Essential Developmental Biology

DATE DUE

Thank you for your purchase. Each new copy includes access to the student website. You will find your unique password printed on the card inserted into this book. If you purchased a used copy of this text, you may purchase a password card to gain access to the animations, review questions, and artwork by visiting **www.blackwellpublishing.com/slack** and clicking on **Order Password Card**.

2nd Edition

Essential Developmental Biology

J.M.W. Slack

Department of Biology and Biochemistry
University of Bath
United Kingdom

Blackwell
Publishing

BLACKWELL PUBLISHING
350 Main Street, Malden, MA 02148-5020, USA
9600 Garsington Road, Oxford OX4 2DQ, UK
550 Swanston Street, Carlton, Victoria 3053, Australia

First edition published 2001 by Blackwell Science Ltd,
Second edition published 2006 by Blackwell Publishing Ltd

1   2006

Library of Congress Cataloging-in-Publication Data

Slack, J.M.W. (Jonathan Michael Wyndham), 1949–
Essential developmental biology / J.M.W. Slack.– 2nd ed.
  p. cm.
Includes bibliographical references and index.
ISBN-13: 978-1-4051-2216-0 (pbk. : alk. paper)
ISBN-10: 1-4051-2216-1 (pbk. : alk. paper)
  1. Developmental biology–Laboratory manuals.   I. Title.

QH491.S6 2006
571.8′1–dc22

2005004145

A catalogue record for this title is available from the British Library.

Set in 9.5/12pt Minion
by Graphicraft Limited, Hong Kong
Printed and bound in Italy
by Rotolito Lombarda, SPA

The publisher's policy is to use permanent paper from mills that operate a sustainable forestry policy, and which has been manufactured from pulp processed using acid-free and elementary chlorine-free practices. Furthermore, the publisher ensures that the text paper and cover board used have met acceptable environmental accreditation standards.

For further information on
Blackwell Publishing, visit our website:
www.blackwellpublishing.com

Contents

Preface

This book presents the basic ideas and facts of modern developmental biology of animals. Special attention has been given to keeping it compact and concise. It should be suitable as a core text for undergraduate courses from the second to the fourth year, and for beginning graduate courses. The first edition has been "road tested" by myself and by many other instructors, and it has been found suitable for both biologically based and medically oriented courses. A basic knowledge of cell and molecular biology is assumed, but no prior knowledge of development, animal structure, or histology should be necessary.

Organization

The book is arranged in four sections and the order of topics is intended to represent a logical progression. The first section introduces the basic concepts and techniques. The second section covers the six main "model organisms," *Xenopus*, zebrafish, chick, mouse, *Drosophila*, and *Caenorhabditis elegans*, describing their early development to the stage of the general body plan. The third section deals with stem cells and organ development, mostly of vertebrates but including also *Drosophila* imaginal discs. The fourth section deals with growth, regeneration, and evolution. To assist readers unfamiliar with the the families of genes and molecules that are important in development, they are listed in the Appendix in the context of a short revision guide to basic molecular and cell biology.

Distinctive approach

This book differs from its main competitors in four important respects, all of which I feel are essential for effective education. Firstly, it keeps the model organisms separate when early development is discussed. This avoids the muddle that arises all too often when students think that knockouts can be made in *Xenopus*, or that bindin is essential for mammalian fertilization. Secondly, I have avoided all considerations of history and

experimental priority because students do not care who did something first if it all happened twenty years ago. Thirdly, on the other hand, I have been careful to stress at all stages *why* we believe what we do. Understanding does not come from simply memorizing long lists of gene names, so I have insisted that students understand how to investigate developmental phenomena and what sorts of evidence are needed to prove a particular type of result. Finally, the work is highly focused. In order to keep the text short and concise I have not wandered off into areas such as the development of plants or lower eukaryotes that may be interesting but are really separate branches of biology.

Changes to this edition

The first edition was very well received by both users and reviewers and I hope that the second edition will make *Essential Developmental Biology* an even more popular choice for undergraduate teaching around the world. The changes made for this edition reflect both the requests of users and the changes in the subject matter over the last few years. Users overwhelmingly wanted color in the illustrations, so this has been provided. There is now a glossary at the end which defines all the key terms shown in **bold** in the text, and each chapter also contains a set of summary bullet points. The web-based materials have been expanded and now include animations in addition to the full set of illustrations used in the book. The users we consulted also tended to want more material on their own favorite topic. This was more difficult to provide as everyone's favorite topic is different, and to please everyone would have led to an explosion in length that would have ended up pleasing nobody. However some modest additions have been made. There is more on mammalian fertilization, which is always of interest to students. There is more on the heart and the gut, as these are such central topics in human embryology, and there is much recent research progress. There is more on stem cells, growth, and aging, all hot research topics with obvious practical significance. Finally, a new chapter on evolution and development now gives this area a higher profile than in the first edition. Otherwise, the text has all been rewritten and updated, the grouping of topics has been reorganized to some extent, the references have been rationalized, and errors have been removed.

Developmental biology has become a very detailed and complex subject and this means that inevitably most of the references in an elementary text have to be to reviews. A consequence of this could be that students would never read any original scientific papers, which would be very undesirable. So I have now included boxes with primary references to some key major discoveries. The choice of these references is of course personal and subjective but I hope they will communicate the excitement involved in research to those who look them up. I have probably been even more subjective in my choices of future priorities in the boxes on "new directions for research," but the object of these is to indicate that the subject is still moving and these boxes may be useful as the starting points for some discussions.

Students sometimes consider developmental biology to be a difficult subject, but this need not be the case as long as certain obstacles to understanding are identified at an early stage. The names and relationships of embryonic body parts are generally new to students, so in this book the number of different parts mentioned is kept to the minimum required for understanding the experiments, and a consistent nomenclature is adopted (e.g. "anterior" is used throughout rather than "rostral" or "cranial"). The competitor texts all mix up species and, for example, would typically consider sea urchin gastrulation, *Xenopus* mesoderm induction, and chick somitogenesis in quick succession. This leaves the student unsure about which processes occur in which organisms. In order to avoid confusion, I have kept separate the animal species in section 2, and for sections 3 and 4 it is made clear to which organisms particular findings apply. Although most students do understand genetics in its simple Mendelian form, they do not necessarily appreciate certain key features prominent in developmental genetics. Among these are the fact that one gene can have several mutant alleles (e.g. loss of function, constitutive, or dominant negative), or that the name of a gene often corresponds to its loss of function phenotype rather than its normal function (e.g. the normal function of the *dorsal* gene in *Drosophila* is to promote ventral development!). Furthermore, pathways with repressive steps, such as the Wnt pathway, cause considerable trouble because of a failure to understand that the lack of something may be just as important as the presence of something. Here, these issues are fully explained in the early chapters, with appropriate reinforcement later on. Finally, I have tried to keep the overall level of detail, in terms of the number of genes, signaling systems and other molecular components, to the bare minimum required to explain the workings of a particular process. This sometimes means that various parallel or redundant components are not mentioned, and the latest detail published in *Cell* is omitted.

Summary of key new features:
• Instructor CD with artwork in downloadable format

• Website including 25 animations, interactive exercises, all text, artwork, and also simple schematic art. Animations are indicated in the margin with the ● icon. Access is free with purchase of new book (access may also be purchased by visiting www.blackwellpublishing.com and searching for ISBN 1-4051-4646-X)
• New chapters on Tissue Organization and Stem Cells (Chapter 13), Development of Endodermal Organs (Chapter 16), and Evolution and Development (Chapter 20)
• Expanded coverage of mammalian fertilization, the heart, growth control, and aging
• "Classic Experiment" boxes with primary references
• "New Directions for Research" boxes
• End-of-book glossary
• End-of-chapter summaries for quick review
• Numerous new figures, including model organism comparison chart (Chapter 6)
• Four-color used throughout

When students have completed a course corresponding to the content of this book they should be able to understand the main principles and methods of the subject. If they wish to enter graduate school, they should be well prepared to enter a graduate program in developmental biology. If they go to work in the pharmaceutical industry, they should be able to evaluate assays based on developmental systems where these are used for the purposes of drug screening or drug development. If they become high school teachers, they should be able to interpret the increasing flow of stories in the media dealing with developmental topics, which are sometimes inaccurate and often sensationalized. Whether the story deals with human cloning, four-legged chickens, or headless frogs, the teacher should be able to understand and explain the true nature of the results and the real motivation behind the work. It is in all our interests to ensure that the results of scientific research are disseminated widely, but also that they are a source of enlightenment and not of sensation.

Acknowledgments

Finally, I should like to thank some people who have been involved with the work: Nancy Whilton who enthusiastically commissioned and guided this edition; Elizabeth Wald who very capably managed the day-to-day details in developing this edition; Debbie Maizels of Zoobotanica for the excellent illustrations; Rosie Hayden, Sarah Edwards, and Brian Johnson who have skillfully handled the complex production; and the numerous reviewers, listed below, who have made many helpful comments on sections of the manuscript. The responsibility for any residual errors is mine and I shall be pleased to hear from readers who discover them.

Reviewers:
Judith E. Heady, University of Michigan-Dearborn

David Heathcote, University of Wisconsin-Milwaukee
Margaret Saha, College of William and Mary
Han Wang, University of Oklahoma
Grant N. Wheeler, University of East Anglia
W.B. Wood, University of Colorado
Lauren Yaich, University of Pittsburgh

Most importantly, I should like to thank my lab members who have put up with a lot of unavailability on my part, and my family whose patience and support during this long writing process was also invaluable.

Jonathan Slack
Bath, 2005

Section 1

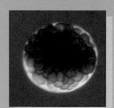

Groundwork

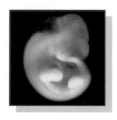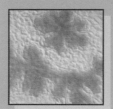

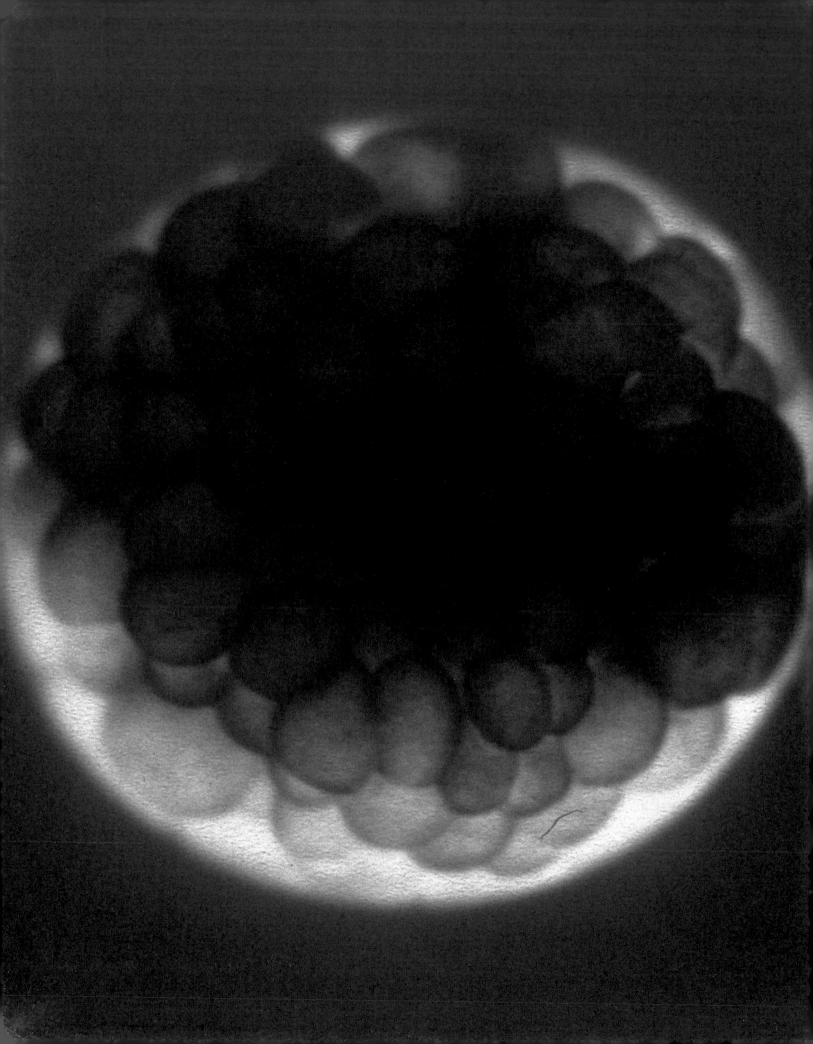

Chapter 1

The excitement of developmental biology

Where the subject came from

One of the most amazing conclusions of modern biological research is that the mechanisms of development are very similar for all animals, including humans. This fact has only been known since it has become possible to examine the molecular basis of developmental processes. As recently as 1980 we knew nothing of these mechanisms, but 25 years later we know a lot and it is possible to write undergraduate textbooks on the subject. Over this period, developmental biology has been one of the most exciting areas of biological research. These dramatic advances came from three main traditions that became fused together into a single world-view: experimental embryology, developmental genetics, and molecular biology.

Experimental embryology had been in existence since the beginning of the twentieth century, consisting mainly of micro-surgical experiments on embryos of frogs and sea urchins. These had demonstrated the existence of **embryonic induction**: chemical signals that controlled the pathways of development of regions of cells within the embryo. The experiments showed where and when these signals operated, but they could not identify the signals, nor the molecular nature of the responses to them.

Developmental genetics has also existed for a long time, but it really flowered in the late 1970s when mass genetic screens were carried out on the fruit fly *Drosophila*, in which thousands of mutations affecting development were examined. These **mutagenesis screens** resulted in the identification of a high proportion of the genes that control development, not just in *Drosophila*, but in all animals.

Molecular biology had started with the discovery of the three-dimensional structure of DNA in 1953, and became a practical science of gene manipulation in the 1970s. The key technical innovations were methods for **molecular cloning** to enable single genes to be amplified to a chemically useful quantity, methods for **nucleic acid hybridization** to enable the identification of DNA or RNA samples, and methods for **DNA sequencing** to determine the primary structures of genes and their protein products. Once this toolkit had been assembled it could be applied to a whole range of biological problems, including those of development. It was used initially to clone the developmental genes of *Drosophila*. This turned out to be of enormous importance because most of the key *Drosophila* genes were found to exist also in other animals, and frequently to be controlling similar developmental processes. Molecular biological methods were also applied directly to vertebrate embryos and used to identify the previously mysterious inducing factors and the genes regulated by them.

The application of molecular biology meant that the mechanisms of development could for the first time be worked out in molecular detail. It also meant that the path of development could be experimentally altered by the introduction of new genes, or the selective removal of genes, or by an alteration of the regulatory relationships between genes. It has turned out that all animals use very similar mechanisms to control their development. This is particularly exciting because it means that we really can learn about human development by understanding how it happens in the fruit fly, zebrafish, frog, or mouse.

Central position in biology

Developmental biology occupies a pivotal position in modern biology. This is because it unites the disciplines of molecular biology, **cell biology**, **genetics**, and **morphology**. Molecular and cell biology tell us about how the individual components work: the inducing factors, their receptors, the signal transduction pathways, the transcription factors. Genetics tells us directly about the function of an individual gene and how it relates to the activities of other genes. Morphology, or anatomical structure, is both a consequence and a cause of the molecular events. The first processes of development create a certain simple morphology which then serves as the basis on which further rounds of signaling and responses can occur, eventually to create a more complex morphology.

So developmental biology is a synthetic discipline in which an understanding of molecular biology, genetics, and morphology is necessary. When thinking about developmental problems it is necessary to be able to use concepts from these three areas simultaneously because they are all necessary to achieve a complete picture.

Impact on society

Certain areas of developmental biology have had a significant impact on society in recent decades. *In vitro* **fertilization (IVF)** is now a routine procedure and has enabled many previously infertile couples to have a baby. Its variants include artificial insemination by donor (AID), egg donation, and storage of fertilized eggs by freezing. It is perhaps less widely appreciated that AID, IVF, embryo freezing, and embryo transfer between mothers is also very important for farm animals. It has been used for many years in cattle to increase the reproductive potential of the best animals.

Developmental biology also led to the understanding that human embryos are particularly sensitive to damage during the period of **organogenesis** (i.e. after the general body plan is formed, and while individual organs are being laid down). The science of **teratology** studies the effects of environmental agents such as chemicals, viral infection, or radiation on embryos. This has led to an awareness of the need to protect pregnant women from the effects of these agents.

Developmental biology is responsible for an understanding of the chromosomal basis of some human **birth defects**. In particular Down's syndrome is due to the presence of an extra chromosome, and there are a number of relatively common abnormalities of the sex chromosomes. These can be detected in cells taken from the amniotic fluid and form the basis of the **amniocentesis** tests taken by millions of expectant mothers every year. Many more birth defects are due to mutations in genes that control development. It is now possible to screen for some of these, either in the DNA of the parents or in the embryo itself, using molecular biology techniques.

Future impact

Although the past impact of developmental biology is significant, the future impact will be much greater. Some of the benefits are indirect and not immediately apparent. Some, particularly those involving human genetic manipulation or cloning, will cause some serious ethical and legal problems. These problems will have to be resolved by society as a whole and not just the scientists who are the current practitioners of the subject. For this reason it is important that an understanding of developmental biology becomes as widespread as possible, because only with an appreciation of the science will people be able to make informed choices.

The human genome is now fully cataloged and sequenced, and so are the genomes of most of the animals used as experimental organisms for studying development. Furthermore techniques are now well advanced for separating and identifying all the proteins in a particular tissue sample (**proteomics**). This means that it has become much easier to identify genes or gene products associated with particular developmental mutations or diseases, and has led to an increased emphasis on understanding their functions. Developmental biology is a central component of these new disciplines of **functional genomics** and **functional proteomics**.

The first main area of practical significance is that an understanding of developmental mechanisms will assist the pharmaceutical industry in designing new drugs effective against cancer or against degenerative diseases such as diabetes, arthritis, and neurodegeneration. As is well known, these conditions cause enormous suffering and premature death. The processes that fail in degenerative diseases are those established in the course of embryonic development, particularly its later stages. Understanding which genes and signaling molecules are involved will provide a large number of potential new **therapeutic targets** for possible intervention. Once the targets have been identified by developmental biology, the new powerful techniques of **combinatorial chemistry** can be applied by pharmaceutical chemists to create drugs that can specifically augment or inhibit their action.

Secondly, and as a quite separate contribution to the work of the pharmaceutical industry, various developmental model systems are important as assays. The *in vivo* function of many **signal transduction pathways** can be visualized in *Xenopus* or zebrafish or *Drosophila* or *Caenorhabditis elegans*, and can be used to assay substances that interfere with them using simple dissecting microscope tests. Genetically manipulated mouse embryos are increasingly being used as **animal models** of human diseases, enabling more detailed study of pathological mechanisms and the testing of new experimental therapies. These are by no means limited to models for human genetic disease as often a targeted mutation in the mouse can mimic a human disease that arises by other means.

Thirdly, there is the possibility of using our understanding of growth and regeneration processes for therapy. This has already been done to some extent. For example the hematopoietic growth factors erythropoietin and granulocyte–macrophage colony-stimulating factor (GM-CSF) have both been used in clinical practice for some years to treat patients whose blood cells are depleted by cancer chemotherapy, or for other reasons. In future other factors may also be developed. For example, something that could make pancreatic β-cells grow would be very useful for the treatment of diabetes, or something that could promote neuronal regeneration would be useful in treating a variety of neurodegenerative disorders.

Fourthly, there is the extension of the existing **prenatal screening** to encompass the whole variety of single-gene disorders. Although this is welcome as a further step in the elimination of human congenital defects, it also presents a problem. The more tests are performed on an individual's genetic makeup, the more likely that individual is to be denied insurance or particular career opportunities because of a susceptibility to some disease or other. It also risks the creation of an underclass of genetically "suspect" persons, contrasted with the screened and supposedly "clean" ones. This is a problem that society as a whole will have to resolve.

Fifthly, and even more controversial, there is the possible application of developmental biology to the production of human tissues or organs for **transplantation**. At present transplantation is seriously limited by the availability of donor organs. There are two conceivable routes to this end. The **tissue engineering** route involves the growth of the tissue or organ *in vitro* either from stem cells or from combinations of mature cells that can be cultivated outside the body. This involves the production of novel types of three-dimensional extracellular matrix, or **scaffold**, on which the cells grow and with which they interact. Tissue engineering will need more input from developmental biology in order to be able to create tissues containing several interacting cell types, or tissues with appropriate vascular and nerve supplies.

The second route to replacement tissues and organs envisages their growth from human **embryonic stem cells (ES cells)**. This may be possible by improvement of culture conditions or it may turn out also to require genetic modification of the cells. In either case there are potential ethical problems connected with genetic modification of human tissues and with the use of human eggs for a purpose other than conventional reproduction. This issue also intersects with the debate about human **cloning**. Although there is virtually universal agreement that human beings should not be "copied" by cloning methods (the procedure called **reproductive cloning**), the majority of scientists do favor the potential use of cloned embryos as a source for tissue grafts. This is called **therapeutic cloning** and involves growing the ES cells from an egg in which the nucleus has been replaced by one from the individual needing the graft. The potential advantage is that this could be a method for creating a limitless supply of grafts with perfect immunological compatibility. The continuing ethical debate on this matter arises because the procedure technically involves the creation of an embryo for a purpose other than reproduction.

Finally, we should not overlook the likely applications of developmental biology to agriculture. With farm animals the possibilities are likely to be limited by a public wish to retain a "traditional" appearance for cows, pigs, sheep, and poultry, but already technologies have been developed to produce pharmaceuticals in the milk of sheep or vaccines in eggs, and other opportunities will doubtless present themselves in the future.

Further reading

Useful web sites

Zygote: http://zygote.swarthmore.edu/
The virtual embryo: http://www.ucalgary.ca/UofC/eduweb/virtualembryo/
Bill Wasserman's developmental biology page: http://www.luc.edu/depts/biology/dev.htm

Textbooks, mainly descriptive

Gilbert, S.F. & Raunio, A.M. (1997) *Embryology: constructing the organism.* Sunderland, MA: Sinauer Associates.
Larsen, W.J. (1997) *Human Embryology*, 2nd edn. New York: Churchill Livingstone.
Hildebrand, M. & Goslow, G.E. (2001) *Analysis of Vertebrate Structure*, 5th edn. New York: John Wiley.
Carlson, B.M. (2004) *Human Embryology and Developmental Biology.* St Louis, MO: Mosby.

Textbooks, mainly analytical

Twyman, R. & Gatherer, D. (2001) *Instant Notes in Developmental Biology.* Oxford: Bios Scientific Publishers.
Wolpert, L. (2002) *Principles of Development*, 2nd edn. Oxford: Oxford University Press.
Gilbert, S.F. (2003) *Developmental Biology*, 7th edn. Sunderland, MA: Sinauer Associates.
Wilt, F.H. & Hake, S.E. (2003) *Principles of Developmental Biology.* New York: W.W. Norton.

Monograph

Martinez-Arias, A. & Stewart, A. (2002) *Molecular Principles of Animal Development.* Oxford: Oxford University Press.

Reproductive technology and ethics

Edwards, R.G. (1997) Recent scientific and medical advances in assisted human conception. *International Journal of Developmental Biology* **41**, 255–262.
Austin, C.R. (1997) Legal, ethical and historical aspects of assisted human reproduction. *International Journal of Developmental Biology* **41**, 263–265.
Braude, P. (2001) Preimplantation genetic diagnosis and embryo research – human developmental biology in clinical practice. *International Journal of Developmental Biology* **45**, 607–611.
Committee on Biological and Biomedical Application (2002) *Stem Cells and the Future of Regenerative Medicine.* Washington DC: National Academy Press.
Maienschein, J. (2003) *Whose View of Life? Embryos, cloning and stem cells.* Cambridge, MA: Harvard University Press.
Hwang, W.S., Ryu, Y.J., Park, J.H. et al. (2004) Evidence of a pluripotent human embryonic stem cell line derived from a cloned blastocyst. *Science* **303**, 1669–1674.

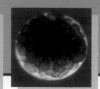

Chapter 2

Common features of development

Developmental biology is the science that seeks to explain how the structure of organisms changes with time. Structure, which may also be called **morphology** or anatomy, encompasses the arrangement of parts, the number of parts, and the different types of parts. Parts may be large, such as whole organs, or small, down to the level of the organization of individual cells.

Development happens most obviously in the course of embryonic development as the fertilized egg develops into a complete organism. This book deals mainly with embryonic development. But it should not be forgotten that development also occurs in postembryonic life. Many animals have life cycles involving a larval stage that is specialized for feeding and/or for dispersal. The larva will at some stage undergo a **metamorphosis** in which the body is remolded to a greater or lesser extent to form the adult. Of the model organisms considered in this book, *Drosophila* shows a drastic metamorphosis during which most of the adult body is formed from **imaginal discs** laid down in the larva. *Xenopus* also undergoes metamorphosis from a tadpole to the adult frog.

Some animals are capable of **asexual reproduction** by forming buds, and this is usually associated with the ability to **regenerate** large parts of the body after loss caused by predators. This is true for example of many **hydroids** and **planarian** worms. Regenerative ability is less evident in higher animals but some amphibians have the ability to regrow limbs and tails after amputation, and even in mammals there is a certain ability to repair tissue damage following wounding. Some developmental events are also associated with cell turnover and differentiation which occur continuously in most types of animal.

Each of these examples of development involves similar problems and they can loosely be classified into four groups:

1 Regional specification deals with how pattern appears in a previously similar population of cells. For example, most early embryos pass through a stage called the **blastula** or **blastoderm** at which they consist of a featureless ball or sheet of cells. Somehow the cells in different regions need to become programmed to form different body parts such as the head, trunk, and tail.

This often involves regulatory molecules deposited in particular positions within the fertilized egg (**determinants**). In addition it always involves some intercellular signaling, called **embryonic induction**, leading to the activation of different combinations of regulatory genes in each region. Processes of regional specification occur in early development, during the formation of individual organs, and in the course of metamorphosis or regeneration.

2 Cell **differentiation** refers to the mechanism whereby different sorts of cells arise. There are more than 200 different specialized cell types in a vertebrate body, ranging from epidermis to thyroid epithelium, lymphocyte, or neuron. Each cell type owes its special character to particular proteins coded by particular genes. The study of cell differentiation deals with the way in which these genes are activated and how their activity is subsequently maintained. Cell differentiation continues throughout life in regions of cell turnover.

3 Morphogenesis refers to the cell and tissue movements that give the developing organ or organism its shape in three dimensions. This depends on the dynamics of the cytoskeleton and on the mechanics and viscoelastic properties of cells. It is less well understood than **1** and **2**. Again, the morphogenetic processes involved in formation of tissue microstructure persist into adult life.

4 Growth refers to increase of size, and the control of proportion between body parts. Although more familiar to the lay person than other aspects of development, it is currently the least well understood aspect in terms of molecular mechanisms.

With a few exceptions, such as the lymphocytes of the immune system, all the different cell types in the animal body retain a complete set of genes. This means that the regulation of gene activity is important for all four processes and occupies a central position in developmental biology. Many techniques for the study of gene expression are described in Chapter 5. The best evidence that all cell types retain a complete set of genes is derived from experiments on the cloning of animals from the nucleus of a single cell.

Genomic equivalence, cloning of animals

In any animal the sperm and eggs and their precursor cells are known as the **germ line**. All other cell types are called **somatic cells**. The germ-line cells have to retain a full complement of genes otherwise reproduction would be impossible. It is generally accepted that virtually all the somatic cells in animals also contain the full complement of genes and that the differences between cells are due to the fact that different subsets of genes are active. The total DNA content of most somatic cell types is the same; and when examined for the presence of a particular gene by standard molecular biological methods, DNA from all tissues gives the same results.

The most persuasive evidence has been obtained from experiments in which a whole animal is created using a single nucleus taken from a somatic cell. Because most genes will be required at some stage of development, the fact that a differentiated cell nucleus can support the whole process implies that all the DNA is still present in that nucleus. The experimental procedure is known as **cloning**, but it should be remembered that the term cloning is also applied to the molecular cloning of genes, and to the cloning of cells in tissue culture, particularly important in the production of monoclonal antibodies.

Many types of animal embryo, including frogs, rabbits, cows, sheep, and mice, can be cloned by transferring a nucleus from an early embryo cell (a blastomere) back into a fertilized egg whose own nucleus has been removed (Fig. 2.1). In such experiments it is important to have some means of distinguishing an animal arising from the donor nucleus from one arising from the original egg nucleus, in case it was not properly removed or destroyed. This is known as a **genetic marker**. For example the donor nuclei in the frog experiments were often taken from albino embryos lacking the gene for pigment synthesis. Only if the embryos arising from the reconstituted eggs are albino can one be sure that they were indeed formed from the donor nucleus. In frogs, nuclear transplantation works progressively less well the more advanced the developmental stage of the donor nucleus. It has been possible to obtain a small number of tadpoles using nuclei from indubitably differentiated epidermal cells from adult frogs, and this strengthened the case for the conservation of DNA in all somatic tissues, but the poor yields led most scientists to think that this was not a practical means for the cloning of animals.

Recent advances in the ability to clone animals have come from those mainly interested in new routes to the genetic modification of farm animals. Using both sheep and cattle it has become clear that it is relatively easy to clone complete animals using nuclei from somatic cells taken from fetuses, and in some cases from adults, even if the cells are grown in tissue culture for some time (Fig. 2.2). The procedure is to isolate a mature **oocyte** (commonly known as an **egg**) from the ovary, and remove the nucleus by sucking it out with a fine glass pipette. The oocyte is

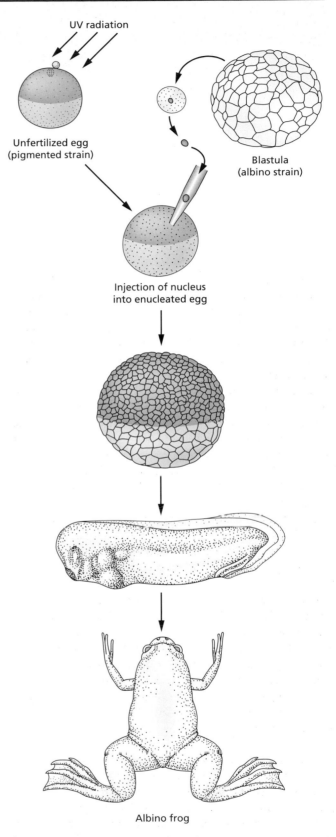

UV radiation

Unfertilized egg
(pigmented strain)

Blastula
(albino strain)

Injection of nucleus
into enucleated egg

Albino frog

Fig. 2.1 Cloning of frogs from blastula nuclei.

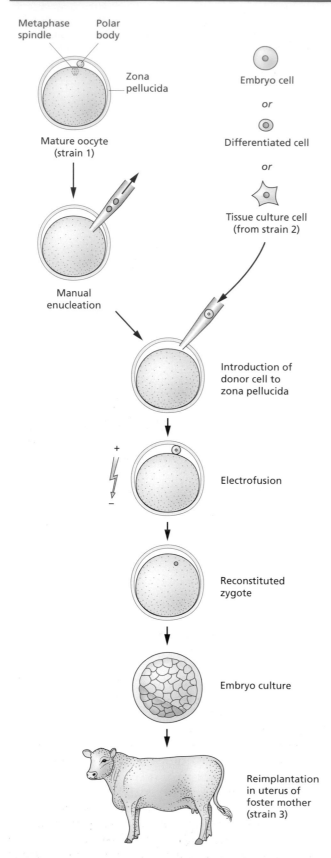

Metaphase spindle

Polar body

Zona pellucida

Mature oocyte (strain 1)

Embryo cell

or

Differentiated cell

or

Tissue culture cell (from strain 2)

Manual enucleation

Introduction of donor cell to zona pellucida

+

−

Electrofusion

Reconstituted zygote

Embryo culture

Reimplantation in uterus of foster mother (strain 3)

Fig. 2.2 Cloning of mammals. The genetic material may come from a somatic cell or a tissue culture cell.

surrounded by a jelly-like coat called the **zona pellucida**, and within this is inserted one cell from the donor tissue culture cell line. The donor cells may be deprived of serum before use to ensure that they are not undergoing DNA replication. An electric pulse is administered which causes fusion of the oocyte and the somatic cell and hence reintroduces a nucleus into the oocyte forming a reconstituted egg. This is cultivated *in vitro* for a while, during which it divides several times and progresses to a stage known as the **blastocyst**. It is then implanted into the uterus of a female who has been hormonally primed to prepare her for pregnancy. Although many eggs do not survive the rigors of cell fusion and reimplantation in the uterus, a percentage of transplanted blastocysts can develop to term and be born. These progeny are genetically clones of the animal which originally donated the nucleus. For such experiments the genetic markers need to distinguish the cell line from both the recipient egg donor, in case the egg nucleus was not properly removed, and the foster mother, in case of accidental pregnancy, and for this purpose DNA polymorphisms distinguishing the relevant animal breeds are detected using the **polymerase chain reaction**.

There are in fact some well-known exceptions to the principle that all somatic cells contain the same genes. The antibody-forming genes of B lymphocytes and the T-cell receptor genes of T lymphocytes are known to undergo rearrangement at the DNA level and lose some genetic information in the process. Certain nematodes, although not *Caenorhabditis elegans*, shed chromosomes from some cell lineages during development. There are also a few examples where total copy number of genes may be altered. **Polyploidy**, where the whole chromosome set is doubled or quadrupled, can occur in some mammalian tissues such as the liver. **Polyteny**, where DNA replication occurs repeatedly without chromosomal division, leading to giant chromosomes, occurs in some tissues in *Drosophila*. However, in general the activity of genes is regulated at the level of transcription and subsequent events, and not at the level of the DNA itself.

Gametogenesis

Sexual reproduction means that the life cycle involves the union of male and female **gametes** to form a fertilized egg or **zygote**. By definition the male gamete is small and motile and called a spermatozoon (sperm), and the female gamete is large and immotile and called an **egg** or ovum. Each gamete contributes a **haploid** (1n) chromosome set so the zygote is **diploid** (2n), containing a maternal- and a paternal-derived copy of each chromosome.

The gametes are formed from **germ cells** in the embryo. The germ cells are referred to collectively as the **germ line**, consisting of cells that will or can become the future gametes, and all other cells are referred to as the **somatic** tissues or soma. The importance of the germ line is that its genetic information can be passed to the next generation, while that of the soma cannot. For example, a germ-line mutation indicates a mutation in germ cells that may be carried to the next generation. By contrast, a somatic mutation may occur in a cell at any stage of development and may be important in the life of the individual animal, but it cannot affect the next generation.

It is often the case that the future germ cells become committed to their fate at an early stage of animal development. In some cases there is a cytoplasmic determinant present in the egg that programs cells that inherit it to become germ cells. This is associated with a visible specialization of the cytoplasm called **germ plasm**. It occurs in *C. elegans* where the cells inheriting the **polar granules** become the P lineage and thereafter the germ cells. It occurs in *Drosophila*, where cells inheriting the **pole plasm** become pole cells and later germ cells. It probably also occurs in *Xenopus* where there is a vegetally localized germ plasm. In other species there may be no visible germ plasm in the egg but germ cells still appear at an early stage of development.

During embryonic development germ cells undergo a period of multiplication and will also often undergo a migration from the site of their formation to the **gonad**, which may be some distance away. The gonad arises from mesoderm and is initially composed entirely of somatic tissues. After the germ cells arrive they become fully integrated into its structure, and in post-embryonic life undergo gamete formation or **gametogenesis**. At some stage in mid-development the key decision of **sex determination** is made and the gonad is determined to become either an ovary or a testis. The molecular mechanism of this is, somewhat surprisingly, different for each of the principal experimental model species, so it will not be described in this chapter. But the upshot is that in the male the germ cells will need to become sperm and in the female they will need to become eggs. Unlike the other model organisms, *C. elegans* is normally a **hermaphrodite** and the germ cells will produce both sperm and eggs in the same individual. However there are also male individuals of *C. elegans*, and the sex-determination mechanism controls the male–hermaphrodite decision rather than a male–female decision.

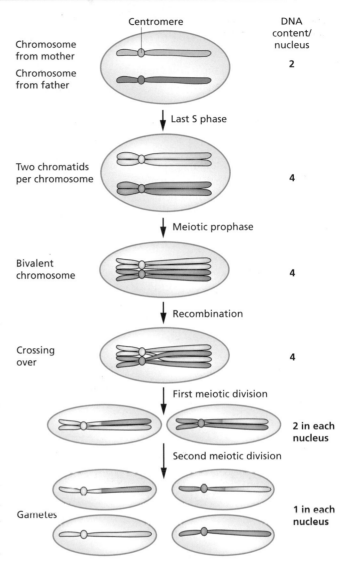

Fig. 2.3 Behavior of chromosomes during meiosis.

Meiosis

The critical cellular event in gamete production is **meiosis**. This is a modified type of cell cycle in which the number of chromosomes is reduced by half (Fig. 2.3). As in mitosis, meiosis is also preceded by an S phase in which each chromosome becomes replicated to form two identical sister chromatids, so the process starts with the nucleus possessing a total DNA content of four times the haploid complement. In mitosis the sister chromatids will separate into two identical diploid daughter cells. But meiosis involves two successive cell divisions. In the first the homologous chromosomes, which are the equivalent chromosome derived from mother and father, pair with each other. At this stage the chromosomes are referred to as **bivalents**, and each consists of four chromatids, two maternal-derived and two paternal-derived. **Crossing over** can occur between these

chromatids, bringing about recombination of the **alleles** present at different loci. Hence, alleles present at two different loci on the same chromosome of one parent may become separated into different gametes and be found in different offspring. The frequency with which alleles on the same chromosome are separated by recombination is roughly related to the physical separation of the loci, and this is why the measurement of recombination frequencies is the basis of genetic mapping. Recombination can also occur between sister chromatids but here all the loci should all be identical since they have just been formed by DNA replication.

In the first meiotic division the four-stranded bivalent chromosomes will separate into pairs which are segregated to the two daughter cells. There is no further DNA replication and in the second meiotic division the two chromatids become separated into individual gametes.

It should be noted that the terms **haploid** (1n) and **diploid** (2n) are normally used to refer to the number of homologous chromosome sets in the nucleus rather than the actual amount of DNA. After DNA replication a nucleus contains twice as much DNA as before, but retains the same ploidy designation.

Oogenesis

The process of formation of eggs is called **oogenesis** (Fig. 2.4). Following sex determination to female, the germ cells become **oogonia**, which continue mitotic division for a period. Follow-ing the last mitotic division, the germ cell becomes known as an **oocyte**. It is a **primary oocyte** until completion of the first meiotic division and a **secondary oocyte** until completion of the second meiotic division. After this it is known as an unfertilized **egg** or ovum. Because in the model organisms considered here fertilization occurs before completion of the second division, it is technically an **oocyte** rather than an **egg** that is being fertilized. However, the term "egg" is often used rather loosely to refer to oocytes, fertilized ova, and even early embryos.

Eggs are larger than sperm and the process of oogenesis involves the accumulation of materials in the oocyte. Usually the primary oocyte is a rather long-lived cell that undergoes a

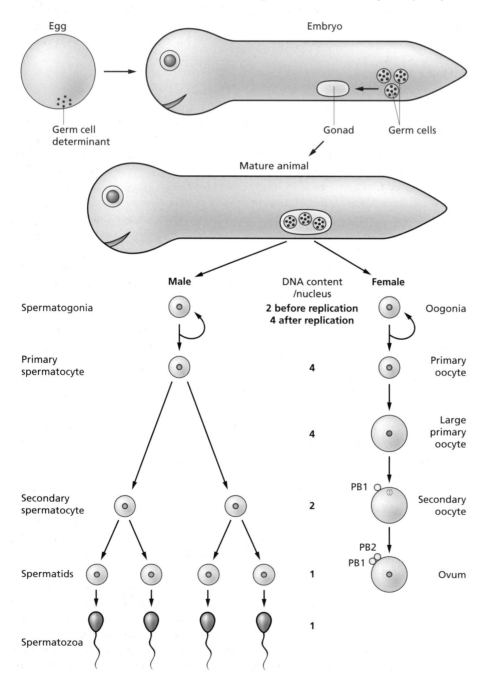

Fig. 2.4 Typical sequence of gametogenesis. The germ cells are initially formed from a cytoplasmic determinant and during development they migrate to enter the gonad. Spermatogenesis generally results in the production of four haploid sperm per meiosis. Oogenesis generally results in the formation of one egg and two polar bodies (PB1 has the same chromosome number but twice the DNA content as PB2).

considerable increase in size. Its growth may be assisted by the absorption of materials from the blood, such as the **yolk** proteins of fish or amphibians which are made in the liver. It may also be assisted by direct transfer of materials from other cells. This is seen in *Drosophila* where the last four mitoses of each oogonium produces an egg chamber containing one oocyte and 15 **nurse cells**. The nurse cells then produce materials that are exported to the oocyte. Animals that produce a lot of eggs usually maintain a pool of oogonia throughout life capable of generating more oocytes. Mammals differ from this pattern as they are thought to produce all their primary oocytes before birth. In humans no more oocytes are produced after the seventh month of gestation, and the primary oocytes then remain dormant until puberty. Although this is the conventional view, there is recent evidence for some postnatal oogenesis in mice.

Ovulation refers to the resumption of the meiotic divisions and the release of the oocyte from the ovary. It is provoked by hormonal stimulation and involves a breakdown of the oocyte nucleus (the **germinal vesicle**) and the migration of the cell-division spindle to the periphery of the cell. The meiotic divisions do not divide the oocyte into two halves, but in each case into a large cell and a small **polar body**. The first meiotic division divides the primary oocyte into a secondary oocyte and the first polar body, which is a small projection containing a replicated chromosome set. The second meiotic division divides the secondary oocyte into an egg and a second polar body, which consists of another small projection enclosing a haploid chromosome set. The polar bodies soon degenerate and play no further role in development.

Spermatogenesis

If the process of sex determination yields a male then the germ cells undergo spermatogenesis (Fig. 2.4). Mitotic germ cells in the testis are known as **spermatogonia**. These are stem cells that can both produce more of themselves and produce cells whose destiny is terminal differentiation into sperm. After the last mitotic division the male germ cell is known as a primary spermatocyte. Meiosis is equal, the first division yielding two secondary spermatocytes and the second division yielding four spermatids, which mature to become motile spermatozoa.

Early development

The process of fertilization differs considerably between animal groups but there are a few common features. When the sperm fuses with the egg there is a fairly rapid change in egg structure that excludes the fusion of any further sperm. This is called a block to **polyspermy**. Fusion activates the inositol trisphosphate signal transduction pathway (see Appendix) resulting in a rapid increase in intracellular calcium. This causes exocytosis of **cortical granules** whose contents form, or contribute to, a fert-

ilization membrane; and also trigger the metabolic activation of the egg, increasing the rate of protein synthesis and, in vertebrates, starting the second meiotic division. The calcium may, in addition, trigger cytoplasmic rearrangements that are important for the future regional specification of the embryo, for example cortical rotation in *Xenopus*, or polar granule segregation in *C. elegans*. The sperm and egg pronuclei fuse to form a single diploid nucleus and at this stage the fertilized egg is known as a **zygote**.

A generalized sequence of early development is shown in Fig. 2.5. A typical zygote of an animal embryo is small, spherical, and polarized along the vertical axis. The upper hemisphere, usually carrying the polar bodies, is called the **animal hemisphere**, and the lower hemisphere, rich in yolk, the **vegetal hemisphere**. The early cell divisions are called **cleavages**. They differ from normal cell division in that there is no growth phase between successive divisions. So each division partitions the mother cell into two half-size daughters. The products of cleavage are called **blastomeres**. Cell division without growth can proceed for a considerable time in free-living embryos without an extracellular yolk mass. Embryos that do have some form of food supply, either mammals that are nourished by the mother, or egg types with a large yolk mass such as birds and reptiles, usually only undergo a limited period of cleavage at the beginning of development. In many species, the embryo's genome remains inactive during part or all of the cleavage phase, and protein synthesis is directed by messenger RNA transcribed during oogenesis (maternal mRNA). This is the stage of **maternal effects** because the properties of the cleavage stage embryo depend entirely on the genotype of the mother and not on that of the embryo itself (see Chapter 3).

Different animal groups have different types of cleavage (Fig. 2.6) and this is controlled to a large extent by the amount of yolk in the egg. Where there is a lot of yolk, as in an avian egg, the cytoplasm is concentrated near the animal pole and only this region cleaves into blastomeres, with the main yolk mass remaining acellular. This type of cleavage is called **meroblastic**. Where cleavage is complete, dividing the whole egg into blastomeres, it is called **holoblastic**. Holoblastic cleavages are often somewhat unequal, with the blastomeres in the yolk-rich vegetal hemisphere being larger (**macromeres**), while those in the animal hemisphere are smaller (**micromeres**).

Each animal class or phylum tends to have a characteristic mode of early cleavage and these can be classified by the arrangement of the blastomeres into such categories as **radial** (echinoderms), bilateral (ascidians), rotational (mammals). An important type is the **spiral** cleavage shown by most annelid worms, molluscs, and flatworms. Here, the macromeres cut off successive tiers of micromeres, first in a right-handed sense when viewed from above, then another tier in a left-handed sense, and so on. Most insects and some crustaceans show a special type of cleavage called **superficial cleavage**. Here only the nuclei divide and there is no cytoplasmic cleavage at the early stages. Thus, the early embryo becomes a **syncytium** consisting

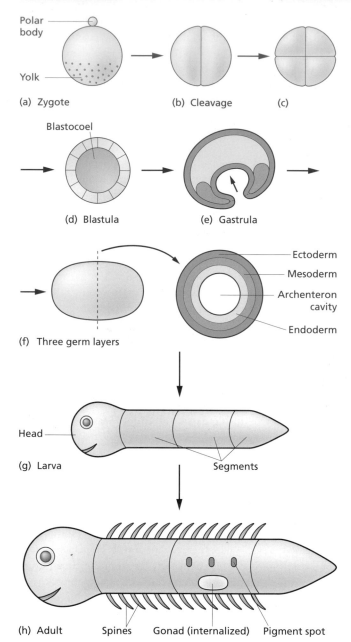

Polar body

Yolk

(a) Zygote

(b) Cleavage

(c)

Blastocoel

(d) Blastula

(e) Gastrula

Ectoderm

Mesoderm

Archenteron cavity

Endoderm

(f) Three germ layers

Head

(g) Larva

Segments

(h) Adult   Spines   Gonad (internalized)   Pigment spot

Fig. 2.5 A generalized sequence of early development. Segments, spines and spots are not necessarily found in all animals but represent a typical external anatomy. The color scheme of ectoderm green, mesoderm orange, and endoderm yellow is used throughout the book.

of many nuclei suspended within the same body of cytoplasm. At a certain stage the nuclei migrate to the periphery and shortly afterwards cell membranes grow in from the outer surface of the embryo and surround the nuclei to form an epithelium.

During the cleavage phase a cavity usually forms in the center of the ball of cells, or sheet of cells in the case of meroblastic cleavage. This expands due to uptake of water and becomes known as the **blastocoel**. At this stage of development the embryo is called a **blastula**. The cells often adhere tightly to

one another, being bound by cadherins, and will usually have a system of **tight junctions** forming a seal between the external environment and the internal environment of the blastocoel.

Following the formation of the blastula, all animal embryos show a phase of cell and tissue movements called **gastrulation** that converts the simple ball or sheet of cells into a three-layered structure known as the **gastrula**. The details of the morphogenetic movements of gastrulation can vary quite a lot even between related animal groups, but the outcome is similar. The three tissue layers formed during gastrulation are called **germ layers**, but these should not be confused with **germ cells**. Conventionally the outer layer is known as the **ectoderm**, and later forms the skin and nervous system; the middle layer is the **mesoderm** and later forms the muscles, connective tissue, excretory organs, and gonads; and the inner layer is the **endoderm**, later forming the epithelial tissues of the gut. The **germ cells** have usually appeared by the stage of gastrulation and are not regarded as belonging to any of the three germ layers.

After the completion of the major body morphogenetic movements most types of animal embryo have reached the general body plan stage at which each major body part is present as a region of committed cells, but is yet to differentiate internally. This stage is often called the **phylotypic stage**, because it is the stage at which different members of an animal group, not necessarily a whole phylum, show maximum similarity to each other (see Chapter 20). For example all vertebrates show a phylotypic stage at the **tailbud** stage when they have a notochord, neural tube, paired somites, branchial arches, and tailbud. All insects show a phylotypic stage at the **extended germ band** when they show six head segments, three appendage-bearing thoracic segments, and a variable number of abdominal segments.

Axes and symmetry

In order that the correct orientation of a specimen can be specified, it is necessary to have terms for describing embryos (Fig. 2.7). If the egg is approximately spherical with an animal and vegetal pole then the line joining the two poles is the animal–vegetal axis. Unfertilized eggs are usually radially symmetrical around this axis, but after fertilization there is often a cytoplasmic rearrangement that breaks the initial radial symmetry and generates a bilateral symmetry. In some organisms, such as *Drosophila*, this may occur earlier, in the oocyte; in others, such as mammals, it may occur later, at a multicellular stage. But even animals such as sea urchins, which are radially symmetrical as adults, or gastropods, which are asymmetrical as adults, still have bilaterally symmetrical early embryos. The change of symmetry means that the animal now has a distinct **dorsal** (upper) and **ventral** (lower) side.

If the animal and vegetal poles are at the top and bottom, then the equatorial plane is the horizontal plane dividing the egg into **animal** and **vegetal hemispheres**, just like the equator of the Earth. Any vertical plane, corresponding to circles of longitude,

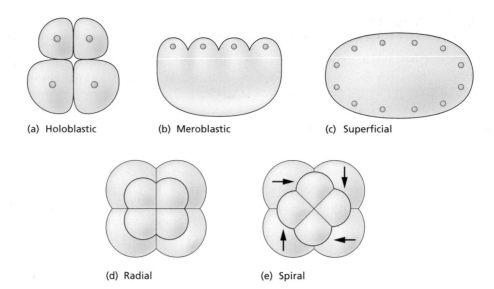

Fig. 2.6 Cleavage types.

(a) Holoblastic   (b) Meroblastic   (c) Superficial

(d) Radial   (e) Spiral

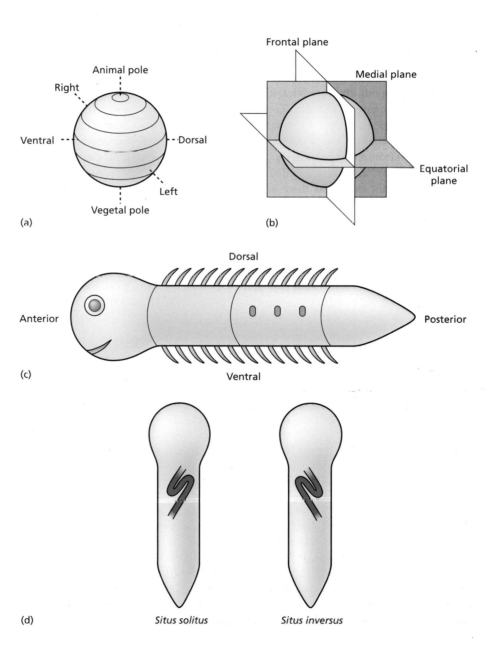

Animal pole

Right

Ventral

Dorsal

Left

Vegetal pole

(a)

Frontal plane

Medial plane

Equatorial plane

(b)

Dorsal

Anterior

Posterior

Ventral

(c)

Situs solitus   *Situs inversus*

(d)

Fig. 2.7 Axes and symmetry. (a) Axes of a fertilized egg after it has acquired a dorsoventral asymmetry. (b) Anatomical planes of an early embryo. (c) Principal axes of an animal viewed from the left side. (d) Ventral view of an animal showing deviation from bilateral symmetry.

is called a meridional plane. Once the embryo has acquired its bilateral symmetry then there is a particularly important meridional plane, the **medial** (= **sagittal**) plane, separating the right and left sides of the body. This is often, but not always, the plane of the first cleavage. The **frontal** plane is the meridional plane at right angles to the medial plane, and is often, but not necessarily, the plane of the second cleavage.

Following gastrulation most animals become elongated. The head end is the **anterior**, the tail end is the **posterior**, so the head-to-tail axis is called **anteroposterior** (= craniocaudal or rostrocaudal). The top–bottom axis is called **dorsoventral** and the left–right axes are called **mediolateral**. In human anatomy, because we stand upright on two legs, the term anteroposterior is normally synonymous with dorsoventral, but the term "craniocaudal" remains acceptable for the head-to-tail axis. The terms **proximal** and **distal** are usually used in relation to appendages, proximal meaning "near the body" and distal meaning "further away from the body."

Generally the principal body parts will become visible some time after completion of gastrulation. Some phyla, including annelids, arthropods, and chordates, show prominent **segmentation** of the anteroposterior axis. To qualify as segmented an organism should show repeated structures that are similar or identical to each other, are principal body parts, and involve contributions from all the germ layers.

Although most animals have an overriding bilateral symmetry, this is not exact and there are systematic deviations which make right and left sides slightly different. For example in mammals the cardiac apex, stomach, and spleen are on the left and the liver, vena cava, and greater lung lobation are on the right. This asymmetrical arrangement is known as *situs solitus*. If the arrangement is inverted, as occurs in some mutants or experimental situations, it is called *situs inversus* (Fig. 2.7). If

the parts on the two sides are partly or wholly equivalent, it is called an **isomerism**.

Morphogenetic processes

Cell shape changes and movements are fundamental to early development as the embryo needs to convert itself from a simple ball or sheet of cells into a multilayered and elongated structure. The processes by which this is achieved are called **gastrulation**. Although the details of gastrulation can differ substantially between even quite similar species, the basic cellular processes are common. At later stages of development the same repertoire of processes is re-used repeatedly in the morphogenesis of individual tissues and organs.

From a morphological point of view most embryonic cells can be regarded as epithelial or mesenchymal (Fig. 2.8). These terms relate to cell shape and behavior rather than to embryonic origin, as epithelia can arise from all three germ layers and mesenchyme from ectoderm and mesoderm. An **epithelium** is a sheet of cells, arranged on a **basement membrane**, each cell joined to its neighbors by specialized junctions, and showing a distinct apical–basal polarity. **Mesenchyme** is a descriptive term for scattered stellate cells embedded in loose extracellular matrix. It fills up much of the embryo and later forms fibroblasts, adipose tissue, smooth muscle, and skeletal tissues. Epithelia and mesenchyme are further described in Chapter 13.

Cell movement

Many morphogenetic processes depend on the movement of individual cells. This is most apparent in the case of migrations,

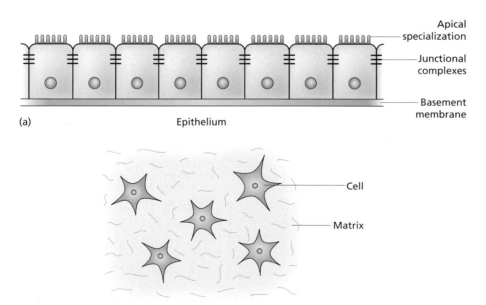

(a) Epithelium

Apical specialization

Junctional complexes

Basement membrane

(b) Mesenchyme

Cell

Matrix

Fig. 2.8 Epithelium and mesenchyme.

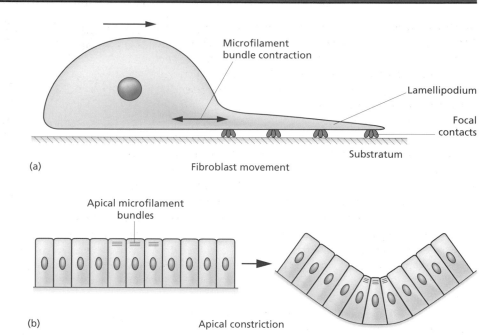

Fig. 2.9 Types of cell movement.

Cell adhesion

for example of the neural crest cells or the germ cells, which may move long distances as individuals. But shorter-range movements are also important for processes such as differential adhesion or shape changes in cell sheets. The mechanism of cell movement is most apparent in fibroblasts moving on a substratum (Fig. 2.9). They extend a flat process called a **lamellipodium** (plural **lamellipodia**) which is rich in microfilaments. This attaches to the substratum by focal attachments containing integrins, which are connected to the microfilament bundles by a complex of actin-associated proteins. The body of the cell is then drawn forward by a process involving active contraction in which myosin molecules migrate towards the plus end of microfilaments. Cells in embryos are thought to move in essentially the same way. Instead of a large flat lamellipodium they may extend multiple thin **filopodia** to form the contacts. Nerve axons also elongate by a similar mechanism. At the tip is a flattened structure called a **growth cone**, which emits the filopodia.

Cell shape can also change because of the activity of microfilament bundles and their associated motor proteins. An apical constriction in a group of epithelial cells will reduce the apical surface area and increase the length of the cells (Fig. 2.9). This is often a preliminary to an **invagination** movement in which the cells leave their epithelium and enter the space below.

Cell adhesion

Vertebrate epithelial cells are bound together by tight junctions, adherens junctions, and desmosomes, the latter two types involving cadherins as major adhesion components (see Appendix). Mesenchymal cells may also adhere by means of cadherins but usually more loosely. The adhesion of early embryo cells is usually dominated by the cadherins as shown by the fact that most types of early embryo can be fully disaggregated into single cells by removal of calcium from the medium.

There is some qualitative specificity to cell adhesion. Cadherin-based adhesion is **homophilic** and so cells carrying E-cadherin will stick more strongly to each other than to cells bearing N-cadherin. The calcium-independent immunoglobulin superfamily-based adhesion systems such as N-CAM (Neural Cell Adhesion Molecule), particularly important on developing neurons and glia, are different again, and also promote adhesion of similar cells. This qualitative specificity of adhesion systems provides a mechanism for the assembly of different types of cell aggregate in close proximity, and also prevents individual cells wandering off into neighboring domains. If cells with different adhesion systems are mixed they will sort out into separate zones, eventually forming a dumbbell-like configuration or even separating altogether (Fig. 2.10a).

In addition to the qualitative aspect of specificity, it is also known that cell-sorting behavior can result simply from different strengths of adhesion of the same system. If cell type A is more adhesive than cell type B then B will eventually surround A (Fig. 2.10b). The process is based on the existence of small random movements of the cells in the aggregates, and the same final configuration will be reached from any starting configuration, for example from an intimate mixture of the two types, or from blocks of the two types pressed together.

Classification of morphogenetic processes

Similar repertoires of morphogenetic processes are re-used repeatedly in different developmental contexts (Fig. 2.11). They

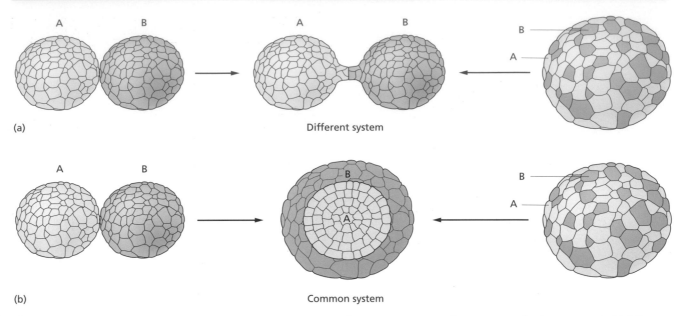

Fig. 2.10 Cell-sorting processes. (a) Cell types A and B each have their own adhesion mechanism, so adhesion A–B<<adhesion A–A or B–B. (b) Cell types A and B have the same adhesion mechanism, but adhesion A–A>B–B and adhesion A–B has the average strength: (A–A + B–B)/2. The center image in (b) represents a section while the others represent surface views.

are driven partly by shape changes of cells due to the contraction of microfilament complexes, partly by changes in gene expression of the cell adhesion molecules (cadherins, CAMs, and integrins), and partly by local changes in cell division rates. These events are in turn under the control of regional specification mechanisms: the inductive signals and regulatory genes.

Condensation of cells to form an aggregate (Fig. 2.11a) is often a prelude to the formation of structures, for example the somites, or the skeletal elements of the limbs. Condensation arises partly by increased local cell division, partly by reduction of local matrix secretion, and partly by increased cell–cell adhesion.

Invagination and involution (Fig. 2.11b) are processes that generate multilayered structures from a simple epithelium. They are found in gastrulation, in neurulation, and in the formation of many other structures such as glands, sense organs, and insect appendages. They are usually initiated from a localized apical constriction that arises from a contraction of the terminal microfilament complexes. In an **invagination** the constriction causes the cell sheet to buckle so that the constricted region of cells forms a protrusion into the interior. Internalization of this protrusion can then occur by several different mechanisms. They will normally involve differential adhesion between the invaginating cells and the surroundings, for example the expression of N-CAM by the forming neural plate and L-CAM (leukocyte cell adhesion molecule) by the epidermis helps to ensure closure and complete invagination of the neural tube. In a distinct mechanism also depending on cell adhesion, the forming archenteron of the sea-urchin embryo is drawn into the embryo by filopodia thrown from cells of the initial invagination across to the far side of the gastrula. If the initial entry of cells arises not

from mechanical buckling of the sheet but from a migration of cells around the edge of the constricted surface then it is called an **involution**. This will involve the formation of a free edge in the involuting tissue. Again the process depends on a differential cell adhesion between the involuting cells and their surroundings. The cell movement may arise from individual migration or by convergent extension (see below) or both.

Invagination is one way of generating a hollow ball or tube of cells. Alternatively a fluid-filled space can be hollowed out from a solid mass of cells by **cavitation** (Fig. 2.11c). This may occur either by cell rearrangement, as in secondary **neurulation**, or by **apoptosis** of the cells in the interior, as found for example in formation of the mouse egg cylinder. However a cavity is produced, it may later be referred to as a **lumen**, and the cells abutting it as being on the lumenal (also spelt luminal) surface.

Epithelial-to-mesenchymal transitions (Fig. 2.11d) occur whenever cells leave an epithelium and move off as individuals or as a mesenchymal mass. This happens in the ingression of cells from the chick epiblast to form the hypoblast, or in the formation of the neural crest from the dorsal neural tube. It requires a local reduction in cell–cell adhesion in the cells that separate out from the epithelium. The reverse mesenchyme-to-epithelium transition (Fig. 2.11e) is also found, for example in the formation of the coelomic lining epithelium or of kidney tubules.

Cell sheets can change their shape quite dramatically as a result of active cell rearrangement. This usually takes the form of **convergent extension** (Fig. 2.11f), in which individual cells intercalate in between each other causing a constriction of the sheet in the direction of intercalation and an elongation of the sheet at right angles to the intercalation. The intercalation movements depend on contractile activity at the termini of the

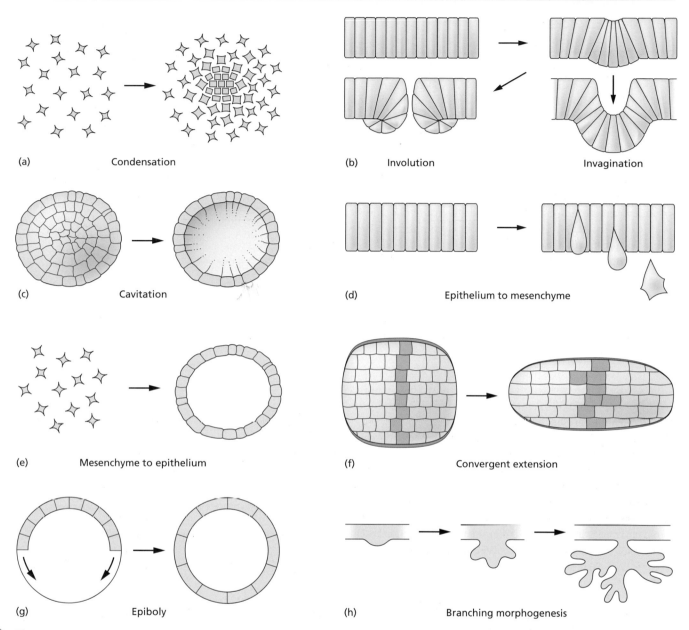

Fig. 2.11 Classification of morphogenetic processes. (a) Mesenchymal condensation. (b) Invagination and involution. (c) Cavitation. (d) Epithelium to mesenchyme transition. (e) Mesenchyme to epithelium transition. (f) Convergent extension. Here a typical file of cells is labeled and the intercalation movements of the cells rearrange the file so that some cells are excluded. (g) Epiboly. (h) Branching morphogenesis.

cells which exert traction on the extracellular matrix and force the cells in between one another. The overall polarity of the process is controlled by boundary regions at which the contractile activity is inhibited. These boundaries will eventually become the long edges of the elongated cell sheet.

An associated process found in gastrulation is **epiboly** (Fig. 2.11g), in which a sheet of cells expands to surround and enclose another population. The mechanisms of this may be somewhat diverse. *Xenopus* and zebrafish gastrulation both show epiboly of the animal hemisphere, which expands to cover the whole embryo. However, in *Xenopus* the process is driven by radial intercalation of cells producing a thinner sheet of greater

area, thus showing some similarity to convergent extension. By contrast, in the zebrafish it is driven by the expansion of the acellular yolk syncytial layer by a microtubule-dependent process.

A type of morphogenetic behavior that is characteristic of organogenesis, rather than early development, is **branching morphogenesis**. Typically an epithelial bud grows into a mesenchymal cell mass and as it does so the number of growing points progressively increases to generate a branched structure (Fig. 2.11h). This process may depend mainly on cell movement, as in the formation of the *Drosophila* tracheal system, or by differential growth at the tips, as is more usual in the formation of vertebrate organs such as the lung or kidney.

Growth and death

A typical animal cell cycle is shown in Fig. 2.12 and some typical patterns of cell division in Fig. 2.13. The cell cycle is conventionally described as consisting of four phases. M indicates the phase of mitosis, S indicates the phase of DNA replication, and G1 and G2 are the intervening phases. For growing cells, the increase in mass is continuous around the cycle and so is the synthesis of most of the cell's proteins. Normally the cell cycle is coordinated with the growth rate. If it were not, cells would increase or decrease in size with each division, as happens for example during embryonic cleavage divisions. There are various internal controls built into the cycle, for example to ensure that mitosis does not start before DNA replication is completed. These controls operate at **checkpoints** around the cycle at which the process stops unless the appropriate conditions are fulfilled.

Control of the cell cycle depends on a metabolic oscillator comprising a number of proteins called cyclins and a number of cyclin-dependent protein kinases (Cdks). In order to pass the M checkpoint and enter mitosis, a complex of cyclin and Cdk (called M-phase promoting factor, MPF) has to be activated. This phosphorylates and thereby activates the various components required for mitosis (nuclear breakdown, spindle formation, chromosome condensation). Exit from M phase requires the inactivation of MPF, via the destruction of cyclin, so by the end of the M phase it has disappeared.

Passage of the G1 checkpoint depends on a similar process operated by a different set of cyclins and Cdks, whose active complexes phosphorylate and activate the enzymes of DNA replication. This is also the point at which the cell size is assessed. The cell cycle of G1, S, G2 and M phases is universal although there are some modifications in special circumstances. The rapid cleavage cycles of early development have short or absent G1 and G2 phases and there is no size check, the cells halving in volume with each division. The meiotic cycles require the same active MPF complex to get through the two nuclear divisions, but there is no S phase in between.

The early embryo cleavage divisions occur in the absence of extracellular **growth factors** (listed in the Appendix), but at later stages of development and particularly in the mature organism most cells are quiescent unless they are stimulated by growth factors. In the absence of growth factors cells enter a state called G0, in which the Cdks and cyclins are absent. Restitution of growth factors induces the resynthesis of these proteins and the resumption of the cycle starting from the G1 checkpoint. One factor maintaining the G0 state is a protein called Rb (retinoblastoma protein). This becomes phosphorylated, and hence deactivated, in the presence of growth factors. In the absence of Rb, a transcription factor called E2F becomes active and initiates a cascade of gene expression culminating in the resynthesis of cyclins, Cdks, and other components needed to initiate S phase.

True growth, meaning increase in size, is not quite the same as cell division. An obvious example of the dissociation between

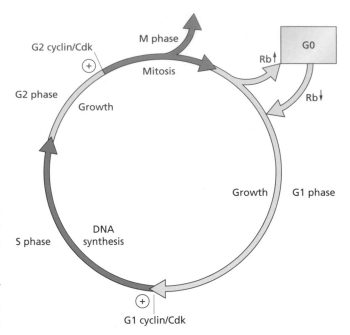

Fig. 2.12 The cell cycle of a typical eukaryotic cell. ⊕ Indicates checkpoints at the entry to M phase and S phase.

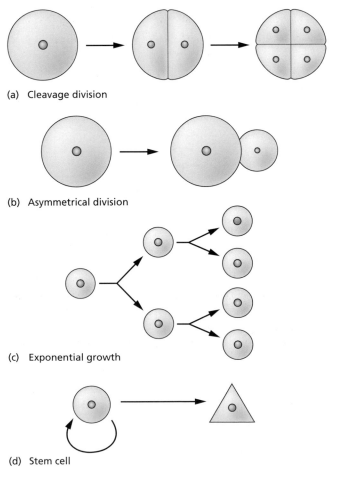

(a) Cleavage division

(b) Asymmetrical division

(c) Exponential growth

(d) Stem cell

Fig. 2.13 Types of cell division.

the two is the embryonic cleavage divisions where cell division occurs without growth (Fig. 2.13a). In later development growth does require an increase in cell number, with doubling of size between divisions, but often also involves an increase in cell size or an increase in the amount of the extracellular matrix. Extracellular material is not extensive in early embryos, but later on a substantial proportion of skeletal tissues such as bone and cartilage is composed of matrix materials. There is no real growth at all in free-living embryo types such as *Xenopus*, zebrafish, or sea urchins. They may expand to some extent because of uptake of water but they do not increase in actual content of matter until they are able to feed themselves. By contrast, those embryo types that do have an external nutrient supply, such as the placenta in mammals, or a large extraembryonic yolk mass, as in birds and reptiles, do grow extensively throughout development. This has implications for experiments, because materials injected into cells of a free-living embryo do not become diluted by growth.

In embryos some cells can be seen to undergo visibly **asymmetric divisions** (Fig. 2.13c). This is often associated with the formation of a cytoplasmic asymmetry prior to their division such that cytoplasmic **determinants** (see below) inherited by the two daughter cells evoke different patterns of gene activity in their nuclei and thus bring about different pathways of development. The nature of determinants is now quite well understood and includes localized mRNAs and self-organizing protein complexes such as the PAR complex (see Chapter 12).

Cells often have the capability for exponential growth in tissue culture (Fig. 2.13b) but this is very rarely found in animals.

Although some differentiated cell types can go on dividing, there is a general tendency for differentiation to be accompanied by a slow down or cessation of division. In postembryonic life most cell division is found among **stem cells** and their immediate progeny called **transit amplifying cells** (see Chapter 13). Stem cells are cells that can both reproduce themselves and generate differentiated progeny for their particular tissue type (Fig. 2.13d). This does not necessarily mean that every division of a stem cell has to be an asymmetrical one, but over a period of time half the progeny will go to renewal and half to differentiation. The term "stem cell" is also used for **embryonic stem cells (ES cells)** of early mammalian embryos. These are early embryo-type cells that can be grown in culture and are capable of repopulating embryos and contributing to all tissue types (see Chapter 10).

Programmed cell death or **apoptosis** is also important in development. Apoptosis is a method of removing cells without spilling their bioactive contents into the surroundings. It involves a molecular pathway which culminates in the activation of proteases called caspases (see Chapter 12). These cause the nucleus to condense, the cell to shrink and to display on its surface a signal for engulfment by other, phagocytic, cells. Apoptosis is often initiated by a withdrawal of growth factors from the cell, but can sometimes be an active response to a signal. In vertebrate development, apoptosis is particularly important in terms of the reduction of motor-neuron pools in response to the availability of growth factors from their target organ (see Chapter 14), and in limb development where digits form because the tissue in between is removed by programmed cell death.

Key Points to Remember

- The main processes in animal development are regional specification, cell differentiation, morphogenesis, and growth.
- Whole animal cloning experiments show that the full set of genes is retained by somatic cells. Development therefore involves the control of gene expression.
- Gametes arise from cells of the germ line by meiosis.
- Events at the earliest stages of development involve components preformed in the egg and so depend on the genome of the mother.
- Animal development normally involves an early cleavage stage leading to the formation of a blastula or blastoderm.

- This early cleavage stage is followed by a phase of morphogenetic movements called gastrulation during which the three germ layers: ectoderm, mesoderm, and endoderm are formed.
- Morphogenetic processes include condensation, involution, invagination, cavitation, transitions between epithelium and mesenchyme, epiboly, and branching processes.
- The cell cycle of G1, S, G2 and M phases, is universal but is modified for specialist developmental processes such as meiosis and cleavage divisions. Growth requires increase in size as well as cell division.

Further reading

Stern, C. ed. (2004) *Gastrulation*. Cold Spring Harbor, NY: Cold Spring Harbor Laboratory Press.
Also see general textbooks cited in Chapter 1.

Cloning
Maienschein, J. (2003) *Whose View of Life? Embryos, cloning and stem cells*. Cambridge, MA: Harvard University Press.

Gametogenesis
Lin, H. (1997) The tao of stem cells in the germline. *Annual Reviews of Genetics* **31**, 455–491.
Wylie, C. (1999) Germ cells. *Cell* **96**, 165–174.
Matova, N. & Cooley, L. (2001) Comparative aspects of animal oogenesis. *Developmental Biology* **231**, 291–320.

Fertilization
Strickler, S.A. (1999) Comparative biology of calcium signaling during fertilization and egg activation in animals. *Developmental Biology* **211**, 157–176.
Hardy, D.M. (2001) *Fertilization*. New York: Academic Press.
Jungnickel, M.K., Sutton, K.A. & Florman, H.M. (2003) In the beginning: lessons from fertilization in mice and worms. *Cell* **114**, 401–404.

Cell cycle and apoptosis
Murray, A. & Hunt, T. (1994) *The Cell Cycle: an introduction*. Oxford: Oxford University Press.
Green, D.R. (2000) Apoptotic pathways: paper wraps stone blunts scissors. *Cell* **102**, 1–4.
Novartis Foundation (2001) *The Cell Cycle and Development*. Novartis Foundation Symposium. New York: John Wiley.
Lawen, A. (2003) Apoptosis – an introduction. *Bioessays* **25**, 888–896.

Morphogenesis
Bard, J.B.L. (1992) *Morphogenesis. The cellular and molecular processes of developmental anatomy*. Cambridge: Cambridge University Press.
De Arcangelis, A. & Georges-Labouesse, E. (2000) Integrin and extracellular matrix functions, roles in vertebrate development. *Trends in Genetics* **16**, 389–395.
Irvine, K.D. & Rauskolb, C. (2001) Boundaries in development: formation and function. *Annual Reviews in Cell and Developmental Biology* **17**, 189–214.
Savagner, P. (2001) Leaving the neighborhood: molecular mechanisms involved during epithelial–mesenchymal transition. *Bioessays* **23**, 912–923.
Schöck, F. & Perrimon, N. (2002) Molecular mechanisms of epithelial morphogenesis. *Annual Reviews in Cell and Developmental Biology* **18**, 463–493.
Tepass, U., Godt, D. & Winklbauer, R. (2002) Cell sorting in animal development: signaling and adhesive mechanisms in the formation of tissue boundaries. *Current Opinion in Genetics and Development* **12**, 572–582.
Affolter, M., Bellusci, S., Itoh, N., Shilo, B., Thiery, J.P. & Werb, Z. (2003) Tube or not tube: remodeling epithelial tissues by branching morphogenesis. *Developmental Cell* **4**, 11–18.
Hardin, J. & Walston, T. (2004) Models of morphogenesis: the mechanisms and mechanics of cell rearrangement. *Current Opinion in Genetics and Development* **14**, 399–406.

Asymmetry
Wood, W.B. (1997) Left–right asymmetry in animal development. *Annual Reviews in Cell and Developmental Biology* **13**, 53–82.
Capdevila, J., Vogan, K.J., Tabin, C.J. & Izpisua-Belmonte, J.C. (2000) Mechanisms of left–right determination in vertebrates. *Cell* **101**, 9–21.
Mercola, M. & Levin, M. (2001) Left–right asymmetry determination in vertebrates. *Annual Reviews in Cell and Developmental Biology* **17**, 779–805.

Chapter 3

Developmental genetics

Developmental mutants

Of all the genes in the genome, approximately 1–2% have functions that are specifically concerned with development. Many more than this are needed in order that normal development should take place, but their primary functions lie in the central areas of cell biology or metabolism. Considerable use is made of **mutants** in developmental biology. The gene **mutations** carried by a mutant organism may be spontaneous, or induced by mutagenic treatments such as chemical mutagens or radiation, or, particularly in the mouse, may be specifically designed "targeted" mutations. Mutations fall into several classes in terms of the molecular basis of the change. Chemical mutagenesis usually results in the creation of point mutations in which a single DNA base is changed to another. These may cause the substitution of one amino acid for another, or, if a new termination codon is created, cause a premature chain termination. Addition or deletion of a single nucleotide will produce a frameshift mutation, changing the entire downstream sequence of amino acids in the protein. X-irradiation often induces deletions of a whole stretch of DNA which may include more than one gene. Spontaneous mutations can be of any of these types, and can, in addition, include the insertion of **transposable elements**.

The different versions of a gene are called **alleles.** There will usually be just one normal allele, called the **wild type**, but there is an almost infinite number of different possible mutant alleles. The totality of nuclear DNA in an organism is known as the **genome**, and the specific combination of alleles carried is the **genotype**. In developmental biology the term genotype is usually used in a specific context to refer to the constitution of just one or a few loci. The totality of characteristics of an organism is known as its **phenotype**, and in developmental biology this usually relates to its visible appearance. The normal, or wild-type, phenotype arises from a wild-type genome, and a mutant phenotype arises from the consequences of one or more mutations carried in the genome. It is not possible to deduce the complete function of a gene simply by looking at a mutant phenotype, although when combined with the primary sequence of the gene

and the normal expression pattern, a mutant phenotype can be very informative.

It is possible to deduce quite a lot about normal gene function by looking at the effects of several mutant alleles of the same gene, but with different phenotypes. The most common type of mutation is a **loss of function**, meaning that the protein product of the mutant gene is less active than the wild type. A complete loss of function is called a **null** mutation and corresponds to a complete lack of active gene product. Sometimes there exists a set of alleles having different degrees of loss of function which can be arranged into an **allelic series**, ordered by the severity of the abnormal phenotype (Fig. 3.1). The set of phenotypes displayed by an allelic series may make the function of the wild-type gene much more apparent than the phenotype of a single

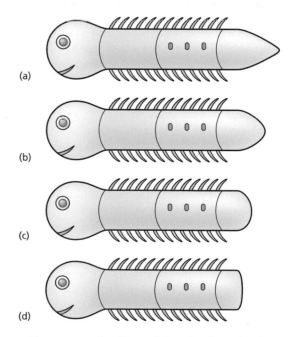

Fig. 3.1 Phenotypes produced by an allelic series of mutations in a gene required for formation of the posterior extremity. (a) Wild type. (b) Slight loss of function. (c) Moderate loss of function. (d) Null mutant.

mutant allele. A weak loss-of-function mutant is more likely to be viable and so survive into adult life than a null. Loss-of-function mutations are usually genetically **recessive**, because their effect will be masked by the presence of a wild-type allele on the other chromosome producing the normal gene product. Sometimes they are genetically **dominant** because the loss of 50% of the product is sufficient to cause an abnormal phenotype. This type of dominance is called **haploinsufficiency**. It has the property that the homozygous phenotype, with 100% loss of function, is much more severe than the heterozygous phenotype with only 50% loss.

There are also dominant mutants showing **gain-of-function** phenotypes. For example, a cell-surface receptor may normally become activated when the ligand is bound to the extracellular domain. But a gain-of-function mutant version may signal all the time regardless of whether the ligand is present or not. Similarly, a transcription factor may normally turn on a target gene in response to some regulatory event, but a gain-of-function mutant form may be active all the time and not respond to the regulation. Gene products that are active all the time are called **constitutive**, and gain-of-function mutants often express constitutive versions of the normal gene product. Such mutants are usually genetically dominant because the gain of function persists even in the presence of wild-type gene product made by the allele on the other chromosome. Another type of mutation that is also genetically dominant, but is not constitutive, is the **dominant negative**. Here the mutant form of the gene product itself has no function, but it interferes with the function of the wild-type form. This may arise, for example, where molecules need to form dimers in order to exert their activity. If the dimer of the dominant negative and the normal form is inactive then the overall activity will be well below the 50% characteristic of a recessive mutation. Recessivity of a mutation is generally indicative of loss of function, but dominance may be due either to a haploinsufficient, or constitutive, or dominant negative effect, and further investigation is required to find which is the case. In principle dominant mutant types can be distinguished by gene dosage analysis since introducing extra copies of a haploinsufficient allele will have little effect, while introducing extra copies of a gain-of-function allele will increase the effect. But such studies are not necessarily easy to perform in all the model organisms.

It should be remembered that genes are often named because of the first-discovered mutant phenotype and this is often a source of confusion on first encounter. If it is a loss-of-function phenotype then the function of the wild-type gene may be opposite to that indicated by its name. For example, the *dorsal* gene in *Drosophila* is responsible for initiating *ventral* development. In the absence of the gene, ventral structures cannot develop and the whole embryo follows the default pathway and becomes dorsal-type all over. Furthermore, homologous genes in different organisms often have different names. This is because they will have been named depending on their mutant phenotypes before the actual gene carrying the mutation was identified. It is often the case that a gene product known from biochemical study of vertebrates is known by the names of its mutations in the model invertebrates used in genetics. For example, the molecule known as beta-catenin in vertebrates is coded by the *armadillo* gene in *Drosophila* and the *wrm-1* gene in *Caenorhabditis elegans*.

Sex chromosomes

The mechanism of sex determination differs quite substantially between animal groups but usually depends on dimorphic sex chromosomes. In mammals, females have two X chromosomes, while males have one X and one Y. In birds it is the female that has two different chromosome types: the male having two Z chromosomes and the female having one W and one Z. In *Drosophila* the sex is determined by the ratio of X chromosomes to autosomes.

Sex-linked mutations are present on a sex chromosome and so their effect depends on the sex of the individual. For example, a single-copy recessive mutation on the mammalian X chromosome will be masked by the wild-type copy in females (XX), but will produce a phenotype in males (XY) because the Y chromosome does not carry the corresponding locus. Chromosomes other than sex chromosomes are called **autosomes**, and any gene or mutation not on a sex chromosome is said to be autosomal.

Maternal and zygotic

Normally in genetics we think of the phenotype as corresponding to the genotype of the same individual. But this is often not the case for the very earliest events of embryonic development. This is because some early developmental events depend on the situation in the *mother* rather than the situation in the embryo itself. For example, if a cytoplasmic determinant is placed in a particular region of the oocyte during oogenesis, then all the genes involved in this process will be those of the mother (Fig. 3.2). If the determinant fails to be formed because of a mutation in the mother's genome, then it will be no use to receive a good copy of the gene from the father because by then it is too late to perform the function. A **maternal-effect** gene is one for which the phenotype of an individual depends not on its own genotype but on that of the mother. A good example is the gene *stella* in the mouse. This encodes a chromosomal protein expressed in germ cells and early embryo blastomeres. If the gene is knocked out, the female −/− mouse is normal but her embryos are defective and die very early.

The period of maternal control of development does not end at fertilization because for most animal types the embryo's own genome, called the **zygotic** genome, remains inactive during the early cleavage stages. Once the zygotic genome has been activated, the normal situation is re-established and the embryo phenotype will thereafter correspond to the embryo genotype.

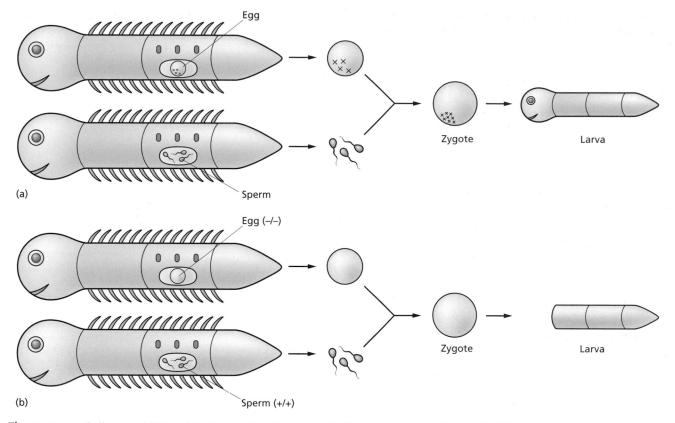

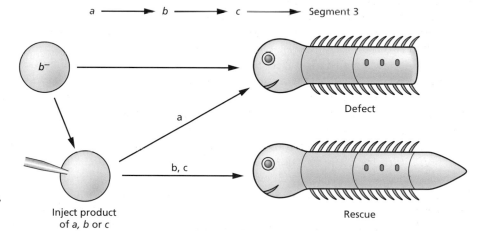

Fig. 3.2 Maternal effect gene. (a) Normal development, in which a maternal effect gene is required to deposit a head determinant in the egg. (b) A mutant mother who produces eggs lacking the determinant, and consequently the offspring are headless.

Fig. 3.3 Elucidation of a genetic pathway by rescue experiments. *a, b, c* are a set of genes that activate each other in a linear pathway and are required for the formation of segment 3. If each of the gene products of *a, b, c* are injected into embryos mutant for *b*, then b and c will rescue a wild-type phenotype but a will not.

The zygotic genome may be activated at different stages in different animal groups, ranging from early cleavage to late blastula.

Genetic pathways

If a set of genes are involved in a single pathway or process it is often possible to deduce the sequence in which they act from genetic data. There are two common situations, one where a group of different mutations have a similar phenotype and the

other where a group of mutations have two phenotypes that are in some sense opposites of one another.

Where several genes have similar mutant phenotypes, the sequence of action can sometimes be established by a rescue protocol. This is illustrated in Fig. 3.3. Imagine that there are three genes in a pathway leading to the formation of the third segment of an animal and that each gene is turned on by the action of the previous one. Loss-of-function mutations in all the genes give the same phenotype, namely loss of the third segment. Now suppose that each of the normal gene products a, b, and c, is

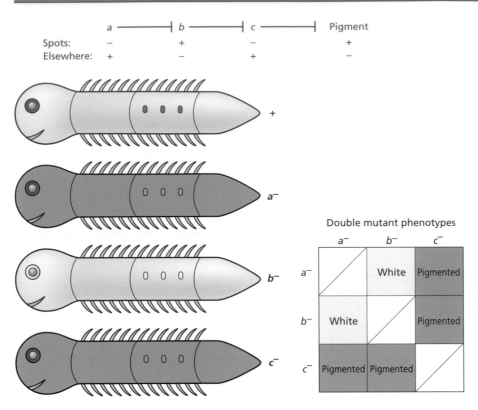

$a \longrightarrow\hspace{-0.8em}| \hspace{0.3em} b \longrightarrow\hspace{-0.8em}| \hspace{0.3em} c \longrightarrow\hspace{-0.8em}| \hspace{0.3em}$ Pigment

	a	b	c	Pigment
Spots:	−	+	−	+
Elsewhere:	+	−	+	−

Double mutant phenotypes

	a^-	b^-	c^-
a^-		White	Pigmented
b^-	White		Pigmented
c^-	Pigmented	Pigmented	

Fig. 3.4 Analysis of a repressive pathway. Normally, a gene a is inactivated in three spots of segment 2, resulting in the formation of dark pigment. In a loss of function mutant of a or c the whole organism is pigmented. In a loss of function mutant of b there is no pigment. The phenotypes of the double mutants show that b must act after a and before c.

injected into embryos which are mutant for gene b. Clearly product a will have no effect because the pathway is blocked at step b. Product b will rescue a mutation in its own gene and give a normal phenotype. But so will c rescue a mutation in b because c lies downstream in the pathway. We could conclude from these data that the pathway goes $a^->(b,c)$, but to order b and c we should need to do a similar experiment to find whether a loss-of-function mutant of c could be rescued by injection of the product b or c. This type of analysis was used to investigate the posterior group mutants in *Drosophila* and to show that *nanos* was the last-acting member of the pathway (see Chapter 11).

If members of a set of mutants have one of two opposite phenotypes then the genes may again code for the successive steps in a pathway but it is likely that some or all the steps will be repressive events rather than activations. Figure 3.4 represents this type of situation. Pigment is made in the spots of segment 2 following the operation of a pathway of three genes in which a represses b which represses c which represses pigment formation. Normally gene a is active everywhere except the spots, so only the spots become pigmented. Two types of loss-of-function mutant can be recovered: b^-, which are unpigmented all over, and a^- or c^-, which are pigmented all over. It is possible to deduce the sequence of action of the genes by examining the phenotype of the double mutants. In each case the phenotype of the double mutant is the same as that produced by the mutant of the *later acting* of the two genes. For example, b^-c^- is pigmented all over because c acts after b. By looking at the phenotype of each double mutant combination, the genes can be arranged into a

pathway. Among many other examples, this type of analysis was used to order the *dorsal* group genes in *Drosophila* (see Chapter 11). Repressive pathways are remarkably common and they can cause much confusion. The best way to understand them is, as in Fig. 3.4, to write out the state of each gene in the pathway under the two possible conditions: that in which the first step is activated, and that in which the first step is repressed. In developmental biology the two conditions often refer two regions within the embryo where the same pathway is under different regulation, for example the dorsal and ventral sides.

These methods are examples of **epistasis** analysis, because if one gene prevents the expression of another it is said to be epistatic to it.

Another method of ordering gene action in development depends on the use of **temperature-sensitive** mutations. In contrast to the pathway situations, this does not depend on any particular relationship between different gene products and can be used to order events in time which are mechanistically quite independent of each other. Temperature-sensitive mutants are those which display the phenotype at a **nonpermissive** (usually high) temperature, and do not show a phenotype at the **permissive** (usually low) temperature. They are frequently weak loss-of-function alleles. They arise from changes in the conformation of the protein product which are sensitive to changes in temperature in the range compatible with embryonic survival. The time of action of a gene may be deduced by subjecting groups of temperature-sensitive mutant embryos to the nonpermissive temperature at different stages of development. If the organism ultimately displays the mutant phenotype, this means that the gene was inactivated at the time of its normal function, in other words that the gene was required during the period of the high temperature exposure. An example would be the time of action of the gene *cyclops*, which is needed for the induction of the floor plate in zebrafish. To form a floor plate the embryos need to be kept at the permissive temperature between the stages of 60 and 90% epiboly, showing that this is the time at which gene function is needed. Temperature-sensitive mutants are more use in poikilothermic organisms such as *C. elegans*, *Drosophila*, and the zebrafish, rather than in homeotherms such as the mouse, because the range of temperatures to which the embryos can be subjected is much wider.

Genetic mosaics

It is sometimes possible to make organisms that consist of mixtures of cells of different genotypes. These are called **genetic mosaics** and can be useful as they provide information about where in the embryo a particular gene is required. For instance an embryo may consist of two territories, A and B, and a particular mutant shows a defect in B. We can consider two informative types of genetic mosaic (Fig. 3.5). One has prospective A cells mutant and prospective B cells wild type, while the other has prospective B cells mutant and prospective A cells wild type. If the organism with B cells mutant shows the abnormal phenotype, then we say that the mutant is **autonomous**; it affects just the region in which the gene is normally active. However, if the organism with A cells mutant shows the abnormal phenotype, then the mutant is **nonautonomous** because it is affecting a structure outside the domain of action of the gene. Nonautonomy means that there must be an inductive signaling step that is affected by the mutation. However, it does not necessarily mean that the mutant gene itself codes for a signaling factor, as failure of the signaling event can be a downstream consequence of the mutation.

Genetic mosaics have been widely used in *Drosophila*. A very useful type is made by pole-cell transplantation and consists of germ cells of one genotype in a host of a different genotype. Such mosaics have enabled the understanding of factors controlling the patterning of the oocyte as a result of interactions with the somatically derived follicle cells of the egg chamber (see Chapter 11). Mosaics have also been used in *C. elegans*, where they arise spontaneously by loss of small duplicated chromosome fragments (see Chapter 12). In zebrafish, mosaics can be created by grafting as there is quite a lot of early cell mixing to disperse the labeled cells. In mammals, the term **mosaic** is usually reserved for a naturally occurring organism composed of two genetically dissimilar cells (e.g. X inactivation mosaic, see Chapter 10) and the term **chimera** is used for embryos made experimentally by cell injection or aggregation of blastocysts.

Genetic mosaics should not be confused with embryos said to show **mosaic** behavior (see Chapter 4). This means that surgical removal of parts causes a defect in the final anatomy corresponding exactly to the fate map. Mosaic behavior is contrasted to **regulative** behavior and has nothing to do with genetic mosaics.

Screening for mutants

The term **forward genetics** is sometimes used to describe investigations that start with the discovery of an interesting mutant phenotype. **Reverse genetics**, by contrast, refers to functional investigations on a known gene. Many interesting mutants have arisen spontaneously, but even more have been recovered in large-scale screens on *Drosophila*, *C. elegans*, zebrafish, and mouse. The details of these screens can be very complex, particularly for *Drosophila* in which there are many selective tricks for reducing the total number of individuals to be dealt with. But the principle is simple and relies just on basic Mendelian genetics. The following description approximates to the procedures used in zebrafish screens, although is still slightly simplified (Fig. 3.6).

A group of males will be mutagenized, for example by treatment with a chemical mutagen. Numerous mutations are induced in the dividing spermatogonial cells, so the animals will thereafter produce sperm containing mutations, potentially in

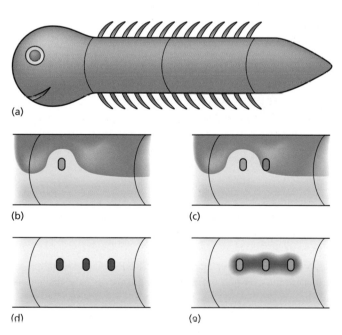

(a)

(b)   (c)

(d)   (e)

Fig. 3.5 Use of genetic mosaic. (a) A mutant of a gene in which the three spots in segment 2 are missing. In (b) and (c) genetic mosaics are made in which the red tissue is null mutant and the green tissue is wild type. In (b) spots appear in the wild-type zone so the gene must have an autonomous function corresponding, for example, to the wild-type expression pattern in (d). In (c) a spot appears in the zone of mutant tissue so the gene must have a nonautonomous function, corresponding, for example, to the expression pattern in (e).

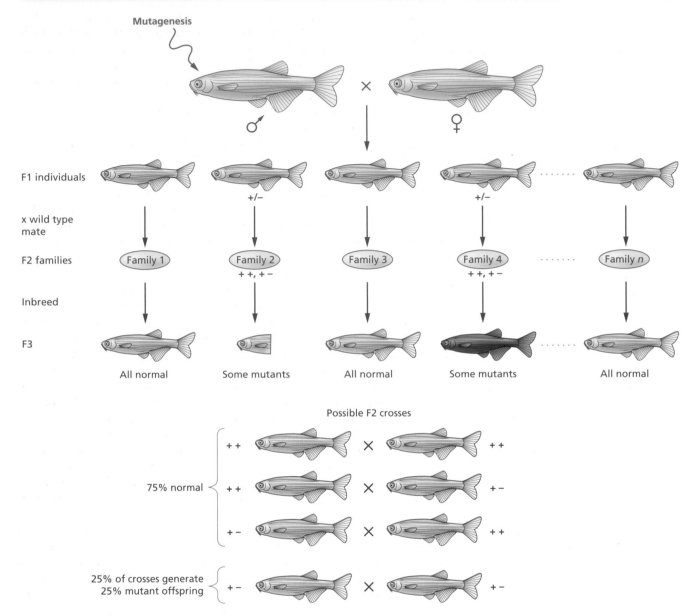

Fig. 3.6 Screen for developmental mutants. This illustrates the simplest possible type of screen for zygotic recessives. Each F1 individual is outcrossed to generate a family at the F2 generation. Pairs of F2 animals are mated and if the family contains a mutation then about one mating out of four will be between two +/− individuals, and such matings will yield 25% −/− progeny which should show an abnormal phenotype.

any gene in the genome. The mutagenized males are mated to normal females, producing an F1 offspring generation. Each of the F1 individuals is likely to carry a mutation in heterozygous form, and each is likely to carry a different mutation from all the others. So each F1 individual is put in a separate container for further mating to a wild-type animal. This produces a family at the F2 generation. If the F1 individual did carry a mutation, then half the F2 individuals will be heterozygous for it. A set of test matings is carried out between pairs of individuals within each F2 family. If there is a mutation present then 1 in 4 matings will be between heterozygotes and yield an F3 generation that is 25% homozygous mutant. The F3 generation is examined and

scored at the embryo stage. By definition developmental mutations are those which perturb the anatomy of the organism, so the homozygous mutants should be visibly abnormal. They may be detected simply by examination of the embryos under the dissecting microscope, or if the screen is more focused, following immunostaining or *in situ* hybridization to display a particular structure or cell type under investigation.

Any mutation that disables a gene essential for early development is quite likely to be lethal and prevent development after the time of normal gene function. So the homozygous mutant F3 embryos may well die at an early stage, and they need to be examined early on before they degenerate. It is obviously not

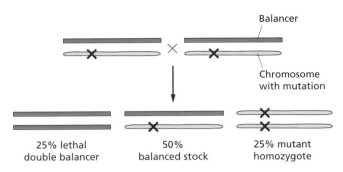

Fig. 3.7 Maintenance of a mutant line by means of a balancer chromosome.

possible to maintain a line of a homozygous lethal mutation, but the mutants can be maintained by keeping the heterozygous parents. These will go on producing batches of embryos in which 25% are homozygous mutant, and when the original F2 animals are too old to go on reproducing they can be replaced by their heterozygous offspring. The overall screening procedure has various elements of randomness. It is quite likely that F2 families will contain no mutation, or none giving an abnormal phenotype, in which case they are discarded. It is also possible for an F3 phenotype to be caused by more than one mutation in the original sperm. If there are two mutations showing independent segregation then the F3 generation will actually be a 9 : 3 : 3 : 1 mix of normal, single homozygotes of each type and double homozygotes. This may not be apparent immediately, but can be resolved by further breeding.

In *Drosophila* there are some very sophisticated methods for reducing the labor involved in screens. The most important is the use of **balancer chromosomes**. These have multiple inversions which mean that there is no recombination between the balancer chromosome and its wild-type homolog. They also carry a recessive lethal mutation, so flies with homozygous balancers are not viable. They also carry some **marker gene** that will enable all flies carrying the balancer in one copy to be easily identified. The uses of balancers are numerous, but one of the most important is in the simple maintenance of a recessive lethal mutant line. This is shown in Fig. 3.7. The line carries one copy of the balancer chromosome and one copy of the homologous chromosome bearing the mutation. In each generation a 1 : 2 : 1 ratio is produced of homozygous balancer, heterozygous and homozygous mutant. The heterozygotes are the only viable offspring and serve to maintain the line. The homozygous mutant embryos are available for experiments. This means that a line can be maintained by repeated mating with no need to test individuals to see whether they are heterozygotes.

Cloning of genes

Developmental genetics existed long before molecular cloning was introduced, but it is now regarded as essential to clone any gene of interest identified by mutagenesis. This caused considerable difficulty in the past but is much easier today with the availability of high-resolution genome maps and genome sequences. A gene is regarded as cloned when the complete coding sequence is incorporated into a bacterial plasmid, or other cloning vector, so that it can be amplified and purified in a quantity suitable for use in any of the types of investigation now described as **reverse genetics**.

Most of the developmentally important genes in *Drosophila* were cloned by inducing mutations with a transposable element, the **P-element**. Once it had been shown that a P-element had integrated into the locus of interest then it could be used as a probe to isolate DNA clones from a genomic library. Nowadays, in most cases for experimental organisms as well as for human genetics, genes are cloned by **positional cloning**. Here a mutation is mapped to very high resolution using microsatellite polymorphisms or restriction fragment length polymorphisms. There are many of these scattered through the genome and so long as the genome has been sequenced all their positions should be known. It is possible to obtain sets of PCR primers that enable each polymorphism to be detected in DNA samples by the presence of a specific band visible on a DNA gel. So long as enough offspring can be produced it is now possible to map a mutation to the specific locus using a single cross. The procedure is shown in principle in Fig. 3.8. In general the test parent will be heterozygous for a recessive lethal mutation of gene g, designated g^+g^-. It is crossed to an individual of another strain in which most of the polymorphic loci are different. The F1 individuals should contain 50% heterozygotes for the mutant allele of g. F1 individuals are crossed together to yield many families, of which about one quarter should contain 25% homozygous lethal g^-/g^- individuals. DNA is isolated from each of the mutants, and also from many of the wild-type siblings, which may be g^+g^+ or g^+g^-. These are then all individually typed for a selection of the polymorphic markers. As shown in Fig. 3.8, a marker that is closely linked to the mutant locus should segregate with it, while others will segregate away by independent assortment of chromosomes or by meiotic crossing over within a chromosome. It will be necessary to go through several cycles of mapping, using the same DNA samples with markers that are more and more closely spaced around the mutant locus. Eventually a small chromosome region will be identified which is known from the genome sequence only to contain a few genes. Each of these is then evaluated as a candidate. One consideration is the putative nature of the protein deduced from the sequence, for example if the mutation has a cell autonomous action it is unlikely to code for a signaling molecule. Another is the expression pattern of the candidate gene relative to the domain of action of the mutation. A gene that is not expressed in the relevant region is unlikely to be a good candidate. The expression pattern can be established using *in situ* hybridization (see Chapter 5) with probes designed from the known sequence. Once a good candidate has been found the mutant DNA can be sequenced at that locus to see if it does, in fact, contain a mutation. Final proof may be obtained by

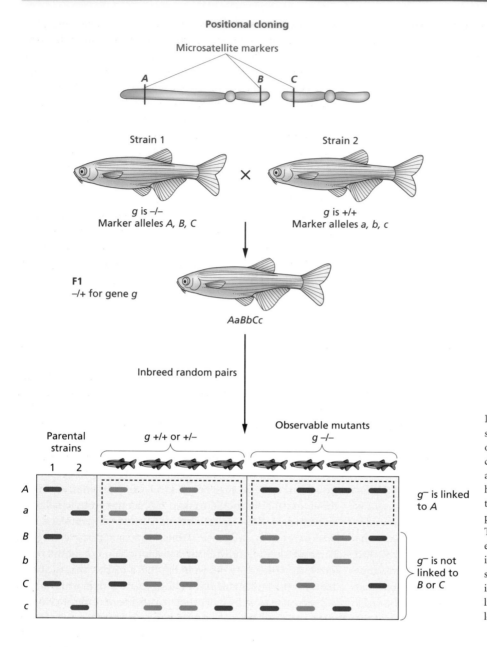

Fig. 3.8 Positional cloning. This is a very simplified presentation of the principle of positional cloning. The mutation to be cloned is in a gene called *g*. Three markers are considered, of which the parental strains have alleles *A*,*B*,*C* or *a*,*b*,*c*. Mutant or wild-type F2 individuals are analyzed using PCR primers for each of the polymorphic loci. The reaction mixtures are separated by gel electrophoresis and each specific allele is indicated by a DNA fragment of a particular size. It may be seen that the g^- mutant allele is linked to *A* and the g^+ wild-type allele is linked to *a*. This indicates that the mutant locus lies near to the *Aa* locus.

introducing a good copy of the gene by transgenesis and finding whether this will rescue the mutant phenotype towards the wild type.

Genome sequencing has finally enabled reasonably accurate estimates to be made of gene numbers, and hence indirectly of the complexity of living organisms. In terms of protein-coding genes, free-living bacteria have about 2000–4000 genes; unicellular eukaryotes like yeast about 6000; invertebrate animals like *Drosophila* and *C. elegans* have up to 20,000; and vertebrates have about 30,000. The number for vertebrates is considerably lower than previous estimates but it should be remembered that complexity arises not only from the number of genes but also from the number of distinct proteins which may be produced by alternative splicing and post-translational modifications. It is

also increased by the complexity of the regulation, and developmentally important genes in particular can have very complex regulatory regions.

Transgenesis

The scope of genetics has been considerably expanded in the molecular era such that many of the genetic variants used in experiments are not produced by mutagenesis, but by more sophisticated and directed methods. Such lines of animals are often called transgenic in popular or legal parlance. In biology it is more usual to reserve the terms transgenic and transgenesis for cases where an extra gene has been introduced into the genome, as opposed to **knockouts** in which a gene has been removed. A **transgene** is simply a gene that has been introduced into an organism by transgenesis. Methods for transgenesis exist for mouse, *Xenopus*, zebrafish, *Drosophila*, and *C. elegans*.

In the mouse and zebrafish, the gene is introduced by injecting DNA into the fertilized egg. In *Xenopus* the DNA is incorporated into the sperm, and in *Drosophila* and *C. elegans* the transgene is incorporated into a transposable element which is injected into the germ cells of the embryo. All methods of transgenesis share the property that the integration site in the genome is random, or at least not controlled. Transgenes are usually designed so that their expression is regulated by a **promoter** within the insert and they are as far as possible immune from the effects of the position within the genome at which they have integrated. However, they are sometimes designed specifically to probe the local environment. An **enhancer trap** consists of a **reporter gene** such as *lacZ* (see Chapter 5) coupled to a minimal promoter that can bind the transcription complex but has very low activity. If the transgene integrates within range of an enhancer then the activity of the promoter will be boosted enough to produce a detectable output of β-galactosidase, and the expression pattern of the *lacZ* will be characteristic of the enhancer controlling it. Enhancer trap lines have found considerable use, particularly in *Drosophila* (see Chapter 17). The **gene trap** is a related method for identifying functional genes by integration and has been used a lot in the mouse (see Chapter 10).

There is no satisfactory method of transgenesis for the chick, but considerable use has been made of **retroviruses** to introduce genes into embryos by infection. Retroviruses are RNA viruses and, after infection of a cell, the genome is reverse transcribed to DNA which becomes integrated in the host genome. If an additional gene is inserted into the viral genome this too will be integrated and expressed under the control of one of the viral promoters. Infected cells make and export virus particles which infect the neighboring cells, and therefore in a few days the infection, and the transgene, spreads through the embryo. As this may eventually have harmful consequences for the embryo, the experiments are usually timed so that just the region of interest has been infected at the critical time. Note that the type of virus used for genetic modification is replication competent, while those used for clonal cell labelling (see Chapter 4) are replication incompetent.

Targeted mutagenesis

The introduction of specifically engineered mutations at a desired site in the genome is a technology that has been brought to a high level in the mouse, but is not generally available for other species. It has mainly been used to produce **knockouts**, or loss-of-function mutations in particular genes selected because they were thought likely to be of developmental or medical importance. Targeted mutagenesis depends on **homologous recombination**, which is the direct replacement of a gene by a

modified version made *in vitro*. If DNA is added to cells or embryos, most of the integration events will occur in the wrong place and so there needs to be a selection step to isolate the homologous recombinants that have replaced the endogenous gene. This has to be done on tissue-culture cells, so the applicability of targeted mutagenesis depends on the availability of cells that can be grown in culture and can then be reintroduced into an embryo. Because tissue culture is not well developed outside the mammals this effectively limits targeted mutagenesis to mammals. In the mouse there are **embryonic stem cells** that can be engineered and then reincorporated into embryos to produce germ-line chimeras from which the new strain can be bred (see Chapter 10). In several species of domestic mammals it has proved possible to clone whole animals by nuclear transplantation from primary cultures of fetal cells into enucleated oocytes (see Chapter 2). This should make it possible to develop targeted mutagenesis also for these species as the necessary selective steps can be done in tissue culture.

Other ways to inhibit gene activity

For the species in which targeted mutagenesis is not possible, other methods for producing specific inhibition of known genes have been developed. They are inherently less specific than mutation and when using them it is important to ensure that experimental tests of specificity have been carried out.

One method involves the design of **dominant negative** reagents. For example a transcription factor lacking its DNA binding domain may often act as a dominant negative because when overexpressed it will sequester all the normal cofactors needed by the wild-type factor, or it may form inactive dimers with the wild-type factor. A dominant negative version of a gene product can "knock out" the normal gene if it is simply introduced as a transgene anywhere in the genome. It does not need to be a homologous recombination. Alternatively the mRNA for the dominant negative protein can be produced *in vitro* and injected into the fertilized egg. This is a technique that is suitable for organisms with large eggs, like *Xenopus* and the zebrafish, because they are easy to inject and the injected mRNA can exert its effect on early developmental events before it is degraded or diluted by growth.

Secondly, there is the **domain swap** method, used extensively for transcription factors. Because transcription factors have a modular design (see Appendix) it is possible to replace an activating region with a repressing domain or vice versa. The domain-swapped factor will still bind to the same site in the DNA but instead of activating its target genes it will repress them (or vice versa). This is not quite the same as a loss-of-function mutation, since there will be an active repression of any gene to which the target factor binds, and these genes would not necessarily be inactive following a simple ablation of the transcription factor. Again this can be introduced either by transgenesis or by injection of mRNA into the fertilized egg. The usual inhibitory domain used in this type of experiment is that from the *Drosophila* gene *engrailed*, and the normal activating domain is that from herpesvirus gene *VP16*.

The third strategy involves the use of **antisense** reagents. If an RNA transcript is made from the noncoding strand of the DNA then it will be an antisense version of the normal messenger RNA. When introduced into the embryo this will form hybrids with the normal mRNA, which are inactive as translation substrates and are often rapidly degraded. There have been fashions for introduction of full length antisense mRNA, and for the external application of antisense oligodeoxynucleotides, but the currently favored methods fall into two groups: the use of morpholinos and the use of RNA interference (RNAi).

Morpholinos are analogs of oligonucleotides, in which the sugar-phosphate backbone is replaced by one incorporating morpholine rings. Unlike normal oligonucleotides they are resistant to degradation by the nucleases that are present in all cells and extracellular fluids, but because the usual four bases are linked to the resistant backbone with the correct spacing, they can still undergo hybridization with normal nucleic acids. Morpholinos are usually synthesized to be about 20 residues long and are designed to be complementary to a region of the mRNA likely to be accessible, such as the translation start region. The hybrid of morpholino and mRNA is not degraded but remains inactive for protein synthesis. Because there is no degradation of the mRNA it is necessary to show that specific protein synthesis has been blocked, which requires the availability of an antibody to the protein that can be used for Western blotting or immunoprecipitation or *in situ* immunostaining. Morpholinos cannot generally penetrate cell membranes and so their main application has been in early embryos of free-living embryos where they can easily be administered by intracellular injection, namely *Xenopus*, zebrafish, sea urchins, or ascidians.

The **RNA interference** (**RNAi**) method depends on normal host defenses against RNA viruses. In some embryo types introduction of double-stranded RNA (**dsRNA**) of the same sequence as the normal mRNA can be a very effective method for specific mRNA destruction. dsRNA is cleaved by an enzyme called dicer into short (21–23 bp (basepair)) length fragments. These enter a silencing complex that can bind to, unwind, and cleave mRNAs that contain complementary sequence. Because the mechanism is catalytic a few molecules of double-stranded RNA can destroy a much larger quantity of mRNA. In mammalian cells the long dsRNA causes nonspecific inhibition of translation, but the short (21–23 bp) length processed fragments do not, and can be used directly to bring about destruction of specific mRNAs. dsRNA does not enter cells readily but can be introduced using the same type of lipid transfection reagent that is used to introduce plasmids into cells. Because it is possible to make large libraries of dsRNA this method is now being used instead of chemical mutagenesis to conduct screens. It is particularly suitable for *C. elegans*.

A final method of specific inhibition is treatment of the embryo with a specific **neutralizing antibody** directed against the protein product of the gene of interest. Antibodies will not penetrate intact embryos and so they must be injected at the site of interest. A common problem with this method is that most antibodies that bind to a particular protein will not neutralize its activity, so it is necessary to have some independent test to show that the antibody does, in fact, neutralize the target protein.

Some of these inhibitory techniques can be used in transgenic mode, but they are very often used as transient, nongenetic procedures. As mentioned above, it is easy to inject substances into *Xenopus* or zebrafish embryos, and it is also easy to treat later organ cultures from mammalian or chick embryos. This can be very useful so long as the inhibitor is able to penetrate to the site of action.

Gene duplication

Gene duplication is probably the major source of evolutionary novelty. If a gene becomes duplicated then the constraints on changes to its sequence become relaxed. At one extreme, one copy could continue to be the functional gene and the other copy could accumulate mutations such that it acquired a novel and advantageous function. Alternatively the second copy could accumulate deleterious mutations until it became nonfunctional, and maybe eventually not even expressed (a pseudogene). More usually, both copies will accumulate some sequence divergence such that they carry out subsets of the original function. Soon after the duplication the overlap in function will be considerable, while after millions of years the functions will diverge. For example the *cyclops* and *squint* genes of the zebrafish arise from duplication of the *nodal* gene which encodes a critically important mesoderm-inducing factor in vertebrate development, but they have diverged in function such that they act at different developmental stages (see Chapter 8).

The extreme case of gene duplication occurs when the entire genome becomes duplicated, with a doubling of chromosome number. This is called tetraploidization, as the resulting organisms are tetraploid instead of diploid. Tetraploidization can produce a vast array of new genes instantaneously and so enormously enlarge the adaptive possibilities for the line of descent. The pattern of multigene families in vertebrates suggests that two tetraploidization events may have occurred at the time of the origin of vertebrates, temporarily boosting their gene number from about 20,000 to about 80,000. This may account for their subsequent adaptive radiation and evolutionary success, although the count of protein-coding genes in extant vertebrates suggests that the number has been much reduced in subsequent evolutionary time back to about 30,000. It also seems that further tetraploidizations have occurred in various lineages. For example *Xenopus laevis* looks as though it underwent a tetraploidization about 30 million years ago, as most genes are recovered in two copies differing in sequence by about 10%. These are known as **pseudoalleles**. They look like alleles, and generally have the similar expression patterns and functions, but they are not alleles because they occupy distinct genetic loci. Bony fish, including the zebrafish, seem to have undergone a tetraploidization at a much more remote time, about 230 million years ago when they first arose as a lineage. Because of the long time interval since this event, each pair of genes is now substantially diverged in sequence and function so bony fish are considered to be diploid organisms today.

Limitations of developmental genetics

Despite its successes there are certain limitations to the standard protocols of developmental genetics. One problem arises directly from gene duplication because this has resulted in the extensive presence of **redundancy** between genes, meaning that there are two or more genes with a significant overlap in function. It is a particular problem in vertebrates because of the repeated gene duplications and divergence. It is unlikely that two genes ever have an exactly identical function unless they are the result of a very recent duplication event. However there are many examples of substantial overlap when gene function is examined at the laboratory level, where the organisms are not subject to all the vagaries of selection in the wild. The consequence is that mutation of a single gene to inactivity may produce no abnormal phenotype, or a minimally abnormal phenotype not consonant with the true function of the gene. Mutagenesis screens are usually designed to look at single mutations and so the presence of widespread redundancy in the genome greatly limits the number of informative phenotypes that can be recovered.

Where targeted mutagenesis is possible a mutant can be made and characterized without the need for a specific phenotype, and in fact numerous mouse knockouts show no abnormal phenotype or a very minimally abnormal phenotype. However, an abnormal phenotype is often found when mutations in several members of a multigene family are combined by breeding. For example, the knockouts of the **myogenic** genes *MyoD* and *myf5* have limited effects individually, but in combination show an almost complete inhibition of myogenesis. Likewise, the knockouts of individual Hox genes (see Chapter 4 and later) often show little effect, but if all members of a homologous group (a **paralog group** see Chapter 20) are knocked out, then the abnormality does become significant.

A distinct type of problem for genetic analysis arises where a developmental gene has several functions at different stages of development. In a null mutant the embryos may die because the first of these functions cannot be carried out and this means that the phenotype will not be informative about any of the later functions. An example is the knockout of the gene for Fibroblast Growth Factor 4 (*fgf4*). Although FGF4 has important functions in gastrulation, brain development, and limb development, the null phenotype is an early lethal because it is needed in preimplantation stages for cell division in the extraembryonic supporting tissues (see Chapter 10). In this case the problem can be circumvented because of the availability of some sophisticated experimental strategies, but it illustrates how the phenotype of a null mutant does not always reveal much about gene function.

Key Points to Remember

- Mutants have been very important for identifying developmental genes and unraveling developmental mechanisms.
- In general mutations are genetically recessive if they lead to loss of function and genetically dominant if they lead to gain of function.
- Genes may be named after their loss-of-function mutant phenotype, and hence the name may seem opposite to the actual function. The same gene in different organisms may have different names.
- Mutations affecting early developmental processes are often maternal-effect, those affecting later processes are zygotic.
- Regulatory or biochemical pathways can be deduced from genetic experiments, especially from epistasis experiments in which the combined effect is determined of two mutations with opposite phenotypes.

- Screens for developmental mutants can be conducted by mutagenesis followed by breeding to homozygosity.
- Once a mutation has been identified the gene is usually cloned by positional cloning.
- Transgenic organisms are those with an extra gene inserted in the genome, and can be made in most of the laboratory model species.
- Targeted mutagenesis based on homologous recombination is mostly applicable only to the mouse.
- Many other experimental methods of inhibiting specific gene activity exist, including introduction of dominant negative reagents, antisense oligonucleotides, and RNAi.
- The existence of widespread gene duplication in evolution means that many genes have overlapping functions (redundancy) so the loss-of-function phenotype may not reveal the full activity of a gene.

Further reading

General

Wilkins, A.S. (1992) *Genetic Analysis of Animal Development*, 2nd edn. New York: Alan R. Liss.

Hartl, D.L. & Jones, E.W. (2001) *Genetics: analysis of genes and genomes*, 5th edn. Sudbury. MA: Jones and Bartlett.

Hawley, R.S. & Walker, M.Y. (2003) *Advanced Genetic Analysis*. Oxford: Blackwell Science.

Hartwell, L.H., Hood, L., Goldberg, M.L., Reynolds, A.E., Silver, L.M. & Veres, R.C. (2004) *Genetics: from genes to genomes*, 2nd edn. New York: McGrawHill.

Classic mutagenesis screens

Driever, W. (1996) A genetic screen for mutations affecting embryogenesis in zebrafish. *Development* **123**, 37–46.

Haffter, P., Granato, M., Brand, M. et al. (1996) The identification of genes with unique and essential functions in the development of the zebrafish, *Danio rerio. Development* **123**, 1–36.

Hirsh, D. & Vanderslice, R. (1976) Temperature sensitive developmental mutants of *Caenorhabditis elegans. Developmental Biology* **49**, 220–235.

Moser, A.R., Pitot, H.C. & Dove, W.F. (1990) A dominant mutation that predisposes to multiple intestinal neoplasia in the mouse. *Science* **247**, 322–324.

Nüsslein-Volhard, C. & Wieschaus, E. (1980) Mutations affecting segment number and polarity in *Drosophila. Nature* **287**, 795–801.

Positional cloning

Wickling, C. & Williamson, R. (1991) From linked marker to gene. *Trends in Genetics* **7**, 288–293.

Genetic pathways

Anderson, K.V., Jurgens, G. & Nüsslein-Volhard, C. (1985) Establishment of dorso-ventral polarity in the *Drosophila* embryo: genetic studies on the role of the Toll gene product. *Cell* **42**, 779–789.

Schüpbach, T. & Wieschaus, E. (1986) Maternal-effect mutations altering the anterior-posterior pattern of the *Drosophila* embryo. *Wilhelm Roux's Archives of Developmental Biology* **195**, 302–317.

Allelic series

Strecker, T.R., Merriam, J.R. & Lengyel, J.A. (1988) Graded requirement for the zygotic terminal gene, *tailless*, in the brain and tail region of the *Drosophila* embryo. *Development* **102**, 721–734.

Maternal inheritance

Payer, B., Saitou, M., Barton, S.C. et al. (2003) *Stella* is a maternal effect gene required for normal early development in mice. *Current Biology* **13**, 2110–2117.

Temperature shift

Tian, J., Yam, C., Balasundaram, G., Wang, H., Gore, A. & Sampath, K. (2003) A temperature-sensitive mutation in the nodal-related gene *cyclops* reveals that the floor plate is induced during gastrulation in zebrafish. *Development* **130**, 3331–3342.

Mosaics

Nagy, A. & Rossant, J. (2001) Chimaeras and mosaics for dissecting complex mutant phenotypes. *International Journal of Developmental Biology* **45**, 577–582.

Transgenesis and targeted mutagenesis

Rubin, G.M. & Spradling, A.C. (1982) Genetic transformation of *Drosophila* with transposable element vectors. *Science* **218**, 348–353.

Palmiter, R.D. & Brinster, R.L. (1985) Transgenic mice. *Cell* **41**, 343–345.

Thomas, K.R. & Capecchi, M.R. (1987) Site-directed mutagenesis by gene targeting in mouse embryo-derived stem-cells. *Cell* **51**, 503–512.

Kroll, K.L. & Amaya, E. (1996) Transgenic *Xenopus* embryos from sperm nuclear transplantations reveal FGF signaling requirements during gastrulation. *Development* **122**, 3173–3183.

Bishop, J. (1999) *Transgenic Mammals*. Harlow: Longman.

Nakamura, H., Katahira, T., Sato, T., Watanabe, Y. & Funahashi, J.I. (2004) Gain- and loss-of-function in chick embryos by electroporation. *Mechanisms of Development* **121**, 1137–1143.

Other inhibitory techniques

Lagna, G. & Hemmati-Brivanlou, A. (1998) Use of dominant negative constructs to modulate gene expression. *Current Topics In Developmental Biology* **36**, 75–98.

Hannon, G.J. (2002) RNA interference. *Nature* **418**, 244–251.

Heasman, J. (2002) Morpholino oligos – making sense of antisense. *Developmental Biology* **243**, 209–214.

Chapter 4

Experimental embryology

Although the techniques of molecular biology and genetics are now essential in the investigation of development, historically it was the experimental embryologists who gave most thought to mechanism and who formulated the basic conceptual framework which is still used today.

Normal development

Normal development means the course of development which a typical embryo follows in standard laboratory conditions when it is free from experimental disturbance. A sound knowledge of normal development is necessary in order to understand the effects of experimental manipulations. In order to describe embryos, a number of standard terms are in use (Fig. 4.1). The front end of an animal is known as the **anterior** or **cranial** end. The rear end is known as the **posterior** or **caudal** end. The upper surface is **dorsal**, the lower surface is **ventral**. For microscope sections, those taken across the long axis of the animal are called **transverse**. Those parallel to the long axis are **longitudinal**. A vertical longitudinal section is known as **sagittal** if it is in the midline and **parasagittal** if it is to one side of the midline. A horizontal longitudinal section, separating dorsal and ventral sides, is called **frontal** or **coronal**.

All of the model species used for laboratory work have published tables of **stage series** which describe the course of development as a number of standard stages which can be identified by external features under the dissecting microscope. Embryonic development is predictable, so if an embryo has reached stage 10 at a particular time then it is possible to be confident that it will reach stage 20 a particular number of hours later. For mammals and birds, development will always occur at a particular physiological temperature, but for free-living embryos such as *Xenopus* or zebrafish, the rate of development will depend on the temperature. The existence of these tables enables investigators to standardize their procedures by using embryos of the same stage, regardless of the temperature in the laboratory that day.

Features of development are referred to as maternal if they are due to components which exist in the egg that have been accumulated during oogenesis. They are said to be **zygotic** if they are due to components newly synthesized by the embryo itself after fertilization.

The fate map

A **fate map** is a diagram that shows what will become of each region of the embryo in the course of normal development: where it will move, how it will change shape, and what structures it will turn into. The fate map will change from stage to stage because of morphogenetic movements and growth, and so a series of fate maps will depict the trajectory of each part from the fertilized egg to the adult. The precision of a fate map depends on how much random cell mixing occurs in development. If there is none, as in the nematode *Caenorhabditis elegans*, then the fate map can be precise down to the cellular level. For most embryo types there is some local mixing of similar cells and therefore the fate maps cannot be quite this precise. Nonetheless, the fate map is a fundamental concept in embryology and the interpretation of nearly all experiments concerned with early developmental decisions depends on knowledge of the fate map.

Fate maps are constructed by labeling single cells or regions of embryos and locating the position and shape of the labeled patch at a later stage of development (Fig. 4.2). The labeling methods used are those described in Chapter 5, and may be either an extracellular label to a patch of cells, injection of an intracellular label to one cell, or grafting of labeled tissue to replace an exactly equivalent piece in the host embryo. The results of many individual labeling experiments will be combined to form the fate map for one particular stage. It is essential to note that a fate map does not indicate anything about developmental **commitment**. All parts of the embryo have a **fate** throughout development, but commitment to form particular structures or cell types is usually acquired through a series of intercellular interactions.

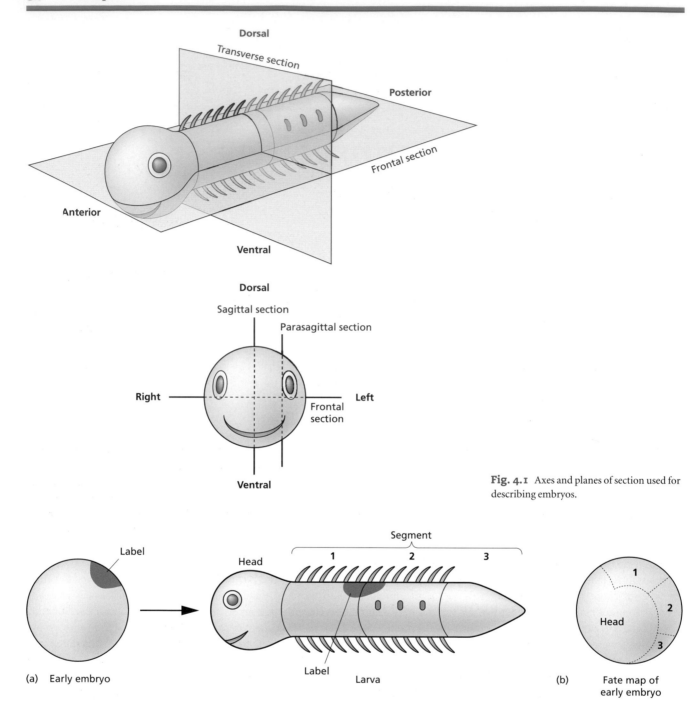

Fig. 4.1 Axes and planes of section used for describing embryos.

Fig. 4.2 Fate mapping. (a) A label placed at a particular position in the early embryo ends up in a reproducible position on the animal. (b) A possible fate map of the early embryo.

In older embryological literature an embryo is referred to as a **mosaic** if experimentally isolated parts develop according to the fate map. It is referred to as **regulative** if an isolated part forms more structures than expected from the fate map. In reality all types of embryo show some aspects of mosaic and of regulative behavior depending on the region of the embryo and the developmental stage concerned. Note that mosaic in this sense is nothing to do with **genetic mosaics**, which are organisms consisting of cells of different genotype (see Chapter 3).

Clonal analysis

Clonal analysis is a form of fate mapping in which a single cell is labeled and the position and cell types of its progeny identified at a later stage. The labeling may be carried out by injection of one cell with a lineage label. This is a simple method where the cells are large and very suitable for organisms that do not grow significantly such as early stage *Xenopus* or zebrafish or sea urchins. For organisms that do grow, such as chick or mouse

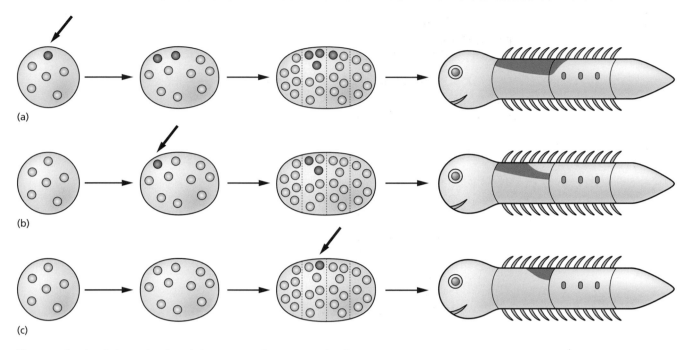

Fig. 4.3 Clonal analysis. No clonal restriction means no determination, but the converse is not necessarily true. (a) Single cell labeled before determination, progeny span later boundary. (b) Cell labeled before determination but progeny fail to span boundary because clone is too small. (c) Cell labeled after determination and cannot cross boundary.

embryos, it is preferable to introduce into a single cell a genetic label that will persist without dilution. This might be insertion by a replication-incompetent retrovirus, or a genetic recombination event that yields a visible marker. The most useful aspect of clonal analysis is to decide whether a cell is committed to form a particular structure or cell type at the time of labeling. If it is committed to form a particular structure called A then its descendants will only be able to populate structure A and nothing else. It follows that if a clone labels both structures A and B then the cell cannot have been exclusively committed to develop into either A or B at the time of labeling (Fig. 4.3). Sometimes it is found that a label applied early will span A and B while a label applied later will populate only A or B. This may be because the cells have become committed in the time between the two labels. However, it may simply be because the later-induced clones are smaller and so have less chance of populating more than one structure. Thus, a clonal analysis can prove *lack* of commitment but cannot prove the *presence* of commitment.

Clonal analysis has been extensively used particularly in the analysis of *Drosophila* segmentation (see Chapter 11) and of vertebrate hindbrain patterning (see Chapter 14). The term **compartment** is sometimes used to indicate a region in an embryo whose boundaries are boundaries of clonal restriction. Once a compartment is established, no cells may enter and none may leave. In other words all the cells within a compartment are the descendants of the founder cells. A compartment usually corresponds to a visible structure such as a segment or an organ rudiment and is maintained either by physical boundaries to cell migration, such as basement membranes, or by a differential

adhesion of the cells of the compartment compared to those outside, such that all cells of the compartment stick together and cannot mix with their neighbors. In a few cases, notably the anterior–posterior compartment boundaries of *Drosophila* imaginal discs (see Chapter 17), the boundary of clonal restriction does not correspond to a visible anatomical boundary.

Developmental commitment

As development proceeds, formerly uncommitted cells become committed to form particular body parts or cell types. We now regard commitment as being encoded as a combination of transcription factors present in the cell and so it can be visualized directly by observing the expression of the relevant genes using *in situ* hybridization. But historically commitment was investigated by embryological experiments. This led to two operational definitions, usually called **specification** and **determination**, which are still useful today.

A cell or tissue explant is said to be specified to become a particular structure if it will develop autonomously into that structure after isolation from the embryo (Fig. 4.4a,b). If a large number of such experiments are performed, and the results combined, it is possible to construct a specification map of the embryo. This shows what the cells have been programmed to do by that particular developmental stage. The specification of a region need not be the same as its fate in normal development. For example the prospective neural plate of a *Xenopus* blastula will differentiate not into a neuroepithelium but into epidermis

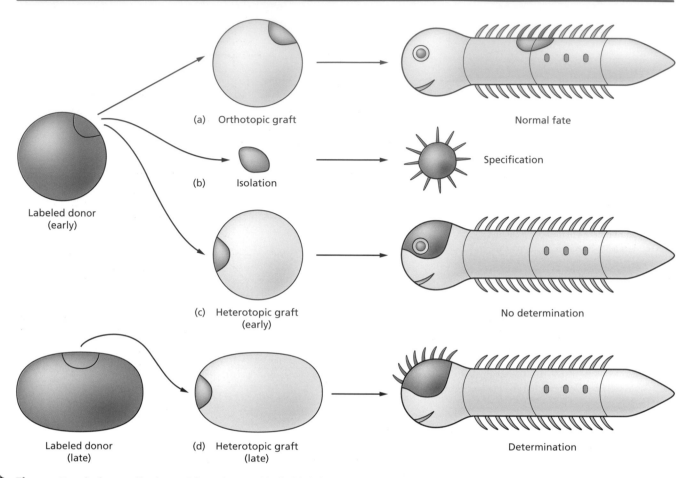

(a) Orthotopic graft

Normal fate

(b) Isolation

Specification

(c) Heterotopic graft (early)

No determination

Labeled donor (early)

Labeled donor (late)

(d) Heterotopic graft (late)

Determination

Fig. 4.4 Tests for fate, specification, and determination. (a) The labeled region will normally contribute to the spiny dorsal part of the animal (fate). (b) When isolated this tissue still forms dorsal spines (specification). (c) When grafted at an early stage to another region it differentiates according to the new position (not determined). (d) When grafted at a later stage to another region it differentiates according to its original position (determined).

when cultured in isolation. In order to form neuroepithelium it needs to receive an inductive signal from the mesoderm (see Chapter 7).

A **determined** region of tissue will also develop autonomously in isolation but differs in that its commitment is irreversible with respect to the range of environments present in the embryo. In other words it will continue to develop autonomously even after it is moved to any other region of the embryo (Fig. 4.4c,d). A very large number of embryological experiments consist of **grafting** a piece of tissue from one place to another and asking whether it develops in accordance with its new position or its old position. Grafts will usually be labeled by one of the methods described in Chapter 5, so that the tissues of graft and host can be distinguished. A graft to the same position of another embryo is called an **orthotopic** graft, and is one of the usual methods of fate mapping. A graft to a different position in the host is called a **heterotopic** graft, and represents the test for **determination**. If the pathway of development is unaltered by such a graft then the tissue is defined as determined. If a heterotopic graft develops according to its new position then it follows that it was not determined, although it may have been specified,

at the time of grafting. A series of such grafts performed at different stages usually show a time at which the tissue becomes determined. For example, the prospective neural plate of a *Xenopus* embryo is not determined at the blastula stage because it will form epidermis or mesoderm if grafted elsewhere in the embryo. It becomes determined to form neural plate during gastrulation, as, after this stage, grafts to other regions of the embryo will always differentiate into neuroepithelium. In a molecular sense, determination means that the cells have lost their responsiveness, or **competence**, to the signals that originally turned on the relevant combination of transcription factors. This may be because the cells have lost receptors or other components of the signaling machinery, or because the transcription factor combination is maintained by other factors than those responsible for turning it on in the first place.

In the development of any embryo there will be a hierarchy of subdivision (Fig. 4.5). For example there will first be formed the three germ layers, ectoderm, mesoderm, and endoderm, then each germ layer will be subdivided, for example the ectoderm into epidermis, neuroepithelium, and neural crest. Later the neuro-epithelium will be subdivided on a smaller scale into subregions

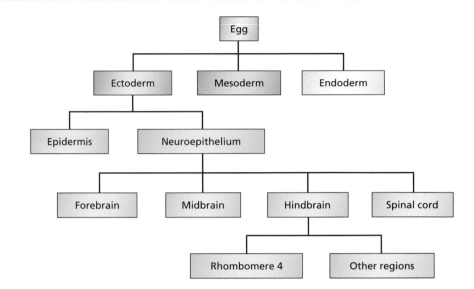

Fig. 4.5 A hierarchy of regional specification in development. Cells that become rhombomere 4 of the hindbrain first need to "decide" to be ectoderm, then to be neuroepithelium, then to be hindbrain, and finally to be rhombomere 4.

such as individual rhombomeres of the hindbrain. This means that any region in the embryos will pass through several states of commitment, each defined by a different combination of transcription factors. The embryological methods for defining specification and determination can be applied at any level of this hierarchy.

Because the early steps of the hierarchy of commitment do not correspond to named body parts they are often referred to by position (e.g. dorsal/ventral, anterior/posterior). This can be very confusing to students beginning the study of developmental biology. Experimental manipulation can alter states of commitment so one might encounter a statement such as "overexpression of X makes dorsal cell population Y ventral." This means that cells in a dorsal position in the embryo have been caused to acquire a state of commitment the same as that normally found in the ventral region. So it is important to be very clear when reading publications about whether positional terms like "dorsal" literally refer to position, or whether they refer to a state of commitment associated with that position in normal development.

The term **potency** is sometimes used to mean the range of possible cell types or structures into which a particular cell population can develop. This is similar to competence, but may include also pathways of development that can be provoked *in vitro* by environments not normally found within the embryo. In general, cell populations at the top of the hierarchy will be **toti-** or **pluripotent**, and those at the bottom will be unipotent.

Acquisition of commitment

Cytoplasmic determinants

A cytoplasmic **determinant** is a substance or substances, located in part of an egg or blastomere, that guarantees the assumption of a particular state of commitment by the cells which inherit it during cleavage (Fig. 4.6). By definition the cell division will be asymmetrical and the two daughters will follow different pathways of development. If cytoplasm containing a determinant for

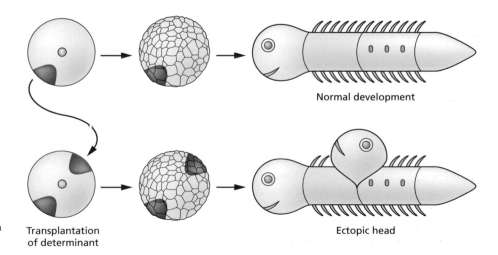

Normal development

Fig. 4.6 Operation of a cytoplasmic determinant coding for the anterior structures of the embryo. In normal development it ensures that the head forms from the cells that inherit it. If the determinant is transplanted elsewhere then it causes formation of an ectopic head.

Transplantation of determinant

Ectopic head

a particular structure is grafted to a different part of the egg, then it will cause formation of that structure from the cells that come to contain it (Fig. 4.6).

Cytoplasmic determinants are of considerable importance for the very earliest stages of embryonic development because they are often responsible for the establishment of the first two or three distinctly specified regions in the embryo. The subsequent complexity of the pattern develops as a result of interactions between these initial domains. Determinants are sometimes mRNAs that are localized to a part of the cell in association with the cytoskeleton, for example the *bicoid* and *nanos* mRNAs in the *Drosophila* egg (see Chapter 11). They may also be proteins. In the early stages of *C. elegans* development a complex containing PAR3, PAR6, and aPKC becomes localized in the anterior and controls the fate of the first two blastomeres (see Chapter 12). A similar system seems to be involved in later events of neurogenesis and epithelial polarization in other animals.

Induction

Most regional specification in development arises from the operation of extracellular signals called **inducing factors**. The families of signaling molecules involved are briefly described in the Appendix. Many of them are also known as growth factors, cytokines, or hormones in other contexts. The ability to respond to an inductive signal is called **competence** and requires not just the presence of specific receptors, but also a functioning signal transduction pathway coupled to the regulation of transcription factors.

To give a concrete example, in the *Xenopus* embryo the mesoderm is induced from the animal hemisphere tissue in response to activin-like signals emitted from the vegetal region (Fig. 4.7 and see Chapter 7). The signals activate the expression of various transcription factors that define the mesodermal state, such as the T box protein brachyury. The remainder of the animal hemisphere becomes ectoderm, as does the whole animal hemisphere in isolation. This interaction can occur between small pieces of the blastula cultured in combination, and so by using pieces taken from embryos of different stages it has been possible to show that the interaction occurs during the blastula stages. This type of interaction is called an **instructive induction** because the responding tissue has a choice before it (either mesoderm or ectoderm), and in normal development the interaction results in an increase in complexity of the embryo.

There are two different types of instructive induction which have somewhat different consequences in terms of regional specification (Fig. 4.8). It may be that the signaling center lies at one end of a cell sheet and is the source of a concentration gradient of the signal substance. The competence of the surrounding tissue embodies different **threshold responses** to different concentrations and hence a series of territories are formed in response to the gradient. It has become common usage to refer to the signal substance in a case of this sort, where there is more

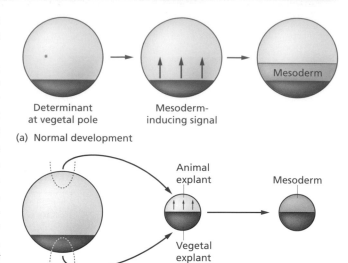

Fig. 4.7 Mesoderm induction in *Xenopus*. (a) As it occurs in normal development. (b) As it occurs in an animal–vegetal combination experiment.

than one positive outcome, as a **morphogen**. Well-established examples of morphogen gradients are the gradients of sonic hedgehog protein in the neural tube, of active BMP (bone morphogenetic protein) in the early *Xenopus* embryo, or of decapentaplegic protein in the *Drosophila* imaginal discs (see later chapters).

The other possibility is that the signaling centers lie in one cell sheet and the responding cells in another. When they are brought together, the appropriate structures are induced as a result of a single threshold response in those parts of the responding tissue immediately adjacent to the signaling centers. This probably happens, for example, in the induction of nasal, lens, and otic **placodes** from the head epidermis of vertebrate neurulae under the influence of the underlying tissues. In the presence of the signal the epidermis forms a placode, in its absence it differentiates as normal epidermis. This is called **appositional induction**. Typically only one threshold response would be made by the responding tissue, and the inducing factor for this reason would not be called a morphogen, even though the same substance might function as a morphogen in another context.

There is a further kind of inductive interaction which is called **permissive**. Here the signal is necessary for the successful self-differentiation of the responding tissue but cannot influence the developmental pathway selected (Fig. 4.8). Permissive interactions are very important in late development. For example in the development of the kidney, the mesenchyme will form tubules on receipt of a permissive signal from the ureteric bud. In the absence of the signal it simply fails to develop, and does not form any alternative tissue. The essential difference between instructive and permissive is that instructive inductions lead to a subdivision of the competent tissue while permissive inductions do not.

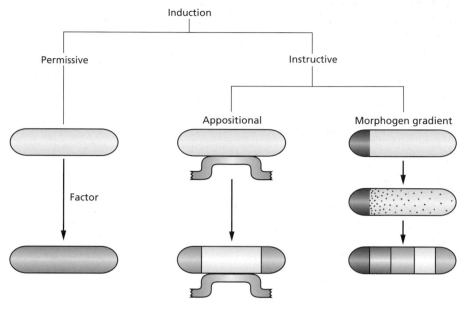

Fig. 4.8 Types of induction: permissive and instructive. Instructive inductions may be appositional or morphogen type.

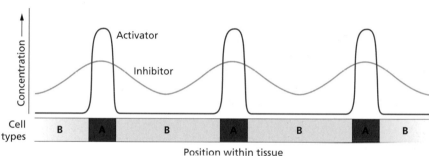

Fig. 4.9 Lateral inhibition. Cell type A produces both the activator and the inhibitor. Where the activator prevails cell type A is stabilized, where the inhibitor prevails cell type A is suppressed.

It has generally been supposed that inductive signals diffuse through the extracellular space. However, many of the factors concerned are growth factors that are tightly bound by extracellular glycosaminoglycans. For this and other reasons, alternative routes for diffusion have been considered, and one possibility is transport through the fine intercellular processes called cytonemes that are found in many developing tissues. Even where inducing factors do travel through the intercellular space their distribution may depend on cellular processes of endo- and exocytosis. The transport of inducing factors is best understood in the *Drosophila* imaginal discs (see Chapter 17).

Lateral inhibition

Another important class of cell communication that could be called induction, but is usually considered separately, is **lateral inhibition**. This is best known in terms of the behavior of the Notch-Delta system that governs many cases where individual cells or cell clusters from a uniform population follow one pathway of development while all around follow another. Examples include **neurogenesis** in the early neural plate (see Chapter 14), or formation of endocrine cells in the epithelia of the gut (see Chapter 16). In principle lateral inhibition systems work by a signaling center becoming established that suppresses the formation of the signal in the surrounding cells. This is shown in Fig. 4.9. There is a field of cells that are committed to become cell type B but are spontaneously progressing toward commitment to form cell type A. The first few cells to become type A produce an activator substance which promotes development of cell type A, and an inhibitor which antagonizes the action of the activator, such that inhibited cells are unable to continue the progress toward type A and become cell type B. The reason that the system generates pattern is that the activator is short range, perhaps only active through intercellular contacts, while the inhibitor is long range, moving freely by diffusion. This means that near the source of activator, the activator will prevail over the inhibitor and guarantee formation and maintenance of the cell type A, which produces both substances. In the surrounding region the inhibitor will prevail over the activator and so suppress formation of cell type A. Beyond a certain range the action of the inhibitor will be insufficient to prevent the formation of further signaling centers of type A cells and so the final result will be the formation of many signaling centers spaced out across the field of cells in an approximately uniform way. How regular the final pattern is will depend on the details of the mechanism.

Stochasticity in development

Development often seems to involve the creation of pattern from homogeneity. This occurs when a lateral inhibition system gets going. It occurs when a cytoplasmic determinant becomes localized at one end of the cell rather than another. It also occurs in any case where multiple cell types appear to differentiate from a single type of progenitor, even when they are cultured *in vitro* in a uniform environment. All of these processes involve **symmetry breaking**. Very small naturally occurring fluctuations in the concentration of particular substances, or the activity of specific genes, become amplified at particular locations as a result of positive feedback, and become repressed in adjacent regions as a consequence of the amplification. What is the original source of these fluctuations? It seems that it is ultimately due to the small number of regulatory molecules in a cell. There may be a few hundred copies of a particular transcription factor within each cell and they have to find their binding sites in the DNA by searching a huge amount of DNA sequence. There are only two copies of each gene, and, at any particular time, each copy of a particular regulatory sequence will either have a transcription factor attached or it will not have one attached. The residence times of the individual molecules is quite long, measured in minutes, so a population of apparently identical cells will actually be heterogeneous in terms of the instantaneous occupation of regulatory sites. Given the existence of positive feedback systems, such fluctuations can easily be amplified to macroscopic and irreversible differences.

Homeotic genes

A **homeotic gene** is one that, when mutated, causes conversion of one part of the body into the likeness of another. Homeotic genes code for transcription factors, often but not necessarily of the homeodomain type, and the presence of a particular combination of these factors encodes the state of commitment of a cell. The expression of homeotic genes may be controlled by cytoplasmic determinants or, more usually, by inducing factors. Homeotic genes are also sometimes called **selector genes** because their activity *selects* a particular developmental pathway for the cell.

Like all transcription factors, the products of homeotic genes work by regulating the activity of other genes, and it is important to note that as much information is encoded by the "off" state as by the "on" state. In molecular language, the absence of a repressor can be equivalent to the presence of an activator. The existence of two discrete states of gene activity is a natural way of ensuring a sharp and discontinuous threshold response to a determinant or inductive signal. One way of ensuring that there are just two discrete states is to have a positive-feedback regulation as shown in Fig. 4.10. This type of system is called a **bistable switch**, because it has two stable states. The gene is off if both the regulator and the gene product are absent. It is initially turned

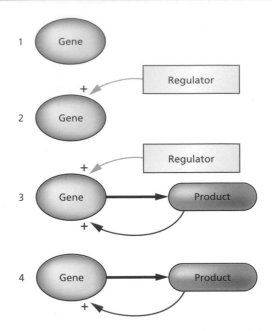

Fig. 4.10 Operation of a bistable switch. A temporal sequence is depicted: in 2 the gene is activated by a regulator; in 3 it is also activated by its product; in 4 it remains on because of the product even though the regulator is now gone.

on by the regulator, which might be a cytoplasmic determinant or a signal transduction pathway activated by an inducing factor. Once the gene product has accumulated, the gene remains on even if the regulator is removed. This model shows three critically important features of gene regulation in development:

1 It can yield a sharp and discontinuous **threshold** in response to the regulator.

2 The system has memory of exposure to the regulator. This is because the gene remains on permanently despite its transient exposure to the regulator.

3 Bistable switches like this are kinetic phenomena. This means that they depend on the continuous production and removal of substances, in this case the gene product. Simple equilibrium thermodynamic properties of binding and dissociation on their own cannot create sharp thresholds or memory.

There are several examples of bistable states of this sort, for example the autoregulation of the Hox gene *Deformed* in early *Drosophila* development, or of *Ubx* in the later visceral mesoderm. Although sharp thresholds of gene activity are very common, they are in most cases maintained by more complex mechanisms than this simple positive-feedback loop.

Properties of morphogen gradients

A stable concentration gradient cannot be produced simply by releasing a pulse of the morphogen. Such a pulse would spread out by diffusion and eventually the concentration would become uniform all over the embryo. A concentration gradient is only set up in a situation where the morphogen is continuously produced in one region (a **source**) and destroyed in another (a **sink**). This will produce a **gradient** of concentration, with a flux of material from the source to the sink. A model that seems to fit the behavior of many natural systems maintains a constant concentration of morphogen in the signaling region, and has destruction at a rate proportional to local concentration throughout the responding tissue. This produces a gradient which is approximately of exponential form when it reaches the steady state.

A concentration gradient has two important properties. It can subdivide the competent field of cells into several states of commitment, by means of threshold responses, and it automatically imparts a **polarity** as well as a pattern to the responding tissue. In Fig. 4.11 are shown three examples to illustrate basic properties of the system. The top figure represents normal development of the anteroposterior pattern of an animal controlled by a morphogen gradient. The competent field is subdivided into four territories by the activation of three homeotic genes in a nested pattern. Since there are just the two states of gene activity, they are represented by binary digits where "1" means "on" and "0" means "off." The action of the gradient will subdivide the field into four territories (head and three body segments) with the codings 000, 001, 011, 111. In the second example, suppose that a graft has been carried out to place a second source at the other end of the field. With a source at both ends and destruction throughout the central region, the concentration gradient will become U-shaped. The same threshold responses will now produce a different pattern 111, 011, 001, 011, 111. This is a type of structure called a mirror-symmetrical duplication, because it consists of two similar halves joined by a plane of **mirror symmetry**. The polarity is inverted in the left half of the field. Mirror-symmetrical duplications arise quite often in embryological experiments following the grafting of signaling centers, for example the double-dorsal *Xenopus* embryo arising from the creation of a second organizer (see Chapter 7), or the double posterior limbs produced by ZPA grafting (see Chapter 15), and their existence strongly suggests the existence of an underlying morphogen gradient. In the lower panel is shown another type of experiment involving the insertion of an impermeable barrier that interrupts the passage of the morphogen. Since no morphogen is being produced on the left-hand side, the concentration soon falls to zero, while on the right-hand side it actually piles up to a higher concentration than normal. This is because the size of the sink has been reduced relative to the source. As the rate of destruction is proportional to concentration, the overall concentration has to increase to re-establish the steady state. Operation of the same threshold responses shows that not only has the left half of the pattern been lost, but so has part of the right half because the elevation of concentration has expanded the size of the most posterior territories. This example shows how the properties of developmental systems may not be obvious at first sight. They may sometimes be counterintuitive, and interpretation of experimental results may require an

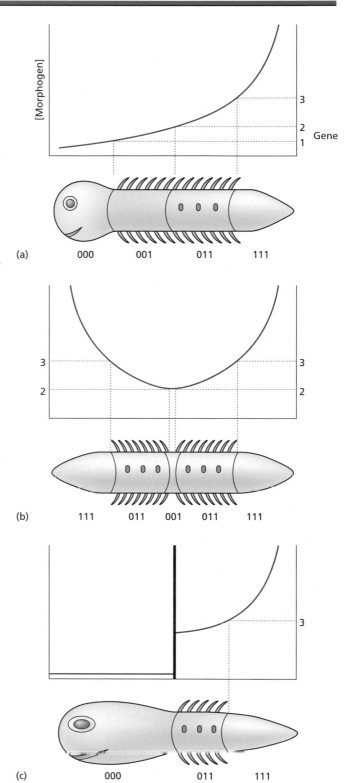

Fig. 4.11 Properties of morphogen gradients. (a) Normal development of an animal with a head and three segments. (b) Graft of the posterior source to the anterior causes formation of a U-shaped gradient and produces a double-posterior animal. (c) Insertion of an impermeable barrier causes formation of a large gap in the pattern. The binary codes indicate the activity in each body region of the genes 3, 2, and 1 respectively.

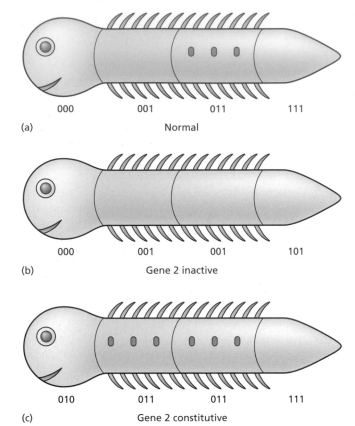

(a) Normal

000   001   011   111

(b) Gene 2 inactive

000   001   001   101

(c) Gene 2 constitutive

010   011   011   111

Fig. 4.12 Homeotic mutants. (a) Normal genotype and phenotype. (b) Loss-of-function mutation of gene 2 causes second body segment to resemble the first. (c) Gain-of-function mutation of gene 2 causes first body segment to resemble the second. This example assumes that the abnormal codings (010 and 101) do not produce homeotic effects.

understanding of the properties of the underlying mechanisms as well as a familiarity with the genes and signaling molecules at work.

Homeotic mutations

Now suppose that the organism is a mutant and that the second gene in the series is permanently inactive and cannot be turned on. In this case the codings will be 000, 001, 001, and 101. In other words the second body segment has now been turned into another copy of the first body segment (Fig. 4.12a,b). What happens to the third, tail, segment cannot be predicted without knowing more about the logical circuitry, since this now has a novel coding not found in the normal organism. This example illustrates the behavior of a loss-of-function homeotic mutation. As mentioned in Chapter 3, such mutations are genetically recessive since function would be restored by a good copy of the gene on either chromosome.

On the other hand, suppose that the mutation in gene 2 caused constitutive activity, in other words the gene is on all the

time everywhere. Now the sequence of codings would become 010, 011, 011, 111. Here the first body segment has become a copy of the second body segment and the head has an abnormal coding (Fig. 4.12c). This is a gain-of-function mutation. It is genetically dominant since inappropriate activation will occur with only one copy of the mutated gene.

This example represents, in a highly oversimplified manner, the operation of the Hox gene system for anteroposterior patterning of animals. In general loss-of-function mutations produce anteriorizations as in Fig. 4.12b and gain-of-function mutations produce posteriorizations as in Fig. 4.12c. It also illustrates the **combinatorial** nature of regional specification in development. Although striking anatomical transformations may sometimes be produced by the manipulation of individual genes, the identity of body parts is actually controlled in normal development by combinations of genes.

Homeotic, homeobox, and Hox

Considerable confusion arises about the relationship between the definition of homeotic behavior and the molecular identity of homeotic genes. In 1984 a common DNA sequence was discovered in the genes of the Bithorax and Antennapedia complexes of *Drosophila* (see Classic experiments box "Discovery of the homeobox" in Chapter 11). Because these genes were homeotic, this motif was called the "**homeobox**." The proteins encoded by these genes are all transcription factors and the homeobox encodes a sequence of 60 basic amino acids which form their DNA binding domain, called the **homeodomain**. Subsequently many homeobox-containing genes were found in all types of eukaryotic organisms. All homeobox-containing genes encode homeodomain-containing transcription factors. Many, but not all, of these are concerned with development, but rather few of them are homeotic when mutated or misexpressed.

The **Hox genes** are a family of genes, found in animals but not in other eukaryotes, that are responsible for specifying **anteroposterior** identity to body levels. They are a subset of the general class of homeobox genes and are specifically the homologs of the *Drosophila* Bithorax and Antennapedia gene clusters. In most animals Hox genes form just one cluster, with the different genes adjacent to each other on the chromosome.

The Hox genes are activated at an early stage in body-plan formation and are generally maximally expressed around the **phylotypic stage** of the group in question (see Chapter 20). Each gene in the cluster is expressed at a particular anteroposterior level, running from a sharp anterior boundary to fade out gradually in the posterior. They are expressed in both central nervous system and mesoderm. Remarkably, the spatial order of expression of Hox genes in the body, from anterior to posterior, is often the same as the order of the arrangement of the genes on the chromosome. Invertebrates have just one cluster of Hox genes but vertebrates have four or more clusters, each situated on a different chromosome.

Criteria for proof

Much research work in developmental biology is concerned with the identification of inducing factors, or of determinants, or of the transcription factors that define the states of developmental commitment. Rigorous proof that a particular molecule really does perform a particular function during normal development requires three independent lines of evidence, concerning *expression, activity,* and *inhibition*:

1 The molecule in question must be there. There must be evidence that it is expressed in the right place, at the right stage, and in a biologically active form. This is usually obtained from **in situ hybridization** or **immunostaining**, but it should be remembered that the presence of mRNA does not guarantee the translation of the polypeptide, nor does the presence of protein guarantee its post-translational processing to an active form. Additional evidence can sometimes be obtained from suitable reporter constructs, for example the presence of biologically active levels of retinoic acid can be detected by introducing a gene consisting of a retinoic acid response element (RARE) coupled to a *lacZ* reporter gene (see Chapter 5). Regions exposed to retinoic acid, and capable of responding to it, should then express β-galactosidase, which can be detected by the X-Gal reaction.

2 The molecule in question must have the appropriate biological activity in a suitable test system. For example, a candidate inducing factor should be able to evoke the correct responses from its target tissue. For a candidate cytoplasmic determinant, it must be possible to inject it into another part of the cell or another blastomere, and cause the injected region to develop along the pathway caused by the determinant. For a transcription factor, introduction by transfection or by transgenesis should lead to the predicted change in developmental pathway of the affected region.

3 If the molecule is inhibited *in vivo* then the process for which it is thought responsible should fail to occur. Where several similar molecules are responsible for a process (redundancy, see Chapter 3) then it may be necessary to inhibit all of them to obtain a result. Inhibition may be achieved at the DNA level by mutation of a gene to inactivity, at the RNA level using morpholinos or RNAi, or at the protein level by introduction of a specific inhibitor of the normal gene product. Extracellular substances may sometimes be successfully inhibited by specific neutralizing antibodies. Although well-characterized mutations are necessarily specific to a single gene, inhibition experiments involving the other methods may not be so specific and need careful evaluation.

Key Points to Remember

- A fate map shows where each part of the embryo will move, and what it will become. It does not, however, indicate the state of commitment of parts of the embryo at the time of labeling.
- Developmental commitment can be labile (specification) or stable (determination). Specification indicates that a particular structure or cell type will be formed by development in isolation. Determination indicates that a particular structure or cell type will be formed following grafting to other regions of the embryo.
- Clonal analysis comprises the deductions that can be made about developmental mechanisms by labeling a single cell. If the progeny of one cell span the boundary between two structures, it shows that the cell was not determined to become either structure at the time of labeling.
- A cytoplasmic determinant will cause commitment to a particular developmental pathway of the cells that inherit it.
- An inducing factor is an extracellular signal substance that can alter the developmental pathway of cells exposed to it. Many inducing factors are known as growth factors or hormones in other contexts. If the factor is simply necessary for continued development of the target cells, it is said to be permissive. If a different developmental pathway is followed in the absence and the presence of the factor, then it is said to be instructive. If there is more than one positive response at different concentrations, the factor is described as a morphogen.
- Lateral inhibition systems generate two mixed cell populations from a single cell sheet. This occurs by amplification of small initial differences between the cells such that differentiating cells inhibit the differentiation of those around them.
- Homeotic genes generally encode transcription factors and show discontinuous threshold responses of expression to cytoplasmic determinants or inducing factors. The developmental commitment of cells is encoded by the combination of homeotic genes that they express.
- Proof that a particular gene product is responsible for a particular process requires evidence of appropriate expression pattern; of appropriate biological activity, and of appropriate consequences of ablation.

Further reading

Also see developmental biology textbooks and websites cited in Chapter 1, and stage series cited in Chapter 6.

General

Meinhardt, H. (1982) *Models of Biological Pattern Formation.* New York: Academic Press.

Slack, J.M.W. (1991) *From Egg to Embryo. Regional specification in early development*, 2nd edn. Cambridge: Cambridge University Press.

Held, L.I. (1994) *Models for Embryonic Periodicity.* Basel: S. Karger.

Fate mapping

Hartenstein, V., Technau, G.M. & Campos-Ortega, J.A. (1985) Fate-mapping in wild-type *Drosophila melanogaster* 3. A fate map of the blastoderm. *Wilhelm Roux's Archives of Developmental Biology* **194**, 213–216.

Dale, L. & Slack, J.M.W. (1987) Fate map for the 32 cell stage of *Xenopus laevis. Development* **99**, 527–551.

Tam, P.P.L. (1989) Regionalization of the mouse embryonic ectoderm: allocation of prospective ectodermal tissues during gastrulation. *Development* **107**, 55–67.

Kimmel, C.B., Warga, R.M. & Schilling, T.F. (1990) Origin and organization of the zebrafish fate map. *Development* **108**, 581–594.

Hatada, Y. & Stern, C.D. (1994) A fate map of the epiblast of the early chick-embryo. *Development* **120**, 2879–2889.

Clonal analysis

Garcia-Bellido, A., Lawrence, P.A. & Morata, G. (1979) Compartments in animal development. *Scientific American* **241**, 90–98 or 102–111.

Kimmel, C.B. & Warga, R.M. (1986) Tissue specific cell lineages originate in the gastrula of the zebrafish. *Science* **231**, 365–368.

Cepko, C., Ryder, E.F., Austin, C.P., Walsh, C. & Fekete, D.M. (1995) Lineage analysis using retroviral vectors. *Methods in Enzymology* **254**, 387–419.

Determinants

Hawkins, N. & Garriga, G. (1998) Asymmetric cell division: from A to Z. *Genes and Development* **12**, 3625–3638.

Knust, E. (2001) G protein signaling and asymmetric cell division. *Cell* **107**, 125–128.

Henrique, D. & Schweisguth, F. (2003) Cell polarity: the ups and downs of the par6/aPKC complex. *Current Topics in Genetics and Development* **13**, 341–350.

Morphogen gradients

Podos, S.D. & Ferguson E.L. (1999) Morphogen gradients: new insights from DPP. *Trends in Genetics* **15**, 396–402.

Teleman, A.A., Strigini, M. & Cohen, S.M. (2001) Shaping morphogen gradients. *Cell* **105**, 559–562.

Vincent, J.P. & Dubois, L. (2002) Morphogen transport along epithelia, an integrated trafficking problem. *Developmental Cell* **3**, 615–623.

Thresholds, stochasticity

Lewis, J., Slack, J.M.W. & Wolpert, L. (1977) Thresholds in development. *Journal of Theoretical Biology* **65**, 579–590.

McAdams, H.H. & Arkin, A. (1999) It's a noisy business! Genetic regulation at the nanomolar scale. *Trends in Genetics* **15**, 65–69.

Elowitz, M.B., Levine, A.J., Siggia, E.D. & Swain, P.S. (2002) Stochastic gene expression in a single cell. *Science* **297**, 1183–1186.

Xiong, W. & Ferrell, J.E. (2003) A positive-feedback-based bistable "memory module" that governs a cell fate decision. *Nature* **426**, 460–465.

Homeosis

Garcia-Bellido, A. (1977) Homoeotic and atavic mutations in insects. *American Zoologist* **17**, 613–629.

Kuziora, M.A. & McGinnis, W. (1989) Autoregulation of a *Drosophila* homeotic selector gene. *Cell* **55**, 477–485.

Lewis, E.B. (1994) Homeosis: the first 100 years. *Trends in Genetics* **10**, 341–343.

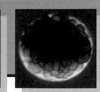
Techniques for the study of development

The basic genetic techniques have already been covered in Chapter 3 and the basic embryological techniques in Chapter 4. Here we shall consider a further set of techniques that are derived from cell and molecular biology but that have particular relevance to the study of development.

Microscopy

Embryos are small and their study inevitably requires the use of microscopes. Experiments often involve manual interventions using a **dissecting microscope**. This is a binocular microscope with a magnification in the range of about 10 to 50 and a good working distance between the objective lens and the specimen (Fig. 5.1). Dissecting microscopes provide a three-dimensional image which allows accurate perception of depth by the observer and assists in the performance of manipulations such as microsurgery or microinjection. Unlike most compound microscopes, a dissecting microscope does not invert the image. If specimens are opaque, such as *Xenopus* or chick embryos, incident lighting is used. This means that the beam is shone down from the light source onto the specimen. For living specimens it is important not to overheat them in a powerful incident light beam, and so a fibreoptic light guide is used, providing a cold but bright illumination. If specimens are transparent, such as embryos of zebrafish, sea urchin, or mouse, then a transmitted light base would usually be used.

The **compound microscope** (Fig. 5.2) is used for the examination of sections, or for whole specimens that are small enough to be transparent, such as embryos of *Drosophila, Caenorhabditis elegans*, or zebrafish. Specimens that are not sections are referred to as **wholemounts**. The compound microscope has a magnification range from about 40 to 1000, and the upper limit of magnification is set by the wave nature of light, which prevents resolution of points closer together than about 0.2 μm. Most compound microscopes invert the image. This is a natural consequence of the optical system and is not normally corrected because to do so would require extra lenses. Under most circumstances the inversion is no problem.

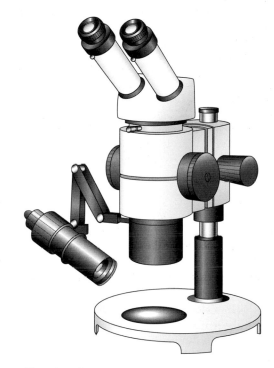

Fig. 5.1 Dissecting microscope.

Optical techniques

There are several different optical techniques used with the compound microscope. Ordinary transmitted light is used for stained sections or wholemounts. Differential interference contrast, otherwise known as Nomarski optics, is used extensively in the study of small wholemount specimens. This is a technique that converts small differences of refractive index into an apparent difference in height when perceived by the observer. It also provides a sharp resolution of one particular optical section within the specimen, so as one focuses the microscope up and down through the specimen, different optical sections come into view. Nomarski optics provides very clear visualization of single cells within the specimen, and it was this technique that

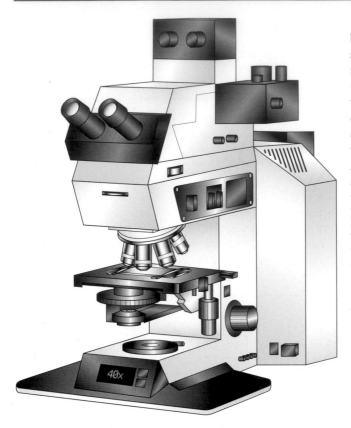

Fig. 5.2 Compound microscope.

Fluorescence microscopy is used for a variety of purposes but they all depend on visualizing the location of a fluorescent substance, or **fluorochrome**, within a specimen. They include fluorescent antibody staining, fluorescent *in situ* hybridization (FISH), and the visualization of fluorochromes introduced into the specimen to label subsets of cells. The principle of fluorescence is that the fluorochrome absorbs light of a particular energy and emits light of a lower energy. Lower energy means longer wavelength or a color shifted toward the red end of the spectrum. A particular fluorochrome will have an **excitation** spectrum showing how the intensity of fluorescence varies with the excitation wavelength. It will also have an **emission** spectrum showing how the intensity of fluorescence is distributed across the wavelength spectrum (Fig. 5.3). The excitation and the emission spectra are characteristics of the substance. A fluorescent microscope has an attachment that shines the excitation beam down onto the specimen. This consists of a powerful lamp, usually a mercury arc lamp, then a filter to select a narrow excitation wavelength band suitable for the fluorochrome in use, then a dichroic mirror that reflects wavelengths below a certain cutoff and transmits them above this cutoff. The dichroic mirror reflects the excitation beam down onto the specimen. Because the emission is of longer wavelength, it will be transmitted and can then be visualized by the observer (Fig. 5.3). A fluorescent microscope will usually contain several filter sets, one for each fluorochrome in use. If a specimen contains two or three fluorochromes it should be possible to visualize each one separately using the appropriate filter set. Because of the need to examine fluorescence in whole specimens arising from the various applications of green fluorescent protein (see below), fluorescent dissecting microscopes are now available.

Confocal and two-photon microscopes

Conventional fluorescence microscopy is limited to use on sections or very thin wholemounts because fluorescence from cells above and below the plane of focus would otherwise swamp the image. However, fluorescence from thicker wholemounts can readily be visualized with the **confocal scanning microscope**.

enabled the elucidation of the entire cell lineage of the worm *C. elegans.*

Dark-field microscopy is a method that depends on illumination from a very oblique angle so that only points within the specimen that scatter light extensively are visible as bright points, and the remainder of the specimen is dark. In developmental biology it is only really used for visualizing radioactive *in situ* hybridizations, in which the signal consists of an accumulation of silver grains in a photographic emulsion coating the section (see below).

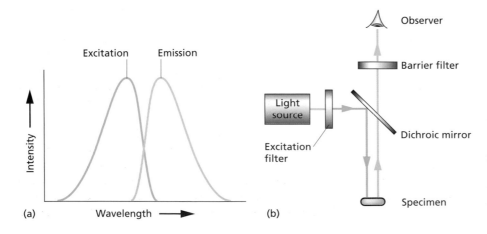

(a) Excitation   Emission

Intensity

Wavelength

Observer

Barrier filter

Light source

Excitation filter

Dichroic mirror

Specimen

(b)

Fig. 5.3 Fluorescence microscopy. (a) Typical spectra for excitation and for emission of a fluorochrome. (b) Arrangement of components in a fluorescence microscope.

This is a device that uses a laser for illumination so excitation is achieved with just the single wavelength characteristic of that laser. Instead of illuminating the whole specimen, it illuminates just one point at a time. All the points in one particular optical section are scanned in turn and the fluorescence from each point is recorded by a detector and then used to reconstruct the image of the section. Because only one point is viewed at a time, the quality of the optical section is very high. Light from points above or below the plane that is in focus will be dispersed and contribute very little to the signal. The image from a confocal microscope is necessarily a digital image stored on a computer.

The two-photon microscope can provide even better optical sectioning of fluorescent specimens than the confocal microscope. Here the light source is of only half the energy required to excite the fluorochrome. Thus, an excitation can only occur if two photons strike one fluorochrome molecule simultaneously, and this can only happen at the point of focus of the excitation beam where the photons are very dense.

Digital images

Digital representation and storage of the image is essential to the operation of the confocal microscope but is by no means confined to it. Until the 1990s all microscopic images were recorded by photography. There are still some situations in which photography is used but most image collection is now done using **charge coupled devices (CCDs)**. The detection chip of a CCD camera contains an array of pixels each of which can be "filled up" with electrons, and the charge each pixel accumulates over a given time is proportional to the light intensity falling on it. An exposure is taken and then the charge on each pixel is read out in sequence by a detector and converted to a digital image. For an 8-bit image, the intensity of each pixel would be represented by a number from 0 to 255. A 24-bit color image would have one 8-bit number for each of three colors red, green, and blue. There are different types of CCD cameras for different purposes. The most sensitive will function only in black and white and can collect long exposures from very faint specimens. Less sensitive ones can collect images fast enough to display on a video screen, and enable the recording of movies in real time.

Once the image is recorded it is transmitted to a computer and displayed on the screen, and then various types of computation can be carried out. For example the contrast and brightness can be altered, the image can be smoothed or sharpened, densitometric measurements can be taken, or multiple black and white fluorescent images obtained through different channels using different fluorochromes can be recombined so that several colors are visualized on the same image. One of the many advantages of digital image collection is that the signal/noise ratio of the image can be improved by taking several exposures and then averaging them. The signal will be the same in each exposure and therefore will be unaffected by averaging, whereas the noise will be different in each exposure and therefore will become low and uniform.

Histological methods

Although wholemounts are used extensively in developmental biology, even with Nomarski or confocal microscopy there is a limit to the degree of resolution with which the structure of a specimen can be resolved. Thus, the need to prepare and examine **histological sections** will always remain. The first step in preparing a specimen for any form of microscopical examination is to fix it. **Fixation** means that it is killed and that it is made mechanically robust enough to withstand osmotic shocks or a certain amount of handling. **Fixatives** work in various ways. The most commonly used is formalin, which is a solution of the gas formaldehyde. This reacts with amino or sulfhydryl groups of macromolecules producing some denaturation and some cross-linking. Glutaraldehyde has two aldehyde groups and is a very effective intermolecular cross-linking agent. Other common components of fixatives are acids and organic solvents. These act as denaturing and precipitating agents.

Once the specimen has been fixed it must be embedded in a solid supporting material that will permit it to be cut into very thin sections. The most common material is paraffin wax. In order to infiltrate the specimen with wax it needs to be **dehydrated**, and this is achieved by passing it through a series of baths of ethanol of increasing concentration. Direct transfer from water to pure ethanol causes too much tissue damage because of the mixing forces exerted by the solvents. Once the water is removed, the specimen is equilibrated in a solvent miscible with wax, such as xylene, then it is placed in molten wax at about 60°C and left until the wax has thoroughly penetrated every part. The wax is allowed to solidify and the resulting block can be stored permanently at room temperature. In order to make sections, the block is mounted on a **microtome** which passes it repeatedly across a very sharp knife, each time advancing the block by a few microns. This results in the formation of a connected set, or ribbon, of sections. These are mounted on microscope slides and can then be stained or processed for immunostaining or *in situ* hybridization.

In embryology the orientation of sections is very important. Because it is often desirable to analyze the disposition of structures in the entire specimen it is necessary to have a complete set of connected, or serial sections. For this reason paraffin wax is a very useful embedding material as the sections naturally form a ribbon as they are cut. Its disadvantage is that the specimen needs to be dehydrated with organic solvents and needs to be heated to 60°C for the embedding period. This can lead to damage to proteins or nucleic acids in the specimen which may compromise immunostaining or *in situ* hybridization. For these techniques it is quite common to use frozen sections. Here the specimen may not even need to be fixed but is frozen rapidly in a medium containing a high concentration of sucrose. This is cut into a block and mounted on a **cryostat**, which is simply a microtome operating in a cooled chamber. The quality of frozen sections is often not quite as good as paraffin wax, and it is very difficult to collect serial sections as they do not form a ribbon.

But it is a very useful technique if only a few representative sections are required.

For some purposes paraffin wax does not provide sufficient quality of sections, as it is difficult to cut them thinner than about 6 μm. There are other embedding materials based on various plastics which can be cut at 1 μm, or even at fractions of a micron for the electron microscope. But these materials may not be compatible with immunostaining and *in situ* hybridization, and cannot provide serial sections. **Electron microscopy** of sections is rarely used in developmental biology as it provides more magnification than is required to identify patterns of cell types in tissues or embryos. But the scanning electron microscope can often provide vivid three-dimensional views of wholemounts.

Study of gene expression by biochemical methods

When studying development it is very important to know the normal expression patterns of the genes under investigation. It is necessary to know at what developmental stages they are active, in which parts of the embryo, and to what level of activity. There are two main classes of method for determining expression patterns: biochemical methods which give a reasonably accurate quantitative measure, but no anatomical information, and *in situ* methods which give accurate anatomical information but limited quantification. In both cases, there are separate methods for studying mRNA and protein. Strictly, it is desirable to do both, as the transcription of a gene does not guarantee its later translation into protein, and the presence of a protein in a particular place does not necessarily mean that it has been synthesized there, as it may have been transported from some other site of synthesis.

With the biochemical methods, a limited degree of regional information can be achieved by dissection of the specimen, but they are all intrinsically techniques for making a semiquantitative measurement of the amount of a specific messenger RNA or protein in the whole specimen. They are often used to obtain a **stage series** for expression of a gene, by examining groups of embryos of different developmental stages (Fig. 5.4).

Only a brief description of the biochemical methods is given here, as they are described in detail in textbooks of molecular biology. In all cases nonradioactive labeling and detection methods are now often used instead of radioactivity, although ^{32}P is still used as it provides the highest level of sensitivity.

Methods for messenger RNA

Northern blotting is the oldest and least sensitive technique. It involves extracting total mRNA from the specimen and separating it by gel electrophoresis on a denaturing agarose gel. After a good separation has been achieved the contents of the gel are transferred ("blotted") onto a hybridization membrane. The membrane is hybridized with the specific probe, which is radio-labeled with ^{32}P. The radioactivity bound to the membrane is visualized by autoradiography or with a phosphorimager. The image will usually consist of a single radioactive band, corresponding to a single mRNA complementary to the probe. Over a certain range, the intensity of the band is proportional to the amount of specific mRNA. If there is alternative splicing leading to the formation of several messages from one gene, or if there are mRNAs with similar sequences that will cross-hybridize, then the pattern may contain several bands.

RNAse protection is more sensitive and more specific than Northern blotting. Again a ^{32}P-labeled probe is prepared. This is

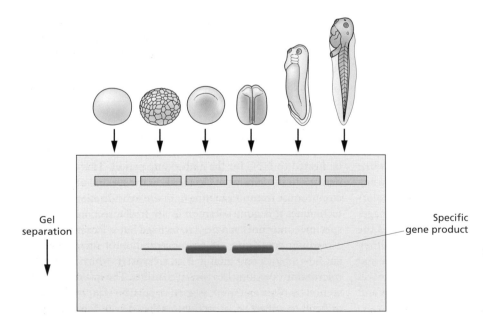

Fig. 5.4 A "stage series" for a particular gene product. This could be a specific mRNA or a protein extracted from whole embryos at the different stages and separated by a suitable technique.

Gel separation

Specific gene product

hybridized in solution with total RNA from the specimen and then digested with ribonucleases. The single-stranded probe will be digested but any probe incorporated into double-stranded hybrid will be protected from digestion. The mixture is then separated on a sequencing gel which provides very high resolution, and the radioactivity on the gel is visualized by autoradiography or phosphorimager. There should be a single band corresponding in length to the protected region of the probe, and over a certain range the intensity is proportional to the content of specific mRNA in the sample. Because the probe is used in excess, all of the complementary mRNA should become hybridized, which accounts for the sensitivity of the method.

Reverse transcription polymerase chain reaction (RT-PCR) is the most sensitive technique. The total RNA is extracted from the sample and reverse-transcribed into complementary DNA (cDNA) using reverse transcriptase. Then two oligonucleotides are added, chosen to correspond to sequences in the target cDNA a certain distance apart. For details of the PCR process, a textbook of molecular biology should be consulted. In essence the sequence between the primers is amplified by repeated cycles of synthesis, melting, and hybridization. Usually a radioactive nucleoside triphosphate is included to increase sensitivity, although it is possible to use the technique without radioactivity. After a suitable number of cycles the reaction mixture is fractionated on a gel and the DNA bands are visualized in the usual way. If the reaction has worked then there should be a band corresponding to the region of cDNA between the two primers. If there is no band then it means that there was no specific target cDNA in the sample. Over a certain amplification range the intensity of the band bears some relation to the initial amount of specific cDNA, and therefore of specific mRNA, in the original sample. More accurate quantitative measurement can be achieved with a **real-time PCR** method. Here the rate of formation of the amplified product is monitored in real time and this bears a more precise relationship to the original mRNA content than does the final amount of product at the end of the process.

Microarrays have been introduced in recent years to enable examination of large numbers of gene products simultaneously. In principle, where the genome sequence is known, it is possible to examine the full inventory of cDNAs corresponding to the entire genome, although in practice microarrays usually only contain a subset of the complete genome. Microarrays are typically used in the early stage of a research project to identify the genes likely to be involved in a particular system or process. They are made using machines derived from the computer industry and so are sometimes referred to as chips. They come in two basic types: the cDNA array and the Affymetrix oligonucleotide array. cDNA arrays consist of a regular arrangement of closely spaced spots, each of a different cDNA, arranged in a rectangular array on a glass slide. Sometimes instead of cDNAs, long synthetic oligonucleotides are used. In the Affymetrix system, which is a proprietary product, a larger number of short oligonucleotides are synthesized directly on the sides. Here several oligonucleotides will represent sequences from one gene. In both cases

the array is used for nucleic acid hybridization. A probe is prepared by extracting RNA from the embryo or tissue sample and reverse transcribing the mRNA into cDNA. This is labeled with a fluorescent dye and hybridized to the microarray under suitable conditions. The array is then scanned by a chip reader, which is a fluorimeter that can measure the fluorescence from the dye bound to each spot. For a given cDNA on the array, the intensity of fluorescence should represent the amount of the cDNA in the probe, and thus the amount of that specific mRNA in the tissue sample. In developmental biology, microarrays are often used to make comparisons, for example between two embryonic stages, or between cells treated and untreated with an inducing factor (Fig. 5.5). In this case there is a probe from each of the two different samples to be compared. These are labeled with different fluorochromes, usually the dyes Cy3 (green emission) and Cy5 (red emission). The probes are mixed before the hybridization is carried out and then the ratio of green to red fluorescence is measured. Genes whose expression does not change will give a yellow signal because both probes will bind, while those whose expression goes up or down will give a green or red signal resepectively. The analyses of both types of microarray depend on very sophisticated software that can match the position of each spot with the identity of the cDNA.

Methods for protein

It is now possible to identify individual unknown proteins from a complex mixture and this is likely to become increasingly important in developmental biology. The set of techniques employed are often collectively referred to as "**proteomics**." As with microarrays, they are most often employed at the beginning of an investigation where it is required to find which proteins change in expression level during a particular developmental event. Total protein extracts are separated by two-dimensional gel electrophoresis, in which the first dimension is an isoelectric focusing gel separating by isoelectric point, and the second dimension is an SDS-polyacrylamide slab gel separating by molecular weight. This gives a pattern of spots, each spot representing one protein. Comparison of preparations from the two samples under investigation will hopefully yield a small number of differences in the spot pattern. Then individual proteins are identified by **mass spectrometry**. A mass spectrometer works by ionizing and volatilizing the substance then determining its time of flight to the detector after acceleration in an electric field. This enables measurement of the molecular weight to very high precision. Where the genome has been sequenced the measurement of protein molecular weight may itself allow an identification, since this can be calculated from the known amino acid composition of each polypeptide represented in the genome sequence. Otherwise the individual protein can be sequenced by tandem mass spectrometry. The protein is digested with trypsin into a number of peptides. These are separated in the first cycle of mass spectrometry, then each peptide is broken up by ion

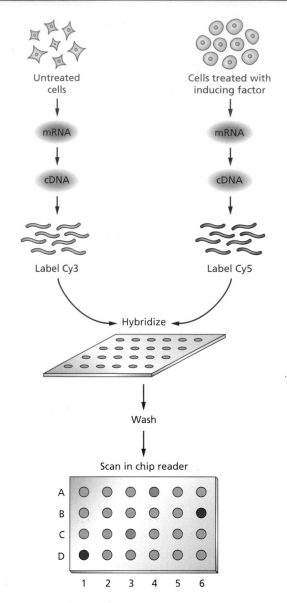

Fig. 5.5 Use of a microarray to compare gene expression in two cell populations. The results show that genes B6 and D1 have been upregulated, and genes A4 and C3 have been downregulated, by the inducing factor treatment.

bombardment and the fragments separated in a second cycle of mass spectrometry. Using sophisticated software it is possible to identify the amino acid sequence of the peptide from the characteristic fragmentation pattern.

Immunochemical methods

The methods of measuring a particular protein in a sample usually depend on having a specific **antibody** directed against the protein. This may be a **monoclonal** or a **polyclonal** antibody and will have been prepared by immunizing an animal with the pure protein or portion of a protein.

An **immunoprecipitation** is not a true precipitation but is a method of isolating the protein recognized by a specific antibody. It is often used to find whether a particular protein is being newly synthesized in an embryo or tissue sample. The live specimen is incubated in radioactive amino acids, usually ^{35}S-methionine or ^{35}S-cysteine, so all of the proteins made during the labeling period become radioactive. The total proteins are extracted and are incubated with the specific antibody. This will bind to its target protein and form an immune complex. The immune complexes are isolated by incubating the mixture with protein A bound to agarose beads. Protein A is a bacterial protein that binds tightly to the constant region of IgG-type antibodies. It will capture the immune complexes and immobilize them on the beads. The beads are then washed to remove contaminating proteins, and are boiled in a highly denaturing sample buffer to release the bound antibody and target protein. This is run on a sodium dodecyl sulfate (SDS) protein gel, which separates by molecular weight; the gel is dried, and then the radioactive band of target protein is visualized by autoradiography or phosphorimager.

A **Western blot** is a method that shows the total content of a specific protein rather than the newly synthesized protein. These may not be the same in a developing embryo where genes are turning on and off. For example the embryo may inherit a large maternal store of a particular protein from the egg, but may not be making it any more. To do a Western blot, the total proteins of the sample are extracted and separated on a protein gel. The content of the gel is transferred ("blotted") onto a membrane. The membrane is then incubated with the specific antibody which should bind only to the band of specific target protein and not to other components separated on the gel. The bound antibody is then visualized with a second antibody which is directed against the constant region of the first antibody. The second antibody is likely to be purchased commercially and it will be modified for easy detection, nowadays usually by being conjugated with **horseradish peroxidase (HRP)**, which has available very sensitive chemiluminescent substrates. A chemiluminescent substrate will produce a phosphorescent reaction product that decays spontaneously and emits light as it does so. After incubation in the substrate mixture, the blot is exposed to X-ray film or placed in a phosphorimager, and the phosphorescence of the reaction product is recorded in the position where the antibody is bound, corresponding to the position and quantity of the target protein in the original sample.

Study of gene expression by *in situ* methods

In situ methods are designed to reveal the spatial domains of gene expression in a specimen (Fig. 5.6). If the specimens are small enough and transparent enough, *in situ* procedures can be performed on wholemounts, which has the advantage of rapid and clear three-dimensional visualization. If specimens are too

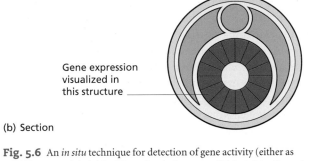

(a) Wholemount

Gene expression visualized in this structure

(b) Section

Fig. 5.6 An *in situ* technique for detection of gene activity (either as mRNA or protein). (a) Staining of a wholemount. (b) Staining of a section.

large or opaque then *in situ* methods can be used on sections, which for embryological work will usually be serial sections.

In situ *hybridization*

In situ **hybridization** reveals the regions of a specimen where a specific mRNA is present. The chemistry is the same as a Northern blot, as an antisense probe is synthesized *in vitro* complementary to the mRNA to be detected. This is hybridized to the specimen and then visualized (Fig. 5.7a). Most *in situ* hybridizations are now performed with nonradioactive probes. For detection, these include an extra chemical group recognizable by a commercially available specific antibody. The group is attached to one of the nucleoside triphosphates used for synthesis of the probe. It is incorporated during synthesis and does not affect the hybridization capacity of the probe. Favorite groups for this purpose are digoxigenin (DIG, a plant sterol) and fluorescein.

For wholemount *in situ* hybridizations, the specimen will usually need to be permeabilized by a short treatment with protease or detergent to enable the large probe molecules to enter the cells. For either wholemounts or sections, the hybridization reaction is conducted overnight, the specimen is washed extensively, then an enzyme-linked antiprobe antibody is added, for example an anti-DIG antibody conjugated to alkaline phosphatase. The location of this can be revealed by placing it in a suitable substrate mixture which yields an intensely colored insoluble precipitate at the site of the reaction. Thus, the color forms where the enzyme is located and this shows where the probe is bound and therefore where the specific mRNA was present.

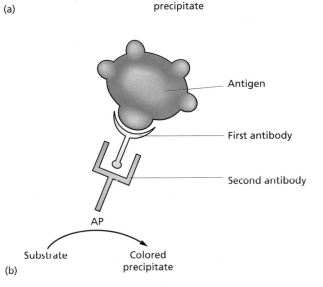

Fig. 5.7 (a) *In situ* hybridization for detection of mRNA. (b) Immunostaining for detection of protein.

Some substrates are fluorescent and produce a signal that can be seen in the fluorescent microscope. Fluorescent *in situ* hybridization is often known by its acronym "FISH." This is very convenient when more than one probe has been used and they are visualized through different fluorescent channels to detect regions of separation or overlap in the expression domains. In order to visualize more than one probe at a time it is necessary to use different substituents in the probes, which can be detected with different antibodies and substrate mixtures. Note that although fluorescein is often used in probes, the number of fluorescein groups present in the probe is not sufficient on its own to provide a fluorescent signal. Visualization of fluorescein in this context still requires antibody binding and amplification by an enzyme–substrate reaction.

In situ hybridization on sections is usually also performed with nonradioactive probes but radioactive probes are still sometimes used. These are labeled with ^{35}S, which is more suitable for autoradiography than the ^{32}P used for Northern blots, because of the shorter path length of the emitted β particles.

Although sulfur atoms do not occur in natural RNA nucleotides, it is possible to incorporate a sulfur atom without affecting the hybridization properties. After the hybridization and washes, the slide containing the sections is coated with a photographic emulsion and put in a light-proof box for some days or weeks. As the ^{35}S atoms decay they release β particles which are recorded by the emulsion. After a suitable time has elapsed, the emulsion is developed and the sensitized points are converted to silver grains. The silver grains will overlay the region of the section in which the original specific mRNA was present. Radioactive *in situ* hybridizations are often visualized using dark-field microscopy as this method causes the silver grains to shine brightly against a dark background. Radioactive *in situ* hybridizations have the advantage of high sensitivity, but the main disadvantage is that they have to be developed at a particular time which requires prior knowledge of the likely optimal exposure time. A color reaction, by contrast, can be watched as it proceeds and be stopped when the color has reached the desired intensity.

Immunostaining

Staining of specimens with specific antibodies is also very important in developmental biology (Fig. 5.7b). It is possible to make an antibody against almost any protein or carbohydrate molecule by injecting it into an animal with a suitable adjuvant. The target of a specific antibody is referred to as its **antigen**, regardless of its chemical nature. The particular parts of an antigen molecule that are recognized by the antibody are called **epitopes**. **Polyclonal** antibodies are often made in rabbits. After repeated immunizations there should be a high concentration, or titer, of specific antibody in the serum. Although the antibody is specific for the particular antigen, it will consist of the products of several clones of B lymphocytes, each making antibody from a different antibody gene. In addition to the specific antibody, the serum will contain thousands of other antibodies recognizing unrelated antigens, and thousands of nonantibody proteins. These other components may cause background or nonspecific staining, and therefore the antibody will usually need to be partially purified from the serum before use. Antibody molecules consist of a variable region, containing the part that recognizes the antigen, and a constant region, which is characteristic of the immunoglobulin class and the animal species. A frequently used partial purification method is a column of protein A. As mentioned above, this is a bacterial protein binding to the constant region of IgG-type antibody molecules. Thus, the protein A column will isolate all the IgG, including the specific antibody, and leave all the other serum components behind. A better method, if enough of the antigen is available, is to make an affinity column carrying the antigen, and select out just the specific antibody from the serum, removing all other antibodies as well as other proteins.

Monoclonal antibodies are usually made by immunizing mice, then fusing their spleen cells with a human tumor cell line called a myeloma. The hybrid cells, called **hybridomas**, are capable both of antibody production and also of growth without limit. Numerous multiwell plates containing clones of the fused cells are screened in order to find the one required. Once the desired cell clone has been isolated, it can be grown without limit and its culture medium harvested as a source of just one particular monoclonal antibody. The advantage of the monoclonal method is that it does not require a pure antigen to start with; in fact wholly uncharacterized tissue extracts can be used for immunization. The antibodies made by individual hybridomas can then be screened for interesting expression patterns and cloned in the molecular sense by screening an expression library with the same antibody. On the other hand, more work and skill is required to make monoclonal rather than polyclonal antibodies, and individual monoclonal antibodies often do not have the same affinity as polyclonal antibodies.

As for *in situ* hybridization, immunostaining can be performed either on **wholemounts** or on **sections**. Wholemounts have the advantage of providing a single three-dimensional view of the location of the antigen in the specimen, while sections provide an intrinsically higher resolution. In both cases, the specimen is incubated with the specific antibody for a suitable period, then washed and incubated with a second antibody. This is a commercially available antibody directed against the species-specific constant region of the first antibody. It will carry a group suitable for detection, either a fluorescent group like fluorescein or rhodamine, or an enzyme such as alkaline phosphatase or horseradish peroxidase. If a fluorescent second antibody was used, the specimen can be examined directly in the fluorescence microscope. If an enzyme-linked second antibody was used, then the specimen is incubated in the substrate to allow the colored precipitate to develop, and can then be mounted and visualized in transmitted light.

Enzyme-linked methods are usually more sensitive than fluorescence, because the enzyme–substrate reaction provides an additional amplification step. Also, once the precipitate is formed it is possible to dehydrate the specimen and use a non-aqueous mounting medium. Nonaqueous media are usually preferable to aqueous media because they have a higher refractive index and render the specimen more nearly transparent. For fluorescent immunostaining it is necessary that the immune complex remains intact, and this makes it impossible to dehydrate the specimen. For the same reason fluorescent specimens may not be permanent because the immune complexes will eventually dissociate. On the other hand, fluorescent methods are simpler, because there are fewer steps, and they provide the best way of looking at more than one antigen in overlapping domains, because each fluorochrome can be examined separately in its own channel of the fluorescence microscope.

Reporter genes

There are numerous circumstances in developmental biology where it is not convenient to monitor the expression of the gene one is actually interested in, but it is possible to achieve the same result by monitoring a **reporter gene**. For example there are many experiments in which the regulatory region of a gene is chopped into short sequences with the object of finding which particular sequence drives expression in different regions or cell types of the embryo. Each short sequence is attached to a reporter gene whose product is easily detected and introduced into embryos as a transgene. As for direct detection of gene products discussed above, it is possible to monitor reporter gene activity either by biochemical methods, which are quantitative but lack spatial resolution, or by *in situ* methods that have good spatial resolution but are not quantitative.

The most popular reporter gene overall is *E. coli lacZ*, coding for β-**galactosidase**. The name arises from the fact that it is the "*Z*" gene of the *lac* operon, subject of classical studies on gene regulation in the 1950s. β-galactosidase is a large, tetrameric enzyme capable of hydrolyzing a whole range of β-galactosides, which are substances consisting of a chemical group joined by a β linkage to the 1-carbon of galactose. Biochemical measurement of β-galactosidase activity is possible with a variety of colorimetric or fluorescent substrates, but in developmental biology this reporter is usually used in *in situ* mode. For this the substrate is 5-bromo-4-chloro-3-indolyl-β-D-galactoside, or X-Gal for short. When hydrolyzed away from the galactose, the X part of the molecule immediately forms a green-blue insoluble precipitate. The sensitivity of the reaction means that very low levels of expression can be detected. β-galactosidase is a very useful reporter both because of its sensitivity and because most animal tissues do not contain cross-reacting enzymes, which means that background staining is usually low. It will work following aldehyde fixation and therefore can be combined with other techniques such as conventional staining or immunostaining.

It should be remembered that the β-galactosidase protein is very stable, so its presence may indicate past as well as present activity of the gene. It is therefore possible to find regions of an embryo in which β-galactosidase enzyme is present but the mRNA encoding it is absent. This phenomenon is known as **perdurance** and may be particularly significant in embryos that do not show much growth. For types that grow rapidly, the protein level will quickly fall by dilution once the gene has been turned off.

β-galactosidase will often remain active as a **fusion protein** when the polypeptide has been fused to some other protein sequence. This is important for some applications, including the use of a construct called β-**geo** in mouse knockout technology. β-geo is a fusion of the β-galactosidase enzyme with the product of the *neomycin resistance* gene and possesses both of the biological activities in the one molecule.

A second very important group of reporter genes is based on **green fluorescent protein (GFP)**, which was introduced during the late 1990s. As its name suggests this is a protein showing intense green fluorescence emission. It originates from the jellyfish *Aequorea victoria*. Like β-galactosidase it often remains active when fused to other proteins. Various alterations have been introduced into the coding sequence of GFP to increase the intensity or alter the color of the fluorescence, and so there are now available several derivative proteins emitting red, yellow, or blue fluorescence. One important advantage of the fluorescent proteins is that they are easy to visualize in living specimens so allow real-time examination of the labeled cells as well as detection in fixed specimens.

For biochemical detection another commonly used reporter is firefly **luciferase**. This is an enzyme that catalyzes breakdown of a substrate, luciferin, in the presence of ATP. The reaction leads to emission of light (**phosphorescence**). Detection in tissue samples is carried out with a luminometer, and very sensitive measurements can be made. This method is not very suitable for *in situ* detection because the phosphorescence is below the detection level of most CCD cameras, although it can be visualized with highly sensitive photon counters. Another biochemical reporter is the enzyme chloramphenicol acetyl transferase (CAT), which catalyzes acetylation of the antibiotic chloramphenical by acetyl CoA. The assay is done by adding the tissue extract to a substrate mixture containing radioactive chloramphenicol, then separating the products by thin layer chromatography and detecting the acetylated product by autoradiography or phosphorimager.

Microinjection

There are various reasons why it is often desirable to introduce a substance into a single cell of an embryo. Most of the experiments on *Xenopus* embryos involving overexpression of genes are carried out by making synthetic mRNA *in vitro* and injecting it into the fertilized egg. The creation of transgenic mice relies on the ability to inject DNA into a pronucleus of a fertilized egg, and to create transgenic *Drosophila* the DNA must be injected at the posterior end of the egg, where the germ cells are about to form. Microinjection methods may also be used to introduce inhibitors such as specific antibodies or antisense oligonucleotides into cells. In addition, microinjection is essential for a whole variety of cell-labeling experiments where a particular cell lineage, or a graft, needs to be identified by the presence of a visible substance called a **lineage label**.

The equipment required for microinjection depends on the size of the target cell. However, it will always be mounted on a microscope to allow visual control of the injection (Fig. 5.8). It will require some form of micromanipulator to hold the injection needle and reduce the manual movements of the experimenter to a scale commensurate with the target cell. The injection needle itself will be made of glass tubing. This is drawn out into a fine-tipped injection needle using a needle-pulling machine that heats the glass to near melting point and then applies an appropriate pull to draw it out. The substance to be injected is introduced into the needle, either sucked into the tip by capillarity or injected with a syringe into the rear end. Then the needle is connected to an injection controller by a flexible

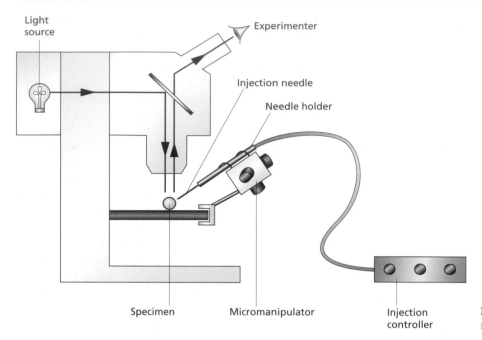

Light source

Experimenter

Injection needle

Needle holder

Specimen

Micromanipulator

Injection controller

Fig. 5.8 Setup for microinjection under the fluorescence microscope.

tube. The controller may be a pressure device that applies sharp pulses of pressure to the needle and forces a small volume out of the sharp end. It may be an iontophoretic device which applies an electric field across the needle and causes a migration of appropriately charged molecules out of the tip. In either event, the needle is filled, is connected to the controller, and attached to the micromanipulator. Watching down the microscope the experimenter impales the cell required and operates the controller to drive a pulse of the substance out of the tip. If the substance is fluorescent, as is often the case for a lineage label, then the microscope will be fitted with a fluorescent attachment to enable immediate visualization of the effectiveness of the injection. *Xenopus* embryos are injected under a dissecting microscope, but smaller specimens will require compound microscope magnification. Some setups will use an upright and others an inverted microscope, but in either case the optics of the microscope will usually be arranged so that the image is not inverted, and its mechanics will be such that the stage holding the manipulators is fixed while the lenses move. These are special features, not standard on the normal type of upright compound microscope.

Cell-labeling methods

Cell-labeling methods are used for a whole variety of different reasons. They are used for **fate mapping** to show the normal destiny of embryo regions in the course of development. They are used to label single cells in order to do **clonal analysis** (see Chapter 6). They are used to label cells into which other substances such as mRNA or antibodies have been introduced. They are used to label whole embryos which can then be used as

donors for labeled grafts. The labels used in developmental biology are normally intended to be **lineage labels**, meaning that they label all progeny of the originally labeled cells and nothing else, although sometimes this ideal is not achieved in practice. The importance of cell labeling lies in the fact that cells in all types of animal embryos can move around considerably and it is impossible to keep track of individual cells by observation alone.

Extracellular labels

The oldest type of extracellular label are the **vital dyes** that were introduced in the early years of the twentieth century for fate mapping. The most commonly used are Nile Blue and Neutral Red. They are called vital dyes because a reasonable intensity of color can be taken up by living cells and does not produce toxic effects. They can be used to label whole specimens by application in the medium, or to label specific regions of an embryo by local application of a small block of agar impregnated with the dye. They are still sometimes used because they are quick, simple, and cheap. However, they are not really lineage labels. It is difficult to label deep within a specimen, and the vital dyes do spread and fade, and therefore only produce an approximate indication of the original site of labeling.

More recently the carbocyanine dyes, **DiI** and **DiO**, have been favored for applications involving small patches of extracellular label. They are applied using an extracellular variant of the microinjection device discussed above. Being very hydrophobic substances they dissolve in the lipid membranes of the labeled cells and are well retained in the original cells and their progeny. They are intensely fluorescent, DiI producing red and DiO producing green emission, so they are visualized using the fluorescence microscope. Although DiI and DiO themselves are removed from the specimen by organic solvents, chemical derivatives are now available that are retained during histological processing so they can be examined in paraffin wax sections.

Intracellular labels

The most commonly used intracellular labels are fluorescent dextrans. Dextran is a polymer of glucose, freely soluble in water and metabolically inert. Water-soluble molecules cannot enter or leave cells by diffusion through the lipid plasma membrane, and although small molecules can move from cell to cell through gap junctions, these are not able to carry molecules of molecular weight above about 1000. Therefore a dextran of about 10,000 molecular weight will remain confined to the cell into which it has been injected. To make the dextran visible it is conjugated to a fluorochrome, such as fluorescein or rhodamine, and to make it fixable it is also conjugated to the amino acid lysine, which bears a free amino group which readily reacts with aldehyde fixatives. Commonly used substances of this group are FDA (fluorescein dextran amine) and RDA (rhodamine dextran amine).

Another commonly used intracellular label is the enzyme horseradish peroxidase (HRP) which is also retained within cells after injection. Peroxidases catalyze the oxidation of various substrates by hydrogen peroxide, and the one normally used is diaminobenzidine, which is converted to a brown insoluble material, stable to paraffin wax histology. HRP can be visualized in formalin-fixed wholemounts, but the enzyme itself is not stable to paraffin wax histology and therefore frozen sections need to be cut if the specimen is too big for wholemount staining.

Genetic labels

Substances such as those described above are at their most useful in embryos that show little or no growth during development, i.e free-living species like *Xenopus* and zebrafish, although they have also been successfully used for limited periods of mammalian or chick development. Without growth the substances do not become diluted and can remain clearly visible for days. However, with growth they will rapidly become diluted and cease to be visible. Under these circumstances the best labels are those incorporated into the genome of the cell such that they become replicated every cell cycle and are not diluted. Much of the early experimentation on mouse and *Drosophila* utilized genetic labels of this kind although they all had severe disadvantages. For example the mouse glucose phosphate isomerase isozyme system depended on the different electrophoretic mobility of the enzyme present in different mouse strains. But this could not be visualized *in situ*, only by biochemical analysis of dissected tissue pieces, so this had very low spatial resolution. The *yellow* and *multiple wing hairs* mutants of *Drosophila* were

widely used but were only expressed in the adult fly and only in the cuticle. They were not suitable for the analysis of larvae or of the internal tissues of the adult.

Nowadays much better genetic labels are available, which are expressed in all tissues, at all developmental stages, and can be visualized by *in situ* methods. For example, there are transgenic mouse lines expressing *Escherichia coli* β-galactosidase (coded by the *lacZ* gene) or human alkaline phosphatase (hPLAP) in all tissues. These can be detected by sensitive histochemical methods, and the hPLAP has the advantage that the enzyme activity is stable to paraffin wax embedding so it can be visualized histochemically in sections. The equivalent in *Drosophila* are transgenic lines which include various easily detectable proteins whose synthesis is directed by a promoter active in all tissues. The protein in question may again be β-galactosidase, or may be some inactive substance for which a good antibody is available, such as the "myc-tag," which is a short sequence from the *myc* oncogene. In any experiment involving a combination of cells from a labeled and unlabeled source, it is possible to stain the specimen and find which cells derived from each of the original components.

It is also possible to introduce a genetic label by means of a replication-incompetent retrovirus. Retroviruses are RNA viruses that carry a reverse transcriptase enzyme. On infection of a cell the reverse transcriptase makes a DNA copy of the viral genome that can then integrate into the chromosome of the cell, and will remain present in all its progeny. A replication-incompetent virus lacks genes that are necessary for the assembly of virus particles, and therefore the viral genome simply stays in the chromosome and is unable to produce further virus or lyze the cell. Replication-incompetent viruses are themselves propagated in special cell lines called packaging cell lines which contain the missing viral genes and are able to complement the functions missing from the virus. A retrovirus used for cell labeling will usually contain a *lacZ* gene driven by the viral **long-terminal repeat** which is a strong promoter sequence. They are mainly used for **clonal analysis** (see Chapter 4), therefore infection is instigated at low multiplicity such that each focus of infection represents the progeny of a single labeled cell.

Cell sorting

Stem cell research in particular has depended on the ability to identify and sort cells from complex mixed populations such as those found in the mammalian bone marrow. The instrument that does this is called a flow cytometer. If it can sort cells as well as analyze them it is called a fluorescence activated cell sorter or **FACS** machine. Cells may be sorted by size or by fluorescence signal depending on the binding of fluorescent antibodies to molecules on the cell surface. To carry out a fractionation, the cells are placed in a reservoir in suspension in a special medium (sheath fluid), then pass through a vibrating nozzle that divides the stream into a series of tiny droplets. The cell density is such

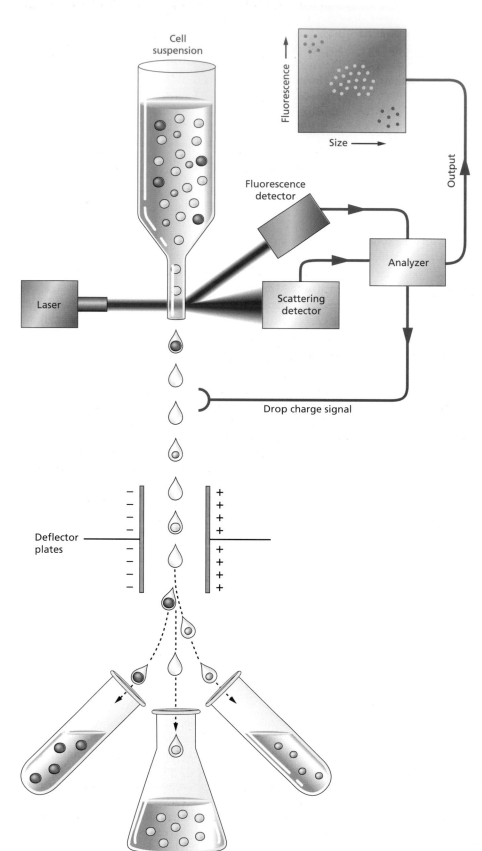

Fig. 5.9 Fluorescence-activated cell sorter. The output shows that the cell population contains minority fractions which are small-high fluorescence (green) and large-low fluorescence (red) cells. These can be sorted into different tubes by appropriate settings of the machine.

that most droplets do not contain any cells and a few contain just one cell. A laser beam is directed at the stream just before it breaks into droplets and as each cell passes through the beam its fluorescence signal is measured by a detector. A forward scattering detector also provides a measure of cell size. The experimenter will program the machine to set criteria for fluorescence and size, and the machine will impart a positive or negative charge to the droplet depending on these criteria and the measurements of the detectors. The droplets then pass between positively and negatively charged metal plates such that droplets carrying a charge become deflected. This means that a pure subset of cells can be purified in viable form from a mixed starting population (Fig. 5.9). Machines may have more than one laser and detector enabling more sophisticated separations.

If it is simply required to measure the proportion of cells with different properties without harvesting them in viable form then the simpler type of analysis-only machine can be used. In this case the cells can be fixed and can be tagged with antibodies recognizing internal molecules as well as those on the cell surface.

Key Points to Remember

- Dissecting microscopes are used to handle and manipulate specimens or to examine large wholemounts. Compound microscopes are used to examine sections or small wholemounts.
- Preparation of standard histological sections involves fixation, dehydration, embedding in paraffin wax, sectioning on a microtome, and removing wax from the sections before staining them.
- Gene expression may be studied by biochemical methods that give good quantification but poor spatial resolution. Microarray analysis can give a semiquantitative overview of which genes are expressed in the sample. Methods for specific mRNAs include Northern blots, RNAse protection, and RT-PCR. Methods for specific proteins include immunoprecipitation and Western blotting.

- *In situ* hybridization for RNA or immunostaining for protein can give good spatial resolution but are not quantitiative techniques.
- Reporter genes, most often *lacZ* or *gfp*, are used for the analysis of promoters in transgenics, and for many other purposes.
- Extracellular labels used for fate mapping comprise the traditional vital dyes and the newer carbocyanine dyes. Fluorescent dextrans are often used as intracellular labels.
- Genetic labels are not diluted by growth of the embryo and so are very useful in mammalian or avian embryos.
- Cells can be counted and sorted with a flow cytometer if they can be labeled in a suitable manner.

Further reading

General practical manuals

Stern, C.D. & Holland P.W.H. (1993) *Essential Developmental Biology. A practical approach.* Oxford: Oxford University Press.

Sharpe, P.T. & Mason, I. (1998) *Molecular Embryology: methods and protocols (Methods in Molecular Biology Series).* Totowa, NJ: Humana Press.

De Pablo, F., Ferrus, A. & Stern C.D., eds (1999) *Cellular and Molecular Procedures in Developmental Biology (Current Topics in Developmental Biology,* vol. 36). New York: John Wiley.

Tuan, R.S. & Lo, C. (1999) *Developmental Biology Protocols (Methods in Molecular Biology Series).* Humana Press.

Gibbs, M.A. (2003) *A Practical Guide to Developmental Biology.* Oxford: Oxford University Press.

Beffa, M. ed. (2004) *Key Techniques in Practical Developmental Biology.* Cambridge: Cambridge University Press.

Histological methods, in situ hybridization, and immunostaining

Nieto, M.A., Patel, K. & Wilkinson, D.G. (1996) *In situ* hybridization analysis of chick embryos in whole mount and tissue sections. *Methods in Cell Biology* **51**, 219–237.

Jowett, T. (1997) *Tissue In Situ Hybridization: methods in animal development.* New York: John Wiley.

Polak, J.M. & McGee J.O'D. (1999) *In Situ Hybridization: principles and practice.* Oxford: Oxford University Press.

Kiernan, J.A. (2000) *Histological and Histochemical Methods: theory and practice.* Oxford: Butterworth Heinemann.

Bancroft, J.D. & Stevens, A., eds (2001) *Theory and Practice of Histological Techniques.* Edinburgh: Churchill Livingstone.

Organism-specific manuals

Bronner-Fraser, M. ed. (1996) *Methods in Avian Embryology (Methods in Cell Biology Series).* New York: Academic Press.

Hope, I.A. ed. (1999) *C.elegans – a practical approach.* Oxford: Oxford University Press.

Sive, H.L. , Grainger, R.M. & Harland, R.M., eds (1999) *Early Development of Xenopus laevis: course manual.* Cold Spring Harbor, NY: Cold Spring Harbor Laboratory Press.

Nagy, A., Gertsenstein, M., Vintersten, K. & Behringer, R. (2003) *Manipulating the Mouse Embryo. A laboratory manual*, 3rd edn. Cold Spring Harbor Laboratory Press.

Ashburner, M., Hawley, S. & Golic, K. (2004) *Drosophila: a laboratory handbook*, 2nd edn. Cold Spring Harbor Laboratory Press.

Detrich, H., Zon, L.I. & Westerfield, M. eds (2004) *The Zebrafish: cellular and developmental biology*, 2nd edn. *Methods in Cell Biology*, vol. 76. New York: Academic Press.

Other techniques

Bradbury, S. (1989) *An Introduction to the Optical Microscope.* Royal Microscopical Handbooks no. 1. London: Oxford University Press.

Lacal, J.C., Perona, R. & Feramisco, J. (1999) *Microinjection (Methods & Tools in Biosciences & Medicine Series).* Basel, Switzerland: Birkhauser Verlag AG.

Darzynkiewicz, Z., Crissman, H.A. & Robinson, J.P., eds (2000) *Flow Cytometry*, 2nd edn. *Methods in Cell Biology*, vols 41, 42. New York: Academic Press.

Section 2

Major model organisms

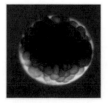
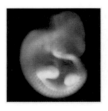

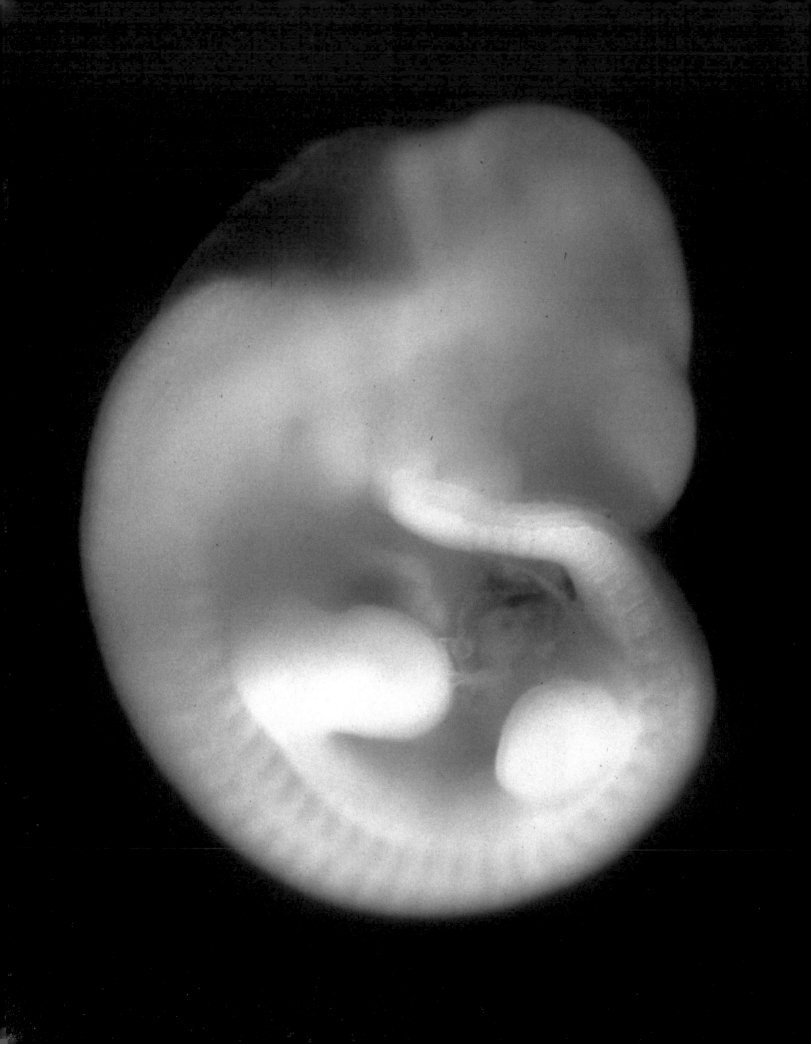

Chapter 6

Model organisms

Out of over 1 million animal species, modern developmental biology has focused on a very small number which are often described as "model organisms." This is because the motivation for their study is not simply to understand how that particular animal develops, but to use it as an example of how all animals develop. Much developmental biology research is supported by medical research funding bodies and their ultimate goal is to understand how the human body develops, even if this is not the immediate goal of the investigators themselves. For this reason there is often an attraction to studying a process that also occurs in humans. In this chapter it will be explained why research activity has focused down onto a small number of species, with some comparative indication of their strengths and weaknesses. In a short book a line has to be drawn somewhere and so just six of the most important models organisms are covered in detail: the mouse, the chick, the frog *Xenopus*, the zebrafish, the fruit fly *Drosophila*, and the nematode *Caenorhabditis elegans*. Other species have also been extensively studied, the next most popular being various species of sea urchin, but worldwide the largest number of active researchers are working with the big six.

The big six

The organisms discussed in detail in this book are listed in Table 6.1 and their positions in the phylogenetic tree of animals are shown in Fig. 6.1. It will be noted that they do not provide a very good coverage of the animal kingdom, as there are just two invertebrates and four vertebrates. However, the evolutionary distance between them is sufficiently large to make it likely that any developmental features possessed by all six are shared by the entire animal kingdom.

The six model species have each been selected because they have some particular experimental advantages for developmental biology research. It is true that sometimes individual scientists choose their organisms just because they like looking at them and working with them. But there are also some more objective practical considerations that govern the selection of organisms for particular purposes which are summarized in Table 6.2.

Availability and cost

All the six species are available all year round. Without this facility they would not have been selected as model organisms at all. It is worth remembering that it can be very difficult to "domesticate" an organism for laboratory life, and to attempt this for a "new" species is a major undertaking. Although some marine invertebrates have been used in developmental research (see below), none of the big six is a marine organism, probably because of the extra difficulties involved in keeping them in the lab. For example it is extremely difficult to breed and rear sea urchins in the lab.

Table 6.1 Organisms discussed in detail in this book.

Species	Common name	Phylum, subphylum, class
Caenorhabditis elegans	Worm	Nematode, phasmida
Drosophila melanogaster	Fruit fly	Arthropod, uniramian, insect
Brachydanio rerio	Zebrafish	Chordate, vertebrate, fish
Xenopus laevis	African clawed frog	Chordate, vertebrate, amphibian
Gallus domestica	Chicken	Chordate, vertebrate, bird
Mus musculus	Mouse	Chordate, vertebrate, mammal

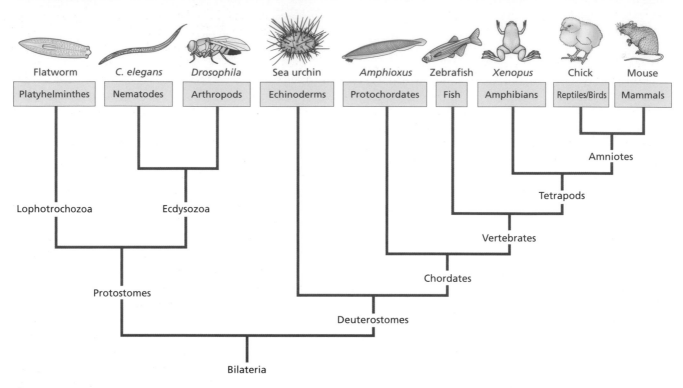

Fig. 6.1 Phylogenetic tree showing the positions of the big six model organisms used in developmental biology.

	C. elegans	*Drosophila*	Zebrafish	*Xenopus*	Chick	Mouse
Numbers of embryos	high	high	high	high	low	low
Cost	low	medium	medium	medium	low	high
Access	good	good	good	good	good	poor
Micromanipulation	limited	limited	fair	good	good	limited
Genetics	good	good	fair	none	none	good
Gene inventory	known	known	known	known*	known	known

Table 6.2 Experimental advantages and disadvantages of six model organisms.

*X. tropicalis

In terms of numbers, it is relatively easy to obtain thousands of eggs from the first four species. Chick eggs are normally not bred in the lab but are purchased from commercial hatcheries. A large incubator will accommodate a few hundred eggs. Mice are less prolific than the other species but a mated pair are likely to produce a litter of, say, 12 embryos, so it is not too difficult to produce moderate numbers.

Cost is an important consideration because the pressure of modern research demands a weekly or even daily supply of embryos. *Caenorhabditis elegans* are very cheap, as they can be grown on agar plates coated with bacteria, and genetic stocks can be stored frozen. *Drosophila* are potentially cheap to keep but may need temperature- and humidity-controlled fly rooms and significant technician time to maintain all the different genetic stocks. They cannot presently be frozen. Zebrafish need an expensive aquarium facility and large-scale stock maintenance. The capital cost of setting up a lab is quite high and the running cost is probably similar to *Drosophila*. *Xenopus* also need an aquarium, although these have traditionally been of a less sophisticated type than those used for zebrafish. The care is relatively cheap since they have "no genetics" and therefore there are few stocks to maintain. Chicks are very cheap to keep as they are purchased from chick hatcheries and just require an egg incubator in the lab. However, if a lab needs to do genetic crosses or have access to very early stages, it will need a chicken-breeding facility, and since the animals are quite large this will be costly. Mice may seem small and cheap but are actually by far the most expensive organisms on the list. The logistics of mammalian breeding and the regulations about space and standards in laboratory animal facilities mean that the technician time and space required for a significant mouse operation is very high. Mouse sperm and embryos can, however, be frozen, reducing the long-term cost of stock maintenance.

Access and micromanipulation

"Access" refers to how easy it is to get at embryos at all stages of development. Just about any embryo can be obtained by dissection, but many experiments require that something should be done to the embryo and it should then be maintained alive until a later stage to observe the consequences. From this point of view the free-living organisms with external fertilization are the most favorable. This means *Xenopus* and zebrafish. *Drosophila* eggs are also laid soon after fertilization. *Caenorhabditis elegans* need to be dissected from the mother during cleavage stages, although they can survive perfectly well on their own. Chicks undergo their cleavages in the reproductive tract of the hen and are already a double-layered structure containing about 60,000 cells when the eggs are laid. After this stage they are easy to get at because it just requires a hole to be cut in the egg shell. They will survive perfectly well if the hole is resealed with adhesive tape and they are kept in a humidified incubator. Mice are the least good from the access point of view. For the first 4 days the embryos can be flushed from the reproductive tract and cultured *in vitro*. After this, they become implanted into the uterus of the mother and depend on the placenta for their nutrition. It is very difficult to grow preimplantation embryos into postimplantation stages *in vitro*, and if an early postimplantation embryo is removed from the mother's uterus it can only be cultured *in vitro* for another 1–2 days. However, individual organ rudiments from mouse and chick embryos can usually be cultured for long periods and show good differentiation *in vitro*. This means that they are often favored for studies of late development.

The other requirement for embryonic experimentation is the ease of microsurgical manipulation. This may mean removing a single cell, or a small piece of tissue, grafting an explant to another position in a second embryo, or injecting individual cells with substances. All this is relatively easy in *Xenopus*. Because of the large size of amphibian eggs, microsurgery can be done freehand under the dissecting microscope, and microinjection requires relatively cheap and simple equipment. A good level of micromanipulation can also be achieved in the chick, particularly at later stages. The other organisms are somewhat less favorable. With zebrafish it is possible to remove or inject cells during the early stages. *Caenorhabditis elegans* and *Drosophila* eggs are small and surrounded by a tough outer coat. Mouse embryos are small at preimplantation stages and hard to culture at postimplantation stages.

The *C. elegans* and fish embryos have the particular advantage of being transparent and so it is easier to follow cell movements *in vivo* than it is for the other species.

Genetics and genome maps

All the organisms under consideration have genomes, but they do not all have "genetics" in the sense of a technology for doing genetic experiments in the lab. *Drosophila* genetics is very sophisticated as it was practiced for many decades before *Drosophila* was adopted as a model for the study of development. The short life cycle of 2 weeks, and the ease of keeping large numbers of animals are both decisive. Also, the existence of balancer chromosomes (see Chapter 11) simplifies the keeping of stocks of mutants that are lethal in the homozygous form. *Caenorhabditis elegans* is also very favorable for genetics, because of a short life cycle and ease of keeping large numbers. Because it is a self-fertilized hermaphrodite, new mutations will segregate to the homozygous state automatically without the need to set up any crosses. Mouse genetics has also been practiced for many decades but it still falls below that of *Drosophila* in sophistication, partly because of the difficulty of handling lethal mutations, and also because of the huge cost of keeping the large numbers of animals required for mutagenesis screens. However, the ability to make "knockouts" of selected genes has become an enormously important technique. The zebrafish has a shorter history as a laboratory organism, and so lacks some of the sophisticated technology that exists for the more longstanding models. In terms of numbers and life-cycle duration (4 months) it is worse than the invertebrates but good for a vertebrate. *Xenopus* has never been taken seriously for genetics because of the long life cycle; it takes at least 9 months to rear an animal to sexual maturity. However, a related species, *Xenopus tropicalis*, will grow to maturity in 4 months.

The large-scale genome-sequencing activity of recent years means that all the model organisms now have more or less complete inventories of genes and high-resolution genome maps. The main importance of this for developmental biology is that it takes most of the labor out of cloning a new gene. In the past, to obtain a homolog of a known gene in your organism you had to clone it yourself and this can take considerable time and effort. Even worse, the positional cloning of a gene known only as a point mutation could take several years. When complete gene inventories and maps are available any particular gene can be obtained, at least to the level of a large piece of genomic DNA, from a central depository. Furthermore the complete gene inventory means that all the members of a gene family are known in advance. This is very helpful because there is frequently considerable redundancy of function between members of a gene family and it is necessary to know all the members to be able to interpret the results of both overexpression and loss of function experiments.

Unfortunately it has turned out that *Xenopus laevis* is **pseudotetraploid**. This means that it doubled the number of chromosomes about 30 million years ago, since when mutations have accumulated that make the gene copies on the duplicated chromosomes somewhat different from each other. But the gene pairs still remain fairly similar with similar functions. They are sometimes called "**pseudoalleles**" because the sequences make them look like alternative alleles at the same genetic locus but they are really not alleles at all. Pseudotetraploidy is unfavorable to the experimentalist because it means more genome to sequence and more redundancy of function. So it is unlikely that *Xenopus laevis* will be sequenced in the near future and attention has focused on *Xenopus tropicalis* which is a true diploid. Fortunately the level of sequence divergence between the two *Xenopus* species is low enough that probes from one species will normally hybridize with the homologous gene from the other.

Bony fish, including the zebrafish, also underwent an extra genome duplication, but this is much more ancient and so the gene pairs have diverged considerably. They have acquired rather different functions, and many have also been lost. Because of this the zebrafish is effectively diploid although does tend to have extra copies of many genes important in development.

Relevance and tempo

Table 6.2 summarizes the advantages and disadvantages of the six model organisms. A quick assessment of this table will show that they all have their strengths and weaknesses and that none of them is ideal in all respects. On balance the mouse and chick score lower than the others, and indeed some of their basic processes have been elucidated more recently because they are technically a little less favorable for experimental work. However, they are among the six favorites because of another important consideration, perceived relevance. Both mouse and chick are **amniotes**, and the mouse is a mammal. This means that they appear much more similar to humans than the other models. This consideration has guaranteed the mouse and chick a good share of medical research funding over the years. It has actually turned out that, at the molecular level, the other organisms are much more similar than previously thought to the human, but since we ourselves are mammals we shall always have a special interest in mammalian development.

In today's competitive world it is not possible to work on a system where each experiment will take a very long time. In this context the rates of development of the big six are shown in Fig. 6.2. This shows that the difference of time span to maturity is enormous, ranging from *C. elegans* at 3 days to *Xenopus laevis* at 9 months. Of course only the fast models are used for genetic research, but the very short generation time of *C. elegans* and *Drosophila* is a distinct advantage compared to the mouse or zebrafish in terms of getting experiments done at moderate cost in time and personnel. The top part of the diagram covers the embryonic rather than postembryonic period and shows that all the models do enable experiments to be conducted in a few days. This is because the "endpoint" is not usually the end of development but rather it is the stage at which the developmental process under study has been completed. So for example most *Xenopus* experiments have concerned early development and are completed in 2–3 days from fertilization, while many mouse experiments are scored in mid-gestation.

Other organisms

A very wide variety of organisms has been used at one time or another for developmental research but have not ended up among the big six. A striking feature of the less popular models is that particular research communities tend to work on many different similar species rather than just one species. This is in fact an important reason why they are not major model organisms. In the molecular age research moves faster when probes can be readily exchanged and a total genomic sequence is a very valuable resource enabling the full set of genes in a particular gene family to be known and enabling the rapid identification of mutants. For different species the primary sequences of genes and their genomic organization is inevitably somewhat different, and without worldwide agreement on a single species none of this can be done. For example the various planarian and urodele amphibian species used for regeneration research (see Chapter 19) are too far apart for cross-hybridization of probes to be possible and this is a significant handicap to progress.

Probably the most important of the other models is the sea urchin. This is a "senior citizen" of embryological research and did have a much more prominent position in the pre-molecular era. It was used for the first experiments in the late nineteenth century that demonstrated embryonic regulation (formation of a whole larva from one blastomere), and in the 1930s for experiments demonstrating the existence of a vegetal to animal gradient controlling the body plan. More recently it has been used to build models of developmental genetic networks. It is easy to obtain large quantities of eggs and to fertilize them *in vitro*. Sea urchin embryos are usually transparent and so morphogenetic processes can be observed *in vivo*. But the life cycle is very long, it is hard to rear the animals through metamorphosis, and they are not suitable for experimental genetics. Although microsurgical experiments have been conducted, the embryos are small (<100 μm diameter) so this type of work is very demanding. Sea urchins had their heyday in the 1960–80 period when it was advantageous to be able to obtain gram quantities of embryos for biochemical studies. This culminated in the discovery of the cyclins, proteins controlling the cell cycle. Also sea urchins are very convenient for studying fertilization, but the successful outcome of this work has shown that the molecular mechanisms have rather little in common with mammalian fertilization.

Ascidians are lower chordates that have had a modest following. Again they have a distinguished history having been used for some of the classic cell lineage studies around the beginning of the twentieth century. Isolation and transplantation experiments demonstrated the existence of cytoplasmic determinants, some of which have now been identified. Various gastropods such as *Patella* and *Ilyanassa* have also appealed for studies of cell lineage and cytoplasmic determinants. But in the end the concentration of attention on *C. elegans* has achieved most of what might have been expected to come from these marine invertebrates in terms of understanding the molecular nature and function of cytoplasmic determinants.

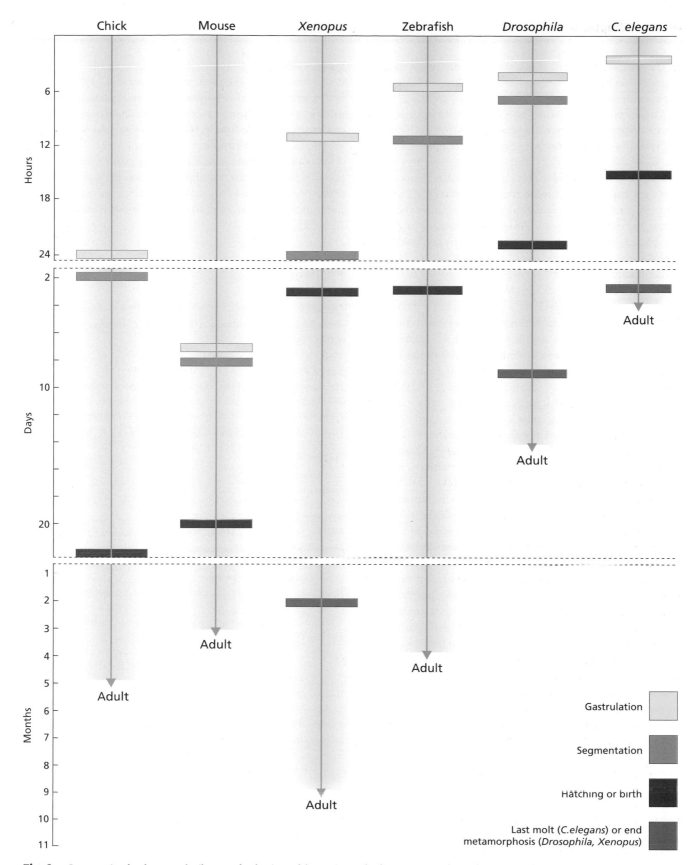

Fig. 6.2 Comparative developmental milestones for the six model organisms. The diagram is spread over three time scales, and times are only approximate since they may depend considerably on temperature for the free-living forms. "Gastrulation" refers to the start of germ layer formation. "Segmentation" refers to the start of visible segmentation. "Adult" refers to reproductive maturity. All times are measured postfertilization, including the chick where this adds approximately 1 day to the usual time scale of days of egg incubation.

The **planarian** worms display some of the most dramatic regeneration behavior in the animal kingdom and might perhaps one day provide us with a seventh model organism. At present however there are rather few labs engaged on the work and several different species are still being studied.

Certain lower eukaryotes including the cellular slime mold *Dictyostelium*, the acellular slime mold *Physarum*, and the alga *Volvox*, have been studied in relation to particular characteristics that they display clearly, respectively chemotactic cell aggregation, synchronous nuclear division, and morphogenetic movements. But all of them are a big evolutionary distance away from the human, substantially more so than even the invertebrate animals.

Then there are the ancillary organisms that lost out historically to the front running models but have some particular technical advantage that keeps them in use in a niche market. The rat is better than the mouse for whole embryo culture over the early postimplantation period and is often used for teratogenicity testing. The quail is easier to keep and breed in the lab than the chicken so can be used for things that require treatment of the mother, like vitamin A depletion studies. It is also used in conjunction with the chick for the labeling of grafts with quail-specific antibody. As mentioned above, *Xenopus tropicalis* offers a simpler genome and the possibility of experimental genetics not shared by *Xenopus laevis*. Also the **urodele** amphibians such as the axolotl and various newts show regenerative behavior greatly surpassing that of *Xenopus*. The medaka is another fish that has been domesticated to laboratory life in Japan and has been better for making transgenics than the zebrafish.

Finally there are some that have been studied not because they are thought to be models for the human but rather from the perspective of trying to understand animal evolution. Here the key issue is not experimental convenience but rather the position in the phylogenetic tree. For example, amphioxus is a cephalochordate thought to resemble to some extent the common ancestor of vertebrates. The features that it displays such as a single Hox cluster are felt to be primitive in the vertebrate lineage. Cnidarians, including the familiar freshwater *Hydra*, have a comparable position insofar as they may resemble the common ancestors of all animals. They also display a high degree of regeneration behavior. Many insects other than *Drosophila* have been used for developmental research, including the dragonfly *Platycnemis*, the cricket *Acheta*, the beetle *Tenebrio*, and the silk moth *Bombyx*. Some of these display a mode of development very different from *Drosophila*, in which the blastoderm grows as it produces posterior parts. Their main interest today is also for evolutionary comparison because of the exquisite level of molecular detail with which *Drosophila* development is understood.

All of these organisms have something to offer, but the big six account for most contemporary work and the convenience of using a standard model organism is such that the research funding agencies expect to see good reasons presented for working on something different.

Further reading

General

Bard, J.B.L. (1994) *Embryos. A colour atlas of development.* London: Wolfe Publishing.

Stage series

Hamburger, V. & Hamilton, H.L. (1951) A series of normal stages in the development of the chick embryo. *Journal of Morphology* **88**, 49–92. Reprinted in *Developmental Dynamics* **195**, 231–272 (1992).

Nieuwkoop, P.D. & Faber, J. (1967) *Normal Table of* Xenopus laevis. Amsterdam: N. Holland. Reprinted by Garland Publishing, London (1994).

Eyal-Giladi, H. & Kochav, S. (1976) From cleavage to primitive streak formation: a complementary normal table and a new look at the first stages of the development of the chick. *Developmental Biology* **49**, 321–337.

Theiler, K. (1989) *The House Mouse. Development and normal stages from fertilization to four weeks of age*, 2nd edn. Berlin: Springer-Verlag.

Hausen, P. & Riebesell, M. (1991) *The Early Development of* Xenopus laevis. Berlin: Springer-Verlag.

Kaufman, M.H. (1992) *The Atlas of Mouse Development.* London: Academic Press.

Hartenstein, V. (1993) *Atlas of* Drosophila *Development.* Cold Spring Harbor, NY: Cold Spring Harbor Laboratory Press.

Kimmel, C.B., Ballard, W.W., Kimmel, S.R., Ullmann, B. & Schilling, T.F. (1995) Stages of embryonic development of the zebrafish. *Developmental Dynamics* **203**, 253–310.

Bellairs, R. & Osmund, M. (1997) *The Atlas of Chick Development.* London: Academic Press.

Campos-Ortega, J.A. & Hartenstein, V. (1997) *The Embryonic Development of Drosophila melanogaster*, 2nd edn. Berlin: Springer-Verlag.

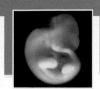

Chapter 7

Xenopus

Although other amphibian species have been used for experimental work in the past, the African clawed frog *Xenopus laevis* has been the world standard for many years. This is because of its ease of maintenance, ease of induced spawning, and robustness of the embryos. The experimental production of *Xenopus* embryos is very simple. The male and female are both injected with chorionic gonadotrophin, are put together overnight, and the next morning there are normally embryos. Nowadays it has become more common to perform *in vitro* fertilization which generates smaller numbers of embryos, but ones whose time of fertilization is known precisely and whose development shows a high degree of synchrony. *Xenopus* embryos are about 1.4 mm in diameter, which is large enough for quite discriminating microsurgery. Because the egg contains a high content of yolk granules and other reserve food materials it is possible for small fragments of embryos to survive and differentiate for several days in very simple media, and the use of embryo fragments (explants) is the basis of many experiments. The introduction of cell lineage labels such as fluorescent dextrans has enabled the accurate identification of cells in graft–host combinations. Methods have been introduced for overexpressing genes by making synthetic mRNA *in vitro* and injecting it into the fertilized egg. It is also possible to make **transgenic** *Xenopus* by incorporating DNA into the sperm and injecting the sperm into the egg.

Many *Xenopus* gene names are prefixed with an "X" to indicate their species of origin, for example *Xbra* for the *Xenopus brachyury* gene, or *Xgsc* for the *Xenopus goosecoid* gene. However, the convention is not uniformly adhered to, and in this book the "Xs" are omitted, partly for simplicity and partly to emphasize the similar developmental functions of homologous genes from different vertebrate species.

Oogenesis, maturation, fertilization

The frog ovary consists of large numbers of **oocytes** surrounded by layers of follicle cells and blood vessels. The **oogonia** in the frog ovary persist and continue to divide throughout life. Oogonia become oocytes following their last mitotic division. Oogenesis then takes several months, during which time the oocytes grow enormously, acquiring the food reserves needed to support the embryo over the several days of development before larval feeding can commence (Fig. 7.1). The oocyte nucleus is very large and is known as the **germinal vesicle**. The chromosomes are the four-stranded **bivalents** characteristic of meiotic prophase, but they remain active during oogenesis and display numerous protruding loops of chromatin. Because of this appearance they are called **lampbrush chromosomes**.

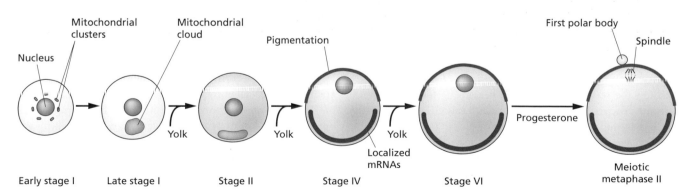

Fig. 7.1 Oogenesis in *Xenopus*. Roman numerals are used for staging. There is considerable increase in size from stage I to VI.

To support protein synthesis in early development, the oocytes need to accumulate a large store of ribosomes and transfer RNA. For this reason the gene cluster coding for ribosomal RNA becomes amplified at an early stage into many additional extra-chromosomal copies, all of which are active as templates for ribosomal RNA synthesis. During the growth phase the oocyte also acquires a large amount of **yolk** proteins. These are not made by the oocytes themselves, but by the liver of the mother, and are absorbed by the oocytes from the bloodstream. The early previtellogenic (= preyolky) oocyte is transparent, but it becomes opaque as the yolk granules begin to accumulate.

During oogenesis the **animal–vegetal polarity** of the oocyte arises. In the previtellogenic stage a special cytoplasmic region becomes assembled. This is rich in mitochondria and called the mitochondrial cloud. It is the precursor of the **germ plasm**, and its location determines the future vegetal pole of the egg. A number of mRNAs including *vegT* (see below) are associated with the cloud, and they become localized to the vegetal cortex as the cloud attaches to the cortex in early vitellogenesis. The hemisphere containing the germinal vesicle is called the **animal hemisphere**. In mid-oogenesis this becomes dark due to an accumulation of pigment granules, while the vegetal hemisphere remains light colored. At about the stage that the pigmentation difference appears, a second group of mRNAs including *vg1* (see below), also become localized to the vegetal cortex.

The fully grown primary oocyte is about 1.2 mm diameter, and its synthetic activity has waned. Maturation is achieved in response to gonadotrophins secreted by the pituitary gland of the mother. These travel through the bloodstream and provoke release of progesterone from the ovarian follicle cells. This binds to steroid receptors in the oocyte and activates translation of an oncoprotein, c-mos, which activates the phosphatase cdc25 and hence activates maturation promoting factor (MPF, also referred to as M-phase promoting factor, the complex of cdk and cyclin required to initiate M phase, see Chapter 2). The germinal vesicle breaks down and the first meiotic division takes place, resulting in the formation of the first polar body. The result is often called an **unfertilized egg**, although it is strictly a secondary oocyte arrested in second meiotic metaphase. The metaphase arrest is due to the fact that cyclin breakdown is inhibited by a complex of c-mos with cdk2, known as cytostatic factor. The eggs are shed into the body cavity, enter the oviducts through the fimbriae at the anterior end, and travel down the oviducts where they become wrapped in jelly.

In a normal mating the male would be clasping the female and would fertilize the eggs as they emerge from the cloaca. On fertil-ization the secondary oocyte/egg becomes a **fertilized egg** or **zygote**. The rise in intracellular calcium caused by sperm entry brings about the destruction of cytostatic factor, leading to the breakdown of cyclin and progression into the second meiotic division, with release of the second polar body. The calcium also causes exocytosis of **cortical granules** near the egg surface whose contents lift the **vitelline membrane** off the egg surface and allow the egg to rotate freely under the influence of gravity to bring the animal hemisphere uppermost.

Embryonic development

Development up to the general body plan stage may be sub-divided into **cleavage**, **gastrulation**, and **neurulation**, and in *Xenopus* these stages are completed within 24 hours at 24°C. A numerical stage series was devised by Nieuwkoop and Faber, according to which stage 8 is the mid-blastula, stage 10 the early gastrula, stage 13 the early neurula, and stage 20 the end of neurulation (Fig. 7.2).

The sperm enters the animal hemisphere and initiates a cyto-plasmic rearrangement called the **cortical rotation** (Fig. 7.3). This is a rotation of the egg cortex relative to the interior which is associated with the transient appearance of an orientated array of microtubules in the vegetal hemisphere. It leads to a reduction in the pigmentation of the animal hemisphere on the prospective dorsal side, opposite the sperm entry point. Internally a dorsal determinant is moved from the vegetal pole to the dorsal side, ensuring that the dorsal structures will develop opposite the point of sperm entry. In some other frog species a similar pigmentation change gives rise to a surface feature known as the **gray crescent**.

Cleavage

The first cleavage is vertical and separates the egg into right and left halves. The second cleavage is also vertical, and at right angles to the first, separating prospective dorsal from ventral halves. The third cleavage is equatorial, separating animal from vegetal halves. Subsequent cleavages vary between individuals, but it is usually possible to obtain some embryos showing four tiers of eight cells at the 32-cell stage. As in other species, the large cells resulting from early cleavage divisions are called **blastomeres**. As cleavage takes place a cavity called the **blastocoel** forms in the center of the animal hemisphere and the embryo is referred to as a **blastula**. The outer surface of the blastula consists of the original oocyte plasma membrane. A complete network of tight junctions around the exterior cell margins seals the blastocoel from the exterior and renders the penetration of almost all substances highly inefficient. This is why radiochemicals and other substances need to be introduced into the embryo by microinjection. Internal cells are connected by cadherins and are readily dissociated by the removal of calcium ions from the medium. Desmosomes are not found until the neurula stages, but all the cells of early cleavage stages are connected by gap junctions.

Cleavage continues rapidly for 12 divisions after which an important transition occurs, known as the **midblastula trans-ition** or **MBT**, although it is actually in the *late* blastula. The rate

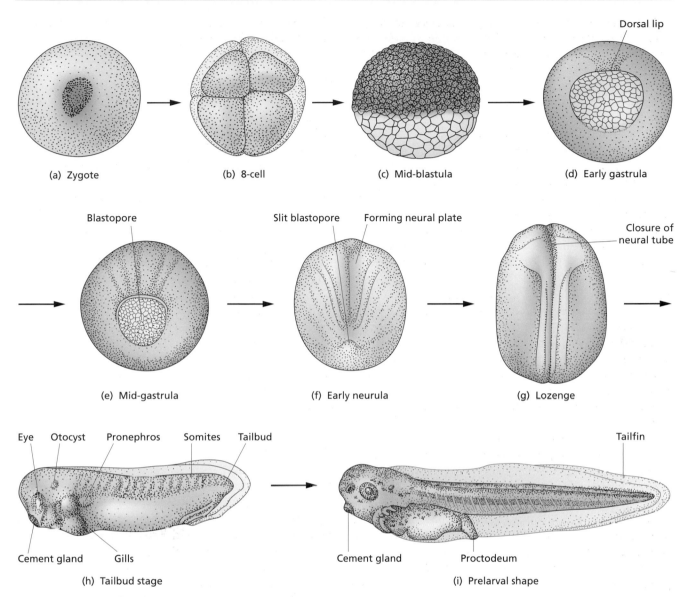

Fig. 7.2 Stages of *Xenopus* development. Here the numbers of the Nieuwkoop–Faber series are used for staging. (a) Zygote, from animal pole; (b) eight cells, stage 4, from animal pole; (c) mid-blastula, stage 8, from side; (d) early gastrula, stage 10, from vegetal pole; (e) mid-gastrula, stage 11, from vegetal pole; (f) early neurula, stage 14, from dorsal side; (g) lozenge, stage 22, about 1 day old, from dorsal side; (h) tailbud, stage 30, about 2 days old; (i) prelarva, stage 40, about 3 days old.

of cleavage slows down, the synchrony of cell divisions is lost, and the strength of intercellular adhesion increases so the blastula appears as a smooth instead of a knobbly ball. The MBT is the time at which significant transcription of the zygotic genome commences, although it is possible to detect low level transcription of some genes during the cleavage stages.

The onset of significant zygotic transcription makes it possible to visualize the early domains of commitment by *in situ* hybridization for specific transcription factors (Fig. 7.4). Only a few examples can be given here of the large number of genes that have been studied. The entire mesoderm can be visualized by the

expression of the T-box gene *brachyury*, which is needed to activate later mesodermal genes and to control gastrulation movements. The dorsal sector, which will form the **organizer** region (= **Spemann's organizer**), is characterized by the expression of a variety of transcription factor genes including *siamois*, *goosecoid*, *not*, and *lim1*. All these are required for axial differentiation and, in addition, *siamois* is involved in the original formation of the organizer, and *goosecoid* is needed for gastrulation movements. The ventral mesoderm expresses the homeobox genes *vent1* and *vent2* in nested pattern. These act in a combinatorial way to specify the lateral plate (vent1+2) and somitic (vent2 only)

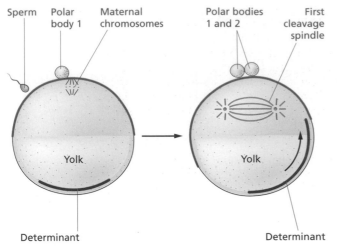

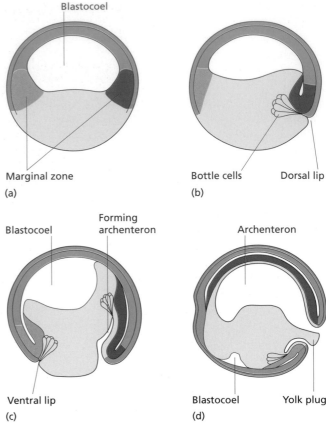

Fig. 7.3 Cortical rotation. The egg cortex moves about 30 degrees relative to the internal cytoplasm, in the direction of the sperm entry point. The vegetally located dorsal determinant moves to the dorsal side.

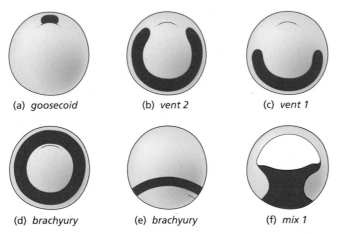

(a) *goosecoid* (b) *vent 2* (c) *vent 1*

(d) *brachyury* (e) *brachyury* (f) *mix 1*

Fig. 7.4 Early domains of zygotic genes. (a–d) From vegetal pole; (e) from side; (f) section from side.

Fig. 7.5 Gastrulation. (a) Blastula; (b) early gastrula; (c) mid-gastrula; (d) late gastrula. All views medial section. In this and subsequent figures, ectoderm is green, mesoderm orange, and endoderm yellow.

regions. The endoderm can be visualized as the domain of expression of several transcription factors including *mix1* and *sox17*; these are later required for activation of various genes of particular endoderm-derived tissues.

Gastrulation

Gastrulation is a phase of morphogenetic movements in the course of which the belt of tissue around the equator, called the **marginal zone**, becomes internalized through an opening called the **blastopore** (Fig. 7.5). This establishes the typical three-layered structure of an animal body, with outer **ectoderm**, middle **mesoderm**, and inner **endoderm**. The start of gastrulation is marked by the appearance of a pigmented depression in the dorsal vegetal quadrant. This is the **dorsal lip** of the blastopore. The blastopore becomes elongated laterally and soon becomes a complete circle. When it is circular the part referred to as the dorsal lip is the dorsal segment of the complete circle and the part referred to as the ventral lip is the ventral part of the complete circle. Elongated cells called bottle cells are present around the blastopore, although their precise mechanical role is not understood. Tissue invaginates all round the circular blastopore but the invagination is much more extensive on the dorsal side where it proceeds until the leading edge of invaginating tissue is well past the animal pole. In the lateral and ventral parts of the blastopore there is only a small extent of invagination.

Several components of the gastrulation movements may be distinguished. The mechanics of these movements are still poorly understood but it is known that they are to some extent independent of each other:

1 Spreading of the animal hemisphere (**epiboly**) which leads to it eventually covering the whole embryo surface.

2 **Invagination** of the marginal zone. This starts on the dorsal side and spreads to the lateral and ventral side until the blastopore is circular. The cavity formed by the invagination is called the **archenteron** and expands at the expense of the blastocoel as gastrulation proceeds. It has the form of a cylinder which is very much longer on the dorsal than the ventral side. When the yolk plug becomes internalized at the end of gastrulation it becomes part of the archenteron floor. In the course of the invagination the leading edge of the endoderm literally crawls up the inside of the blastocoel, and requires a layer of fibronectin on the blastocoelic surface to do so.

3 The prospective mesoderm is internal from the start of gastrulation. As the invagination proceeds, the mesoderm separates from the endoderm and **involutes** as a separate tissue layer between the ectoderm and endoderm.

4 Elongation of the dorsal axial mesoderm in the anteroposterior direction. This occurs by an active process of cellular intercalation in all three germ layers, called **convergent extension**, which helps drive the internalization of the marginal zone and the closure of the blastocoel.

5 Expansion of the ventrolateral mesoderm towards the dorsal midline.

The cellular mechanisms of gastrulation are still poorly understood. Most attention has been devoted to convergent extension. This requires the small GTP exchange proteins Rho and Rac, as introduction of dominant negative versions of these proteins will block convergent extension. These are thought to be activated by the Wnt planar polarity pathway (see Appendix and Chapter 17). In addition there is evidence that the Wnt Ca pathway regulates cell adhesion during gastrulation. As explained in Chapter 2, differential cell adhesion is critical for cell sorting behavior. In *Xenopus* embryos it is known that stimulation of this pathway does reduce adhesiveness. In addition, in intact embryos, the dorsal cells undergoing convergent extension show an increase in intracellular calcium. Finally, if synthesis of the Wnt receptor Frizzled7 is blocked using an antisense **morpholino** (see Chapter 3), then the tissue layers adhere to each other instead of remaining separate.

By the end of gastrulation the former animal cap ectoderm has covered the whole external surface of the embryo and the yolky vegetal tissues have become a mass of endoderm in the interior. The former marginal zone has generated a cylindrical layer of mesoderm extending on the dorsal side from the slit-shaped blastopore right the way to the anterior end, and on the ventral side just a limited distance from the blastopore. The mesodermal layer remains incomplete in the anteroventral region for some time.

The fates of the three germ layers in terms of tissue type are as follows:

Ectoderm becomes epidermis, nervous system, lens and ear, cement gland;

Mesoderm becomes head mesoderm, notochord, somites, kidney, lateral plate, blood, blood vessels, heart, limbs, gonads;

Endoderm becomes epithelial lining of gut, lungs, liver, pancreas, bladder.

New Directions in Research

The main opportunity in early *Xenopus* is the study of *morphogenetic movements*, which remain poorly understood. The advantage is the ability to set up simple *in vitro* systems such as the animal cap or the dorsal marginal zone, which can be observed in real time and which can easily be modified by injection of mRNA into the fertilized egg, or by treatment with biologically active substances.

Later stage *Xenopus* can now be used for research into *organogenesis*, *metamorphosis*, and *regeneration*, because transgenic methods can be used to modify gene expression, and can be combined with microsurgical procedures such as isolation or grafting experiments.

In the course of the invagination of the marginal zone, the archenteron has become the principal cavity at the expense of the blastocoel, and the embryo has rotated so that the dorsal side is uppermost. It now has a true **anteroposterior** axis which runs from the leading edge of the mesoderm at the **anterior** to the residual blastopore at the **posterior**.

If embryos are placed in too strong a salt solution, the gastrulation movements are seriously deranged and instead of invaginating into the interior, the endomesoderm evaginates from the ectoderm to form a dumbbell-like structure (Fig. 7.6). This is known as **exogastrulation**. Both the ectoderm and the endomesoderm of the exogastrula is remarkably normally patterned, although the central nervous system is substantially defective and there is no tail.

Although superficially similar, the gastrulation movements of *Xenopus* differ somewhat from the urodele species that were used for classical studies on amphibian embryology. This means that older textbook accounts may differ somewhat from this one.

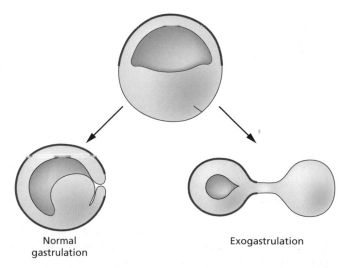

Normal gastrulation

Exogastrulation

Fig. 7.6 Exogastrulation.

Neurulation and later stages

The next stage of development is called the **neurula**, in which the ectoderm on the dorsal side becomes the central nervous system. The **neural plate** becomes visible as a keyhole-shaped region delimited by raised neural folds and covering much of the dorsal surface of the embryo. Quite rapidly the folds rise and move together to form the **neural tube** which, after closure, becomes covered by the ectoderm from beyond the folds, now known as **epidermis** (Fig. 7.7). During and after neurulation there is a

striking elongation of the body, which means that the whole trunk and tail region are derived from the posterior quarter of the neurula. This is driven by a continuing process of cellular intercalation both in the notochord and in other tissues. The neural tube, notochord, and somites are collectively known as the **axis**, not to be confused with the geometrical axes used for anatomical description.

By the **tailbud** stage all major body parts are in their final positions (Fig. 7.8a,b). The **notochord** forms from the dorsal midline of the mesoderm, rows of segmented **somites** appear on either side, the lateral plate mesoderm later gives rise to limb buds, kidney and coelomic mesothelium, and in the ventroposterior region of the mesoderm is a string of blood islands which provide the early tadpole with its erythrocytes. Another population of blood cells later form in the dorsolateral mesoderm (equivalent to the AGM region of higher vertebrates; see Chapter 9). A complex region at the posterior end, comprising the posterior neural plate and the mesoderm beneath it, becomes the **tailbud** (Fig. 7.8c), which generates the notochord, neural tube, and somites of the tail over the next 1–2 days. Because the neural folds close over the blastopore there is a connection created between the neural tube lumen and the gut, called the **neuroenteric canal**. This persists for about a day and is then blocked. The epithelia of the gut, comprising pharynx, lungs, stomach, liver, pancreas, and intestine, develop from the endoderm, although differentiation occurs much later than for the

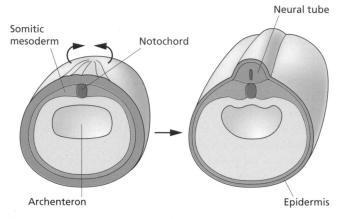

Fig. 7.7 Neurulation. Section through mid-body region.

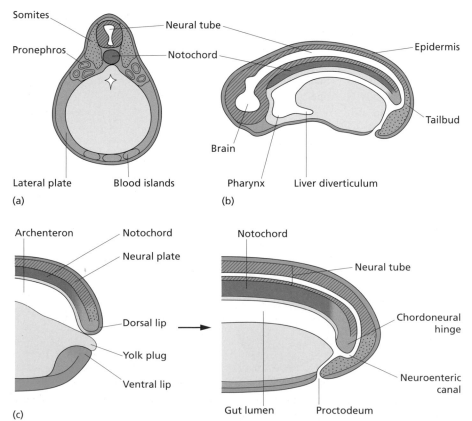

Fig. 7.8 Tailbud stage. (a) Transverse section through trunk; (b) median section; (c) formation of the tailbud by closure of the neural plate over the residual blastopore.

ectodermal and mesodermal tissues. The **proctodeum**, terminating in the anus, forms from the channel between neuroenteric canal and exterior formed by the process of neural tube closure. The mouth develops somewhat later from a new aperture formed at the anterior end.

In the head the anterior neural tube forms three vesicles that become the **forebrain, midbrain,** and **hindbrain.** The epidermis forms various columnar thickenings called **placodes.** These include the nasal placode, consisting of sensory cells from which originate the axons of the olfactory nerve connecting to the telencephalon; the lens placodes forming the lens of the eye; and the otic placodes that form the ear. The eyes develop as outgrowths of the forebrain (optic lobes) that invaginate into a cup shape such that the **pigment epithelium** forms from the outer layer and the retina forms from the inner layer. The lens invaginates from the surface epidermal lens placode and becomes surrounded by the **optic cup.** The optic nerve grows back from the retina down the optic stalk and projects to the **optic tecta** in the midbrain (see also Chapter 14).

Some cells from the folds of the neural plate which come to lie on the dorsal side of the neural tube later become the **neural crest.** This is a migratory tissue that forms a variety of tissue types. In the head the neural crest forms most of the skeletal tissues of the skull. In the trunk it forms the dorsal root ganglia, the sympathetic ganglia, and the **melanocytes** (pigment cells). In the vagal and sacral regions of the trunk it also forms the parasympathetic (= enteric) ganglia.

Anteriorly, the head mesoderm forms part of the jaw muscles and branchial arches. A structure known as the cement gland develops from the anterior epidermis, ventral to the future mouth. This is a prominent external feature of the head from the late neurula. The heart is formed from the ventral edges of the anterior lateral mesoderm that move down and fuse in the ventral midline around the end of neurulation.

Fate maps

Numerous **fate maps** have been published for amphibian embryos at stages from the fertilized egg to the end of gastrulation. Until 1983 all studies were by localized vital staining with the dyes neutral red or nile blue, applied to the embryo surface from a small fragment of impregnated agar.

More recently, injectable **lineage labels** have been preferred such as horseradish peroxidase (HRP) or fluorescein-dextranamine (FDA) (see Chapter 5). These can fill whole cells and do not diffuse. For surface marking **DiI** is now preferred. Because there is little increase in size of the early embryo, passive labels of these sorts do not become diluted and so remain clearly visible for several days. The lineage labels have revealed a certain degree of local cell mixing and this means that the fate map cannot be quite precise since the mixing causes some overlap between prospective regions. As this feature is not so apparent with vital dyes, maps shown in older textbooks imply a spurious degree of

precision which does not exist in reality. A modern fate map is shown in Fig. 7.9a and b. Figure 7.9c shows results of filling a dorsal blastomere (C1) and a ventral blastomere (C4) with a lineage label.

Important features of the fate map are:
1 The neural plate arises from the dorsal half of the animal hemisphere and the epidermis from the ventral half.
2 The mesoderm arises from a broad belt around the equator of the blastula, much from the animal hemisphere.
3 The endoderm arises from the vegetal hemisphere.
4 The somitic muscle arises from most of the marginal zone circumference, much from the ventral half of the blastula.
5 Both dorsal and ventral structures at the anterior end of the body come from the dorsal side of the blastula.

Experimental methods

As discussed in Chapter 4, when attempting to establish the function of any gene product in development it is necessary to know at least the following:
1 the expression pattern;
2 the biological activity;
3 the effect of specific inhibition *in vivo.*

The biological activity and inhibition experiments are particularly easy in *Xenopus* because of the ease of injecting materials into embryos and the use of ancillary techniques such as microsurgical isolation of explants or UV irradiation.

For biological activity measurement the material to be injected will usually be a mRNA made *in vitro* (Fig. 7.10a,b). The RNA synthesis is performed using plasmids carrying promoters for RNA polymerases of bacteriophage such as Sp6, T3, or T7, and a poly(A) addition site to ensure *in vivo* addition of poly(A) to stabilize the message. The mRNA can be injected into a whole fertilized egg, or into a specific blastomere during cleavage, which gives control over its location. The mRNA will be translated by the protein synthesis machinery of the egg and is likely to persist and remain active throughout early development, although it will eventually be degraded. Usually the same plasmid can be used for preparation of the *in situ* probes required for expression studies.

As many developmentally active molecules have different effects at different times, it may be desirable to inhibit activity of the introduced gene product until a desired stage. A useful method for doing this for transcription factors involves adding the hormone-binding domain from the glucocorticoid receptor to the protein of interest. This then causes it to be sequestered in the cytoplasm by binding to the cytoplasmic protein hsp90, until such time as a glucocorticoid, ususally dexamethasone, is added. As a lipid-soluble substance, dexamethasone can penetrate into the embryo and will bind to the receptor, liberating it from the hsp90, and allowing it to move to the nucleus where the transcription factor part of the molecule can exert its biological activity (Fig. 7.10c).

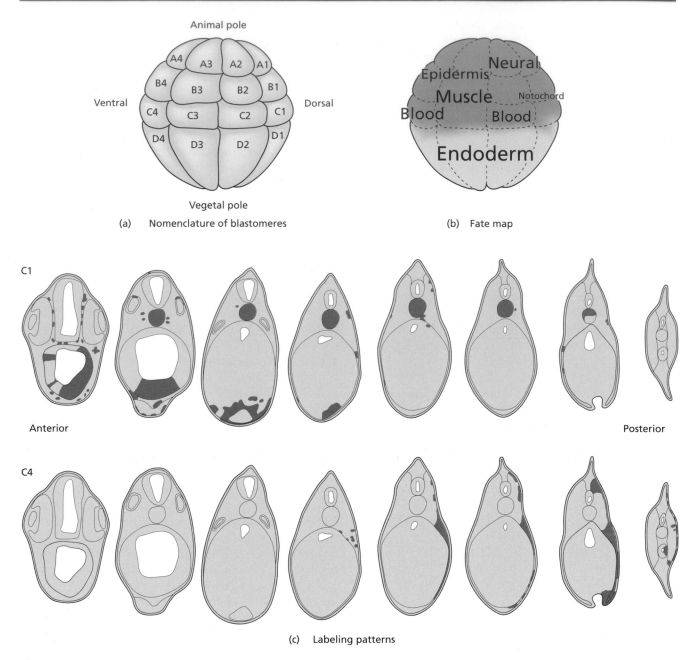

Fig. 7.9 A *Xenopus* fate map for the 32-cell stage. (a) Nomenclature of blastomeres at the 32-cell stage; (b) fate map of tissue types projected onto the 32-cell stage; (c) reconstruction of labeling pattern from blastomeres C1 and C4. The figure shows transverse sections taken at equal intervals along the body from anterior to posterior.

It is also possible to introduce genes by **transgenesis**, particularly important for studying events in late development by which time injected RNA may have been degraded. To make transgenic embryos the transgene DNA is added to sperm heads that have been decondensed in egg extract to make their own DNA accessible, and these are then injected into unfertilized eggs (Fig. 7.10d). It is usual to incorporate a green fluorescent protein (GFP) coding sequence into the transgene so that the transgenic embryos can be identified by their green fluorescence. Each individual transgenic embryo will have a different inser-

tion site and copy number, and this may cause some variability of biological behavior.

In *Xenopus* it is not practical to study the effects of mutating genes to inactivity, but for inhibition experiments there are three standard protocols which are very effective. The most commonly used is the injection of **antisense morpholino** oligos into fertilized eggs (Fig. 7.11a). These hybridize to their complementary mRNA and block translation. It is necessary to have an antibody to the target protein in order to show that the morpholino has been effective and really prevented synthesis of the

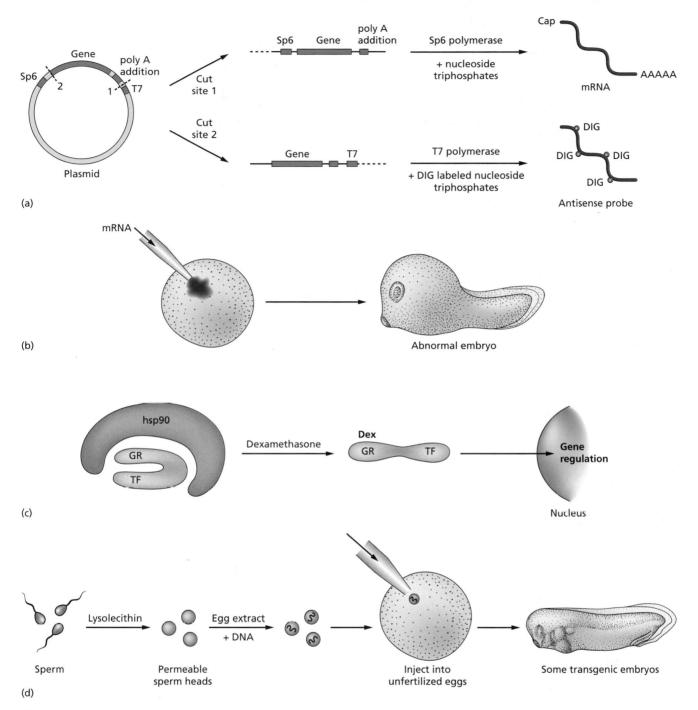

Fig. 7.10 (a) *In vitro* preparation of sense mRNA or antisense hybridization probe from a plasmid. The poly A tail becomes added *in vivo*. (b) Injection of synthetic mRNA into a fertilized egg. (c) Activation of a glucocorticoid receptor fusion protein by dexamethasone. (d) Introduction of genes by transgenesis.

protein. If it is desired to inhibit a maternally acting gene, then antisense deoxy-oligonucleotides are also effective (Fig. 7.11b). It is often very important to be able to deplete the oocyte of a specific mRNA since it is during these stages that various maternal components are laid down that are essential to later, zygotic, development. The antisense oligo is injected into the oocyte

and, like the morpholinos, hybridizes to the target messages. But in this case the resulting RNA–DNA hybrid is a target for the nuclease RNAseH which destroys the message. It is of course necessary to confirm destruction of the specific mRNA by one of the biochemical methods described in Chapter 5. In order to establish later developmental effects of these experiments it is

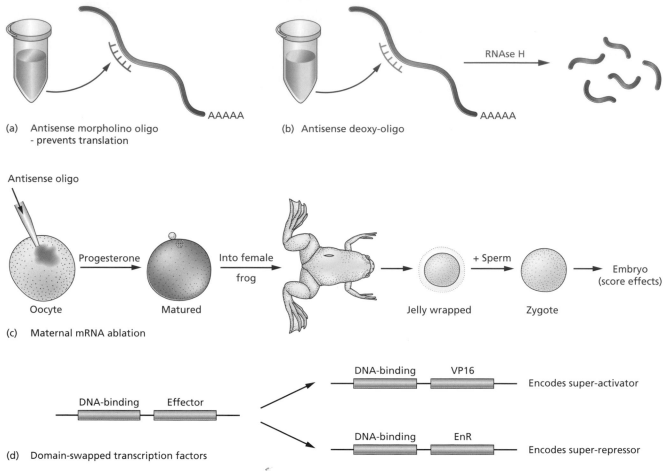

(a) Antisense morpholino oligo
 - prevents translation

(b) Antisense deoxy-oligo

(c) Maternal mRNA ablation

(d) Domain-swapped transcription factors

Fig. 7.11 Methods for inhibition of specific gene activity. (a) Antisense morpholino oligo. (b) Antisense deoxy-oligo. (c) Making an oocyte into an embryo after ablation of a specific maternal mRNA. (d) Domain swapped transcription factors.

then necessary to make the treated oocytes into embryos. This is achieved maturing the oocytes *in vitro* with progesterone, coloring them with a vital dye, and reimplanting them into the abdomen of an ovulating frog so that they pass down the oviduct and become wrapped in jelly. They will then be laid by the female, can be fertilized, and the course of development of the colored embryos observed (Fig. 7.11c).

The third protocol involves the design of **dominant negative** versions of gene products which can specifically inhibit the normal protein on overexpression. There are many possible types of dominant negative reagent. In *Xenopus*, particularly extensive use has been made of **domain swapped** versions of transcription factors (Fig. 7.11d). Here, the effector domain of the factor is replaced by a strong activator (e.g. VP16) or a strong repressor (e.g. engrailed repressor). mRNA is made from each of these constructs and is injected into fertilized eggs. If the effect of, say, the VP16 fusion is similar to the normal factor this shows that the normal factor must be an activator. The phenotype of the engrailed repressor fusion, in which the normal targets of the transcription factor are repressed, then gives an indication of

the functions for which the normal factor is needed. Conversely if the EnR version gives a normal phenotype then the transcription factor is likely to be a repressor, and the phenotype obtained with the VP16 fusion will indicate what functions require the normal gene product.

The effects of overexpression of a specific RNA on the development of the whole embryo are often not very informative because of the nonspecific nature of the defects observed. But two procedures combining overexpression with another technique have been particularly useful. One is the animal cap autoinduction assay (Fig. 7.12a). If an embryo is injected with a component of the mesoderm-inducing or neural-inducing systems and then the animal pole region of the blastula, called the **animal cap**, is explanted, it will autonomously undergo the induction. This is because some or all of the cells in the cap make and secrete the factor and all the cells are competent to respond to it. Untreated animal caps develop as spherical balls of epidermis. Those induced to form axial tissues undergo a convergent extension process and become very elongated. Those induced to form ventral-type tissues may elongate a little, but will swell to

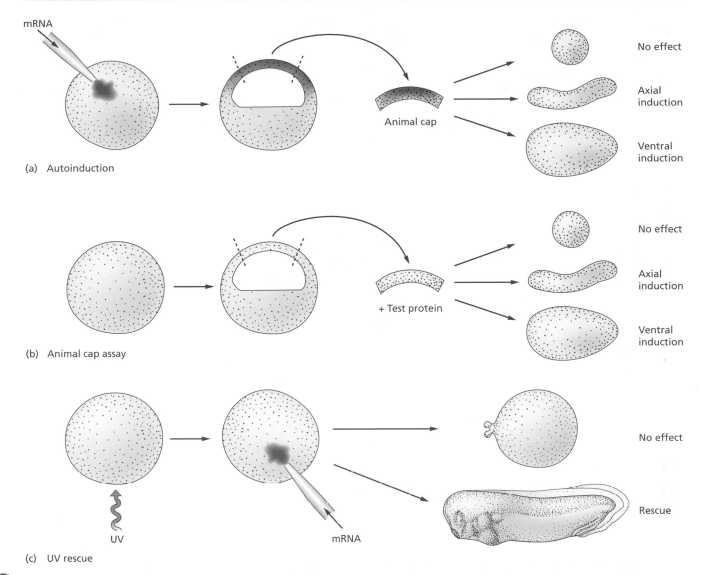

(a) Autoinduction

(b) Animal cap assay

(c) UV rescue

Fig. 7.12 Use of animal caps and UV embryos.

form translucent bodies on the second and third day of culture. As these changes can be observed under the dissecting microscope, this offers a very simple and convenient assay method.

For the study of extracellular factors it is possible simply to apply proteins to explants from embryos and observe the change in developmental pathway (Fig. 7.12b). Since the outer surface of the embryo is impermeable, these treatments must be applied to explants before they round up and become resealed. Whether the assay depends on mRNA injection or on treatment with a protein, the visual scoring of the results can be supplemented by conventional histology, *in situ* hybridization or immunostaining for specific markers, or by biochemical methods for specific mRNAs (see Chapter 5).

Another important combined protocol is the UV rescue method (Fig. 7.12c). As we shall see, it is possible to create embryos lacking all axial structures by UV radiation of the zygote. Injection of an mRNA coding for any component of the

axial induction system can restore part or all of the axis to such embryos, and this is easily scored by visual inspection. Because the head is formed mainly from the dorsal part of the blastula, the degree of head formation provides a semiquantitative measure of the degree of dorsal rescue.

Regional specification

Summary of processes

The fertilized egg contains two determinants, a **vegetal** and a **dorsal** determinant (Fig. 7.13). The vegetal determinant becomes established during oogenesis as a result of mRNA localization to the vegetal cortex and causes formation of the **endoderm.** This is the source of a **mesoderm-inducing** signal that induces a ring of tissue around the equator to become the **mesoderm.** The dorsal

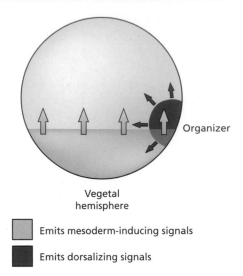

Emits mesoderm-inducing signals

Emits dorsalizing signals

Fig. 7.13 Two signaling centers in the early embryo. The vegetal hemisphere emits a mesoderm-inducing signal (yellow arrows) and the organizer emits a dorsalizing signal (red arrows).

determinant causes formation of the **organizer** in the region that will become the **dorsal lip** of the blastopore. It is initially localized at the vegetal pole, and shifts to the dorsal side during the cortical rotation. The organizer is the source of later dorsalizing signals that both induce the neural plate, and pattern the mesoderm into zones forming different tissues.

Half embryos

Evidence for the existence of the two determinants comes from the effects of early ablation and isolation experiments, which show that formation of a complete body pattern requires the presence of both the vegetal and the dorsal regions. Complete twins can be produced by separation of the first two blastomeres, or the equivalent lateral subdivision of a blastula into right and left halves (Fig. 7.14a). In this case, some vegetal and some dorsal material is present in each half. However, if early cleavage stages or blastulae are divided frontally (separating prospective dorsal and ventral halves) then the dorsal half forms a slightly hyperdorsal whole embryo while the ventral half forms a "belly piece" of extreme ventral character (Fig. 7.14b). Hyperdorsal embryos have a large head and small tail, while ventralized ones have a small head and large tail. Although these effects may seem to

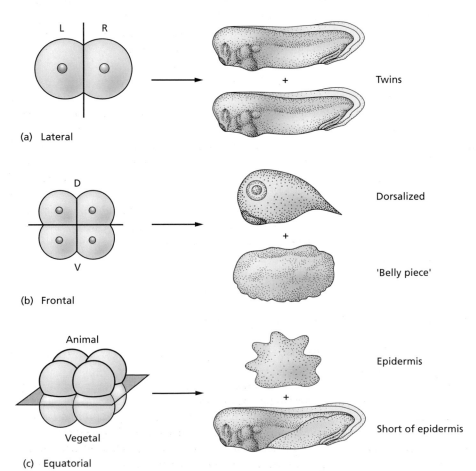

Fig. 7.14 Separation of an early embryo into two halves.

concern the anteroposterior rather than the dorsoventral pattern, it should be remembered that in the fate map the dorsal part of the blastula projects largely to the anterior of the later body and the ventral part projects mainly to the posterior of the later body, so these are the expected results from changing the dorsoventral proportions of the blastula. In the frontal separation, the dorsal half has both vegetal and dorsal tissue, but the ventral half does not. Finally, an equatorial subdivision, into animal and vegetal halves, carried out on the eight-cell to early blastula stage, yields a bag of epidermis from the animal hemisphere and an ectoderm-poor but otherwise well-patterned embryo from the vegetal half (Fig. 7.14c). Again, the explanation is that the vegetal half will contain both dorsal and vegetal material but the animal half will not.

Early dorsoventral patterning

The dorsal determinant is moved from the vegetal pole to the dorsal side by the cortical rotation occurring after fertilization. This depends on an array of parallel microtubules that forms transiently in the sense of the rotation. The microtubules grow from the sperm aster and when they meet the vegetal cortex become anchored to the cell surface by kinesin-related proteins and are aligned with their + ends in the direction of the future cortical rotation. The rotation movement is probably due to the relative displacement of the surface-tethered microtubules relative to internal dynein protein. The cortical rotation can be prevented by microtubule-depolymerizing drugs such as nocodazole or by injection of antibodies to the kinesin-related proteins. It can also be prevented if the vegetal hemisphere is irradiated with UV light before the rotation starts. The embryos that subsequently develop following any of these treatments are radially symmetrical and extreme ventral in character, with the mesoderm mainly consisting of blood islands (Fig. 7.15a,b), showing the importance of the cortical rotation for the development of dorsal structures. The fact that the dorsal side forms opposite the site of sperm entry is explained by the relationship of the microtubule array to the sperm aster.

There is now good evidence that the dorsal determinant consists of inhibitors of glycogen synthase kinase 3 (gsk3), which is a repressive component of the canonical Wnt signal transduction pathway leading to stabilization of β-catenin (see Appendix). Gsk3 is present in eggs and is constitutively active. It phosphorylates, and thereby inhibits, β-catenin. If gsk3 is inhibited, then β-catenin becomes stabilized and active and is free to enter the nucleus and combine with the transcription factor Tcf-3 to activate the target genes that are needed to form dorsal structures. Injection of additional gsk3 mRNA into normal embryos will inhibit formation of the dorsal axis, and, conversely, injection of a dominant negative version of gsk3 can rescue the formation of a dorsal axis in UV embryos or induce a second axis in normal embryos. Evidence for the essential role of β-catenin has been obtained by antisense oligonucleotide-mediated ablation of the β-catenin mRNA from oocytes. When such oocytes are matured and fertilized they develop as ventralized embryos. Although the intracellular components of the Wnt pathway are essential for axis formation, there is no evidence that the extracellular Wnt factors themselves are involved, which is why the pathway in Fig. 7.15e is shown as commencing with dishevelled. In normal development, gsk3 becomes locally inhibited just on the dorsal side as a consequence of the cortical rotation. This means that β-catenin enters the nuclei just on the dorsal side, and this can be observed by immunostaining for β-catenin protein. It extends to some extent above the equator where it is important in the initial formation of the neural plate. One of the targets of the β-catenin-TCF3 complex is the gene encoding the transcription factor siamois, a homeodomain factor important in the subsequent formation of the organizer.

The dishevelled protein, and a protein called GBP (gsk3 binding protein), both inhibit gsk3. Both of these proteins are found in eggs. If fusion proteins containing the dishevelled or GBP sequence joined to GFP are introduced into the egg, then they can be visualized *in vivo* by fluorescence microscopy, and it can be seen that they become sequestered into small vesicles and become moved to the dorsal side by the cortical rotation. GBP binds to kinesin, providing a direct mechanism for this transport process. Both dishevelled and GBP can rescue formation of an axis if injected into UV embryos. It has also been shown that depletion of GBP from oocytes, by the injection of specific antisense olignucleotide followed by conversion of the oocytes into embryos, results in the production of ventralized embryos in a similar way to removal of maternal β-catenin.

It is possible to produce **hyperdorsalized** embryos by treating early blastulae with lithium salts. These have a structure that is the opposite of the UV embryo and resemble radially symmetrical heads (Fig. 7.15a). They arise because the entire mesoderm has been caused to develop as organizer tissue. UV-ventralized embryos can be rescued back to a normal pattern by a localized injection of lithium (Fig. 7.15d), and a localized injection of lithium on the ventral side of a normal embryo will induce a secondary axis (Fig. 7.15e). Lithium is an inhibitor of gsk3, and may also exert additional effects through inhibition of the inositol phosphate pathway (see Appendix).

Inductive interactions

Isolation experiments

Explantation studies on small pieces taken from **blastulae** show three important features (Fig. 7.16):

1 Explants from anywhere in the animal hemisphere form epidermis but not mesodermal or neural tissues.

2 Ventral or lateral explants containing marginal zone form extreme ventral structures (mesenchyme, blood cells).

3 Only a restricted dorsovegetal region, about 60° of circumference, will form axial structures (notochord, somites, neural tube).

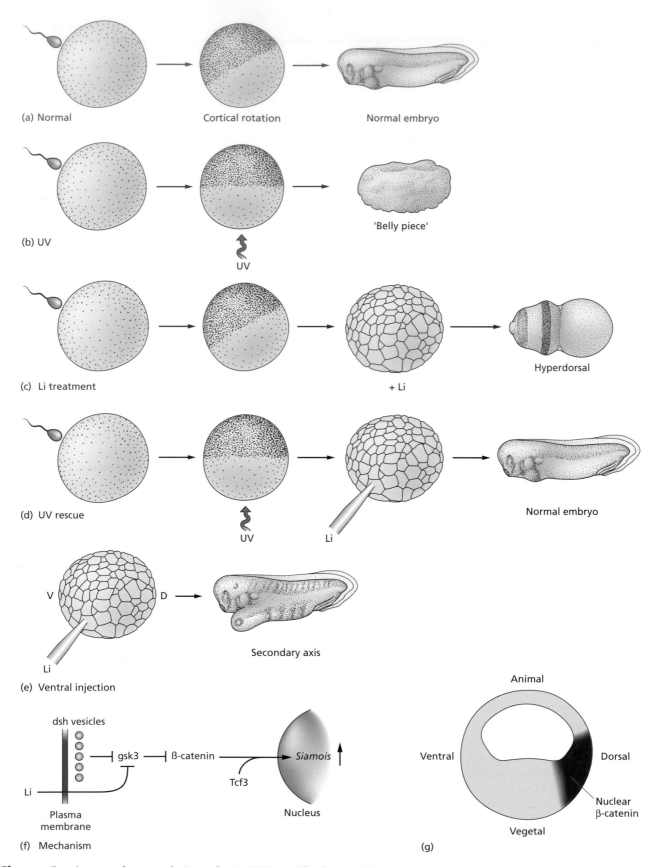

Fig. 7.15 Experiments on dorsoventral axis specification. (a) Normal development. (b) Ventralization by UV irradiation. (c) Formation of hyperdorsal embryo by lithium treatment. (d) Rescue of UV embryo by localized injection of lithium ion. (e) Induction of secondary axis in normal embryo by localized injection of lithium ion. (f) The Wnt pathway in the *Xenopus* egg. gsk3 will normally be repressed by the determinant on the dorsal side, but will be active elsewhere. Lithium can also inhibit gsk3. (g) Location of nuclear β-catenin in the late blastula.

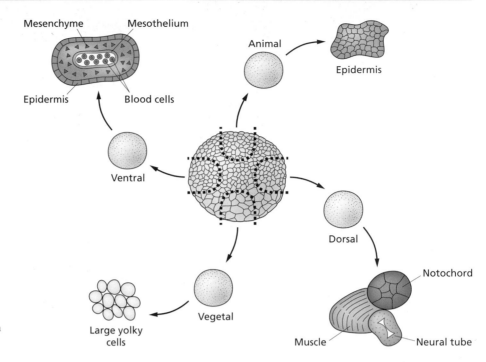

Fig. 7.16 Differentiation of explants from different parts of the blastula.

The differentiation of explants represents the specification of the tissues at the time of isolation. The results differ from the fate map, showing that inductive interactions must be necessary for achieving the final fates, particularly with regard to the formation of the neural plate and the somites.

Germ-layer formation

The fate map shows that the three germ layers arise from different animal–vegetal levels: the endoderm from the vegetal region, the mesoderm from the equatorial region, and the ectoderm from the animal region. The primary cause for this pattern is the localization of maternal mRNAs to the vegetal hemisphere during oogenesis (see Figs 7.1, 7.3). Among these is the mRNA for a T-box type transcription factor called *vegT* (= *brat, apod, ombi*). This appears to act as a **determinant** for the **endoderm**. The evidence is:

1 *VegT* mRNA is localized to the prospective endoderm.
2 It will induce endodermal markers if injected into other parts of the embryo.
3 Antisense oligonucelotide ablation of maternal *vegT* mRNA in oocytes, followed by maturation and fertilization, will produce embryos lacking endoderm.

The endoderm is defined by the expression of a group of transcription factors, all of which depend directly or indirectly on vegT. Some, including the SRY-related factor sox17a and the homeodomain factor mix1, are induced directly by vegT. The genes encoding these factors are activated from MBT even if the cells have been dissociated so that no intercellular signaling can take place. Expression of other endodermal transcription factors, including the homeodomain factor mixer and the zinc finger factor GATA4, depend on the signals emitted from the vegT-containing cells and are not induced if these signals are blocked by cell dissociation. Later on endoderm becomes regionalized into zones that will form the different tissue types of the gut and respiratory system, and this requires the activation of transcription factor genes of the Parahox cluster: *XlHbox8 (=Pdx1)* in the future foregut and *cdx* genes in the future intestine.

The **mesoderm** arises from the equatorial region of the blastula and is characterized by the expression of various transcription factors including brachyury. The part on the vegetal side of the equator may be formed by the endodermal determinants acting at lower concentration than required to form endoderm. However at least the part of the mesoderm arising from the animal hemisphere is formed by **induction** (Fig. 7.17a). Explants from the animal hemisphere develop into **epidermis** after isolation. If an **animal cap** (an explant from the center of the animal hemisphere) is combined with endoderm, then substantial amounts of mesodermal tissues are induced in the cap (Fig. 7.17b).

Only a small dorsovegetal region of the endoderm will induce organizer mesoderm, characterized by expression of transcription factors such as goosecoid, lim1, and not. The remainder of the endoderm induces **ventral-type** mesoderm characterized by expression of transcription factors such as vent1 and vent2. The region of the endoderm with organizer-inducing ability is sometimes called the **Nieuwkoop center**, after the great Dutch embryologist of that name. It is the region in which expression of *siamois* has been activated by the dorsal determinant.

There is also a difference of **competence** within the animal hemisphere such that it is much easier to obtain axial inductions

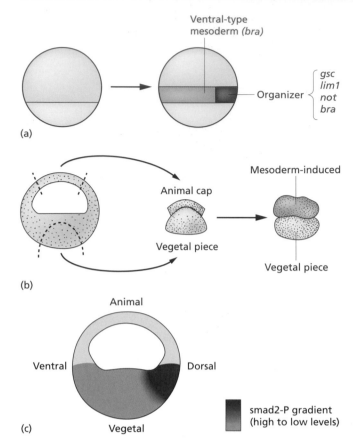

(a)

(b)

(c)

smad2-P gradient
(high to low levels)

Fig. 7.17 Mesoderm induction: (a) in normal development;
(b) mesoderm induced in animal cap combined with vegetal explant;
(c) location of phosphorylated smad 2 in late blastula.

from the dorsal region. This is because the β-catenin activation caused by the dorsal determinant extends into the animal hemisphere. The competence of the ectoderm to respond to the mesoderm-inducing signals rises in the early blastula and falls in the early gastrula. The signaling capacity of the vegetal hemisphere is mostly exerted following MBT, as shown by immunostaining for the activated form of smad2 in intact embryos.

The factors responsible for mesoderm induction are members of the transforming growth factor (TGF)-β superfamily, especially the nodal-related factors (see Chapter 10 for the original *nodal*). This subset of factors activates cell surface receptors that phosphorylate smad2 and smad3 proteins in the cytoplasm, and the activated smads can then move into the nucleus and activate target genes (see Appendix). The evidence for the importance of nodal-related factors is as follows:

1 Nodal-related factors will induce mesoderm if applied to animal caps.

2 They are expressed in the vegetal hemisphere after MBT.

3 Overexpression of a dominant negative receptor that inhibits this group of factors will prevent mesoderm formation.

4 Overexpression of Cerberus-short, a peptide from the Cerberus molecule (see below) that inhibits only the nodal-related factors, will also inhibit mesoderm formation.

5 Smad2 becomes activated after MBT, in the vegetal and equatorial region. This can be visualized by using antibodies specific for the phosphorylated forms (Fig. 7.18c).

6 Reporter constructs, containing a TGF-β-signaling sensitive promoter linked to a *luciferase* gene, are activated when injected into blastomeres in vegetal or equatorial positions, but not blastomeres in animal positions. This shows that the signal *in vivo* affects the equatorial region but does not reach to the animal pole.

Transcription of some of the nodal-related factors is activated directly by vegT, and of others indirectly via the initial signaling process. The nodal-related factors acting by themselves induce pan-mesodermal genes such as *brachyury*, and ventral genes such as the *vents*. In the dorsal quadrant where β-catenin is also activated, there is a synergistic response to both pathways resulting in the activation of genes for goosecoid, not, lim1, and other transcription factors whose activity defines the organizer region.

Dorsalization and neural induction

Mesoderm induction leads to the creation of a belt of mesoderm around the equator of the blastula with a large ventrolateral region of extreme ventral character, and a small organizer region at the dorsal side. The organizer region is often called **Spemann's organizer** after the great German embryologist who originally discovered its properties. It is a key signaling center for the subsequent stages of development, and it acts by secretion of factors that *inhibit* the action of bone morphogenetic proteins (BMPs, see Appendix).

During gastrulation the ventrolateral mesoderm becomes partitioned into zones forming, respectively, somites, kidney, lateral plate, and blood islands, as a function of distance from the organizer (Fig. 7.18a). Isolates from the ventrolateral mesoderm retain an extreme ventral character. But if they are experimentally combined with organizer tissue, they will be dorsalized and form large muscle blocks and pronephric tubules (Fig. 7.18b).

Also during gastrulation the neural plate becomes induced from the ectoderm under the influence of the organizer. When gastrula ectoderm is brought into contact with the organizer it will form **neuroepithelium**, characterized by expression of various transcription factors including sox2, and the cell adhesion molecule N-CAM. There are various ways of achieving this experimentally. The most straightforward is a combination of competent ectoderm with organizer tissue (Fig. 7.19b). Another is the Einsteckung procedure, where bits of organizer tissue are inserted into the blastocoel of an early gastrula and become pressed against the ventral ectoderm by the gastrulation movements (Fig. 7.19c). Finally, in the organizer graft (see below) a secondary neural plate is formed from the ectoderm overlying the secondary mesodermal axis.

If blastulae are placed in a salt solution isotonic to the embryo rather than the usual very dilute solution, they will **exogastrulate**: the endomesoderm becomes extruded from the animal hemisphere instead of invaginating into it (Fig. 7.6). The elongation

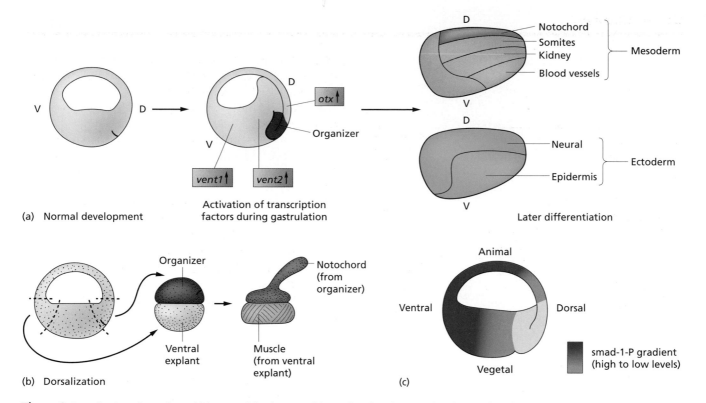

(a) Normal development

Activation of transcription factors during gastrulation

Later differentiation

(b) Dorsalization

(c)

smad-1-P gradient (high to low levels)

Fig. 7.18 Dorsalization of mesoderm: (a) in normal development; (b) muscle induced in ventral explant combined with organizer; (c) location of phosphorylated smad1 in early gastrula.

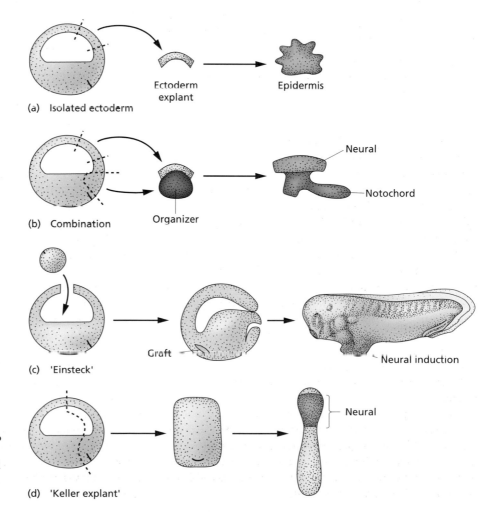

(a) Isolated ectoderm

(b) Combination

(c) 'Einsteck'

(d) 'Keller explant'

Fig. 7.19 Neural induction: (a) isolated ectoderm from prospective brain turns into epidermis; (b) ectoderm is neuralized in combination with the organizer; (c) ventral ectoderm can be neuralized in the Einsteck procedure; (d) Keller explant.

Classic Experiments

DISCOVERY OF INDUCING FACTORS

The inducing factors that control embryonic development were discovered in the 1980s using *Xenopus*. The key step was the ability to apply purified protein solutions to isolated animal caps and to observe the consequences down the dissecting microscope by changes in shape of the explant. This assay led to discovery of the inducing properties of the FGFs and the activin-like factors (now known to be nodals).

The other vital technique was the injection of messenger RNA, prepared *in vitro*, into fertilized eggs, or into UV-ventralized eggs. This led to the discovery of the role of the Wnt pathway in the formation of the primary axis. Further, it allowed the introduction of expression cloning protocols whereby a whole library of expression clones could be divided into subsets, transcribed, and the mixed mRNA assayed by injection. This led to the discovery of noggin and eventually to the mode of action of the organizer.

First inducing factors

Kimelman, D. & Kirschner, M. (1987) Synergistic induction of mesoderm by FGF and TGF-beta and the identification of a messenger-RNA coding for FGF in the early *Xenopus* embryo. *Cell* **51**, 869–877.

Slack, J.M.W., Darlington, B.G., Heath, J.K. & Godsave, S.F. (1987) Mesoderm induction in early *Xenopus* embryos by heparin-binding growth-factors. *Nature* **326**, 197–200.

Smith, J.C., Price, B.M.J., VanNimmen, K. & Huylebroeck, D. (1990) Identification of a potent *Xenopus* mesoderm-inducing factor as a homolog of activin-A. *Nature* **345**, 729–731.

Axis duplication from Wnt

Smith, W.C. & Harland, R.M. (1991) Injected XWnt-8 RNA acts early in *Xenopus* embryos to promote formation of a vegetal dorsalizing center. *Cell* **67**, 753–765.

Sokol, S., Christian, J.L., Moon, R.T. & Melton, D.A. (1991) Injected Wnt RNA induces a complete body axis in *Xenopus* embryos. *Cell* **67**, 741–752.

Discovery of Noggin

Smith, W.C. & Harland, R.M. (1992) Expression cloning of noggin, a new dorsalizing factor localized to the Spemann organizer in *Xenopus* embryos. *Cell* **70**, 829–840.

1 The prospective epidermis makes BMPs and smad1 and 5 becomes phosphorylated over the whole ventrolateral region of the embryo (Fig. 7.18c).

2 The organizer region secretes the BMP inhibitors.

3 Treatment of isolated animal caps with BMP inhibitors causes neuralization.

4 Treatment of isolated ventral mesoderm explants with BMP inhibitors will dorsalize them.

5 Injection of mRNA for dominant negative BMP receptors on the ventral side induces secondary axis formation (i.e. dorsal-type mesoderm on the ventral side).

6 Morpholinos to chordin will suppress dorsal pattern, both in normal embryos and in lithium-dorsalized embryos.

Chordin, noggin, and follistatin act by direct binding to and inhibition of BMP protein and thereby bring about a gradient of BMP activity from ventral to dorsal. The *vent1* gene is activated at high BMP and the *vent2* gene at lower BMP concentration (see Figs 7.4, 7.18a). The proteins coded by the *vent* genes are themselves transcriptional repressors as may be shown by the fact that VP16 fusions have a dorsalizing effect (i.e. opposite to the native factors). The region in which both *vent* genes are on later becomes the lateral plate. The region in which *vent2* but not *vent1* is on later becomes the somites, and soon shows activation of genes for **myogenic** factors such as *Myf5*. In the mid-ventral region that forms the blood islands, BMP activates SCL (stem cell leukemia factor), a bHLH-type transcription factor which later activates hematopoietic differentiation.

Experimentally, neuralization can be provoked by disaggregation of gastrula ectoderm. This is because when cells are disaggregated the ERK signaling pathway becomes activated as a general "wounding" response. This leads to inhibition of the action of smads 1 and 5 by phosphorylation at a second site (the linker region – see Appendix).

In addition to the inhibition of BMP, an element of Wnt inhibition is also involved in the normal dorsalization process. After MBT, the *wnt8* gene becomes activated in the nonorganizer part of the mesoderm and the Wnt pathway now has a *ventralizing* effect. Although the Wnt pathway has a dorsalizing function in the egg, after MBT the transcription factor Tcf3 becomes replaced by Lef1. Both of these factors cooperate with β-catenin, but they show a different spectrum of specificity in relation to

---

of the embryo proceeds normally but the mesoderm is joined to the ectoderm end-to-end rather than being apposed as layers. Another situation in which mesoderm and ectoderm are adjacent, but not apposed as layers, is the **Keller explant**. This is an explant of dorsal tissue extending from the dorsal lip to the animal pole and displays the convergent extension movement without the other features of gastrulation (Fig. 7.19d). In both these situations where the axial mesoderm does not underlie the ectoderm there is still some neural induction, resulting in a correct anteroposterior patterning of structures. This shows that the signals can propagate in the plane of the tissue (tangential induction) as well as from one layer to the other (**appositional induction**).

In normal development, neuralization and mesodermal dorsalization are both caused by inducing factors secreted from the organizer which function as inhibitors of the BMPs. They are chordin, noggin, and follistatin. BMP4 is produced by the embryo in all regions except those dorsal regions with high nuclear β-catenin. The intracellular signal transduction pathway for BMPs is similar to that for the nodal-related factors, but uses different receptors and smads (smad1 and 5; see Appendix). Evidence that inhibition of BMP signaling is the cause of neural induction and mesoderm dorsalization is as follows:

their target genes: Tcf3 is dorsalizing while Lef1 is ventralizing. This can be demonstrated because dominant negative constructs for these two transcription factors show the opposite effects when overexpressed in embryos: dominant negative Tcf3 is ventralizing and dominant negative Lef1 is dorsalizing. This change of competence explains the hitherto puzzling fact that treatment with lithium after MBT is ventralizing rather than dorsalizing. The dorsalizing signals from the organizer include at least one Wnt inhibitor called Frzb (pronounced "frizzbee"), which resembles the extracellular part of the Wnt receptors, and will inhibit the action of Wnt-8.

Anteroposterior patterning

During gastrulation not only is the dorsal to ventral pattern of territories specified, but also the anterior to posterior pattern. Neural-inducing activity is shown both by the organizer and by the axial mesoderm into which the organizer develops. In both cases the inductions show regional specificity. Anterior organizer or anterior axial mesoderm induces brain structures, while posterior organizer or posterior axial mesoderm induces both brain and spinal cord. Since a complete neural axis can be induced by a posterior inducer it is thought that a posterior signal controls anteroposterior patterning during the formation of the central nervous system (CNS).

The initial anteroposterior pattern exists in the form of two domains within the organizer prior to gastrulation, which are derived from the different ratio of β-catenin and nodal-related signaling arising from cortical rotation and mesoderm induction respectively. The anterior part will later become the anterior endoderm and the prechordal mesoderm. During gastrulation this moves towards the animal pole by crawling up

the blastocoelic surface of the ectoderm. It is characterized by expression of transcription factor genes including *goosecoid* and *hex*. It will induce expression of anterior-type genes from ectoderm, such as *otx2* (fore/midbrain) and *XAG1* (cement gland); and its anterior inductive activity is due to secreted factors including Cerberus, which is an inhibitor of Wnts, BMPs, and Nodal-related factors, and Dickkopf, which is a Wnt inhibitor. The posterior part will later become the notochord and somites and, during gastrulation, it elongates considerably by convergent extension. It is characterized by expression of a different group of transcription factor genes including the homeobox gene *not* and the T-box gene *brachyury*. It will induce both anterior- and posterior-type genes from the ectoderm (e.g. both *otx2* and **Hox genes**), and its posteriorizing activity is at least partly due to secretion of fibroblast growth factors (FGFs). These act through the ras–raf–ERK signal transduction pathway (see Appendix) and activate a group of homeobox transcription factors coded by *cdx* genes and these in turn activate the posterior **Hox genes** of paralog groups 6–13 (Fig. 7.20). This subset of Hox genes specify the trunk–tail part of the pattern and they are normally turned on sequentially during gastrulation.

Evidence that FGF signaling is required to induce the trunk–tail region is as follows:

1 Several FGFs are expressed around the blastopore (i.e. the posterior) during gastrulation.

2 The ras–raf–ERK signal transduction pathway is activated in the prospective trunk in the late gastrula, as shown by immunostaining for the activated form of ERK.

3 If animal caps are treated with the BMP inhibitor noggin, then only anterior-type neural genes are induced, but noggin+FGF will also induce Hox gene expression.

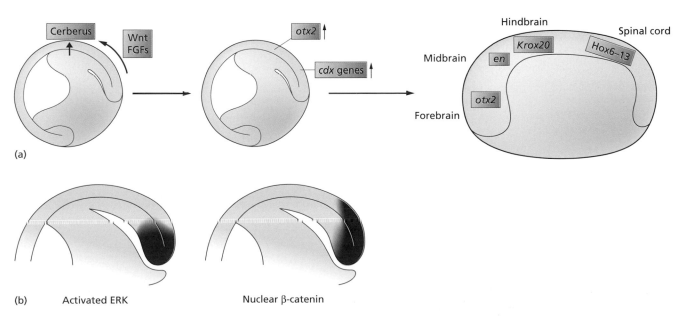

Fig. 7.20 Anteroposterior patterning of the CNS. (a) Action of Cerberus from the anterior part of the organizer, and of FGFs and Wnts from the posterior part. (b) Location of phosphorylated ERK (FGF target) and nuclear β-catenin (Wnt target) in late gastrula.

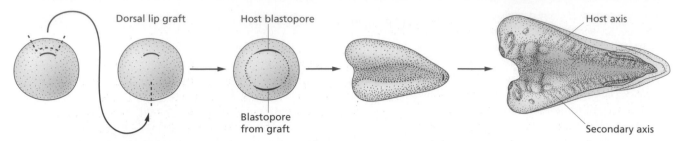

Fig. 7.21 The organizer graft. The graft forms the notochord and head mesoderm of the secondary axis and induces the other parts from the host.

4 Overexpression in embryos of a dominant negative FGF receptor that inhibits endogenous FGF signaling will prevent formation of the trunk and tail.

5 The same phenotype arises from overexpression of a domain-swapped, repressive, version of a *cdx* factor, cdx-EnR, designed to inhibit expression of *cdx* target genes.

Additional components that have been postulated as part of the posteriorizing signal include the Wnt factors and retinoic acid. Evidence for a posteriorizing role for Wnt factors is also good, especially in relation to the hindbrain. A posterior to anterior gradient of nuclear β-catenin can be observed in the early neural plate, and Wnt factors will induce hindbrain markers in animal caps that have been neuralized.

The organizer graft

All three of the processes, dorsalization, neural induction, and anteroposterior patterning, are shown in the organizer graft first performed by Spemann and Mangold in 1924. This is the most famous experiment in embryology, although its true significance has only recently become clear now that the various component processes have become understood. In the graft, a piece of tissue from above the dorsal blastopore lip is implanted into the ventral marginal zone (Fig. 7.21). It leads to the formation of a double dorsal embryo in which the notochord of the secondary embryo is derived from the graft and the remainder of the axis from the host. The ectoderm above the graft becomes induced to form a second neural tube. As gastrulation proceeds, both host and graft axes form progressively more posterior parts. In favorable cases the final result is a symmetrical pair of embryos joined belly to belly.

The movements of the grafted organizer are autonomous and preprogrammed. Therefore it invaginates and undergoes active elongation by convergent extension. The graft emits dorsalizing signals (BMP and Wnt inhibitors). These diffuse to the neighboring mesoderm, and the reduction in BMP signaling suppresses activation of *vent1* and so causes activation of myogenic genes to form a file of somites on either side of the graft. The same substances diffuse to the neighboring ectoderm and activate pan-neural genes. During gastrulation the graft emits FGFs and Wnts which induce *cdx* and Hox genes needed for development of the trunk–tail region. The secondary axes arising from organizer grafts often lack a head, but if the graft includes the deep anterior region of the organizer then this will emit Cerberus and Dickkopf and induce a head in the overlying ectoderm. The net result is the formation of the secondary embryo.

Key Points to Remember

• The descriptive embryology of *Xenopus* is typical of animals generally although details are specific. Cleavages are rapid and synchronous leading to blastula formation. Gastrulation movements lead to the formation of the three germ layers with regional pattern in dorsoventral and anterioposterior axes.

• Accurate fate maps have been constructed by injection of fluorescent dextrans into individual cells.

• Overexpression of gene products is usually carried out by injection of mRNA into fertilized eggs. Inhibition of specific gene products is usually carried out by injection of antisense morpholinos or of mRNA for dominant negative constructs.

• The pattern of the embryo is specified in relation to two determinants in the egg. The vegetal determinant includes the mRNA for the T-box transcription factor vegT which becomes localized to the vegetal cortex during oogenesis. The dorsal determinant comprises dishevelled protein and other Wnt pathway components which become moved from the vegetal to the dorsal side in the sperm-induced cortical rotation.

• VegT protein activates expression of genes encoding endoderm-specific transcription factors (e.g. Sox17) and also the genes encoding nodal factors. The nodal factors are secreted and induce the expression of mesoderm-specific transcription factors (e.g. brachyury) in an equatorial belt of neighboring cells.

• On the dorsal side the Wnt pathway activates expression of dorsal-specific transcription factors (e.g. siamois). The region containing both mesodermal and dorsal transcription factors becomes the organizer. Here genes for BMP inhibitors (noggin, chordin) are activated. These are secreted and create a gradient of BMP activity from ventral to dorsal. Inhibition of BMP activity induces transcription factors such as Sox1 in the animal (ectodermal) region, leading to formation of the neural plate. It also induces transcription factors such as myf5 in the equatorial (mesodermal) region, leading to the formation of the myotomes.

• Anteroposterior patterning depends on a posteriorizing signal emitted from the blastopore region during gastrulation. This consists of FGFs, Wnts, and retinoic acid, and causes activation of Hox genes in a nested manner such that each gene is activated at a particular anteroposterior level and remains on posterior to this.

Further reading

Website

http://www.xenbase.org/

Normal development

Nieuwkoop, P.D. & Faber, J. (1967) *Normal Table of* Xenopus laevis. Amsterdam: N. Holland. Reprinted Garland Publishing, London (1994).

Hausen, P. & Riebesell, M. (1991) *The Early Development of* Xenopus laevis. Berlin: Springer-Verlag.

Oocyte maturation

Tunquist, B.J. & Maller, J.L. (2003) Under arrest: cytostatic factor (CSF)-mediated metaphase arrest in vertebrate eggs. *Genes and Development* **17**, 683–710.

Gastrulation

Keller, R.E., Danilchik, M., Gimlich, R. & Shih, J. (1985) The function and mechanism of convergent extension during gastrulation of *Xenopus laevis. Journal of Embryology and Experimental Morphology* **89**(Suppl.), 185–209.

Keller, R., Shih, J. & Domingo, C. (1992) The patterning and functioning of protrusive activity during convergence and extension of the *Xenopus* organiser. *Development* (Suppl.) 81–91.

Beetschen, J.C. (2001) Amphibian gastrulation: history and evolution of a 125 year old concept. *International Journal of Developmental Biology* **45**, 771–795.

Winklbauer, R., Medina, A., Swain, R.K. & Steinbeisser, H. (2001) Frizzled-7 signaling controls tissue separation during *Xenopus* gastrulation. *Nature* **413**, 856–860.

Wharton, K.A. (2003) Runnin' with the Dvl: proteins that associate with Dsh/Dvl and their significance to Wnt signal transduction *Developmental Biology* **253**, 1–17.

Fate maps

Keller, R.E. (1975) Vital dye mapping of the gastrula and neurula of *Xenopus laevis* I. Prospective areas and morphogenetic movements of the superficial layer. *Developmental Biology* **42**, 222–241.

Keller, R.E. (1976) Vital dye mapping of the gastrula and neurula of *Xenopus laevis* II. Prospective areas and morphogenetic movements of the deep layer. *Developmental Biology* **51**, 118–137.

Dale, L. & Slack, J.M.W. (1987) Fate map for the 32 cell stage of *Xenopus laevis*. *Development* **99**, 527–551.

Bauer, D.V., Huang, S. & Moody, S.A. (1994) The cleavage stage origin of Spemann's organiser: analysis of the movements of blastomere clones before and during gastrulation in *Xenopus*. *Development* **120**, 1179–1189.

Walmsley, M., Ciau-Uitz, A. & Patient, R. (2002) Adult and embryonic blood and endothelium derive from distinct precursor populations which are differentially programmed by BMP in *Xenopus*. *Development* **129**, 5683–5695.

Early dorsoventral polarity

Cooke J. & Smith E.J. (1988) The restrictive effect of early exposure to lithium upon body pattern in *Xenopus* development studied by quantitative anatomy and immunofluorescence. *Development* **102**, 85–99.

Gerhart, J., Danilchik, M., Doniach, T. et al. (1989) Cortical rotation of the *Xenopus* egg: consequences for the anteroposterior pattern of embryonic dorsal development. *Development* (Suppl.) 37–51.

Miller, J.R., Rowning, B.A., Larabell, C.A. et al. (1999) Establishment of the dorsal–ventral axis in *Xenopus* embryos coincides with the dorsal enrichment of dishevelled that is dependent on cortical rotation. *Journal of Cell Biology* **146**, 427–437.

Heasman, J., Crawford, A., Goldstone, K., et al. (1998). From cortical rotation to organizer gene expression: toward a molecular explanation of axis specification in *Xenopus*. *Bioessays* **20**, 536–545.

Weaver, C. & Kimelman, D. (2004) Move it or lose it: axis specification in *Xenopus*. *Development* **131**, 3491–3499.

Endoderm

Stennard, F. (1998) *Xenopus* differentiation: VegT gets specific. *Current Biology* **8**, R928–R930.

Dale, L. (1999) Vertebrate development: multiple phases to endoderm formation. *Current Biology* **9**, R812–R815.

Yasuo, H. & Lemaire, P. (1999) A two-step model for the fate determination of presumptive endodermal blastomeres in *Xenopus* embryos. *Current Biology* **9**, 869–879.

Mesoderm induction

Gotoh, Y. & Nishida, E. (1996) Signals for mesoderm induction. *BBA Reviews in Cancer* **1288**, F1–F7.

Agius, E., Oelgeschlager, M., Wessely, O., Kemp, C. & De Robertis, E.M. (2000) Endodermal nodal-related signals and mesodermal induction in *Xenopus*. *Development* **127**, 1173–1183.

Schier, A.F. & Shen, M.M. (2000) Nodal signaling in vertebrate development. *Nature* **403**, 385–389.

Schohl, A. & Fagotto F. (2002) beta-catenin, MAPK and smad signaling during early *Xenopus* development. *Development* **129**, 37–52.

Xanthos, J.B., Kofron, M., Tao, Q.H., Schaible, K., Wylie, C. & Heasman, J. (2002) The roles of three signaling pathways in the formation and function of the Spemann Organizer. *Development* **129**, 4027–4043.

The organizer

Lemaire, P. & Kodjabachian, L. (1996) The vertebrate organizer: structure and molecules. *Trends in Genetics* **12**, 525–531.

Graf, J.M. (1997) Embryonic patterning: to BMP or not to BMP, that is the question. *Cell* **89**, 171–174.

Harland, R. & Gerhart, J. (1997) Formation and function of Spemann's organizer. *Annual Reviews of Cell and Developmental Biology* **13**, 611–667.

Dale, L. & Wardle, F. (1999) A gradient of BMP activity specifies dorsal–ventral fates in early *Xenopus* embryos. *Seminars in Cell and Developmental Biology* **10**, 319–326.

Niehrs, C. (2004) Regionally specific induction by the Spemann–Mangold organizer. *Nature Reviews Genetics* **5**, 425–434.

Anteroposterior pattern

Doniach, T. (1992) Induction of anteroposterior neural pattern in *Xenopus* by planar signals. *Development* (Suppl.) 183–193.

Blumberg, B. (1997) An essential role for retinoid signaling in anteroposterior neural specification and neuronal differentiation. *Seminars in Cell and Developmental Biology* **8**, 417–428.

Sasai, Y. & De Robertis, E.M. (1997) Ectodermal patterning in vertebrate embryos. *Developmental Biology* **182**, 5–20.

Piccolo, S., Agius, E., Leyns, L., et al. (1999) The head inducer Cerberus is a multifunctional antagonist of Nodal, BMP and Wnt signals *Nature* **397**, 707–710.

Gamse, J. & Sive, H. (2000) Vertebrate anteroposterior patterning: the *Xenopus* neurectoderm as a paradigm. *Bioessays* **22**, 976–986.

Kiecker, C. & Niehrs, C. (2001) A morphogen gradient of Wnt/beta-catenin signaling regulates anteroposterior neural patterning in *Xenopus*. *Development* **128**, 4189–4201.

Chapter 8

The zebrafish

This chapter may seem rather short because the overall course of early development in the zebrafish (*Danio rerio*) is quite similar to *Xenopus* and the chapter focuses on the zebrafish-specific features rather than repeating the description of the common features. In addition to the study of early development described here, the zebrafish is becoming increasingly important in organogenesis research and will feature again in later chapters. In contrast to *Xenopus* the main approach to research with the zebrafish has been through mutagenesis, and this means that mutations are available in many of the important genes. A short list of some developmental genes is provided in Table 8.1. These are homologs of the corresponding genes in *Xenopus* and the mouse, but because most were originally discovered by mutation, and named before the identity of the gene was known, the names are specific to the zebrafish.

Normal development

Zebrafish eggs are about 0.7 mm in diameter and surrounded by a transparent chorion. The eggs and sperm are shed by the parents into the water and fertilization is external. As in *Xenopus*, the animal hemisphere is cytoplasm-rich and the vegetal hemisphere is yolk-rich. The course of development is shown in Fig. 8.1. Cleavage is **meroblastic** as it involves only the animal pole region. The first three cleavages are all vertical, generating an eight-cell stage composed of two rows of four blastomeres. The outer parts of the cell contacts are enriched in the proteins vasa and nanos (homologs of the corresponding *Drosophila* proteins, see Chapter 11) which may be determinants for germ cell formation. During this early phase the blastomeres remain connected to the main yolk mass by cytoplasmic bridges. This yolk

Table 8.1 Some developmental genes in the zebrafish.

Gene	Homolog	Developmental function	Gene product*
headless	*tcf3*	early dorsalization	HMG TF
cyclops	*nodal*	mes-endoderm induction	IF
squint	*nodal*	mes-endoderm induction	IF
one-eyed pinhead	*cripto*	needed for nodal action	EGF-CFC factor
dharma/nieuwkoid/bozozok		defines organizer	paired homeo TF
notail	*brachyury/T*	defines posterior mesoderm	T-box TF
spadetail	*thr6/vegT*	defines trunk mesoderm	T-box TF
acerebellar	*fgf8*	posteriorising	IF
swirl	*bmp2b*	ventralizing	IF
snailhouse	*bmp7*	ventralizing	IF
vent	*vent1/PV1*	defines ventral	homeodomain TF
vox	*vent2/xom*	defines ventral	homeodomain TF
bonnie-and-clyde	*mixer*	defines endoderm	T-box TF
faust	*gata5*	heart/endoderm	Zn finger TF

*TF, transcription factor; IF, inducing factor.

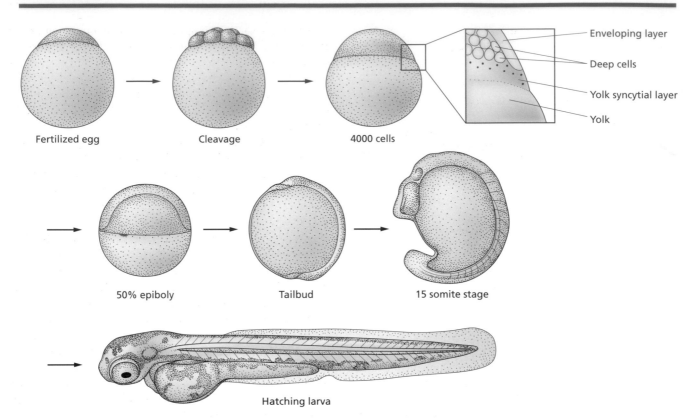

Fig. 8.1 Normal development of zebrafish.

mass becomes known as the **yolk** cell. Succeeding cleavages occur about every 15 minutes, until after about 10 divisions (approximately 3 hours) the **midblastula transition (MBT)** occurs. This is similar to *Xenopus* in that the synchrony of divisions breaks down, the average cell cycle duration increases, cells become more motile, and the transcription of the zygotic genome commences.

By this stage the embryo consists of four regions. The majority of cells form a mass in the animal pole region and are called **deep cells**, which are covered by a thin epithelial **enveloping layer**. The vegetal part of the embryo is occupied by the yolk cell, and the region near the blastoderm contains a number of nuclei and is called the **yolk syncytial layer**. These nuclei arise from the cells at the edge of the blastoderm after MBT, which fuse with the adjacent yolk cell to form the syncytium. The overall phase of morphogenetic movements is quite similar to that in *Xenopus*, although the mechanics may be somewhat different. Starting shortly after MBT, the blastoderm commences an active expansion called **epiboly** such that the margin moves down progressively to cover the yolk cell. Stages of zebrafish development are identified as percentage of epiboly, depending on how much of the yolk cell has been surrounded. This movement is driven by the yolk syncytial layer and depends on the activity of microtubules within the yolk cell. At 50% epiboly (approximately 5.5 hours) the margin of the blastoderm begins to involute. In the zebrafish the term **gastrulation** is normally reserved specifically for the involution movement of the mesoderm as distinct from the epibolic spreading process. Involution takes place all around the blastoderm margin, but, as in *Xenopus*, the dorsal involution is much more pronounced than the ventral. The involution means that the blastoderm is thicker around the margin than elsewhere and this thickening is called the **germ ring**. The involution movement is carried out only by the deep cells. The outer enveloping layer cells do not participate and go on to become the outer layer of the larval epidermis, or periderm. Simultaneous with the involution, the dorsal region starts to elongate in an anteroposterior sense by **convergent extension**, drawing in cells from more lateral regions. This process causes the dorsal marginal zone to thicken relative to the remainder of the circumference and it then becomes known as the **embryonic shield**. Epiboly reaches completion at about 9.5 hours when the yolk cell is completely covered by the blastoderm. The outer layer of the shield becomes the **neural plate**. This sinks into the interior as a solid mass of cells, and a lumen forms by cavitation to create a neural tube. The mesodermal layer forms a midline notochord and paraxial somites, segmenting from anterior to posterior. The basic body plan structures becomes visible by about 14 hours, and the axis straightens by 24 hours. Hatching occurs after about 48 hours, and feeding commences about 5 days after fertilization.

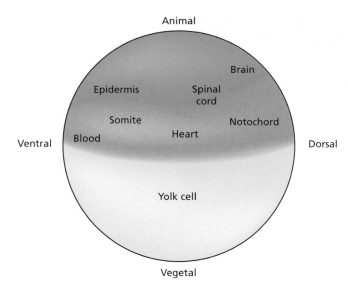

Animal

Brain

Epidermis Spinal
cord

Somite Notochord

Blood Heart

Ventral Dorsal

Yolk cell

Vegetal

Fig. 8.2 Fate map of zebrafish at 50% epiboly.

Fate map

The fate map for the zebrafish has been constructed by injection of individual blastomeres with fluorescent dextrans. As there is considerable mixing of the deep cells, the fate maps from the cleavage stage are only statistical. Each individual blastomere contributes to a very wide region of the later embryo. By the beginning of gastrulation the range of random cell movement is much reduced and a fate map can be produced with similar resolution to that of *Xenopus* (Fig. 8.2). The fate map shown is for the beginning of gastrulation and concerns just the deep cells. Individual cells may populate more than one tissue type or even more than one germ layer if labeled in the early stages of epiboly. But by the start of involution all clones have become restricted to tissue types such as neuroepithelium or somite. This suggests that the timing of the inducing signals responsible for regional specification is similar to *Xenopus*.

Mutagenesis

Recessive screen

The zebrafish has been the subject of several mass mutagenesis screens to isolate mutations in developmental genes. The most common protocol is the identification of recessive mutations, often lethal ones, by breeding for three generations. Mutations are induced in a small number of founder males by treatment with the chemical **mutagen** ethyl nitrosourea (ENU). This induces point mutations at high frequency in dividing cells, up to about one mutation per 500 gametes for a particular gene. The males are allowed to recover and during this period the sperm already undergoing spermatogenesis are lost and new ones are produced from the mutated spermatogonia. The type of breeding schedule used to generate and identify homozygous mutants is described in Chapter 3 (see Fig. 3.6).

The initial screens were carried out by simple dissecting microscope examination of the F3 embryos with the intention of isolating all possible mutations in all developmentally significant genes. But it has become clear that more mutations can be identified if the screen is more focused, so more recent screens have tended to examine one particular organ system in detail. Some of these have used immunostaining or *in situ* hybridization to highlight the structures of interest, and in other cases the parent line is a transgenic line in which the structures of interest express GFP and can easily be visualized under the fluorescence dissecting microscope. When an interesting-looking phenotype is found among the F3 generation, the F2 parents are outcrossed to wild-type fish to reduce the burden of other, nondevelopmental, mutations and to set up a permanent line. The majority of mutations in developmentally important genes are lethal in the homozygous state, therefore the line has to be maintained by identifying heterozygotes and mating them together when mutant embryos are required.

The results of a screen are initially classified according to the phenotype and then **complementation analysis** is carried out for mutations with similar phenotypes to find whether they are different alleles of the same gene. This is done by mating the heterozygotes of the two lines. If the mutations are indeed in the same gene then there will be 25% offspring with the phenotype, whereas if they are in different genes the offspring will not show the phenotype at all. Then the mutation will be mapped by conventional genetic mapping. The map position will be the starting point for **positional cloning** of the gene. Positional cloning used to be a slow and arduous procedure, but nowadays with a near complete genome sequence and a large number of molecular polymorphisms available to use in mapping, it is relatively easy (see Chapter 3, Fig. 3.8). However, many zebrafish mutants have also been identified without positional cloning by the testing of likely candidates based on the similarity of the phenotype to those found in the other model organisms.

Classic Experiments

Because of the simplicity of the genetic methods employed, the original mutagenesis screen performed in the early 1990s represents a paradigm for a zygotic mutagenesis screen for developmental mutants.

Haffter, P. and 16 others (1996) The identification of genes with unique and essential functions in the development of the zebrafish, *Danio rerio*. *Development* (Suppl. 123), 1–36.

Evidence from the results of genome sequencing suggests that the zebrafish, and probably all bony fish, contain up to two similar genes for every one found in higher vertebrates. These related copies have a high level of sequence divergence suggesting a rather ancient duplication of the entire genome. It is estimated that this would have occurred in the ancestor of bony fish shortly after it had separated from the vertebrate main line of descent about 230 million years ago. The effect is that a gene in mouse that has more than one expression domain or developmental function may be represented by two genes in the fish, and these often share out the domains and functions between them such that their combined expression and function resembles that of the single mouse gene. This fact has advantages and disadvantages to the experimenter. On the one hand there are somewhat more genes to sift through and analyze. On the other hand, a gene whose null mutant is an early lethal in mouse may appear as two genes each of whose null phenotype is milder, allowing development to a more advanced stage or even long-term viability.

Other methods

Apart from the normal type of recessive screen, the zebrafish allows two other techniques for mutational screening, which have advantages in particular circumstances (Fig. 8.3). One is the generation of **haploid** embryos. Female F1 fish are taken and squeezed gently to obtain unfertilized eggs. These are fertilized *in vitro* with sperm heavily irradiated with UV light to render them inviable. Such sperm can activate development but the sperm nucleus does not participate and so the embryo is a haploid, arising from the maternal pronucleus alone. In haploid embryos the effects of recessive mutations will be immediately apparent and so an interesting mutation can be identified without the need for the F2 family. However, the haploid embryos are not viable in the long term. They die after a few days and so a mutation needs to be maintained by conventional breeding starting from the mother. Moreover, haploid embryos themselves are not normal even in the absence of mutations, and therefore screens are only feasible for characteristics that would not be masked by the typical haploid syndrome.

The haploid syndrome can be avoided if the eggs are converted into **gynogenetic diploids**. Again, the eggs are squeezed from the female fish and fertilized with inviable sperm. Then they are subjected to a pressure pulse to drive the second polar body back into the egg and cause it to fuse with the maternal pronucleus. This is now a diploid composed entirely of maternal genetic material. However, loci will only be homozygous if there has been no meiotic recombination. If recombination has occurred at the bivalent stage then one mutant copy can be lost into the first polar body and the gynogenetic diploid will potentially be heterozygous for all loci distal to the crossover point. Thus, this technique can be quite useful for genes located near the centromeres, which are unlikely to be lost by recombination, but it does not have a general validity.

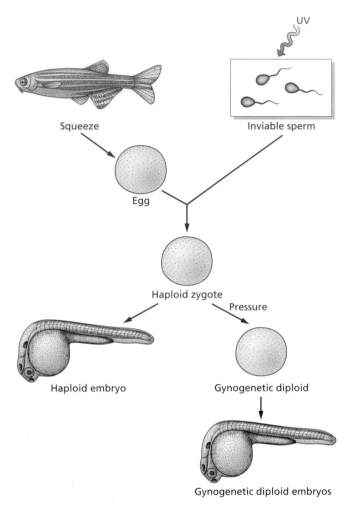

Fig. 8.3 Production of haploid embryos and gynogenetic diploids.

The cloning of genes is rendered easier if mutations are induced by a procedure that results in the insertion of some molecular probe. There are methods for **insertional mutagenesis** based on the introduction of defective retroviruses, but the frequency of mutation is much lower than with ENU, so the number of F2 families to screen is correspondingly higher.

Regional specification

Techniques

As discussed in Chapter 4, the association of a particular molecule with a particular function requires assessment of the expression pattern, the biological activity, and the effects of ablation. Once a zebrafish gene is cloned, the expression pattern is established by *in situ* hybridization or immunostaining. Biological activity is often established, as in *Xenopus*, by injection of specific mRNA into the fertilized egg or into specific blastomeres. Because the starting point for the study of a gene is often a null

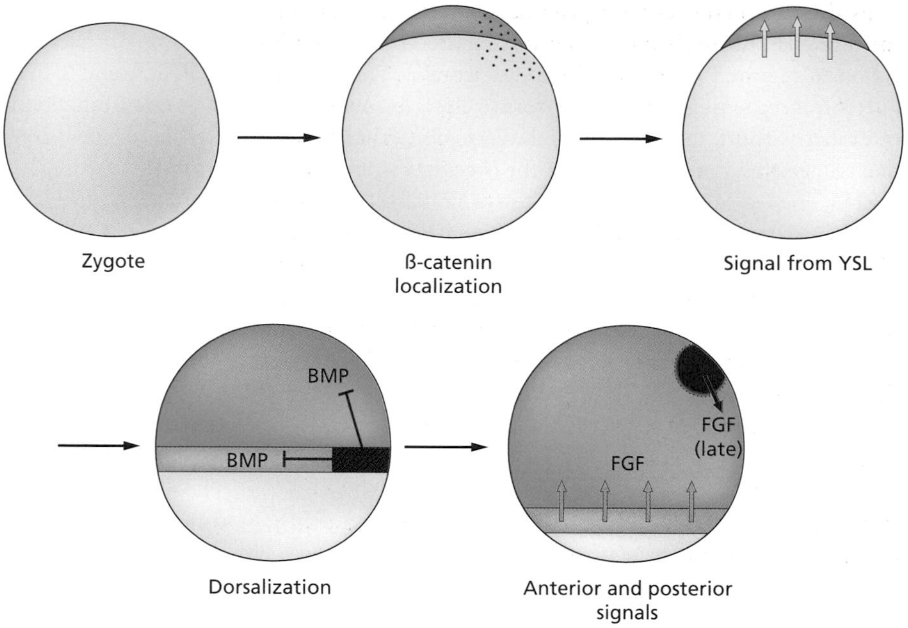

Fig. 8.4 Sequence of inductions in the zebrafish.

mutant, the effect of ablation may already be known. If it is not, then similar methods can be used as in *Xenopus*, involving the injection of morpholino antisense oligos or of mRNA coding for dominant negative variants of the gene under study. The micro-surgical possibilities in zebrafish fall short of those in *Xenopus* because the embryo is considerably smaller. However, using micromanipulation equipment, it is possible to isolate "animal caps" from gastrulae and to transplant groups of cells from one region to another. Transgenesis can be achieved by injection of DNA into the fertilized egg. Normally it is injected into the yolk and cytoplasmic streaming carries it into the nuclei of the early blastomeres, which remain attached by large cytoplasmic bridges. Transgenesis often yields mosaic individuals in which only some cells carry the transgene, although germline integration can be achieved and it has now become routine to produce transgenic lines of fish.

Dorsoventral polarity

There is no obvious cortical rotation in the zebrafish, but treatment with the microtubule depolymerizing drug nocodazole does suppress axis formation, suggesting a role for microtubules in the establishment of the dorsal center. The Tcf3 transcription factor is encoded by the gene *headless*, whose loss of function phenotype is anterior reduction, suggestive of a role for the Wnt pathway. Formation of the dorsal axis can be suppressed by overexpression of gsk3, and a dominant negative gsk3 can induce a secondary axis. There is a migration of β-catenin to the nuclei on the dorsal side, visible by immunostaining. Overexpression of β-catenin will activate transcription of a gene of the paired-homeobox class, *dharma* (also called *nieuwkoid* or *bozozok*).

dharma is necessary for formation of the organizer as the loss-of-function mutant lacks notochord, prechordal plate, and neural tube. This set of evidence suggests that events are comparable to *Xenopus*, with an activation of the intracellular Wnt pathway on the dorsal side .

Meso-endoderm induction

The sequence of inductive events that build up the body pattern also seems very similar to those in *Xenopus* with a few variations (Fig. 8.4). There is an initial mesoderm induction, involving the formation of a ring of mesoderm, with an organizer region (the embryonic shield) on the dorsal side. The organizer region then emits signals leading to neural induction of the ectoderm and to dorsalization of the lateral mesoderm.

One variation concerns the action of vegT. In zebrafish this is encoded by the gene *spadetail*, which is expressed zygotically in the mesoderm and whose mutant phenotype is a defect in the trunk rather than loss of all mesoderm and endoderm. *spadetail* is not expressed maternally. However the yolk cell, which is the zebrafish equivalent of the *Xenopus* vegetal hemisphere, does emit a meso-endodermal inducing signal. If the yolk cell is removed from the embryo before the 16 cell stage, then formation of the mesoderm is prevented. If the yolk cell, with its associated yolk syncytial layer, is recombined with an animal cap, then mesoderm is induced around the junction, as shown by expression of markers like *notail* (the zebrafish homolog of *brachyury*) in the responding tissue.

There are two zebrafish homologs of *nodal*: *cyclops* and *squint*. The double loss-of-function mutant has no germ ring and forms little mesoderm. A similar phenotype is shown by loss-of-function mutants of the gene *one-eyed-pinhead*. This encodes the zebrafish homolog of cripto, an extracellular factor whose action is required for nodal signaling (see also Chapter 10). Furthermore, a similar loss of mesoderm follows overexpression of Cerberus-short, the fragment of the *Xenopus* Cerberus factor that antagonizes nodal. Conversely, the overexpression of nodal mRNA or of a mRNA for a constitutive nodal receptor, will induce mesoderm in zebrafish animal caps. These experiments make up good evidence that mesoderm induction is carried out by nodal signals.

Transcription factor genes characteristic of the early endoderm can be also activated by high concentration of nodal. These include *bonnie and clyde* (=*mixer*), *faust* (=*GATA5*), and *casanova* (*sox* related).

The organizer

The organizer region in zebrafish is called the embryonic shield. Grafting of the shield to a ventral position can induce a secondary axis containing somites and a neural tube. As in *Xenopus*, dorsalization of the mesoderm by signals from the organizer involves inhibitors of BMPs.

Two BMP genes, *swirl* (BMP2b) and *snailhouse* (BMP7), are expressed in the ventral part of the gastrula. Their loss-of-function mutants show dorsalization, showing that BMP signaling is needed for ventral development. The transcription factor dharma, expressed in the organizer region, normally suppresses expression of *swirl*. The loss of function mutant of *dharma* is ventralized, and this can be rescued by injection of mRNA for the EnR domain swap version of dharma, showing that the factor normally acts as a transcriptional repressor. The action of dharma also leads to activation of the *chordin* gene, presumably indirectly since dharma itself is a repressor. The zebrafish chordin homolog of *chordin* is called *dino* or *chordino*, and, as expected, the loss-of-function mutant of *chordino* causes ventralization.

The ventral state is defined by expression of two homeodomain transcription factors, Vent and Vox, which are the homologs of Vent 1 and 2 in *Xenopus*. These have a redundant action, but loss-of-function mutants of both genes together produce a dorsalized phenotype. Their normal expression is dependent on BMP signaling, and overexpression of Vent or Vox has a ventralizing effect.

Thus, as in *Xenopus*, it seems that the dorsalizing action of the organizer arises from inhibition of BMPs by direct transcriptional inhibition in the organizer itself and by secreted BMP inhibitors in the surrounding regions.

Anteroposterior patterning

The anteroposterior patterning mechanism shows some similarity and some difference to that in *Xenopus*. During gastrulation, the blastopore region emits a posteriorizing signal in a similar way. There is good evidence that FGFs are an important constituent of the signal. Overexpression of FGFs causes anterior truncation, or induction of posterior markers in anterior explants. Conversely overexpression of a dominant negative FGF receptor causes posterior truncation. In terms of loss-of-function mutation, if the *fgf8* (*acerebellar*) mutant is combined with injection of a morpholino oligo directed against *fgf24*, then formation of posterior structures is inhibited. This suggests that these two FGFs make up most of the normal posteriorizing signal.

As in *Xenopus*, there is also some evidence for a role for Wnt and retinoic acid signaling since both types of factor will posteriorize on overexpression. It has been argued that the ventral marginal zone is a "tail organizer," as it will form a tail if combined with an animal cap, much of the tail arising from the animal cap cells. This effect can be mimicked by injection of a combination of mRNAs encoding nodal + BMP + Wnt, and since the combination of nodal + BMP might be expected to generate ventral mesoderm, this is further evidence for the posteriorizing activity of Wnt.

There is also a group of anterior ectoderm cells, fated to form anterior telencephalon, pituitary, and nasal placodes, which has an anteriorizing influence when grafted to other parts of the neural plate. Removal of the embryonic shield ablates the notochord and prechordal plate, together with the ventral midline structures of the central nervous system (see Chapter 14), but it does not greatly affect the oveall anteroposterior pattern of forebrain–midbrain–hindbrain–spinal cord. This is now known to be a source of FGF at a later stage.

New Directions in Research

The zebrafish will probably make most contribution in the following two areas:

1 For research into late, organogenesis, stages of development, there will be further mutagenesis screens targeted to particular organs. Individual organs or cell types can be highlighted by transgenic markers or by antibody staining. Such screens should reveal previously unknown genes with key roles in the formation of specific organs.

2 For research into morphogenetic movements, the transparency of the embryo will enable high-resolution studies of cell movement in real time. Again, individual cell populations can be labeled by transgenic markers or microsurgical methods.

Key Points to Remember

- The zebrafish is a vertebrate well suited to genetic experimentation. Embryonic development is rapid, the generation time is short, and large numbers of fish can be kept in a facility. Transparency of the embryos enables visualization of cell behavior *in vivo*.
- Many developmental mutants have been isolated from mutagenesis screens. The genes within which the mutations lie have been identified either by positional cloning or by testing candidates.
- In general the fate map and sequence of inductive steps in early development is similar to *Xenopus*. The Wnt signaling pathway is required for dorsal patterning, nodal factors are required for mesoderm induction, and BMP inhibitors for the effects of the organizer. Anteroposterior patterning involves FGFs, Wnts, and retinoic acid.
- However there are also some differences. In particular *spadetail*, the homolog of *vegT*, is not expressed maternally in the yolk cell but is expressed at blastula stage in the mesoderm. So this is unlikely to be playing a similar role as a vegetal determinant.

Further reading

Website

http://zfin.org

General

Kimmel, C.B., Warga, R.M. & Schilling, T.F. (1990) Origin and organization of the zebrafish fate map. *Development* **108**, 581–594.

Warga, R.M. & Kimmel, C.B. (1990) Cell movements during epiboly and gastrulation in zebrafish. *Development* **108**, 569–580.

Ho, R.K. (1992) Cell movements and cell fate during zebrafish gastrulation. *Development* (Suppl.), 65–73.

Kimmel, C.B., Ballard, W.W., Kimmel, S.R., Ullmann, B. & Schilling, T.F. (1995) Stages of embryonic development of the zebrafish. *Developmental Dynamics* **203**, 253–310.

Solnica-Krezel, L., ed. (2002) *Pattern Formation in Zebrafish*. Berlin: Springer-Verlag.

Genetics

Driever, W., Stemple, D., Schier, A. & Solnica-Krezek, L. (1994) Zebrafish: genetic tools for studying vertebrate development. *Trends in Genetics* **10**, 152–159.

Haffter P. & 16 others (1996) The identification of genes with unique and essential functions in the development of the zebrafish, *Danio rerio*. *Development* **123**, 1–36.

Fishman, M.C. & Chien, K.R. (1997) Fashioning the vertebrate heart: earliest embryonic decisions. *Development* **124**, 2099–2117.

Udvadia, A.J. & Linney, E. (2003) Windows into development: historic, current and future perspectives on transgenic zebrafish. *Developmental Biology* **256**, 1–17.

Inductive interactions

Kodjabachian, L., Dawid, I.B. & Toyama, R. (1999) Gastrulation in zebrafish: what mutants teach us. *Developmental Biology* **213**, 231–245.

Schier, A.F. (2001) Axis formation and patterning in zebrafish. *Current Opinion in Genetics and Development* **11**, 393–404.

Poulain, M. & Lepage, T. (2002) Mezzo, a paired-like homeobox protein is an immediate target of Nodal signaling and regulates endoderm specification in zebrafish. *Development* **129**, 4901–4914.

Schier, A.F. (2003) Nodal signaling in vertebrate development. *Annual Reviews in Cell and Developmental Biology* **19**, 589–621.

Leung, T.C., Bischof, J., Soll, I., et al. (2003) Bozozok directly represses bmp2b transcription and mediates the earliest dorsoventral asymmetry of bmp2b expression in zebrafish. *Development* **130**, 3639–3649.

Dougan, S.T., Warga, R.M., Kane, D.A., Schier, A.F. & Talbot, W.S. (2003) The role of the zebrafish nodal-related genes squint and cyclops in patterning of mesendoderm. *Development* **130**, 1837–1851.

Draper, B.W., Stock, D.W. & Kimmel, C.B. (2003) Zebrafish fgf24 functions with fgf8 to promote posterior mesodermal development. *Development* **130**, 4639–4654.

Agathon, A., Thisse, C. & Thisse, B. (2003) The molecular nature of the zebrafish tail organizer. *Nature* **424**, 448–452.

Chapter 9

The chick

The visible course of development of the chick is superficially very different from the lower vertebrates and is much closer to the mammalian type. But even though the visible morphogenetic movements in the early embryo can appear quite different, at a molecular level it is clear that essentially the same basic processes are taking place in all vertebrates. Because the chick has been extensively used for work on later stages of development, a general outline of vertebrate organogenesis is given in this chapter. In section 3 this will be revisited in greater detail and with incorporation of relevant evidence from other vertebrate species.

Fertilized eggs are usually obtained from commercial hatcheries, and, at the stage of laying, the embryo is a flat **blastoderm** of about 60,000 cells. Development is arrested at low temperatures and therefore eggs can be stored for some days at 10°C during which time embryonic development remains suspended. It recommences when the eggs are incubated at 37.5°C. Unlike *Xenopus* and the zebrafish, the chick undergoes extensive growth during embryonic life because it has at its disposal the food reserves of the egg.

For the experimentalist the chick has the great advantage over the mouse that the embryo is accessible at all stages following laying of the egg. Early blastoderms can be cultured *in vitro* for long enough to form a recognizable primary body plan. Alternatively, a hole can be cut in the egg shell and the embryo can be manipulated *in ovo*, then the hole is resealed with adhesive tape and the whole egg incubated until the embryo has reached a later stage of development. Furthermore it is possible to explant small pieces of tissue onto the **chorioallantoic membrane (CAM)** of advanced embryos, where they become vascularized and will grow and differentiate in effective isolation. Culture of some organ rudiments is also possible *in vitro*. For the labeling of grafts, extensive use has been made of interspecies combination between chick and quail. Quail embryos are anatomically very similar to chick although they are slightly smaller and develop a little faster. Originally they were used because all quail cells possess a condensation of **heterochromatin** associated with

their nucleolus, and this is easily visualized as a dark blob by staining for DNA using the Feulgen histochemical reaction. Nowadays the quail cells are normally visualized by staining with a species-specific antibody.

The chick is not well suited to genetic work. The life cycle is long, the existing mutants limited in number, and there is no routine protocol for transgenesis or targeted mutagenesis. However there are now several useful methods for overexpression of genes in chick embryos. Much use has been made of **retroviruses** carrying the gene in question. These are often called **RCAS viruses** for **r**eplication-**c**ompetent, **a**vian-**s**pecific. The virus can be injected locally in the region to be modified, and the effect will spread as new virus particles are produced by the infected cells. If the range of infection is to be limited, this can be achieved by making an orthotopic graft of tissue from a sensitive strain of chick into an embryo of a resistant strain. Then just the tissue of the graft will become infected and overexpress the gene. More recently, **electroporation** has been used. This involves injecting DNA into the region of interest and then subjecting the embryo to a series of low voltage electric pulses which can drive DNA into the cells without doing too much damage. The DNA is negatively charged and so moves towards the anode, hence the tissue on the cathode side of the injection will be untransformed and can serve as an internal control for the effect of the gene. Localized treatment of embryos with **inducing factors** was originally achieved by implantation of pellets of **tissue culture** cells expressing and secreting the factor in question. With the greater availability of pure factors, they are now usually absorbed to affinity chromatography beads which are implanted into the embryo in the desired position. Such beads can bind a large amount of the factor and then release it slowly for 1–2 days.

As for *Xenopus*, gene ablation studies need to be performed by overexpression of dominant negative reagents. Morpholino oligos are much less useful than for *Xenopus* or zebrafish, because they do not readily penetrate cell membranes and in the chick there are no large blastomeres to inject.

Normal development

The hen's egg is a familiar object, with its shell, albumen layer ("white"), and yolk. But the true egg consists of just the yolk, and, in a fertilized egg, an inconspicuous **blastoderm** of cells, which is surrounded by a vitelline membrane. The yolk corresponds to the mature **oocyte**. It was formerly supposed that avian, like mammalian, oogenesis occurred only during fetal life and that the complement of oocyte present in the newborn represented a lifetime's supply. A question has arisen about whether this is really true for mammals, but in birds the issue has not been recently reinvestigated. Up to about 1–2 weeks before ovulation, the hen's oocyte remains quite small, but it then puts on a tremendous growth spurt and over a few days acquires a weight of about 55 g. On ovulation it is released from the ovary and enters the **oviduct** where, if the hen has recently mated, it will be fertilized. Passage down the oviduct takes about 24 hours, in the course of which the egg is invested successively with the albumen layer, the shell membranes, and the shell itself.

Cleavage is highly **meroblastic** and involves just the patch of cytoplasm 2–3 mm in diameter which is present in the zygote (and oocyte) at the edge of the yolk mass (Fig. 9.1a). The early cleavages take place in the oviduct producing a circular blastoderm initially one cell thick, and later several cells thick. The cleavage pattern is very variable from one embryo to the next and the blastomeres at the ventral and lateral faces of the sheet

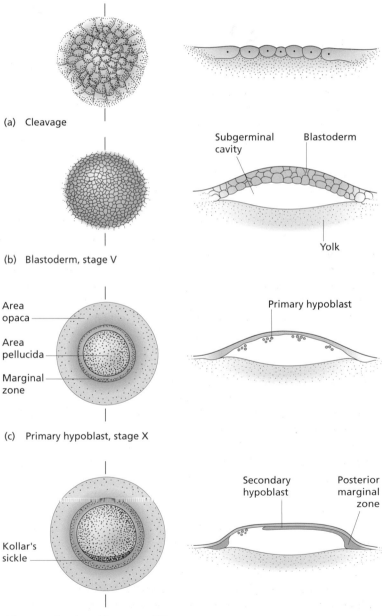

(a) Cleavage

(b) Blastoderm, stage V

Subgerminal cavity   Blastoderm

Yolk

(c) Primary hypoblast, stage X

Area opaca

Area pellucida

Marginal zone

Primary hypoblast

(d) Secondary hypoblast, stage XII

Kollar's sickle

Secondary hypoblast   Posterior marginal zone

Fig. 9.1 Development of the chick blastoderm up until the time of egg laying. Vertical lines indicate planes of section of right side figures.

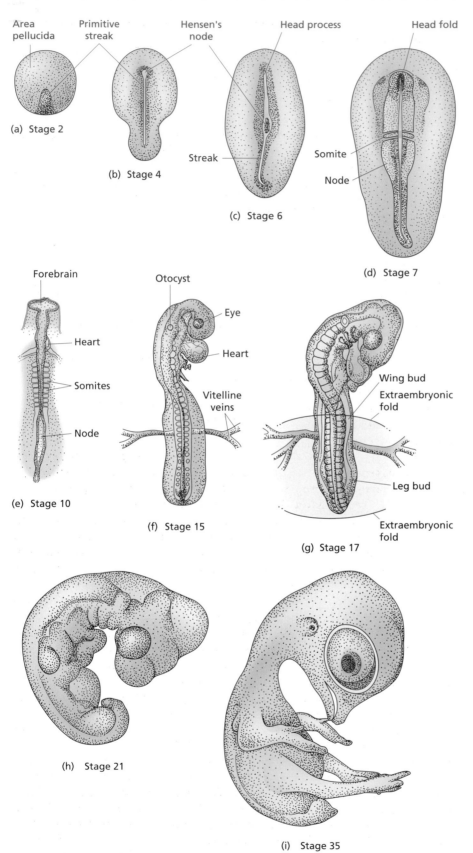

Fig. 9.2 Normal development of the chick. Stage 7 is reached after about 1 day of egg incubation, stage 12 after 2 days, stage 17 after 3 days, stage 24 after 4 days, and stage 35 after 9 days. Time from fertilization is about 1 day more than egg incubation time.

remain connected to the yolk for some time by large cytoplasmic bridges. The egg spends about 20 hours in the lower part of the oviduct, called the uterus, undergoing slow rotations driven by uterine peristalsis while the calcareous shell forms around it. When the blastoderm consists of a few hundred cells, a space called the subgerminal cavity opens beneath it (Fig. 9.1b). Cells are shed from the lower surface of the blastoderm into this cavity and probably die, so that by the end of the uterine period the central region of the blastoderm has thinned to an organized epithelium one or a few cells thick. Because of its translucent appearance this is known as the **area pellucida**. The outer, more opaque, part of the blastoderm is called the **area opaca** and the junctional region the **marginal zone**. Note that the region called the "marginal zone" of the chick embryo is not a homologous structure to the marginal zone of an amphibian embryo.

A lower layer of cells, the **hypoblast**, then develops, partly by ingression of small groups of cells all over the area pellucida (the primary hypoblast) and partly by spreading of cells from the deep part of the posterior marginal zone (the secondary hypoblast). A thickening of the epiblast at the posterior margin is known as **Kollar's sickle** (Fig. 9.1c,d). The hypoblast contributes only to extraembryonic structures and may perhaps be homologous to the visceral endoderm of the mouse egg cylinder (see Chapter 10). The upper layer of cells now becomes known as the **epiblast**. At the time the egg is laid it will usually just have commenced secondary hypoblast formation, and the total blastoderm consists of approximately 60,000 cells. The early developmental stages are described by the stage series of Eyal-Giladi and Kochav which uses Roman numerals. In this series, I represents the fertilized egg, X represents the single-layered blastoderm stage, and XIII represents the complete two-layered blastoderm stage. The stage of egg laying, after which the embryos become accessible for experimentation, is about stage X–XI. Subsequent development is described by the stage series of Hamburger and Hamilton, which uses Arabic numbers and is indicated in what follows.

The stages of body plan formation are depicted in Fig. 9.2 showing top views, and Fig. 9.3 showing transverse sections. A condensation of cells called the **primitive streak** arises at the posterior edge of the area pellucida (stage 2) and elongates until it reaches the center (Figs 9.2a, 9.3a). The streak expresses the T-box transcription factor gene *brachyury*, already encountered in *Xenopus* and zebrafish. Cells from the epiblast migrate into the streak and pass through it to become the **mesoderm** and the **definitive endoderm** part of the lower layer. This process is regarded as **gastrulation** in the chick, although the future gut lumen, or **archenteron**, is not a new cavity but is the pre-existing space below the endoderm. The area pellucida gradually changes from a disc to a pear shape and a further condensation called **Hensen's node** appears at the anterior end of the primitive streak (stage 4; Fig. 9.2b). This expresses various transcription factor genes characteristic of the organizer region in the lower vertebrates, including *goosecoid*, *not*, and *FoxA2(HNF3β)* (chick genes are sometimes given a C prefix, as in *C-goosecoid* or

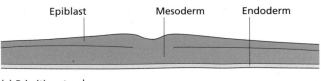

(a) Primitive streak

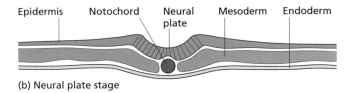

(b) Neural plate stage

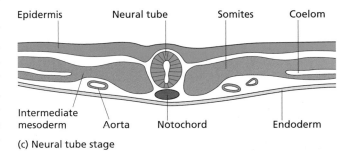

(c) Neural tube stage

Fig. 9.3 Normal development of the chick. Transverse sections during formation of main axial structures.

C-not). The node contains the presumptive notochord cells, some of which migrate anteriorly to form the head process, or part of the notochord lying within the head (stage 6; Fig. 9.2c). The remainder of the node moves posteriorly and as it does so the principal structures of the body plan appear in its wake: the **notochord** in the midline, the **somites** on either side of it, and the **neural plate** in the epiblast (Fig. 9.3b). The **primordial germ cells** appear at the extreme anterior edge of the area pellucida, outside the embryo proper.

By about 1 day of incubation the anterior end of the embryo is marked by an uplifting of the blastoderm called the head fold, and one somite and the anterior neural folds have appeared in the track of the regressing node (stage 7; Fig. 9.2d). From this stage the embryo proper becomes progressively separated from the surrounding extraembryonic tissue. This is achieved by the appearance of folds involving all three germ layers that appear around the embryo and undercut it such that initially the head, and later the tail and trunk, project above the surface of the extraembryonic tissue (Fig. 9.4).

Early on the second day, blood islands appear in the outer extraembryonic part of the blastoderm, and the heart primordium forms by fusion of the rudiments on the right and left side of the anterior mesoderm (Fig. 9.2e). The heart is able to form in this anterior mid-ventral position because the formation of the head fold has now lifted the head above the level of the surroundings (Fig. 9.4). The formation of the head fold has also caused the foregut to become enclosed as a pocket, while the rest

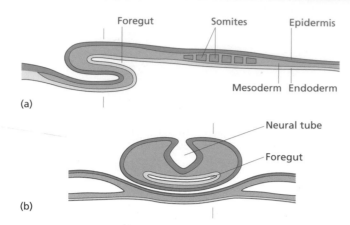

(a)

(b)

Fig. 9.4 The head fold at stage 8. (a) Parasagittal section; lines indicate plane of section of (b). (b) Transverse section; lines indicate plane of section of (a).

of the presumptive gut endoderm is still a lower layer of cells facing the yolk. The **neural tube** closes first over the midbrain and then progressively in both directions. **Somites** continue to arise in anteroposterior sequence from the segmental plates of mesoderm which flank the notochord and neural tube. By about 36 hours there are 10 somites and the neural tube has closed to form forebrain, midbrain, and hindbrain vesicles (stage 10). The **lateral plate mesoderm** becomes divided into a **somatic** layer, adhering to the epidermis, and a **splanchnic** layer, adhering to the endoderm. The space in between is the **coelom** (pronounced "see-loam": Fig. 9.3c). A further subdivision of mesoderm appears as a longitudinal strip in between the presomitic mesoderm and the lateral plate. This is the **intermediate mesoderm** that later forms the kidney, adrenals, and gonads. Although

node regression and the formation of the posterior part of the body continues for some time, this stage marks approximately the junction between early and late development since the general body plan has been laid down and the formation of individual organs is about to begin.

Extraembryonic membranes

In embryos that have an external food supply the formation and arrangement of the **extraembryonic membranes** is essential to their survival. Amniotes (reptiles, birds, and mammals) are so called because they all have an **amnion**, and in fact many of the extraembryonic structures are obviously homologous. The arrangement in the chick embryo is shown diagrammatically in Fig. 9.5.

From the time of gastrulation onwards, the outer **area opaca** expands over the surface of the yolk as a membrane consisting of extraembryonic ectoderm and endoderm. The mesoderm has a more restricted spread and is coincident at any one time with the region of extraembryonic vasculature (see below). Initially the coelom is continuous between the embryonic and extraembryonic regions. The inner extraembryonic layer, which is composed of splanchnic mesoderm, blood-forming tissue, and endoderm, is called the yolk sac. This gradually surrounds the entire yolk mass and serves as a digestive organ, the products of digestion of the yolk being absorbed into the blood vessels and conveyed to the embryo. The outer extraembryonic layer consists of somatic mesoderm and ectoderm and is called the **chorion**. During the third day, a fold of the chorion starts to cover the anterior end of the embryo, and a corresponding fold also grows from the posterior. These folds are shown as lines on

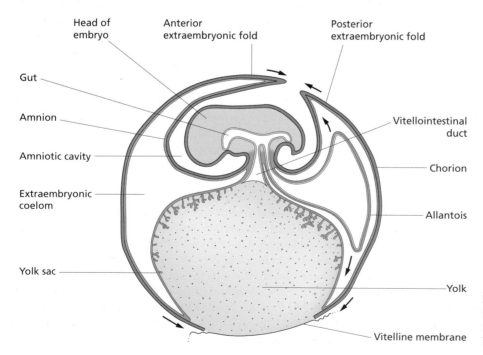

Fig. 9.5 Formation of the extraembryonic membranes in the chick. (After Hildebrand 1995. *Analysis of Vertebrate Structure.* New York: Wiley.)

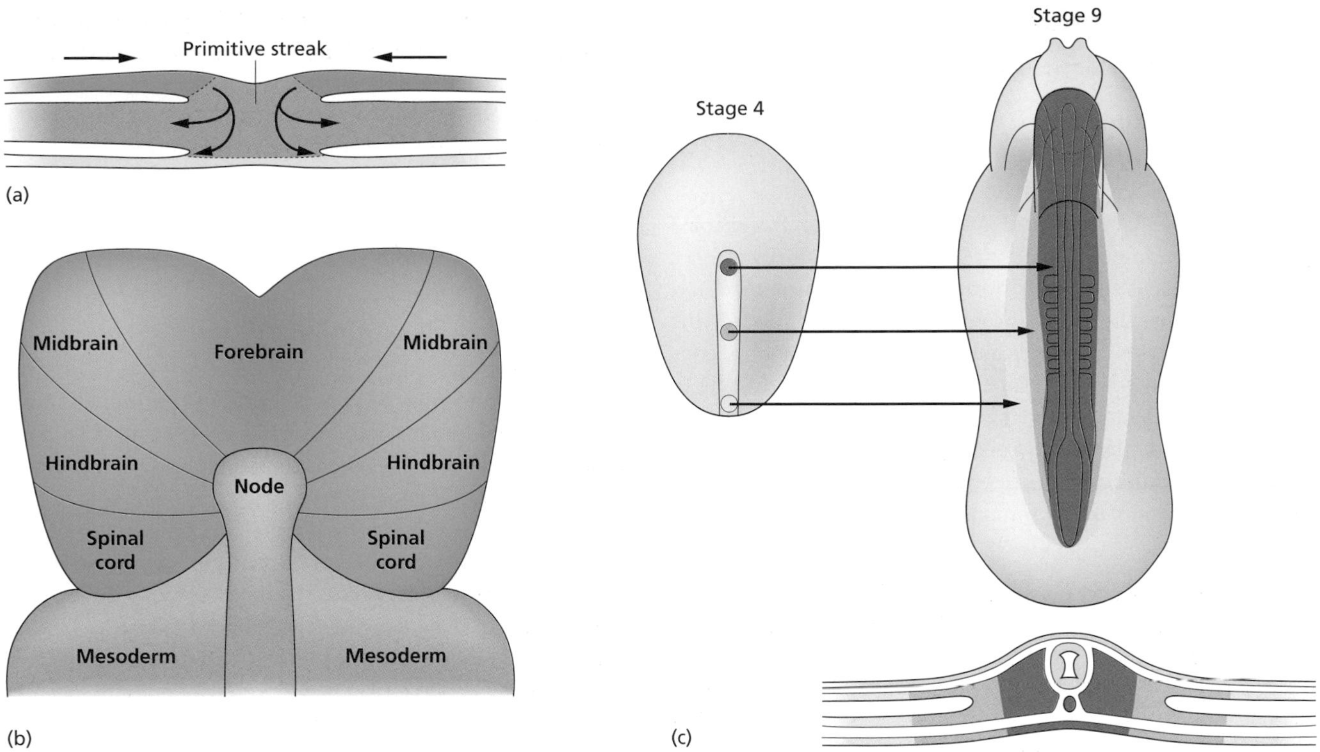

Fig. 9.6 Fate map of the chick embryo. (a) Movement of cells through the primitive streak. (b) The prospective region for the central nervous system about stage 4. (c) The origin of mesodermal and endodermal levels from the primitive streak.

Fig. 9.2g. The folds fuse in the middle to form two complete membranes covering the embryo, the outer still being called the **chorion** and the inner being called the **amnion**.

The **allantois** consists of a layer of endoderm covered by splanchnic mesoderm. It grows out from the hindgut into the extraembryonic coelom and expands to fuse with the undersurface of the chorion. The allantois serves two functions, one being as an excretory receptacle in which uric acid accumulates, the other as the main respiratory organ of the embryo. The **chorioallantoic membrane**, as it is called after fusion, is located just under the shell and is richly supplied with blood vessels for gaseous exchange. It is often used by experimentalists as a site for "*in ovo*" culture of isolated organ rudiments taken from the embryo.

Fate map

Only a small proportion of the blastoderm cells become incorporated into the embryo itself, the rest forming extraembryonic structures. The **fate map** of the early embryo has been extensively studied, most recently using small localized marks of **DiI** and **DiO** which can be visualized at later stages. These studies show that the primitive streak originates from the most posterior part of the area pellucida and from the posterior marginal zone. Like the area pellucida itself, this region contains two cell layers; the upper layer making a substantial contribution to the ectoderm of the streak, and the lower layer to the secondary hypoblast. The streak extends to the center of the area pellucida by active stretching. Although some cells remain resident in the streak, it consists mainly of cells in the process of moving through it, as there is a migration of epiblast cells from both sides which enter and move through to form both the mesoderm and endoderm (Fig. 9.6a). As a result of the entrance of new cells into the midline of the lower layer, the original hypoblast cells become pushed to the outer rim.

Labeling experiments show that the node, like the rest of the streak, has a continuous flux of cells moving in and out. During the phase of node regression the node itself is the prospective region for the notochord along the entire length of the body. The neural plate arises from epiblast around the node, with a conservation of anteroposterior levels (Fig. 9.6b), the somites arise from the region just around the node, and the lateral plate from more posterior parts of the streak. The origin of the endodermal layer closely follows the mesoderm, with medial parts from the anterior streak and lateral parts from the posterior streak (Fig. 9.6c) It is important to note that the primitive streak does not, as is often thought, map in a one-to-one manner onto the later anteroposterior body plan. Instead the anteroposterior axis of the streak maps to the mediolateral axis of the later embryo and this means that the posterior half of the streak is destined to become extraembryonic tissues.

Regional specification of the early embryo

Anteroposterior polarity and fragmentation of the blastoderm

As long ago as 1828, von Baer propounded a rule that enabled the anteroposterior axis to be predicted in the majority of eggs. This states that if the egg is horizontal with the pointed end to the right then the tail of the embryo should be towards the observer. The rule arises because the egg undergoes a continuous rotation when it is in the uterus and this is usually in the same direction relative to the sharp and blunt ends of the egg (Fig. 9.7). The embryo and yolk do not rotate along with the outer surface of the egg but are nevertheless tipped in the direction of rotation and, in fact, a simple tipping of the blastoderm at

the critical period of 14–16 hours of uterine life is enough to establish the polarity such that the posterior end forms at the uppermost end of the blastoderm.

The early blastoderm can be cut into two or three parts, each of which will form a complete embryonic axis. So long as such fragments contain part of the original posterior of the blastoderm, then this will remain the posterior, showing the importance of the early polarization.

Early inductive interactions

It is possible to identify processes of mesoderm induction and the organizer effect by embryological experiments. To some extent the same molecules seem to be involved as in *Xenopus*, but there may also be some differences.

The primitive streak expresses the T-box transcription factor *brachyury*, and its formation corresponds to **mesoderm induction** in *Xenopus*. Normally the streak arises from the extreme posterior part of the area pellucida epiblast. If an anterior piece of epiblast is combined with a piece of posterior marginal zone (PMZ), then a new streak can be induced with polarity such that its posterior end abuts the PMZ (Fig. 9.8). Competence for streak induction by the PMZ is lost at stage XI, when formation of the secondary hypoblast is commencing, although a streak can still be induced to a substantially later stage (Hamburger & Hamilton (H&H) stage 3) by the secondary hypoblast itself. Interestingly, the secondary hypoblast and germ ring show a nuclear localization of β-catenin, like the dorsal side of the *Xenopus* egg, and lithium treatment can induce axial structures from the epiblast. It is possible to induce a streak from the area pellucida by application of vg1 (a nodal-like molecule) together with Wnt. This suggests a similar mechanism to *Xenopus*, but the evidence is incomplete as there is no satisfactory inhibition experiment at present.

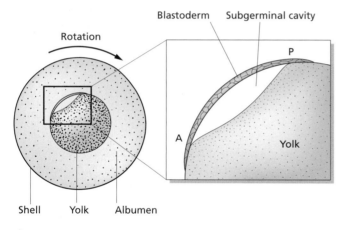

Fig. 9.7 Acquisition of anteroposterior polarity by the chick blastoderm, as a result of intrauterine rotation of the egg. A = anterior; P = posterior.

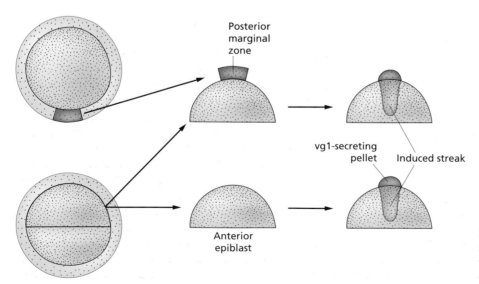

Fig. 9.8 Induction of a primitive streak by the posterior marginal zone or a vg1-secreting cell pellet (vg1 is a nodal-like substance).

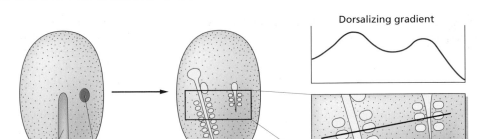

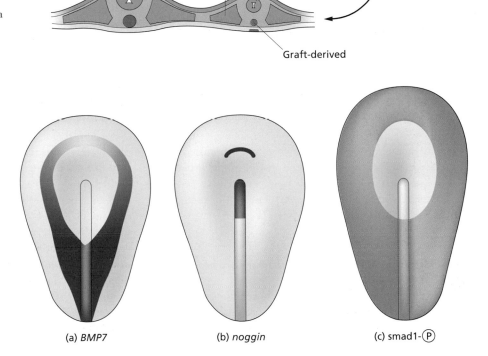

Fig. 9.9 A node grafted to the area pellucida will induce a partial second axis in which the notochord is graft-derived and the neural tube and somites are host-derived.

Fig. 9.10 Expression domains of: (a) a BMP; (b) a BMP inhibitor; (c) visualization of BMP signaling by detection of smad1-phosphate.

(a) *BMP7*   (b) *noggin*   (c) smad1-ⓅP

Interestingly this method can be used to show that formation of one streak represses formation of any others. An isolated anterior half blastoderm will usually form one streak at the posterior margin. If a pellet releasing vg1 is grafted laterally then it will induce a streak locally and suppress the host streak. If two pellets are grafted then they will both induce streaks unless the time interval between them exceeds 5 hours, in which case only the first induces a streak. This shows clearly that a new streak emits a signal inhibiting the formation of additional streaks in its vicinity.

The next inductive interaction is the "organizer effect": the regionalization of the ectoderm and mesoderm under the influence of Hensen's node. A node grafted into the area pellucida not too far from the host streak can induce a secondary axis in which the notochord is derived from the graft but the neural tube and somites are derived from the host (Fig. 9.9). This closely resembles the behavior of the *Xenopus* organizer and shows that

there is an inductive signal emitted by the node. The activity of the node persists until about H&H stage 8 (four somites), with younger nodes inducing anterior neural structures and older nodes inducing posterior structures. The competence of the epiblast disappears before this, after H&H stage 4 (definitive streak). It is tempting to suppose that the organizer effect is due to BMP inhibition, as several BMPs are expressed in the peripheral part of the blastoderm and the BMP inhibitors noggin and chordin are expressed in the node. There is also a reduction of smad1-phosphate detectable by immunostaining in the prospective neural plate region (Fig. 9.10). However, there has been some controversy over this because, in general, the BMP inhibitors do not have the expected neuralizing activity when applied to explants of epiblast. Probably more than one factor is involved, as it is also known that FGFs have some neuralizing activity, FGF inhibitors can suppress neural induction, and Wnts are also antagonistic. As far as effects on the mesoderm are concerned, applied BMPs can suppress somite formation in favor of the lateral plate, suggesting that mesodermal patterning does depend on a graded inhibition of BMP activity, as in *Xenopus*.

Note that it is unwise to refer to the organizer effect in the chick as "dorsalization." Because the chick blastoderm is flat, the axis homologous to the dorsoventral axis of *Xenopus* and zebrafish runs, at this stage, from medial to lateral. In the early chick embryo the "dorsoventral axis" is often referred to as the axis running through the blastoderm from the dorsal surface to the subgerminal cavity. This is homologous to the animal–vegetal axis and not to the dorsoventral axis of *Xenopus*. After ventral closure, the notochord, somites, and lateral plate do indeed run from dorsal to ventral and the nomenclature becomes the same for the chick and the amphibian.

A complete axis can still be produced following extirpation of Hensen's node at an early stage. This is a good example of embryonic regulation (see Chapter 4) whose mechanism remains unknown. The node reforms from the posterior margin of the hole and the process requires the presence of the mid-primitive streak. Once a node has been formed it suppresses formation of other nodes in the vicinity.

Formation of the trunk-tail parts of the anteroposterior pattern may be due to FGF, as in *Xenopus* and zebrafish. *Fgfs* are expressed in the primitive steak, and an FGF-impregnated bead will induce posterior neural tube from the epiblast. FGFs will induce expression of Cdx genes and these will in turn induce expression of the posterior Hox genes (Hox 6–13).

Left–right asymmetry

As mentioned in Chapter 2, vertebrates are not exactly bilaterally symmetrical. In the chick, the deviation arises at an early stage with a tilt of Hensen's node to the left. This is soon followed by the S shape of the heart tube and the flexion of the

whole embryo, with the head lying to the right when viewed from above. When the stomach and intestine develop, they are markedly asymmetrical. Several of the gene products involved in this process have been identified in the chick because of their asymmetrical expression patterns, and the sequence of events has been worked out from experiments in which these factors are applied locally, either as RCAS virus or as protein on beads, and the effects on expression of other factors is observed. The process involves four steps. Firstly a breakage of the basic bilateral symmetry of the embryo in the node or midline structures. Secondly an amplification to create different regimes of gene expression on either side of the midline. Thirdly a spread of the information out to the lateral mesoderm which is the tissue layer most involved in the formation of the asymmetrical organs; and finally the control of the events of cell adhesion and movement that actually bring about the asymmetrical morphology.

In the chick the original symmetry-breaking event is still unclear, although in the mouse there is good evidence that it depends on the asymmetrical structure of the cilia in the node (see Chapter 10). The key player among the asymmetrically expressed genes is believed to be *nodal*, since it is preferentially expressed on the left side in all types of vertebrates, and application of nodal to the right side will randomize the situs of multiple organ systems. The components upstream and downstream of *nodal* do, however, vary between vertebrates. In the chick, *nodal* appears in a small domain on the left of the regressing node at about H&H stage 6 and then spreads to a much wider domain in the left lateral plate mesoderm. Its expression on the left side is preceded by *sonic hedgehog*, and on the right by *activinβB fgf4* and *fgf8* (Fig. 9.11). Shh will activate nodal expression when applied to the right side and activin or FGFs will repress nodal if applied to the left side, indicating that they are upstream regulatory components.

Spread of the *nodal* zone to the lateral plate mesoderm seems to involve partly an autocatalytic loop whereby nodal signaling activates *nodal* transcription, and partly a Cerberus-like BMP inhibitor called Caronte whose transcription is activated by nodal. BMPs are expressed on both sides and can suppress *nodal* expression. But Caronte suppresses BMP signaling on the left and thereby allows the spread of nodal expression on this side. The limitation of spread is controlled by another TGF-β-like factor called lefty, which becomes expressed on either side of the nodal domain. The lefty protein is an inhibitor of nodal and reduces its signaling activity. The end product of the gene cascade is expression of the homeodomain transcription factor pitx2 on the left side. This is controlled by nodal signaling through

Classic Experiments

LEFT–RIGHT ASYMMETRY

Vertebrate embryos are more or less bilaterally symmetrical, and so for many years the nature of the mechanism producing asymmetry was mysterious. In 1995 it was shown that some key genes (*nodal*, *sonic hedgehog*, and *activin receptor IIA*) have asymmetrical expression patterns in the early chick, and regulated each other to form a linked pathway from initial breaking of lateral symmetry to final morphological asymmetry of the heart and viscera.

Levin, M., Johnson, R.L., Stern, C.D., Kuehn, M. & Tabin, C. (1995) A molecular pathway determining left–right asymmetry in chick embryogenesis. *Cell* **82**, 803–814.

Subsequently it was shown in the mouse that the node bears cilia which, because of their inherent molecular asymmetry, drives fluid preferentially to the left.

Nonaka, S., Tanaka, Y., Okada, Y., Takeda, S., Harada, A., Kanai, Y., Kido, M., & Hirokawa, N. (1998) Randomization of left–right asymmetry due to loss of nodal cilia generating leftward flow of extraembryonic fluid in mice lacking KIF3B motor protein. *Cell* **95**, 829–837

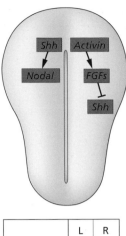

	L	R
Activin ßB	−	+
FGFs	−	+
Shh	+	−
nodal	+	−

(a) Early events

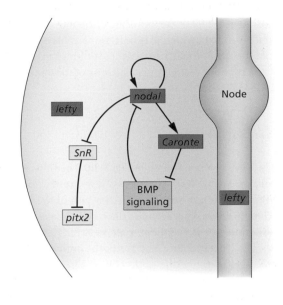

(b) Later events

Fig. 9.11 Development of left–right asymmetry in the chick embryo.

transcriptional repression of snail-related, a zinc finger transcription factor, which itself represses *pitx2* transcription. The sequence of events is shown in Fig. 9.11.

Although some steps of this mechanism are not shared by other vertebrates, nodal and pitx2 are always expressed on the left and seem to represent the principal signaling step and final controller of cell differentiation.

Description of organogenesis in the chick

The later stages of chick embryo development have been very well known for a long time and for this reason serve as the basic resource for our knowledge of vertebrate organogenesis in general. As the experimental work on several organ systems is described in more detail in later chapters, this section contains just a brief summary of some of the major morphological events.

Whole embryo

By the second day the heart is bent to the right and the **optic vesicles** appear from the forebrain (see Fig. 9.2e). The head turns to face the right and at this time (stage 13) an anterior extraembryonic fold rises to cover the head. This fold moves progressively posteriorly and will later become the chorion and amnion (see Fig. 9.5). The head becomes sharply flexed between the region of the forebrain and the hindbrain. The first three **pharyngeal pouches** appear, the optic vesicles invaginate, and the lenses appear in the adjacent epidermis. On day 3 the limb

buds appear in the lateral plate mesoderm (stage 17) and after a further day are as long as they are broad. A posterior extraembryonic fold appears and moves anteriorly, eventually meeting the anterior fold and fusing shortly after the appearance of the limb buds to enclose the embryo within an amniotic cavity (see Fig. 9.5). Eye pigmentation appears from about 3.5 days.

By the third day the original head fold has deepened into an anterior body fold, such that the anterior half of the body has become elevated above the surroundings (see Figs 9.4, 9.5). This results in the formation of a closed tube of **foregut** running anteriorly from an anterior intestinal portal connecting it to the subendodermal space. Over the fourth day, a corresponding posterior body fold lifts the rear part of the embryo and results in the formation of a **hindgut**. The residual ventral opening of the gut becomes progressively smaller and eventually becomes narrowed to a **vitellointestinal duct** joining the yolk mass to the midgut (see Fig. 9.5). The **allantois** arises from the hindgut and expands rapidly into the space between chorion and amnion.

A mouth is formed at the anterior end of the foregut. The face is formed by the fusion of a set of paired processes: above the mouth are the frontonasal and maxillary processes and below the mouth the mandibular processes. Each pair fuses in the midline to make up the face. In birds, a beak appears from about 5.5 days, the upper part from the maxillary and the lower part from the mandibular processes. The outer **epidermis** of the embryo is often called **ectoderm** for several days because of its undifferentiated appearance, but this is a misnomer as it is no longer able to form neural tissues after the primitive streak stage. In the chick, the feather germs start to appear in the epidermis from about 6.5 days.

From 3 to 4 days the posterior extremity of the embryo consists of a **tailbud**. This consists of a juxtaposition of the various axial tissue types: the notochord, neural tube, somites, and hindgut. In the chick it is only responsible for producing the most posterior four to five somites, although in other vertebrates it can produce many more. Hatching of the chick occurs on the 20th or 21st day.

Nervous system

The early brain is shown in Fig. 9.12. The three primary brain vesicles, visible from the second day, are the **forebrain**, **midbrain**, and **hindbrain**. The anterior part of the forebrain is the **telencephalon**, later forming the cerebral hemispheres, and the posterior part is the **diencephalon**, which produces the optic vesicles. The midbrain later forms the **optic tecta**, which are the receptive areas for the optic nerves. The hindbrain forms the **cerebellum**, controlling the body's movements, and the medulla oblongata, site of control centers for various vital functions. The remainder of the neural tube forms the spinal cord. Ten pairs of **cranial nerves** leave the brain to innervate various muscles and sense organs, and the spinal cord produces pairs of spinal nerves in between the vertebrae.

The dorsal part of the neural tube gives rise to a migratory population of cells called the **neural crest**. In the head this contributes to the cranial nerve ganglia and a large proportion of the skeleton of the skull. In the trunk it forms the spinal ganglia, autonomic ganglia, adrenal medulla, and pigment cells.

Further details of development of the nervous system will be found in Chapter 14.

Pharyngeal arch region

At the level of the hindbrain the body has an obviously segmental arrangement (Fig. 9.13). This is made up of elements from different germ layers. In the endoderm of the pharynx, paired lateral pouches develop. These are the famous "gill slits," or branchial clefts, which all vertebrate embryo possess in rudimentary form, although in the chick they do not become fully patent. The hindbrain itself is divided into seven **rhombomeres**. Rhombomere 1 becomes the cerebellum. Each pair of rhombomeres 2–7 produces the neural crest cells that migrate to form one cartilaginous **branchial arch** surrounding the pharynx. The first of these, lying anterior to the first cleft, is the mandibular arch, which subsequently becomes the mandibular and maxillary processes which form the lower half of the face. The second arch is called the hyoid arch. Each pair of rhombomeres also produces one cranial nerve running into its associated arch. The cranial ganglia are composed partly of cells from the neural crest and partly of the corresponding epibranchial placodes, which form in the adjacent epidermis. Each pharyngeal arch is associated with a vascular aortic arch connecting the ventral aorta

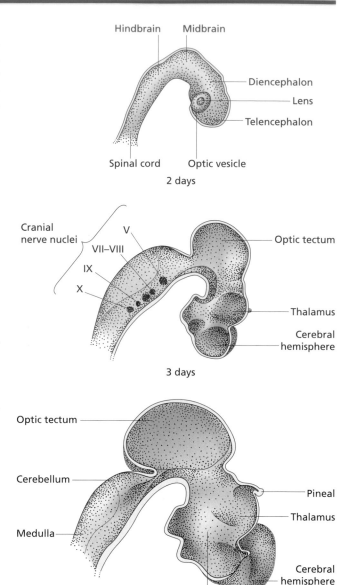

Fig. 9.12 Change in shape of the chick embryo brain from 2 to 4.5 days of development.

from the heart with the paired dorsal aortas (see below). In the endoderm, the thyroid forms in the ventral midline from second pouch tissue, and paired thymus and parathyroid rudiments form from the third and fourth pouch. This whole segmental arrangement is transitory, so the clefts are only open for a short time and the aortic arches are not all functional simultaneously.

Heart and circulation

The heart originates at the hindbrain level from paired endothelial condensations between the splanchnic mesoderm and the

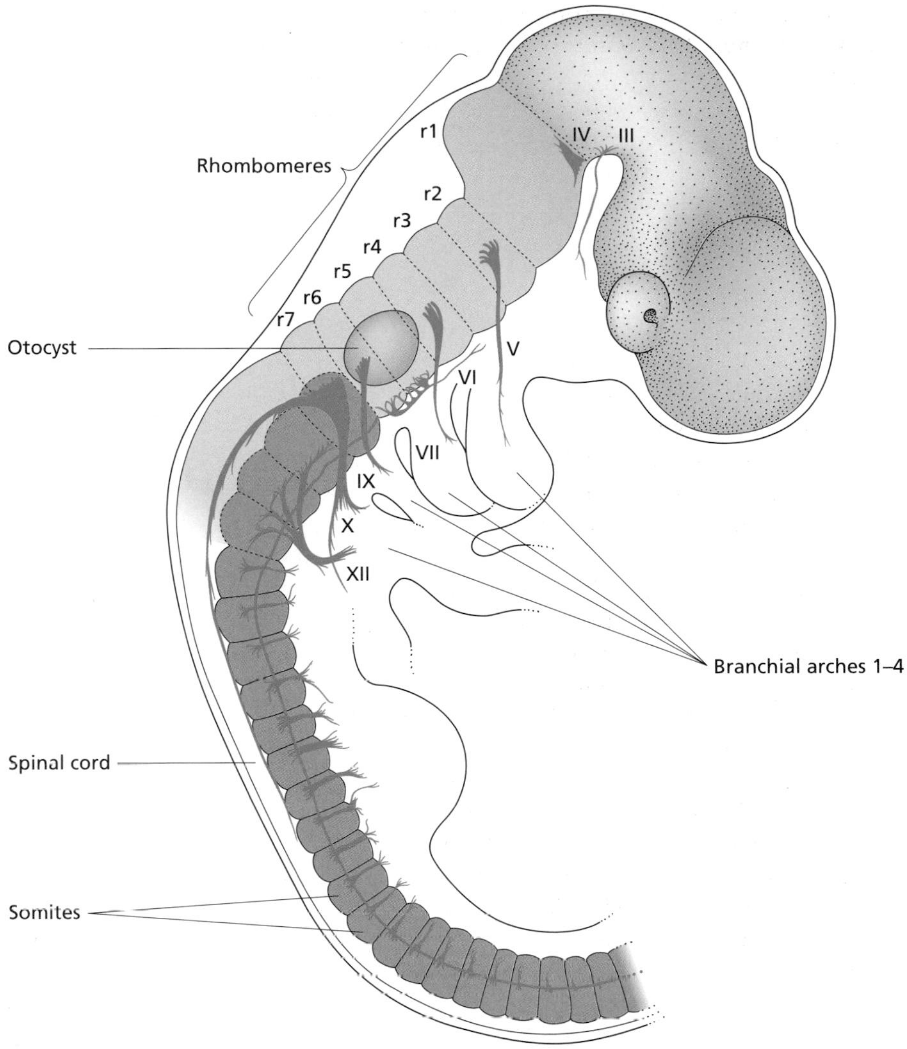

Fig. 9.13 Pharyngeal arch region of a 3-day chick embryo. The rhombomeres are labeled r1–r7 and the associated cranial nerves V–XII. Each pair of rhombomeres innervate one arch. The otocyst is at the level of r5–6. (After Lumsden 1991. *Philosophical Transactions of the Royal Society* **331**, 281–286. Figure 1.)

gut. These initially form separate tubes and then early on the second day, from about seven somites, they fuse in the ventral midline to form a single tube of **endocardium**. The **myocardium**, or muscular wall of the heart, originates from the adjacent splanchnic mesoderm. As the fusion is taking place, the heart moves posteriorly, following the progress of the anterior body-fold. By 48 hours the heart is a coiled tube consisting of, from posterior to anterior, sinus venosus, atrium, ventricle, and out-flow tract.

During the early stages of heart formation, a system of blood islands and blood vessels appears in the area opaca. The vessels grow into the area pellucida and join up with vessels arising from the mesoderm within the embryo. The heart starts to beat about the middle of the second day, and by the third day establishes a blood circulation between the embryo and the yolk mass. From the heart, blood flows into the short ventral aorta, through the aortic arches, initially one for each of the first three pharyngeal arches, into the dorsal aortas (Fig. 9.14). These are initially paired but become progressively united, from anterior to

posterior, into a single aorta. From these, blood flows to all parts of the embryo and also out of the embryo through paired vitelline arteries located at the level of the trunk. It then becomes oxygenated in the extraembryonic capillary bed and returns via the vitelline veins, which approach the embryo from both anterior and posterior directions, and, with the embryonic venous system, join up at the sinus venosus. As the body folds reduce the connection between the embryo and the yolk mass, the vitelline arteries and veins are moved together into the **umbilical tube**. Later on, from the sixth day, most of the blood flow becomes directed through the **allantois** (see Fig. 9.5), as this becomes the principal respiratory organ.

Mesoderm

Somites arise from the mesoderm on either side of the noto-chord, often called the **paraxial mesoderm**. From the end of day 1 the somites appear in anterior-to-posterior sequence. The

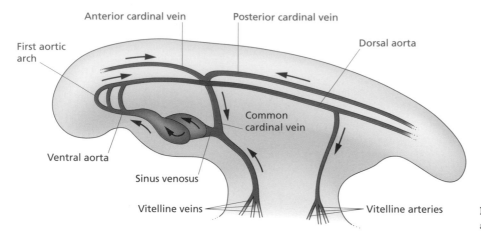

Fig. 9.14 The basic circulation of an amniote embryo.

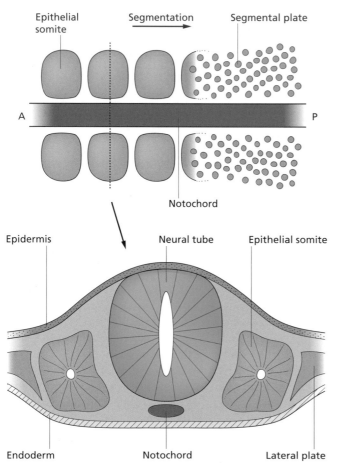

Fig. 9.15 Somitogenesis. Somites are formed sequentially from anterior (A) to posterior (P).

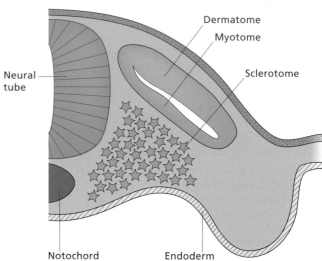

Fig. 9.16 Formation of dermatome, myotome and sclerotome from the somite.

the **sclerotome** and later becomes the vertebrae. The formation of vertebrae is out of phase with the somites, so each vertebra is formed from the posterior sclerotome of one somite and the anterior sclerotome of the next. The middle part of the somite is the **myotome**, which will form segmental striated muscles. The outer part is the **dermatome**, which will contribute to the dermis of the skin. In the chick, the most anterior two somites disperse shortly after their formation and the next four are occipital somites, which contribute to the occipital part of the skull rather than forming vertebrae.

The kidney develops from the intermediate mesoderm (Fig. 9.17). First the **pronephros** develops on day 2–3 at the level of the seventh to 15th somite. A nephric duct grows posteriorly from the pronephric area down to the cloaca. In fish and amphibians the pronephros is functional, but in amniotes it soon degenerates as a **mesonephros** appears from the 16th to 27th somite. In the chick this develops on the third and fourth

visible event of segmentation corresponds to a transition of each somite from a mesenchymal morphology to epithelial spheres of tightly apposed cells (Fig. 9.15). Somites continue to form until there are 45 by 4 days. They will later form three types of structure (Fig. 9.16). The inner part, flanking the notochord, is called

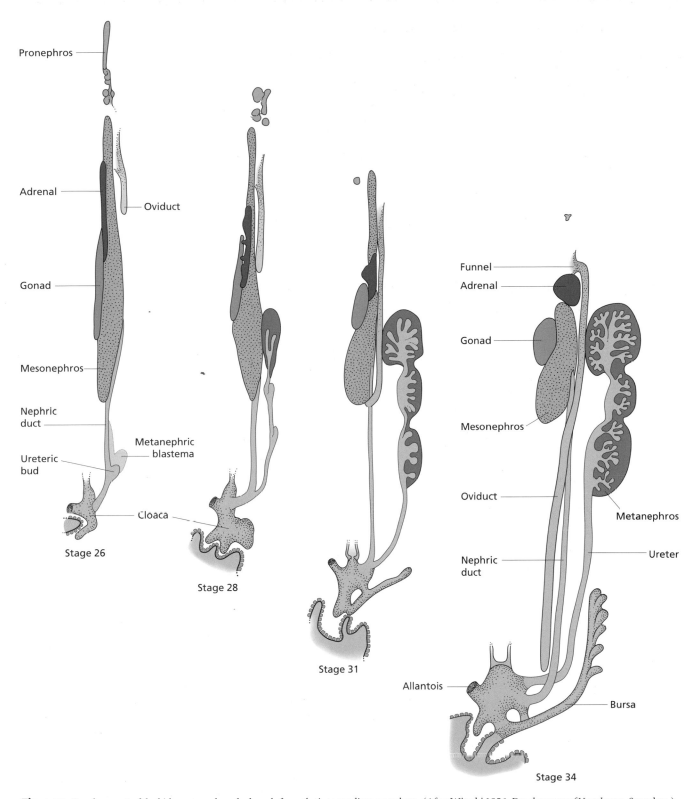

Fig. 9.17 Development of the kidney, gonads and adrenals from the intermediate mesoderm. (After Witschi 1956. *Development of Vertebrates.* Saunders.)

New Directions in Research

The early chick offers a good opportunity to study the as yet mysterious property of embryonic regulation. In most embryo types regulation requires the persistence of the signaling centers but in the chick the main signaling center, Hensen's node, can itself be replaced following extirpation.

The later chick embryo will continue to provide material for research on organogenesis, because of the ease of micromanipulation. Research programs often combine microsurgical experiments on the chick with the use of tissues from knockout mouse lines.

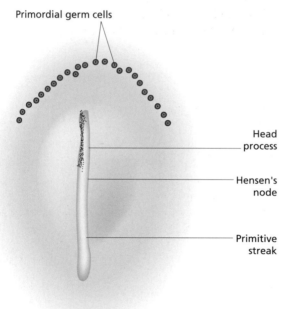

Fig. 9.18 Extraembryonic origin of the primordial germ cells.

day, and consists of glomeruli and tubules that attach to the nephric duct. The mesonephros is the functional kidney in birds during embryonic life. It is superseded after hatching by the **metanephros**. This develops from about 5 to 8 days from the posterior end of the intermediate mesoderm. A branch from the nephric duct, the **ureteric bud**, grows into the neighboring mesoderm and provokes the formation of the metanephric tubules in a mesenchymal to epithelial transition.

The strip of intermediate mesoderm ventromedial to the kidney forms the adrenal gland and the gonads. The adrenal gradually becomes a compact body over days 4–8, with the outer cortex formed from intermediate mesoderm and the inner medulla formed from the neural crest. The gonads arise both from the mesenchyme and from the overlying coelomic epithelium, which together form a protrusion into the coelom called the **genital ridge**. The **primordial germ cells** arise from an anterior extraembryonic position (Fig. 9.18) and enter the germinal ridge after a long migration. On the fourth day the gonads are still of similar appearance in males and females, but subsequently they differentiate as testes or ovaries, respectively. The region of mesoderm forming the aorta, gonads, and mesonephros is sometimes called the **AGM region** and is important in hematopoiesis.

The limbs appear on day 3 as buds formed from the lateral plate mesoderm and overlying epidermis. They elongate, and from the fourth day start to differentiate in proximal to distal sequence. In birds the forelimb buds become the wings and the hindlimb buds the legs.

Key Points to Remember

- The chick is an amniote with a generally similar morphology to mammalian embryos.
- Early development occurs as a flat blastoderm. The primitive streak is induced at the posterior margin and elongates to the anterior. During gastrulation cells pass through the streak and become the major body parts of the embryo.
- The chick possesses a set of extraembryonic membranes needed to support the embryo and to transfer nutrients and oxygen to the embryo. These are the yolk sac, the chorion, the amnion, and the allantois.
- The fate map of the primitive streak shows that the anteroposterior axis of the streak becomes the mediolateral axis of the primary body plan.
- The sequence of inductive interactions is comparable to *Xenopus*. The induction of the primitive streak corresponds to mesoderm induction in *Xenopus* and the behavior of Hensen's node is similar to the *Xenopus* organizer. However some of the molecular components underlying these processes may be different.
- The formation of left–right asymmetry of the embryo depends on the expression of nodal on the left side of the axis.
- The chick is very important in research on organogenesis (see later chapters).

Further reading

General

Hamburger, V. & Hamilton, H.L. (1951) A series of normal stages in the development of the chick embryo. *Journal of Morphology* **88**, 49–92. Reprinted in *Developmental Dynamics* **195**, 231–272 (1992).

Eyal-Giladi, H. & Kochav, S. (1976) From cleavage to primitive streak formation: a complementary normal table and a new look at the first stages of the development of the chick. *Developmental Biology* **49**, 321–337.

Bellairs, R. & Osmund, M. (1997) *The Atlas of Chick Development*. London: Academic Press.

Brown, W.R.A., Hubbard, S.J., Tickle, C. & Wilson, S.A. (2003) The chicken as a model for large scale analysis of vertebrate gene function. *Nature Reviews Genetics* **4**, 87–98.

Stern, C., ed. (2004) The chick in developmental biology. *Mechanisms of Development* **121** (special issue 9).

Fate maps

Psychoyos, D. & Stern, C.D. (1996) Fates and migratory routes of primitive streak cells in the chick embryo. *Development* **122**, 1523–1534.

Fernández-Garre, P., Rodríguez-Gallardo, L., Gallego-Diaz, V., Alvarez, I.S. & Puelles, L. (2002) Fate map of the chicken neural plate at stage 4. *Development* **129**, 2807–2822.

Lawson, A. & Schoenwolf, G.C. (2003) Epiblast and primitive streak origins of the endoderm in the gastrulating chick embryo. *Development* **130**, 3491–3501.

Early development

Lemaire, L. & Kessel, M. (1997) Gastrulation and homeobox genes in chick embryos. *Mechanisms of Development* **67**, 3–16.

Bachvarova, R.F., Skromne, I. & Stern, C.D. (1998) Induction of primitive streak and Hensen's node by the posterior marginal zone in the early chick embryo. *Development* **125**, 3521–3534.

Joubin, K. & Stern, C.D. (1999) Molecular interactions continuously define the organizer during the cell movements of gastrulation. *Cell* **98**, 559–571.

Smith, J.L. & Schoenwolf, G.C. (1998) Getting organized: new insights into the organizer of higher vertebrates. *Current Topics in Developmental Biology* **40**, 79–110.

Wilson, S.I., Rydström, A., Trimborn, T., et al. (2001) The status of Wnt signaling regulates neural and epidermal fates in the chick embryo. *Nature* **411**, 325–330.

Faure, S., de Santa Barbera, P., Roberts, D.J. & Whitman, M. (2002) Endogenous patterns of BMP signaling during early chick development. *Developmental Biology* **244**, 44–65.

Stern, C.D. (2002) Induction and initial patterning of the nervous system – the chick embryo enters the scene. *Current Opinion in Genetics and Development* **12**, 447–451.

Left–right asymmetry

Levin, M. (1998) Left–right asymmetry and the chick embryo. *Seminars in Cell and Developmental Biology* **9**, 67–76.

Capdevila, J., Vogan, K.J., Tabin, C.J. & Izpisua-Belmonte, J.C. (2000) Mechanisms of left–right determination in vertebrates. *Cell* **101**, 9–21.

Mercola, M. & Levin, M. (2001) Left–right asymmetry determination in vertebrates. *Annual Reviews in Cell and Developmental Biology* **17**, 779–805.

Organogenesis, general

Also see references in Chapters 14, 15, and 16

Witschi, E. (1956) *Development of Vertebrates*. Philadelphia: W.B. Saunders.

Balinsky, B.I. & Fabian, B.C. (1981) *An Introduction to Embryology*, 5th edn. Philadelphia: W.B. Saunders.

Hildebrand, M. (1995) *Analysis of Vertebrate Structure*, 4th edn. New York: John Wiley.

Kardong, K.V. (2002) *Vertebrates: comparative anatomy, function, evolution*. New York: McGraw-Hill.

Chapter 10

The mouse

Unlike the other model organisms considered here, mammals are **viviparous** and the resulting inaccessibility of the post-implantation stages of development means that microsurgical procedures are used less than in *Xenopus*, zebrafish, or chick. For this reason the developmental biology of the mouse has depended to a much greater extent on genetic manipulation (transgenesis or knockout). There are many laboratory strains of mice that have been inbred to almost total homozygosity at all loci and each have their own advantages and disadvantages for different types of experiment. Each strain has a characteristic coat color (e.g. black, albino, agouti), and in experiments involving embryos of more than one strain these differences can serve as a visible indication of the genetic constitution.

Mice mate in the night and so the age of the embryos is often expressed as days and a half, for example a 7.5-day embryo (= E7.5 embryo) is recovered on the eighth day after the mice mated. A solid white deposit, or "plug," that is formed in the female vagina after mating allows determination of the mating night. There is a stage series by Theiler, in which stages 1–5 are preimplantation, 6–14 are early postimplantation, comprising body plan formation and turning, and 15–27 cover organogenesis and fetal growth phases up to birth, which occurs at about 20 days after fertilization.

Ovulation occurs a few hours after mating and fertilization takes place at the upper end of the oviduct. For the early, pre-implantation, stages the embryos are located first in the oviduct and then the uterus of the mother. During this time they can be collected and kept *in vitro* in reasonably simple media and it is possible to do microsurgical manipulation of these early stages, including those required to make transgenics and knockouts. In order to turn a modified preimplantation embryo into a late embryo or into an adult mouse, it must be transplanted into the uterus of a **foster mother** which has been made **pseudopregnant**, and thus receptive to the embryos, by previous mating with a vasectomized, and therefore sterile, male. So long as the transplanted embryos implant, they should develop and be born in the normal way.

At late stages it is possible, as in the chick, to explant individual organ or tissue rudiments from embryos into *in vitro* culture, where they are accessible to manual intervention. The scope of these experiments is greatly increased in the mouse by the ability to use tissues from transgenic and knockout strains.

Because of their viviparity, mammalian embryos have a considerable external nutrient supply and undergo extensive growth during development. In this respect the

Classic Experiments

The main contribution of the mouse to developmental biology has been through transgenic and knockout technology. These techniques are now fundamental to virtually every investigation of early development or organogenesis. They have also been used to create many mouse models for human diseases which are used to investigate both the molecular mechanisms of the disease and to test potential therapies.

First transgenesis
Gordon, J.W., Scangos, G.A., Plotkin, D.J., Barbosa, J.A. & Ruddle, F.H. (1980) Genetic-transformation of mouse embryos by micro-injection of purified DNA. *Proceedings of the National Academy of Sciences USA* **77**, 7380–7384.

Discovery of ES cells
Evans, M.J. & Kaufman, M.H. (1981) Establishment in culture of pluripotential cells from mouse embryos. *Nature* **292**, 154–156.
Martin, G.R. (1981) Isolation of a pluripotent cell line from early mouse embryos cultured in a medium conditioned by teratocarcinoma cells. *Proceedings of the National Academy of Sciences USA* **78**, 7634–7638.

Invention of basic knockout procedure
Thomas, K.R. & Capecchi, M.R. (1987) Site directed mutagenesis by gene targeting in mouse embryo-derived stem cells. *Cell* **51**, 503–512.
Mansour, S.L., Thomas, K.R. & Capecchi, M.R. (1988) Disruption of the protooncogene int-2 in mouse embryo-derived stem cells: a general strategy for targeting mutations to non-selectable genes. *Nature* **336**, 348–352.

mouse resembles the chick and differs from *Xenopus* and the zebrafish.

Mammalian fertilization

From a biological standpoint, the fertilization mechanism needs to bring the male and female gametes together in a productive union, while avoiding both cross-species fertilization (**hybridization**) and fertilization of the egg by multiple sperm (**polyspermy**). Fertilization is also a topic of great practical importance in human reproduction. The mechanisms have been studied in a variety of animal models, including especially marine invertebrates such as the sea urchin. However, the molecular mechanisms of fertilization are much more diverse than those of other developmental processes, and it has turned out that there are rather few features in common between the sea urchin and mammals. So in this case the value of invertebrate models for building a general picture is somewhat reduced. To maintain coherence the following account relates just to the mouse, although most features apply also to other mammals. Some of the evidence depends on the use of mice in which specific genes have been "knocked out," or "knocked in" (i.e. replacing an endogenous gene). The techniques for making such strains are described later in the chapter.

The sperm is a highly specialized cell (Fig. 10.1). The nucleus is haploid and the DNA is highly condensed with protamines instead of histones making up much of the protein content of the chromatin. In front of the nucleus lies a large Golgi-like body called the **acrosome**. Behind the nucleus lies a centriole, a mid-piece rich in mitochondria, and the tail which is a reinforced flagellum having the usual 9+2 arrangement of microtubules. The swimming movements of the sperm are driven in an ATP-dependent process by dynein arms which are attached to the microtubules.

In most mammals the sperm are not capable of fertilization immediately after release. They need to spend a period in the female reproductive tract during which they become competent to fertilize, a process known as **capacitation**. The details of capacitation are not fully understood but it can be brought about *in vitro* in simple synthetic media containing albumin, calcium, and bicarbonate. One element of capacitation is the loss of cholesterol, as a medium rich in cholesterol will inhibit capacitation. It is thought that the loss of cholesterol makes the membrane permeable to the Ca^{++} and HCO_3^-, which can directly activate an adenylyl cyclase, resulting in the production of cAMP and activation of protein kinase A. This has various consequences including an increase of membrane potential from about −30 to −50 mV, which may make the subsequent opening of Ca^{++} channels easier. Capacitation also involves the loss of glycoproteins that prevent the sperm–zona interaction (see below).

The "egg" of most mammals, including the mouse and the human, is strictly speaking an **oocyte** arrested in second meiotic metaphase. It is released from the ovary, together with ovarian follicle cells (**cumulus** cells), as an oocyte–cumulus complex. The oocyte itself is surrounded by a transparent layer of extracellular material called the **zona pellucida**, secreted by the follicle cells. Outside this lie cumulus cells, which are embedded in an extracellular matrix rich in hyaluronic acid. The oocyte–cumulus complex is picked up by the funnel (infundibulum) at the entrance to the **oviduct**, a process which depends on adhesion of the cilia of the infundibulum to the extracellular matrix of the complex. It is then "churned" to compress the matrix and allow it through the narrow neck (ostium) into the oviduct itself.

Although chemotaxis of sperm towards eggs occurs in some other types of animal, the evidence for this in mammals is not clearcut. The transport of the sperm certainly depends to some extent on muscular movements of the female reproductive tract which assist its passage up the vagina, through the uterus and into the oviducts. The main steps of sperm–egg interaction are shown in Fig. 10.2. The sperm carry a membrane-bound hyaluronidase that assists their passage through the extracellular matrix of the oocyte–cumulus complex. The next stage is the binding of sperm to the zona pellucida and this is the stage at which species specificity is controlled. If the zona is removed, then cross-species fertilization is possible, and this is the basis for the routine assay of the effectiveness of human sperm using hamster eggs from which the zona has been removed. The zona is composed of three glycoproteins called ZP1, ZP2, and ZP3, which share a common "ZP" peptide sequence motif at the C-terminus. Of these, ZP3, with its specific O-linked oligosaccharide, is the specific sperm receptor. Low concentrations of ZP3, or the oligosaccharide alone, will prevent sperm-zona binding. Female mice in which the gene for ZP3 has been knocked out will form normal oocytes but they do not have a zona and the mice are infertile. If the human ZP3 gene is "knocked in" to replace the mouse gene then zona formation and fertility is restored. Interestingly, eggs from these mice do not acquire a

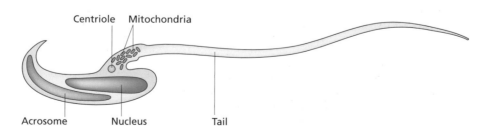

Fig. 10.1 Diagram of a mouse sperm.

Centriole   Mitochondria

Acrosome   Nucleus   Tail

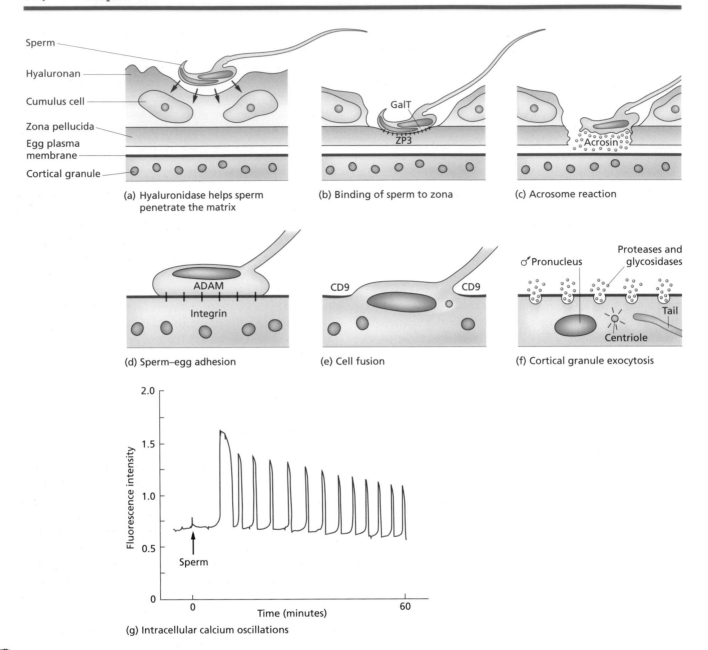

(a) Hyaluronidase helps sperm penetrate the matrix

(b) Binding of sperm to zona

(c) Acrosome reaction

(d) Sperm–egg adhesion

(e) Cell fusion

(f) Cortical granule exocytosis

(g) Intracellular calcium oscillations

Fig. 10.2 Events of fertilization in the mouse.

competence to be fertilized by human sperm. This is because the species specificity resides in the carbohydrate attached to the ZP3 polypeptide, and this has the murine structure even though the polypeptide is of human origin, because it is assembled by the murine glycosyl transferases.

The recognition protein for ZP3 on the sperm is a cell surface β-1,4-galactosyl transferase (GalT) and binding of ZP3 to this provokes the acrosome reaction. The evidence for this is that GalT will bind tightly to the ZP3 oligosaccharide, and that sperm from mice in which the GalT gene is knocked out show reduced binding to the zona and are not provoked into the acrosome reaction. The acrosome reaction is a rapid exocytosis of the acrosomal vesicle and the released products are needed for the

later events of fertilization. The coupling between GalT and the exocytosis is thought to proceed by the sequence: G-protein activation, less negative membrane potential, opening of voltage-gated Ca^{++} channels which cause a rise in intracellular Ca^{++}. There is also a rise in intracellular pH. GalT itself is a single pass membrane protein with a G-protein activation domain in the cytoplasm. If GalT is expressed in *Xenopus* oocytes, then they will bind ZP3, and ZP3 will stimulate G-protein activation and activation events such as cortical granule exocytosis. Here the *Xenopus* oocyte is used simply as an experimental test system, because it is a large cell into which it is easy to inject mRNA. The fact that it is an oocyte is not relevant since the normal location of GalT is the sperm.

The materials released by the acrosome reaction include hydrolytic enzymes, such as the serine protease acrosin, that help digest a path through the zona and enable the sperm to reach the egg surface. At this stage there is a second recognition process. The sperm–egg recognition is thought to be carried out by ADAM proteins on the sperm binding to integrins on the egg surface. ADAM stands for a disintegrin and metallprotease domain, and proteins of this class will bind tightly to integrins. The sperm contains three ADAM proteins: fertilin α, fertilin β, and cyritestin. All probably play some role in the interaction. Peptides from fertilin β will block sperm–egg binding, and sperm from mice in which the genes for fertilin β or cyritestin have been knocked out show greatly reduced fertility. The best candidate for the target integrin on the egg is integrin α6 because binding is prevented by a monoclonal antibody to this protein. However as the knockout mouse for this integrin has normal fertility there must be some other components also involved which have a redundant function. The binding of sperm to egg is the first stage in fusion of the plasma membranes, which occurs at a domain on the side of the sperm head. Fusion requires a four-pass membrane protein on the egg called tetraspanin or CD9. Evidence for this is that female knockout mice for CD9 show no fusion and are infertile; but the fusion ability of the eggs can be restored by injection of CD9 mRNA. In the course of fusion the whole sperm, including the tail, enters the egg.

Cell fusion causes a rise of intracellular calcium ion concentration which is responsible for all the subsequent events of fertilization. This calcium rise is also the aspect of fertilization that does seem to be universal across the animal kingdom. In many mammals including the mouse the initial calcium spike is followed by a series of others making up an oscillatory pattern over several hours (Fig. 10.2f). Calcium transients of this sort are observed by loading the eggs with calcium-sensitive reagents such as the phosphorescent protein aequorin or the fluorescent dye fura2 and measuring the resulting light emission or fluorescence. It is thought that the Ca^{++} release is caused by activation of the inositol trisphosphate (IP_3) pathway by a specific phospholipase C introduced by the sperm. The evidence for this is as follows:

1 Injection of IP_3 will provoke Ca^{++} release.
2 Inhibitors of phospholipase C or of IP_3 receptor will inhibit Ca^{++} release.
3 Injection of whole sperm, or demembranated sperm heads, or sperm extracts, will provoke Ca^{++} release.
4 Sperm contains a specific phospholipase C (PLCζ) which itself will cause Ca^{++} release.
5 Immunodepletion of this PLCζ from sperm extract will abolish activity.

Injection of Ca^{++}, or treatment with calcium ionophore, that allows entry of Ca^{++} from the medium, will both cause the same events of egg activation as fertilization by sperm. These events comprise the exocytosis of **cortical granules**, the completion of the second meiotic division, the resumption of DNA synthesis, and a general metabolic activation (Fig. 10.2g). The cortical granules are found just under the plasma membrane and their contents include glycosidases and proteases that modify the zona receptors so that they can no longer bind sperm. In the mouse this process is the main factor preventing polyspermy. The completion of the second meiotic division results in expulsion of the **second polar body** containing the surplus chromosomes. The sperm nucleus itself decondenses, assisted by reduction of protamine disulfide bonds by the – SH containing peptide glutathione in the egg. In addition to the nucleus, the sperm also introduces some mitochondria and a centriole. The mitochondria degenerate and do not participate in later development, but the centriole becomes the **microtubule organizing center** for the sperm **aster**, and later divides to form the first mitotic spindle. The two pronuclei migrate slowly towards each other and undergo DNA replication. In mammals they do not fuse to form a true zygote nucleus, instead the pronuclear envelopes break down as they meet, and the chromosomes become aligned on the mitotic spindle ready for the first cleavage. Athough a variety of treatments that elevate intracellular calcium can cause egg activation, such eggs are **parthenogenetic** (i.e. contain no paternal nucleus) and because of this they cannot develop far (see imprinting below).

Normal embryonic development

Preimplantation

The course of preimplantation development is shown in Fig. 10.3. Following fertilization the first few cleavages are very slow, and in contrast to *Xenopus* and the zebrafish, expression of the zygotic genome commences as early as the two-cell stage. The first cleavage occurs after about 24 hours and the second and third cleavages, which are not entirely synchronous, follow at intervals of about 12 hours. This slow tempo of early development may be related to the time required for the uterus to prepare for implantation. In the early eight-cell stage the shapes of individual cells are still clearly visible but they cease to be visible when the whole embryo acquires a more nearly spherical shape in a process called **compaction** (Fig. 10.3d). This consists of a flattening of blastomeres to maximize intercellular contacts and is mediated by the calcium-dependent adhesion molecule E-cadherin (also called leukocyte cell adhesion molecule (L-CAM) or uvomorulin). At this stage the cells become polarized in a radial direction. This is most obviously apparent from the appearance of **microvilli** on the outer surfaces, but it also involves a variety of changes in the cell interior. **Gap junctions** are also formed at this stage and allow diffusion of low molecular weight substances throughout the embryo.

The embryo is called a **morula** from compaction until about the 32-cell stage. During this period, desmosomes and tight junctions appear, creating a permeability seal between the inside and outside of the embryo, and a fluid-filled **blastocoel** begins to form in the interior. This is about 3 days after fertilization and around the time that the embryo moves from oviduct to uterus. The cavity expands the embryo into a **blastocyst** which

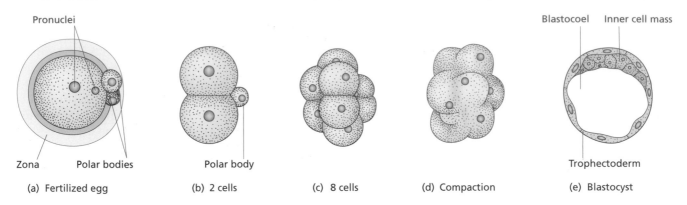

Fig. 10.3 Preimplantation development. The zona remains present but is not shown in (b) to (e).

consists of an outer cell layer of epithelial morphology called the **trophectoderm** and a clump of cells attached to its interior, the **inner cell mass (ICM)** (Fig. 10.3e). At the 60-cell stage, about one-quarter of the cells are found in the ICM and three-quarters in the trophectoderm. The ICM expresses a POU domain transcription factor Oct4, and secretes FGF4. The trophectoderm contains an FGF receptor, FGFR2.

From E3.5 to E4.5 both the ICM and the trophectoderm diversify (Fig. 10.4a). The ICM delaminates a layer of **primitive endoderm** on its blastocoelic surface. This layer contributes to extraembryonic tissues but not to the **definitive endoderm** of the embryo itself. The trophectoderm becomes divided into a polar component, which overlies the ICM, and a mural component which makes up the remainder. While the polar trophectoderm continues to proliferate, the mural trophectoderm becomes transformed into **polyploid** giant cells in which the DNA continues to be replicated but without mitosis.

Early postimplantation stages

At about this stage the embryo hatches from the zona and becomes implanted in a uterine crypt. The uterus is only competent to receive embryos during a short period about 4 days from mating. The uterus is attached to the body wall by a membrane called the **mesometrium**, which carries the uterine blood vessels. When the embryos implant, they are orientated such that the ICM lies away from the mesometrial side of the uterus, which is the side on which the placentas will form.

The course of early postimplantation development is shown in Fig. 10.4b and c. After implantation the trophectoderm becomes known as the **trophoblast** and soon stimulates proliferation of the connective tissue of the uterine mucosa to form a **deciduum** (plural **decidua**). From this stage onward the embryo receives a nutrient supply from the mother and can begin to grow in size and weight. As in the chick, the zygote does not just form an embryo but also a complex of extraembryonic membranes and the whole is referred to as the **conceptus**.

The next stage of development is known as the **egg cylinder**. The cylinder itself can be regarded as homologous to the area

pellucida of the chick embryo and consists of an "upper" layer of **primitive ectoderm** or **epiblast**, homologous to the chick epiblast, and a "lower" layer of **primitive endoderm**, homologous to the chick hypoblast. The terms "upper" and "lower" are in quotes because the embryo is actually the shape of a deep cup with the epiblast on the inside and the endoderm on the outside. It appears U shaped in a sagittal section or O shaped in a transverse section. Cells derived from the primitive endoderm move out to cover the whole inner surface of the mural trophectoderm and start to secrete an extracellular basement membrane known as Reichert's membrane containing laminin, entactin, and type IV collagen. These cells are called the **parietal endoderm**. The remainder of the primitive endoderm remains epithelial and forms a layer of **visceral endoderm** around the epiblast. The cells of this layer somewhat resemble the later fetal liver being characterized by the synthesis of α-fetoprotein, transferrin, and other secreted proteins. In addition to the epiblast, the inside of the egg cylinder also contains extraembryonic ectoderm derived from the polar trophectoderm. This extraembryonic region has now become the ectoplacental cone and as it proliferates it produces further layers of giant cells which move around and reinforce the trophoblast. In contrast to the differentiated extraembryonic tissues, the primitive ectoderm from which the entire embryo will be derived remains visibly undifferentiated.

At about E6.5 the anteroposterior axis of the future embryo becomes apparent with the formation of the **primitive streak** at one edge of the epiblast. The streak marks the posterior end of the future embryonic axis which will extend across the ectoderm toward the distal tip of the cup. It is a region of cell movements similar to that found in the chick embryo and these movements result in the formation of the **definitive endoderm** and mesoderm. By this stage the egg cylinder has become somewhat compressed so that the longer axis, as seen in transverse section, coincides with the anteroposterior axis of the embryo. The streak elongates until it has reached the distal tip of the egg cylinder and the **node** appears at its anterior end. This is homologous to Hensen's node in the chick and consists of two cell layers while the remainder of the streak has three. By E7.5 a head process appears anterior to the node consisting of a forming

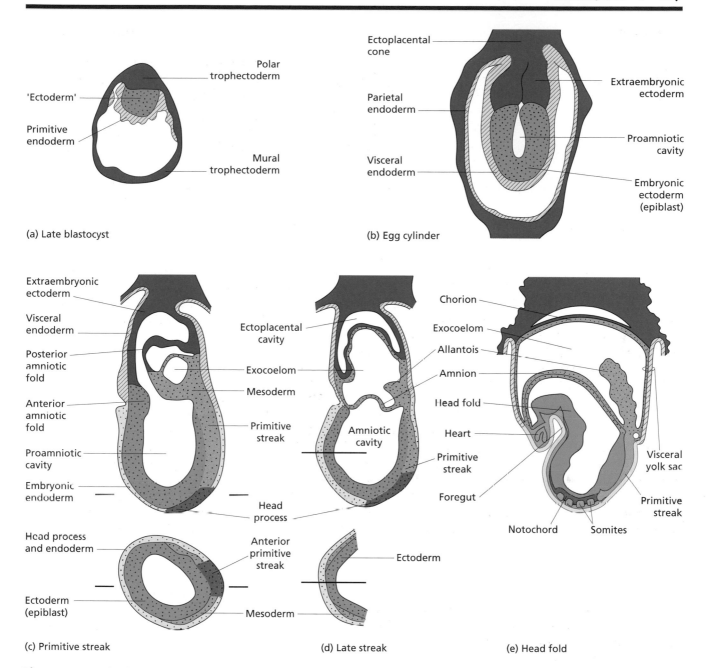

Fig. 10.4 Peri- and early postimplantation development. (c) and (d) show both sagittal and transverse sections at levels indicated by the lines.

notochord flanked by definitive endoderm in the lower layer and **neural plate** in the upper layer. As in the chick the node then moves posteriorly and the axial parts appear in anteroposterior sequence in its track. By E8.5 the embryo has elongated somewhat in length and a massive head fold has formed at the anterior end, mainly composed of the anterior neural tube. The **somites** begin to form from E8 in anteroposterior sequence, at the rate of about one somite per 1.5 hours.

Although at first sight the morphology of the mouse embryo and the lower vertebrates seems rather different, the homology of parts is clearly displayed by the gene expression patterns during gastrulation. The primitive streak and notochord express *brachyury* (usually called *T* in the mouse), with the posterior

part expressing *cdx* genes. The node expresses *goosecoid*. The node, notochord, and floor plate express *FoxA2* (*hnf3β*). As far as inducing factors are concerned *nodal* is expressed in the node, *bmp4* in the surrounding ectoderm, and the BMP-inhibitor *follistatin* in the streak (Fig. 10.5).

The amniotic fold forms at about 7 days as an outpushing of the ectoderm and mesoderm at the junction of posterior primitive streak and extraembryonic ectoderm (Fig. 10.4c,d). The side of this fold nearer the embryo becomes the **amnion** and the side nearer the ectoplacental cone becomes the **chorion**. The fold pushes across the proamniotic space and divides it into three: an amniotic cavity above the embryo, an exocoelom separating amnion and chorion, and an ectoplacental cavity lined with

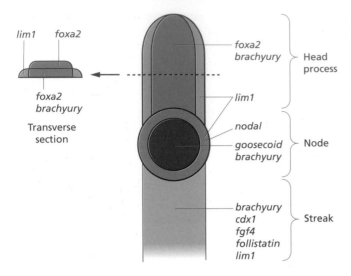

Fig. 10.5 Expression of genes at head process stage.

extraembryonic ectoderm. Into the exocoelom grows the **allantois**, consisting of mesoderm from the posterior end of the primitive streak. This grows to contact the chorion and later forms the embryonic blood vessels of the placenta. The placenta itself forms from the ectoplacental cone region which is directed toward the mesometrium. From the maternal side the mature placenta consists of the following layers: maternal decidual tissues; giant trophoblast cells with maternal blood sinuses; diploid trophoblast with fetal blood vessels; and finally crypts lined with extraembryonic endoderm (Fig. 10.6). The placenta is not only important as the organ supplying nutrients to the fetus, but also has endocrine functions, producing progesterone and estrogen to maintain the pregnancy as well as a number of other important hormones.

Around 8.5 days a most remarkable process takes place known as **turning**, which brings the germ layers into the proper orientation within the embryo. This is best described as a rotation of the embryo around its own long axis. It starts as a U-shaped structure with the dorsal side concave and the rotation

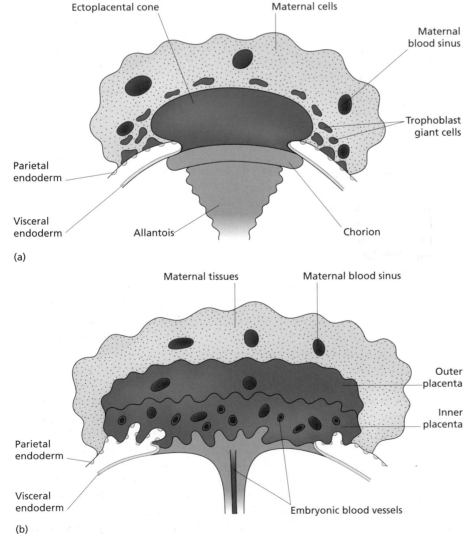

Fig. 10.6 Schematic view of placenta: (a) E8.5; (b) E14.5.

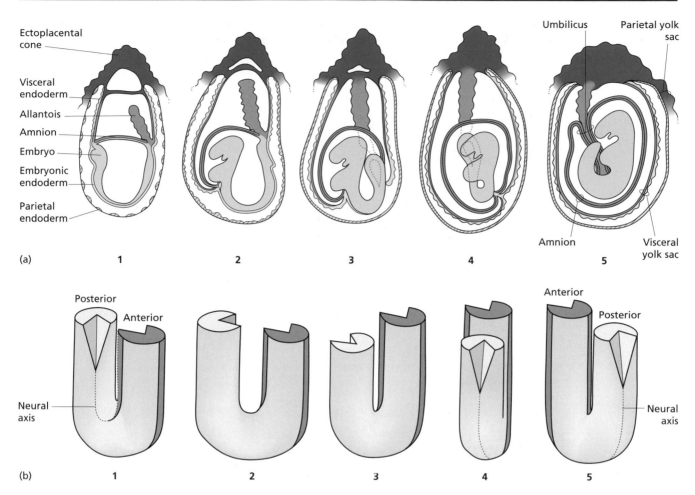

Fig. 10.7 Turning. (a) From about E7.5 to E9.5 the embryo becomes rotated around its long axis leading to ventral closure of the gut. (b) Diagrammatic representation of turning. (Both after M. Kaufman. In: Copp & Cockroft (eds) 1990. Postimplantation Mammalian Embryos. IRL Press, pp. 88–9.)

converts it into an inverted U with the dorsal side convex (Fig. 10.7). This change of orientation has drastic consequences for the arrangement of the extraembryonic membranes. The amnion becomes enlarged so that it surrounds the whole embryo instead of just covering the dorsal surface. The membrane lining the exocoelom, which consists of visceral endoderm and mesoderm, also becomes stretched to cover the entire embryo, and becomes known as the visceral yolk sac. The final arrangement of membranes after turning has the amnion on the inside, the visceral yolk sac next and the parietal yolk sac, formed from the trophectoderm lined with parietal endoderm, on the outside. Turning leads to a rapid closure of the midgut, which starts as a large area of endoderm exposed on the ventral side and becomes constricted to a small **umbilical tube** containing the vitellointestinal duct, the vitelline vessels, and the allantois.

Organogenesis

By about E9.5, when the axis has formed, the embryo has turned, and the gut has closed, the embryo has reached the junction between the phase of body plan formation and of **organogenesis**. Organogenesis is in most respects very similar to that in the chick so only a brief sketch is given here with Fig. 10.8 showing exterior views of the embryo. Further details of organogenesis will be found in Chapters 14–16.

Neural tube closure takes place simultaneously with turning. It commences in the hindbrain and proceeds both anteriorly and posteriorly, with the anterior neuropore closing at E9 and the posterior neuropore at E10–10.5. The **optic vesicles** form at E9.5 and the lens has been incorporated into the eye by E11.5. The **neural crest** emerges from the neural tube over the period E8.5 to E10.5. As in the chick, it forms the skeletal structures of the head, the dorsal root ganglia, Schwann cells, sympathetic ganglia, pigment cells, adrenal medulla, and enteric ganglia. **Somites** continue to form until E14 by which time there are about 65, many being in the tail which is much longer than that of the chick. As in other vertebrates, the somites form the vertebrae and myotomes and contribute to the dermis. The paired heart primordia fuse at E8.5 and the heartbeat begins at E9. The left and right atria become separated at E11.5 and the ventricles at E14. The gut originates from fore- and hindgut pockets as in

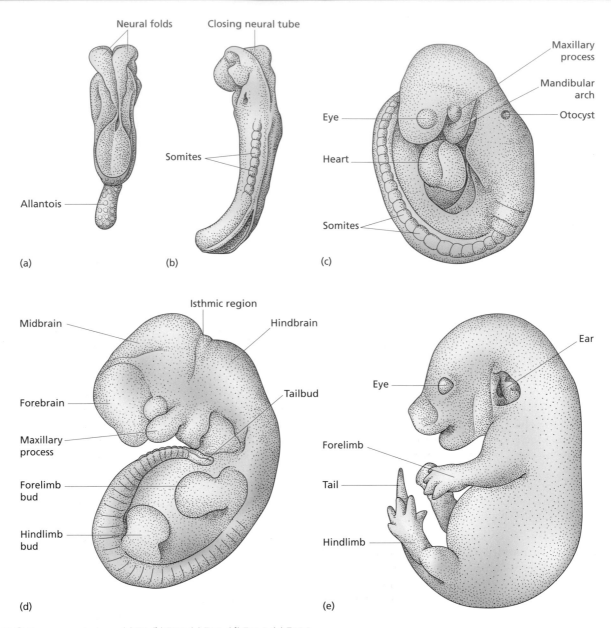

Fig. 10.8 Organogenesis stages: (a) E8; (b) E8.5; (c) E9.5; (d) E11.5; (e) E15.5.

the chick, although the closure of the midgut is faster because it is driven by the turning process. There are six **pharyngeal arches** forming from E9, although the fifth arch is vestigial. The mouth forms at E9.

The mesoderm lateral to the somites is the **intermediate mesoderm**, which becomes the kidney and gonads. The pronephric primordium arises about E9, but has no function. The **genital ridges** become visible from about E10 from mid-trunk to mid-tail. The lateral part of these ridges forms the mesonephros, although in mammals this too is vestigial and without function. The nephric duct produces the **ureteric bud** at E11.5 and this grows into the posterior nephrogenic mesoderm to produce the **metanephros**, which is the functional kidney. The medial part of the urogenital ridges produces the **gonads**. The

germ cells originate from the extraembryonic mesoderm in the posterior amniotic fold. They enter the hindgut at E10 and migrate up the mesentery, reaching the gonads between E11 and E13. Lateral to the intermediate mesoderm is the **lateral plate**, which is divided by the coelom into the outer **somatopleure** (epidermis + somatic mesoderm) and the inner **splanchnopleure** (endoderm + splanchnic mesoderm). The limb buds arise from the somatopleure at E9.5–10 with the forelimb bud at the level of somites 8–12 and the hindlimb buds at the level of somites 23–28. The outer epidermis of the mouse, like other mammals, is covered with hair follicles, arising from E14 onwards.

It should be stressed that although the course of organogenesis is similar in most vertebrates, the arrangement of extraembryonic membranes can differ considerably even between

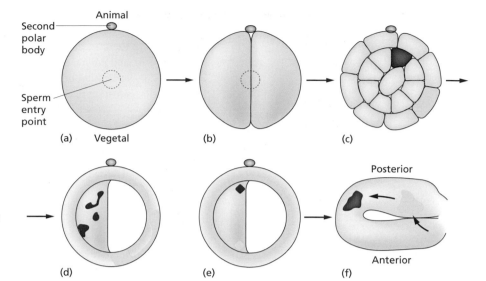

Fig. 10.9 Fate mapping of early stages. (a) The second polar body marks the animal pole of the zygote. (b) The first cleavage is meridional. (c,d) In the compacted blastocyst, labeling of inner cells shows that they enter the inner cell mass, while the outer cells enter the trophectoderm. (e,f) In the inner cell mass, labeled animal cells become distal and labeled vegetal cells become proximal in the egg cylinder.

different groups of mammals. In particular, rodents are unusual in having their deep cup-shaped egg cylinder and the associated turning process. In most mammals the embryo develops more like the chick, from a flat plate of epiblast. Particular features of human development that differ from the mouse are that the amniotic cavity arises precociously within the ICM shortly after the blastocyst has commenced implantation, and that the allantois is vestigial, the embryonic blood vessels being formed from extraembryonic mesoderm lining the chorion.

Fate map

It used to be believed that mammalian embryos, with their high regulative capacity, had no definable fate map at the stage of the fertilized egg. However it is now recognized that the fertilized egg does possess a polarity which is retained through the early embryonic stages (Fig. 10.9a–d). This has been established by a combination of careful observation and the labeling of portions of the egg surface with fluorescent lectin-coated beads and other markers. The original **animal pole** of the egg can be identified by the position of the second polar body. The first cleavage is approximately meridional and separates blastomeres that will preferentially become the embryonic (i.e. the end with the inner cell masss) and abembryonic (the end of the mural trophectoderm) poles of the blastocyst. There is also some evidence for a coincidence of the first cleavage plane with the site of sperm entry.

Labeling of individual blastomeres of early preimplantation embryos by injection of horseradish peroxidase (HRP) or fluorescent dextrans shows that the trophectoderm arises from the polar cells on the exterior of the morula, while the inner cell mass arises from the apolar cells in the interior. By the time the blastocyst expands, there is no further interchange of cells between these two populations. In blastocysts, single labeled cells of the polar trophectoderm give rise to clones of postmitotic

cells in the mural trophectoderm, showing that the polar region is a proliferating zone feeding the mural trophectoderm. The ultimate fate of ICM and trophectoderm was shown by reconstitution of blastocysts from ICM and trophectoderm of genetically distinguishable mouse strains followed by reimplantation into foster mothers. This showed that the ectoplacental cone, the giant cells, and the extraembryonic ectoderm of later stages were derived solely from the trophectoderm. The entire fetus, the amnion, allantois, and extraembryonic endoderm were derived from the inner cell mass.

There is also a predictable relationship between the axes of the **blastocyst** and the later **egg cylinder** stage (Fig. 10.9e,f). If the cells on the animal pole end of the blastocyst are labeled by injection of mRNA for green fluorescent protein (GFP) it is found that they end up distally in the visceral endoderm of the egg cylinder. Conversely if ICM cells at the vegetal pole end are labeled, they end up proximally in the egg cylinder. This distal–proximal distribution of cells arises from the early stages of the morphogenetic movements which form the primitive streak of the embryo. It follows from this that the original animal pole of the zygote is likely to end up as the posterior side of the embryo, i.e. the side of the forming primitive streak, when it becomes visible in the egg cylinder stage.

Fate mapping of postimplantation stages mostly involves labeling the embryo by injection of cells with HRP, or extracellularly with DiI, then culturing them *in vitro* for up to 48 hours (Fig. 10.10). These studies show that the primitive streak and the mesoderm of the amnion and allantois all arise from the posterior edge of the epiblast. The streak pushes distally and the extraembryonic tissue expands into the proamniotic space, mainly driven by growth but also by some lateral movement of cells around the cup of epiblast. The anterior part of the streak becomes the node, which forms the notochord and part of the somites along the entire length of the body. Cells also enter the somites and the neural plate from positions lateral to the midline. The middle part of the streak populates mainly lateral plate

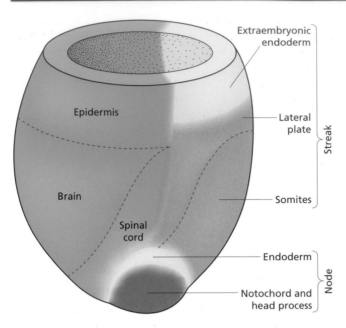

Fig. 10.10 Fate map of the late streak stage. The boundaries are much fuzzier than shown, because of cell mixing.

the surface layer moving laterally towards and then through the streak to become definitive endoderm and mesoderm. But all the studies show a substantial degree of mixing between labeled and unlabeled cells showing that the fate map is a statistical construction and that the boundaries between prospective regions are not precise.

Technology of mouse development

Transgenic mice

The creation of mice with an engineered genetic constitution has become a substantial industry, with applications in many branches of biology. Sometimes the term **transgenic** is used to include all forms of modification, here it will be used more narrowly to indicate introduction of new genes into the germ line. The standard method is to inject the DNA directly into one **pronucleus** of the fertilized egg (Fig. 10.11). This leads to a reasonable yield of transgenics with a good probability that the germ line will be transgenic. Normally integration is at a random position in the genome and comprises many copies of the transgene joined head to tail in a tandem array. Better levels of gene expression are achieved if the gene to be injected is purified away from plasmid DNA as the prokaryotic sequences are usually inhibitory to transgene expression. It is also important to use genomic DNA containing introns, as this gives better expression

mesoderm in the posterior half of the body. The posterior part of the streak populates mainly mesoderm of the amnion, visceral yolk sac, and allantois. It seems that the primitive streak in the mouse works in a similar way to that of the chick with cells from

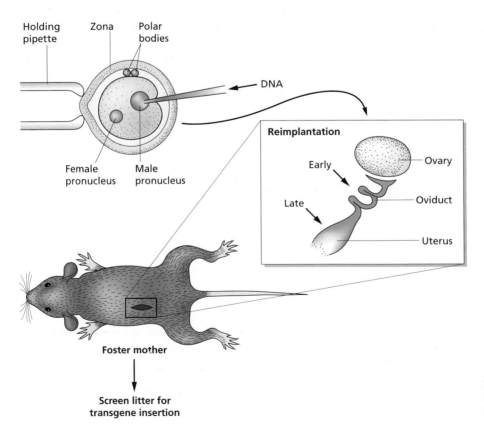

Fig. 10.11 Making transgenic mice by injection of DNA into a pronucleus of the fertilized egg.

levels than complementary DNA (cDNA; i.e. DNA copied from mRNA, without introns). Normally, linear rather than circular DNA is preferred for injection. In addition to the simple pronuclear injection method, it is possible to introduce genes into embryos via embryonic cells, following the methods used to knock out genes (see below), and gene introduction has also occasionally been achieved using retroviruses.

The newborn mice arising from a transgenesis experiment will be screened for incorporation of the gene, by performing a genomic **Southern blot** or **polymerase chain reaction** (**PCR**) on a DNA sample from the tail tip of each individual. A transgene will usually behave as a simple dominant gene in subsequent breeding. It may be desirable to breed it to homozygosity so that all embryos from the line are known to possess the gene.

It is known that, despite the abnormal chromosomal location of transgenes, they can show correct temporal and regional expression so long as sufficient flanking DNA sequences are included. A very large amount of work has been performed in which the flanking sequences of the genes are modified in order to identify the particular regulatory sequences responsible for control of their expression. For this type of work it may be sufficient to use transient transgenics. This means that the injections are done, the embryos reimplanted, and then later recovered and analyzed directly without any breeding to set up a permanent line of animals. Instead of monitoring the normal gene product, such studies are usually done using a **reporter gene** whose product is more easily detected. Commonly used reporter genes are *lacZ* and *luciferase*. The β-galactosidase coded by the *lacZ* gene can easily be detected in wholemounts or sections using the X-Gal reaction. Luciferase is an enzyme from an insect that will convert a specific substrate, luciferin, into a phosphorescent product. It can be measured in tissue samples with very high sensitivity using a luminometer.

By making the appropriate molecular constructs it is possible to express the coding region of one gene under the control of the promoter of another, and this enables **ectopic** expression of specific genes, and the resulting modification of the course of development. Sometimes uniform expression is required, which can be achieved using promoters for housekeeping genes such as *cytoskeletal (β) actin* or *histone H4*. Sometimes a particular stage- or region-specific promoter is required to drive expression in just that tissue or position. It is also possible to ablate a particular region of the embryo by expressing a toxin, such as diphtheria toxin, under the control of a suitable promoter. Transgenes may also be made **inducible**, so they can be activated at a particular stage by a treatment given by the experimenter (see below).

Mosaicism and chimerism

Among mammals an animal composed of cells which are genetically dissimilar is called a **mosaic** or a **chimera**. Usually the term "mosaic" is used if the animal has arisen from a single zygote, and "chimera" if it has arisen from some experimental or natural mixture of cells from different zygotes. One type of naturally occurring mosaic is the **X-inactivation** mosaic, described below. Genetic mosaics should not be confused with **mosaic** development, an entirely different concept discussed in Chapter 4.

There are two experimental methods for making a chimera (Fig. 10.12). **Aggregation chimeras** are made by removing the zona pellucida of four- to eight-cell-stage embryos, then gently pressing them together so that the cells adhere. The fused embryo is reimplanted into the uterus of a pseudopregnant recipient. Such embryos show normal development and consist of a mixture of the cell types of the two components. Injection chimeras are made by injecting cells into the blastocoelic cavity of expanded blastocysts, with zona intact. Again they are implanted in foster mothers for further development. It is often possible to detect chimerism by examination of the coat pattern, if the two mouse strains used have different coat colors. Although the pigment cells themselves are derived from the neural crest, the expression of pigment in the hair follicles will depend also

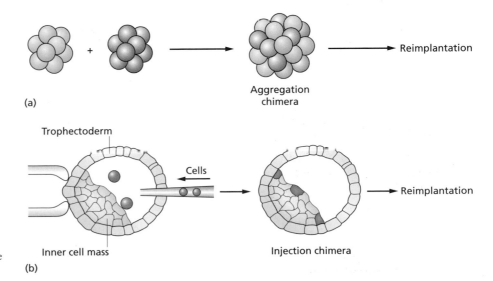

Fig. 10.12 Making chimeras: (a) by aggregation; (b) by injection of cells into the blastocyst.

on the local genetic composition of the epidermis. Therefore chimeras will often be blotchy in coat pattern, or at least have some admixture of the hair colors of the two contributing strains. Injection chimeras have enormous practical importance because they represent an important step in the production of knockout mice.

In addition both types of chimera, and also X-inactivation mosaics, can be used for the **clonal analysis** of structures in the mature mouse. If a particular type of structure is always entirely composed of one of the two genotypes, then it must normally be formed by a single cell. If it contains patches of the two genotypes, then it is possible to calculate the probable number of progenitor cells, allowing for the fact that some similar clones will be adjacent. This type of analysis has been carried out for hair follicles and for intestinal crypts (see Chapter 13).

Clonal analysis can also be carried out using transgenics for a modified form of the *lacZ* gene called *laacZ*. This gives an inactive product due to the insertion of a repeated sequence into the coding region. Intragenic recombination can restore the functional *lacZ* sequence and this happens spontaneously at low frequency. Once the recombination event has occurred all progeny cells will express β-galactosidase permanently so this can be used as a method for random labeling of clones.

Embryonic stem cells

Cell lines showing developmental **pluripotency** can be produced by putting mouse blastocysts into culture. After the blastocyst hatches from the zona it will adhere to the substrate and, in a suitable medium, the ICM cells will proliferate. The resulting cells are **embryonic stem cells (ES cells)**. These resemble the ICM and the epiblast of the normal embryo, but microarray analysis shows that they are slightly different from both. They can be cultured for many passages and be frozen for later use. They are grown on **feeder layers** of irradiated fibroblasts or in the presence of leukemia inhibitory factor (LIF). When the feeder cells or the LIF are removed, the ES cells will differentiate into **embryoid bodies** in which the outer layer of cells resembles the primitive endoderm. In the normal embryo, LIF is expressed in the trophectoderm and the receptors LIFR and gp130 are present in the inner cell mass.

ES cells can be established from any strain of mouse, although most are from strain 129. The stem-cell property depends on the transcription factors Oct4 (POU class) and Nanog (homeodomain), as knockouts of their genes result in cells that can form differentiated extraembryonic cell types but will not self-renew. ES cells are usually of normal karyotype, although this can become abnormal on prolonged culture. When implanted into immunologically compatible adult mice, ES cells form tumors containing several differentiated tissue types; and, more significantly from the developmental point of view, when ES cells are injected into blastocysts they will colonize the resulting embryos giving a high frequency of chimerism. In many cases the chimerism extends to the **germ cells**, making it possible to breed intact mice from cells that have been grown in culture (Fig. 10.13). The existence of ES cells shows that it is possible to disengage growth from developmental commitment, as lines have been passaged as many as 250 times without loss of the ability to repopulate embryos.

The main practical importance of ES cells is that they offer a sophisticated route for the reintroduction of genes into an embryo. Although genes can be introduced by simple injection of DNA into the zygote, this offers no control over the number of copies introduced or the location of the insertion site. By contrast, with ES cells in culture it is possible to select for rare events, particularly **homologous recombination** of exogenous DNA into the complementary site in the genome. This has mainly been used to **knock out** individual genes by replacing them with an inactive variant, but there are many other possibilities such as replacement of one gene by another, or the assembly of complex binary and conditional gene regulation systems.

Knockouts

In developmental biology, the main object of knocking out individual genes is an expectation that the null phenotype will reveal the gene's function. Usually the product of the gene in question is believed to be important for some reason, such as its biochemical activity, or expression pattern, or the existence of a developmental function for the homolog in another organism, such as *Drosophila*. But knockouts are also useful to create mice mutant for particular gene variants responsible for human genetic diseases. This can create an **animal model** of the human disease, so that further information can be gained about the pathology of the disease, and strategies for therapy can be tested. An example would be the creation of mice mutant for the cystic fibrosis transmembrane conductance regulator (CFTR protein) to serve as a model of human cystic fibrosis. Another is a mouse lacking the gene for apolipoprotein B, which develops atherosclerosis over 2–3 months. A further application in medical research is the creation of mice especially susceptible to cancer, for example by removal of tumor-suppressor genes such as the *p53* gene.

Most genes have been knocked out using the positive–negative method, which is a selection procedure for homologous rather than random integrations. A targeting construct is assembled that consists of genomic DNA for the region to be replaced, with an essential functional region of the gene replaced by an antibiotic-resistance gene, usually neomycin resistance (*neo*^{r}). Flanking this is a copy of a viral gene coding for thymidine kinase (*tk*). The construct is transfected into the ES cells and if it integrates by homologous recombination, only the *neo*^{r} will be incorporated (Fig. 10.14a). If it integrates at random, in the wrong place, both *neo*^{r} and *tk* will probably be incorporated (Fig. 10.14b). Then the cells are subject to selection using two drugs. **Neomycin** will kill the cells that have not incorporated the targeting vector at all, as the host ES cells are sensitive to it (the feeder cells are resistant). **Ganciclovir** will kill cells that have incorporated the construct in such a way that the *tk* gene is

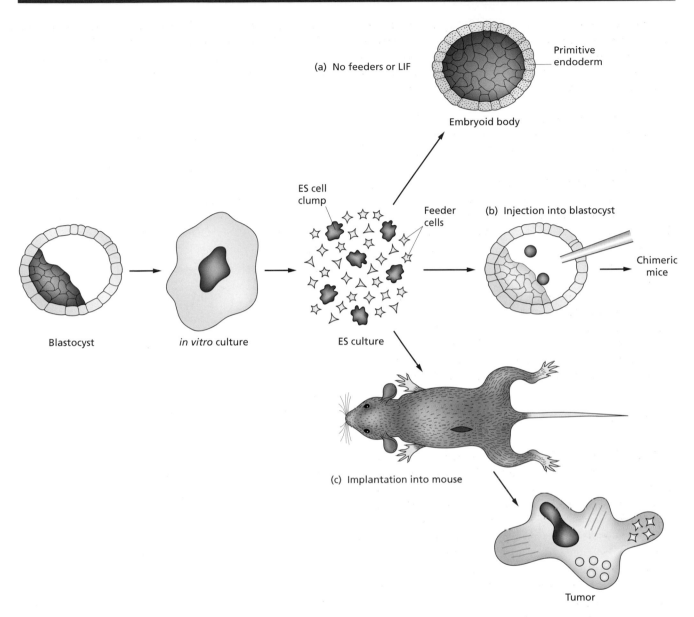

Fig. 10.13 ES cells. (a) Removal of feeder cells will cause differentiation *in vitro* into "embryoid bodies." (b) Injection into blastocysts can generate chimeric mice. (c) Implantation under the kidney capsule can produce a teratocarcinoma.

present. This is because the thymidine kinase converts the drug into a cytotoxic product. The net effect is that the surviving cells are often ones that have undergone homologous recombination such that the target gene has been exactly replaced by the inactive version. These cells are grown up as individual clones and are screened by Southern blotting or PCR to ensure that the targeting construct has indeed integrated at the predicted position (Fig. 10.15). Then the cells are injected into host blastocysts and the embryos reimplanted into foster mothers. The strain of the host embryos will be chosen such that the coat color differs from that of the parent strain of the ES cells. Most ES cells are from strain 129 mice, which have an agouti coat color (mixed black/yellow, like many wild mammals), so a black or albino strain is preferred for the host blastocysts. The chimerism should then be indicated by a patchy coat color, and this is confirmed by analysis of DNA from tail tips. Chimeric mice may or may not have any of the donor cells in their germ line, so they are mated and the resulting offspring also subject to tail-tip DNA analysis. If the F1 generation do contain the targeting construct, then they can be mated together to generate homozygotes, and the phenotype of the null mutation can be established.

Exactly the same technology can be used to perform "**knock-ins**." Here instead of replacing the endogenous gene with an inactive version, it is replaced by another gene, maybe a modified version of itself. The knock-in method has an advantage over traditional transgenesis because the insertion site is defined and so there should be no unpredictable consequences of random insertion.

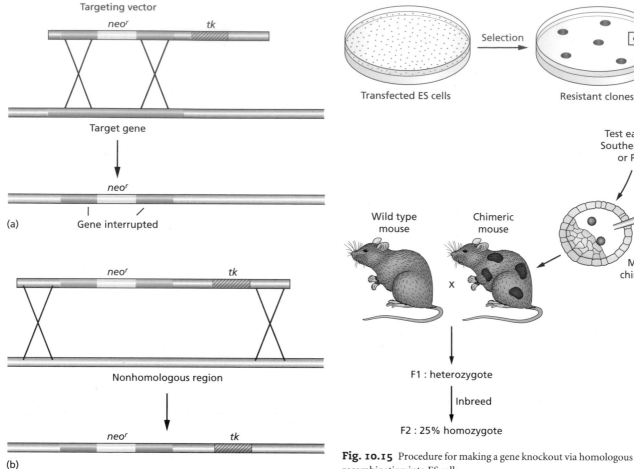

Fig. 10.14 Gene targeting. (a) Recombination at the homologous site disrupts the host gene and introduces *neo*^{r} but not *tk*. (b) Recombination at a nonhomologous site introduces both *neo*^{r} and *tk*.

Fig. 10.15 Procedure for making a gene knockout via homologous recombination into ES cells.

Various problems have been encountered with the use of knockout mice. Firstly, there may be no abnormality of phenotype, or a barely perceptible abnormality. This is usually ascribed to **redundancy**, or the presence of other genes in the genome with overlapping functions, which is very common in vertebrates because of the large number of multigene families (see also Chapter 3). In order to establish the function of a set of genes showing significant redundancy it is necessary to knock them all out and then assemble a multiple homozygous null by mating of the individual knockout lines together. Considerable work of this sort has been done, for example on **Hox gene paralog** groups and on the various retinoic acid receptors. Secondly, the effects of the knockout may vary considerably depending on the genetic background, or strain of mouse into which the mutation is introduced. Although it may complicate the functional analysis this can be a useful property as it can help to identify other interacting genes important for the process under investigation. Finally, the phenotype may be so severe that the embryos die at an early stage which precludes any examination of later functions of the same gene. There are various solutions to

this problem. One is to use a **conditional knockout** strategy as discussed below. Another is to make **chimeras** in which the gene is knocked out in only part of the conceptus.

Chimeric knockouts

The usual application of the chimera method is the case where a gene is required at an early stage in the extraembryonic tissues and at a later stage in the embryo itself, so the early death is due to failure of nutritional support from the extraembryonic tissues. It turns out that when ES cells are injected into a **tetraploid** blastocyst, the ES cells mainly form the embryo and the tetraploid cells mainly form the extraembryonic tissues. Tetraploid embryos are made by electrofusion of the blastomeres at the two-cell stage and are not themselves viable in the long term. To establish the postimplantation function of a gene, homozygous null ES cells, or ICM cells from a homozygous null blastocyst, are injected into tetraploid blastocysts, which are implanted into foster mothers. The tetraploid-derived extraembryonic structures support development until the gene is needed in the embryo itself, and from this stage an identifiable defect will arise which will hopefully reveal something of the normal gene function. An example is the case of *fgf receptor 2*. If knocked out, this causes preimplantation death, due to failure of proliferation of the polar trophectoderm. But a chimeric embryo, composed of *fgfr2⁻* cells in a wild-type tetraploid blastocyst, will develop to about E10.5, and at this stage it is clear that the embryos lack both lungs and limbs, showing that FGF signaling through this receptor is needed for the formation of both these organs.

Conditional systems

In a conditional system the desired genetic change occurs only when the appropriate *condition* is achieved. The most commonly used conditional system is the **cre-lox** system. Cre (pronounced "cree") is a recombinase enzyme from phage P1 which can excise segments of DNA flanked by binding sequences called *loxP* sites. In order to knock out a gene in just one tissue, and thereby circumvent problems of early lethality, two mouse lines need to be created. The first is a line in which the target gene has *loxP* sites inserted on either side (the gene is said to be "**floxed**"). The second is a transgenic line in which the Cre recombinase is driven by a tissue-specific promoter. When the two types of mice are mated together, the target gene should be excised in the offspring, but only in the tissue containing the Cre (Fig. 10.16). In practice the Cre does not give a 100% effective excision, so it is best if the parent with the floxed target gene is a heterozygote with the second copy of the gene inactive.

The utility of the *cre-lox* system is not confined to tissue-specific knockouts. It can also be used for a tissue-specific activation of a gene if the Cre is used to excise an inhibitory segment. For example, a short polypeptide with termination codon can be inserted at the beginning of the coding region and be flanked with *loxP* sites. When this is removed by the Cre, the gene will become active. A similar effect could be achieved more simply by driving the gene off a tissue-specific promoter in a conventional transgenic, but the *cre-lox* system enables greater sophistication. For example, if it is desired to know what will normally be formed by a particular group of cells that activate a particular promoter at a particular stage, then a mouse can be assembled in which the promoter in question drives the Cre, and the Cre activates a reporter gene, such as *lacZ*, by excision of an inhibitory segment (Fig. 10.17). In such a mouse, the descendant cells of those in which the promoter was turned on will continue to express β-galactosidase thereafter, regardless of subsequent changes in the state of the tissue-specific promoter. An example of this strategy is the proof that all cell types of the pancreas come from endoderm expressing the transcription factor Pdx1 (see chapter 16).

Inducibility

Another important element of mouse embryo technology that can be incorporated both into the conventional transgenic, or into a binary combination, is inducibility. Early experiments

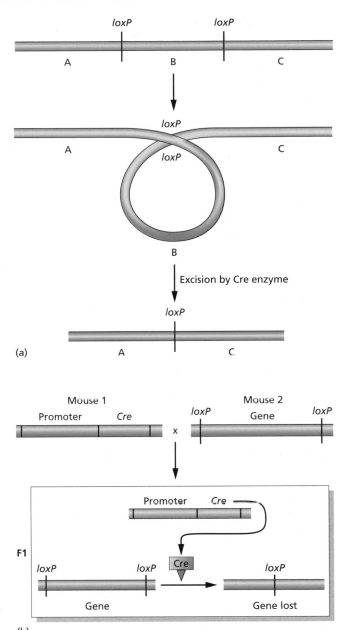

Fig. 10.16 *Cre-lox* system. (a) The *cre* recombinase will excise DNA between two *loxP* sites. (b) A binary mouse. When mouse 1 and 2 are mated, the *cre* recombinase is expressed according to the specificity of the promoter to which it is coupled and will drive excision of the gene between the *loxP* sites.

used the *metallothionein* promoter, which is activated by heavy metals such as zinc or cadmium. But this is not tissue specific, and it is often desirable to confer inducibility on an otherwise noninducible, but tissue-specific, promoter. The preferred system uses elements of the tetracycline system from *Escherichia coli*. The enzymes that degrade tetracycline are coded by genes of the *tet* operon. In the absence of tetracycline the operon is repressed, but in the presence of tetracycline it becomes activated. The system works because the tetracycline combines

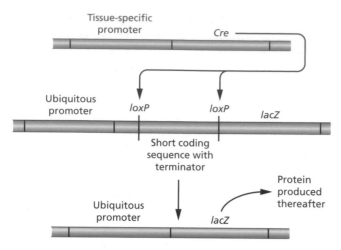

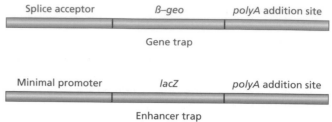

Fig. 10.19 Gene trap and enhancer trap.

Fig. 10.17 Use of the *cre-lox* system to label permanently a population of cells in which the tissue-specific promoter was active only transiently.

sequestered in the cytoplasm by binding to heat shock proteins. When their ligand is supplied this will displace the heat shock protein and allow migration of the ER fusion protein to the nucleus. Cre-ER can still be driven by a tissue-specific promoter but the time at which it becomes active is controlled by injecting the mother with a suitable receptor ligand, usually the synthetic estrogen **antagonist** tamoxifen.

Gene and enhancer traps

The knockout technology is very powerful, but it can only be applied to identified genes. New genes need to be identified by mutagenesis. This may be chemical mutagenesis, as described in Chapters 3 and 8, followed by positional cloning of the genes. It can also be useful to be able to carry out **insertional mutagenesis** with a suitable DNA vector, as this allows easy identification of the modified locus. The relevant constructs are called **gene traps** and contain a splice acceptor site, a reporter gene, and a selectable marker (Fig. 10.19). Often the reporter and selectable marker are a single gene called **β-geo**, which is a fusion of *E. coli lacZ* and the neomycin-resistance gene, having the activities of both. The construct is transfected into ES cells. If it integrates outside a gene it remains inactive because it has no promoter of its own. If it integrates within a gene it should function as the last exon, with the splice acceptor enabling incorporation into the coding sequence. Because the endogenous gene is disrupted by the insertion, it is likely to be mutated to inactivity or reduced activity. The β-geo will be expressed as a fusion on the C-terminal end of the truncated endogenous protein, and if the splicing is in frame then the cells should be selectable by

with a Tet repressor protein (TetR), and thereby causes it to dissociate from the *tet* operator (*tetO*) sequence in the DNA, enabling transcription to take place.

This system has been modified, and improved in specificity, by converting the TetR into an activator (TetA) in which its normal repression domain is replaced by the VP16 activation domain (Fig. 10.18). In this variant, called "**tet-off**," the gene is inactive in the presence of tetracycline but, when the tetracycline is withdrawn, the TetA will bind to the *tetO* and activate transcription. In a transgenic situation, the *tetO* will be combined with the tissue-specific promoter driving the target gene and the TetA will be expressed from a constitutive transgene elsewhere in the genome. Tetracycline is supplied continuously to the pregnant female mice by including it in their drinking water, so the TetA is sequestered and the target gene remains off. If it is desired to activate the target gene on a particular day of development, then the tetracycline is withdrawn. It becomes cleared from maternal and fetal circulation within a few hours, the TetA is able to bind the *tetO*, and the target gene becomes activated.

It is possible to make the *cre* system inducible by using a fusion of the Cre enzyme with the estrogen receptor (Cre-ER). In the same way as discussed for the glucocorticoid receptor in Chapter 7, the nuclear hormone receptors like ER are

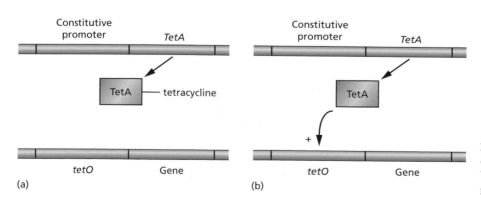

Fig. 10.18 Tet-off system. In the presence of tetracycline the target gene is inactive. When tetracycline is removed, the target gene is activated.

neomycin resistance, and should also express β-galactosidase. There are more sophisticated versions of this method that will work even if the insertion is not in frame.

The gene trap is transfected into ES cells, then introduced into mice by the same procedure as used for knockouts. It is often possible to examine the normal expression pattern of the gene by doing X-Gal staining on the chimeras themselves. Although only some of the cells will carry the construct, it may be enough for this purpose. If the expression pattern looks interesting, the chimeras will be bred to establish an F1 generation heterozygous for the gene trap, and these will be mated with each other to establish an F2 with 25% homozygosity. At this stage it will be clear whether the trap has produced an interesting recessive phenotype. If so the gene will be cloned by making a genomic DNA library from the gene trap line and probing it for the gene trap vector sequence. The 5′ junction sequence should be within an intron of the trapped gene and enable rapid cloning of the remainder.

One important gene trap mouse strain is called **Rosa26**. In this case the trapped gene encodes an untranscribed RNA of unknown function. But it has the property of being reliably expressed in a ubiquitous manner, at all stages and in all tissues. For this reason mouse lines designed for constitutive ubiquitous expression of a transgene are now often made by "knocking in" the required gene to the *rosa26* locus, using the techniques of homologous recombination. For example a ubiquitous Cre expressing line is of this type. Also the locus has been modified by insertion of a *lacZ* gene containing a floxed termination sequence (R26R strain of mice). This is useful for testing transgenic Cre lines. Once the Cre line has been made, it can be mated to R26R and the offspring will express β-galactosidase permanently in all regions where the Cre was active. This enables the tissue specificity and level of expression of the Cre line to be established before experiments are undertaken with functional target genes.

An **enhancer trap** is a gene trap that lacks the splice acceptor and carries its own minimal promoter, which provides a basal RNA polymerase II binding site but is not sufficient for detectable transcription (Fig. 10.19). If it enters the genome within range of an endogenous promoter or enhancer then this will complement the minimal promoter and activate significant transcription of the *lacZ*. Enhancer traps are usually not mutagenic as they may be activated anywhere within effective range

of an endogenous enhancer. Because they may be at some distance from any endogenous gene they are also not so useful for cloning purposes. Their main use is providing lines of mice in which particular tissues or cell types are highlighted by expression of β-galactosidase and are therefore very easy to visualize.

Regional specification in development

Embryo vs. extraembryonic structures

The brief sketch given here concerns only body-plan formation, as mammalian organogenesis will be considered further in Section 3.

The early blastomeres of a mouse embryo are known to be **totipotent**. It is possible to obtain formation of a complete blastocyst from each single blastomere isolated from the two- or four-cell stage. These blastocysts tend to have a higher proportion than normal of trophectoderm but they will form complete normal embryos after reimplantation. From the eight-cell stage it is no longer possible to obtain a complete embryo from one blastomere, although blastomeres from the eight-cell stage can still integrate into host blastocysts and contribute to all tissues, both embryonic and extraembryonic.

In normal development the formation of ICM and trophectoderm depends on the cell polarization that occurs at the eight-cell stage (Fig. 10.20). It is known that the polarization depends on cell contact and that the microvillous region always appears at the external surface, but the identity of cytoplasmic determinants responsible for initiating the genetic programs for the two cell types is still not known. Although the normal specification commences at the eight-cell stage, it is still possible for a period to force polar cells to become ICM, or apolar cells to become trophectoderm, by putting them, respectively, on the inside or the outside of a cell aggregate. By the 64-cell stage the two cell types have stabilized and are no longer interconvertible.

After implantation the ICM becomes divided into an outer layer of primitive endoderm and an inner core of epiblast. This probably depends on the cell layer in contact with the blastocoel being induced to form primitive endoderm. The trophectoderm becomes divided into the polar trophectoderm, neighboring the

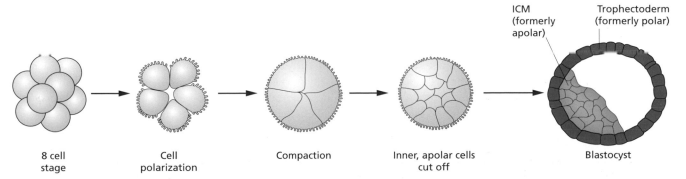

Fig. 10.20 Origin of ICM and trophectoderm from cell polarization at the eight-cell stage.

New Directions in Research

Mice will continue to be used to make transgenic and knockout lines to address a huge range of biological issues, by no means confined only to mechanisms of development. For example these include problems in cell biology, immunity, and neurobiology, as well as the study of disease mechanisms.

The phenomena of imprinting and X-inactivation will continue to be used as models to try to understand the mechanisms of **epigenetic** inheritance. The role of chromatin structure in gene regulation, and of inheritance of chromatin modifications through cell division are still poorly understood.

ICM, and the mural trophectoderm surrounding the rest of the blastocyst. The maintenance of cell division in the polar region depends on the proximity of the ICM, and, among other factors, requires FGF4. It is possible to grow trophectoderm cells *in vitro* in the presence of FGFs. A knockout of *fgf4*, or of *oct4*, expressed in the ICM, or of *fgfr2*, expressed in the trophectoderm, will terminate development at this stage.

The primitive endoderm becomes divided into visceral endoderm, in contact with the epiblast, and parietal endoderm, in contact with the mural trophectoderm. There is good evidence from isolation and recombination experiments that these distinctions are also due to a difference in cellular environment, but the molecular nature of the signals is not known.

Embryonic body plan

The whole embryo derives from the epiblast of the egg cylinder stage. Even until late primitive streak stage it is possible to induce twinning by treatment with cytotoxic drugs, presumably by killing large numbers of cells and causing regulation to occur from small nests of survivors. This shows that regional determination has not become irreversible before this stage. By comparison, it is thought that human identical twins usually arise from division of the ICM (70–75%), less often from blastomere separation (25–30%), and most infrequently from division of the primitive streak (1%). These figures are arrived at on the basis of whether the twins share a common placenta and amnion. In cases of blastomere separation the placentas will be separate, in the case of ICM division the placenta will be common but the amnions separate, and in the case of primitive streak division, both placenta and amnion will be common. The fact that cells are still able to contribute to more than one individual up to primitive streak stage is the basis for the law in the UK permitting some experimentation on human embryos up to this stage, which is reached at about 14 days.

As it is very difficult to do microsurgical experiments in mice, much of the investigation of early postimplantation development relies on the interpretation of knockouts. These have underlined the critical importance of the nodal factor, a member of the TGF-β superfamily. Nodal was originally discovered by retroviral mutagenesis of the mouse, and found to be expressed in the node, although it is in fact expressed earlier throughout the epiblast. The homozygous null embryos are unable to form any embryonic pattern, and many of the developmentally important genes active during formation of the primitive streak and the node are not expressed. As we have seen, *nodal* homologs are also important for mesoderm formation and patterning in *Xenopus*, zebrafish, and chick. The knockout of the so-called *activin receptor IIA* and *B*, which are receptors for nodal, or of *smad2*, which is required for signal transduction of nodal-like factors, or *foxh1*, which is a partner of smad2 in transcriptional regulation, all abolish the anteroposterior polarity of the embryo, such that the mesoderm that forms is entirely extraembryonic.

The anteroposterior pattern of the embryo derives initially from the proximodistal pattern in the early egg cylinder. The transcription factor gene *hex* and the inducing factor genes *cerberus-like, dickkopf*, and *lefty1* are expressed at the distal tip in the visceral endoderm and a number of genes including that encoding the T-box transcription factor brachyury (usually called *T* in the mouse) are expressed at the proximal end in the epiblast (Fig. 10.21). This proximo–distal pattern is thought to be due to signals emitted from the extraembryonic ectoderm. Two good candidates for components of this signal are Wnt3 and BMP4. Genes for both these inducing factors are expressed in the proximal part of the egg cylinder, abutting the future epiblast. Knockouts for both are early lethals, the *wnt3* knockout having no primitive streak or mesoderm, and the *bmp4* knockout having no allantois and being defective in embryonic mesoderm.

As the cup-shaped egg cylinder elongates and the proamniotic cavity forms, morphogenetic movements shift the *hex* domain to one side and the *T* domain to the other. These, respectively, become **anterior** and **posterior** ends of the embryo. As indicated above, there is a statistical association between the position of the original animal pole of the zygote, and the posterior side of the early embryo, so it is possible that the breakage of the radial symmetry of the egg cylinder arises because of small asymmetries derived from the fertilized egg. The anterior visceral endoderm (**AVE**) then expresses a group of genes for transcription factors (including *otx2, foxA2*, and *lim1*) and genes for secreted factors (including *nodal* and *cerberus-like*) that are associated with anterior development. A little later the same genes are activated in the overlying epiblast that becomes the head fold. By this time

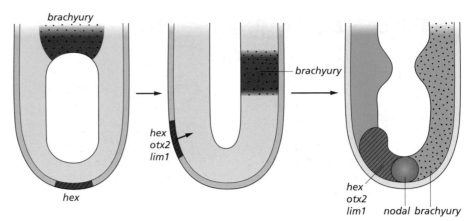

Fig. 10.21 Anteroposterior patterning of epiblast in the egg cylinder to primitive streak stage.

the primitive streak has formed in the posterior, and the node lies in between the head domain and the streak. Microsurgical recombination experiments show that the head structures can be induced in the epiblast by the action of the AVE. Knockouts of the genes expressed in the AVE tend to lead to anterior truncations. The relative requirement for gene function in the AVE and in the future head itself can be established by the chimera protocol described above. When chimeras are made of mutant ES cells in wild-type 4n blastocysts, the phenotype will reflect the requirement for the gene in the epiblast. If chimeras are made by injecting normal ES cells into mutant blastocysts, then the phenotype will reflect the requirement in the visceral endoderm. Such experiments have shown that some of the anterior genes, such as *otx2* and *foxA2*, are required just in the visceral endoderm at this stage, and others, such as *lim1 (=lhx1)*, are required in both the visceral endoderm and in the epiblast. At the same time the signals from the AVE suppress formation of a streak in the adjacent epiblast. This is probably through the inhibitory action of Cerberus-like and Lefty-1, as a double knockout of both factors leads to the formation of ectopic streaks.

The node behaves like the organizer in *Xenopus*, as it is able to induce a second axis containing host-derived neural tube and somites if transplanted to another part of the epiblast. Mouse nodes will also induce second axes in chick blastoderms. It is currently unclear whether BMP inhibitors are the principal signals involved, although both chordin and noggin are expressed in the node. Moreover the double knockout of chordin and noggin prevents formation of the forebrain and gives defects in notochord and sclerotome, indicating at least some role in neural induction and mesodermal dorsalization.

As in *Xenopus*, the FGF–Cdx–Hox pathway probably controls patterning of the posterior region as *fgf4* and *fgf8* genes are expressed in the streak, activated ERK can be detected by immunostaining, and knockouts of *fgf8* or *fgf receptor 1* or *cdx* genes show posterior defects or anteriorizations associated with reduction of Hox activity. A gradient of FGF8 protein can be seen by immunostaining in all three germ at tailbud stages. This arises by transcription of *fgf8* in the tailbud followed by gradual decay of the mRNA after the cells have left the tailbud.

This mechanism depends on growth from a posterior zone, and would not work in *Xenopus* or zebrafish where there is little growth at early stages. However the end result of a gradient of FGF activity seems similar. Some *wnt* genes are also important in posterior body formation in the mouse. *wnt3A* in particular is expressed in the posterior and the knockout for *wnt3A* lacks posterior parts.

The available information suggests that body patterning in the mouse is comparable to *Xenopus*, although not identical. In both species, the initial regionalization is into just two parts, a head and a trunk region, and depends originally on the pattern in the original animal–vegetal axis of the egg. In the mouse, the head pattern depends on a group of genes activated initially in the anterior visceral endoderm, with an inductive signal required to induce a corresponding anterior territory in the epiblast. The trunk pattern depends on the signaling center associated with gastrulation movements (the dorsal lip or node), which is also responsible for neural induction and for dorsalization of the mesoderm. In both species the nodal factors have a key role in mesoderm formation, and the FGF and Wnt systems are needed for formation of posterior parts. But the relative importance of the BMP inhibitors for neural induction in the higher vertebrates is still under debate.

Left–right asymmetry

As in the chick, *nodal* becomes expressed preferentially on the left side of the node from the two- to three-somite stage, along with another TGF-β superfamily member, *lefty2*. Expression then spreads to the left lateral plate mesoderm. The sequence of subsequent gene activations is different from the chick but eventually these factors bring about the asymmetries of organogenesis by differential activation of downstream targets on the two sides. For example, within the early heart there are two bHLH type transcription factors, dHAND being mainly on the right side and eHAND mainly on the left.

The original cause of the asymmetry comes from the action of the nodal cilia. Each node cell bears a single cilium and they are motile in the period just before the onset of asymmetric gene

Wild type   iv⁻; KIF3B⁻   inv⁻

50%   50%

Situs solitus   *Situs solitus*   *Situs inversus*   *Situs inversus*

Fig. 10.22 Asymmetry mutants in the mouse.

expression, driving a flow of fluid from right to left. This can be visualized by addition of fluorescent dyes to embryos cultured *in vitro*. The fluid flow is causally related to the later gene expression as may be shown by experimentally reversing its sense, which brings about nodal expression on the right instead of the left. The exact mechanism remains unclear but it is likely that the fluid flow stimulates sensory cilia on the left side provoking an increase of intracellular calcium, which affects gene expression.

Cilia are intrinsically asymmetrical structures based on the standard 9+2 arrangement of microtubules together with motor proteins to generate the movement. In the mouse there are numerous mutants which perturb the normal left–right asymmetry and a high proportion of these cause defects in cilia. Three distinct classes of mutant phenotype can be distinguished (Fig. 10.22). First are mutants that cause randomization of *situs*. This means that half the embryos have the normal ***situs solitus***, and half have ***situs inversus***. Such an outcome suggests that the breakage of bilateral symmetry still occurs but that the bias that normally guarantees the usual *situs solitus* has been removed. In such mutants the expression of *nodal* and *lefty2* may be on either side, but not both, indicating that the mutant genes are upstream of these factors. The mutations are in **motor proteins**, for example the knockouts of the kinesin components KIF3A and KIF3B, or the naturally occurring *iv* mutant, which is in a dynein (lrd). These motor proteins are needed either for the assembly of the cilia, in the case of kinesin, or for ciliary motion, in the case of dynein. The second class of mutants leaves the embryo in a state of bilateral symmetry. This is known as an **isomerism** (not, of course, to be confused with isomerism of molecules). An example is the *TG737* mutant. Here, the gene product, called polaris, is also involved in the assembly of cilia and the mutant has no cilia and shows symmetrical *nodal* expression. The third class of mutant has a reversed asymmetry (*situs inversus*). An example is *inv*, which arises from a transgene insertion but, like *iv* and *kif3b⁻*, is genetically recessive. Its homozygotes all have *situs inversus*. In *inv⁻* embryos, *nodal* and *lefty* are expressed on the right side. The *inv* product, inversin, is a cytosolic protein with ankryn repeats. It is expressed early and affects the direction of the ciliary beat. In the mutant the fluid flow is reduced but rightwards instead of leftwards and expression of nodal is on the right side.

The human genetic disease Kartagener's syndrome involves a randomization of *situs*. Genetically it is somewhat heterogeneous but many cases involve dynein mutants. Individuals also suffer from male infertility, because their sperm are immotile, and from lung diseases because the cilia of their bronchial epithelia are immotile and cannot remove the accumulated mucus from the lungs.

Hox genes

In the mouse the four **Hox gene** clusters contain a total of 39 genes. Most of the paralog groups have two or three members. As in other vertebrates, the genes are maximally expressed at the **phylotypic** stage, they tend to have sharp anterior expression boundaries in the central nervous system (CNS) and mesoderm and to fade out in the posterior, and members of the same paralog group have similar anterior boundaries (Fig. 10.23). Extensive work has been done on Hox function by making knockouts of the Hox genes, or by expressing them ectopically by making transgenics in which a Hox gene is driven by a foreign promoter.

In general a knockout results in an anterior transformation. The effects are usually strongest at the anterior expression boundary of the gene in question, because it is there that the particular gene distinguishes the combination of active transcription factors from that in more anterior positions. Often the knockout of just one Hox gene has only a modest effect, but when the whole paralog group is knocked out then the effect becomes substantial. For example, knockout of the paralog group 10 causes all lumbar vertebrae to convert to thoracic character, and knockout of the paralog group 11 causes all sacral vertebrae to convert to lumbar. This is because the members of a paralog group are quite similar and are expressed in similar domains, so there will normally be some redundancy of function between them.

Ectopic expression of Hox genes generally results in a posterior transformation. For example if *hoxa7*, normally with its anterior boundary in the thoracic region, is expressed in the head, then the basal occipital bone of the skull is transformed into a pro-atlas type vertebra. Activation of Hox genes can also be provoked by treatment of the mothers with retinoic acid,

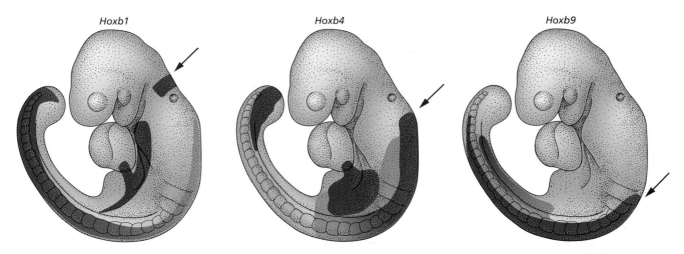

Hoxb1   *Hoxb4*   *Hoxb9*

Fig. 10.23 Expression of three Hox genes, showing the different anterior boundaries.

and this produces multiple posteriorization events due to simultaneous ectopic expression of several Hox genes.

Other topics in mouse development

Nuclear transplantation and imprinting

Nuclei transplanted from early blastomeres to enucleated eggs can support development, although the ability to do so falls off rapidly with stage, perhaps associated with the early onset of zygotic transcription in the mouse. An essential condition for normal development is for the egg to contain one paternally derived and one maternally derived pronucleus. **Parthenogenetic** diploid embryos, made by activation of the egg and suppression of second polar body formation, develop poorly because of inadequate formation of extraembryonic tissues. **Androgenetic** diploids, made by removing the female pronucleus and injecting a second male pronucleus, have abundant extraembryonic tissue but the embryo arrests at an early stage. This suggests that there is some difference between the state of homologous genes on chromosomes from the two parents such that the paternal copy is more readily activated in the trophoblast and the maternal copy in the embryo.

The reason for this is the existence of **imprinted** genes on several chromosomes. These are genes that are only expressed from either the maternal or the paternal chromosome and there are probably about 100 such loci altogether. Many are concerned with growth control and one explanation for the phenomenon is an evolutionary one, as natural selection will always favor traits that maximize the number of offspring. For the father the maximum offspring are achieved by mating with many females and by the embryos achieving the maximum growth rate relative to any other male's offspring. For females the maximum offspring are achieved by devoting equal resources to each embryo and by surviving the birth so that another reproductive cycle can take place. There is thus a potential evolutionary conflict between the traits that will be favored in males and females. Put crudely, the males "want" fast growth of the embryos while the females "want" uniform and controlled growth. An example supporting this principle is the insulin-like growth factor (IGF)-2 system. IGF2 is a growth factor promoting prenatal growth of the embryo and its gene is active only on the paternal chromosome. There is an inhibitor of IGF2 called (misleadingly) the IGF2 receptor, and this is active only on the maternal chromosome. In such cases the effects of genetic heterozygosity depend on which chromosome (maternal or paternal origin) the mutation lies on. If an inactive allele is present on the nonexpressed chromosome then there will be no effect, while if it is present on the expressed chromosome the effect will be similar to that of a homozygous mutant. Conversely, it is possible to have mutations that lead to loss of imprinting on the normally nonexpressing chromosome and this may have effects by doubling the normal gene dosage. For example in humans the Beckwith–Widemann syndrome is a condition involving embryonic overgrowth and predisposition to cancer. It involves the disruption of the gene for a voltage-gated K⁺ channel, but also a high proportion of the cases show loss of imprinting of IGF2. IGF2 also shows loss of imprinting in most cases of the pediatric Wilm's tumor.

Although there are many other imprinted genes, the *igf2* system itself seems the most significant in terms of prevention of parthenogenetic development. If an oocyte is reconstructed to contain two haploid female nuclei, one of which is modified to enable expression of *igf2* like a paternal nucleus, then the embryo can develop to term. The establishment of imprints can also be assayed by nuclear transplantation into eggs, followed by analysis of whether both maternal and paternal alleles at an imprinted locus are expressed. This has shown that the imprints are set up during gametogenesis (Fig. 10.24). Whole embryo clones made with nuclei from primordial germ cells show an erasure of the original imprints from E11.5, probably associated with a genome-wide demethylation of DNA in these cells. The

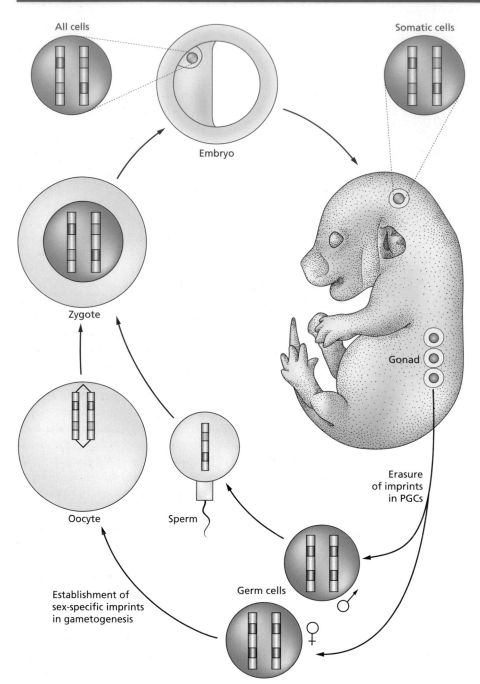

All cells

Somatic cells

Embryo

Zygote

Gonad

Erasure
of imprints
in PGCs

Oocyte

Sperm

Germ cells

Establishment of
sex-specific imprints
in gametogenesis

Fig. 10.24 "Life cycle" of imprinting. Two loci are shown, with red for an expressed gene and blue for an imprinted, inactive, gene. Imprints are erased in the primordial germ cells and reset during gametogenesis.

imprints become re-established late in female gametogenesis, during the growth phase of the oocyte, and earlier in male gametogenesis. It remains unclear how they are initially established, although their maintenance generally correlates with the DNA methylation state of the genes (see Appendix). Several experimentally introduced transgenes have been found to become imprinted, and are differentially methylated in sperm and eggs. In the early stages of embryonic development there is another general demethylation of the genome but this does not affect the imprints, which must be protected in some way.

X-inactivation

In all female mammals, one copy of the **X-chromosome** becomes **heterochromatic** and inactive, and is visible as a Barr body within the nucleus. This occurs at random at the primitive streak stage such that either the maternally derived or the paternally derived X becomes inactive. The inactivation of one copy of the X-chromosome produces a **mosaic** with respect to any heterozygous locus, since some cells will express the gene in question and others will not, and this ensures that all female

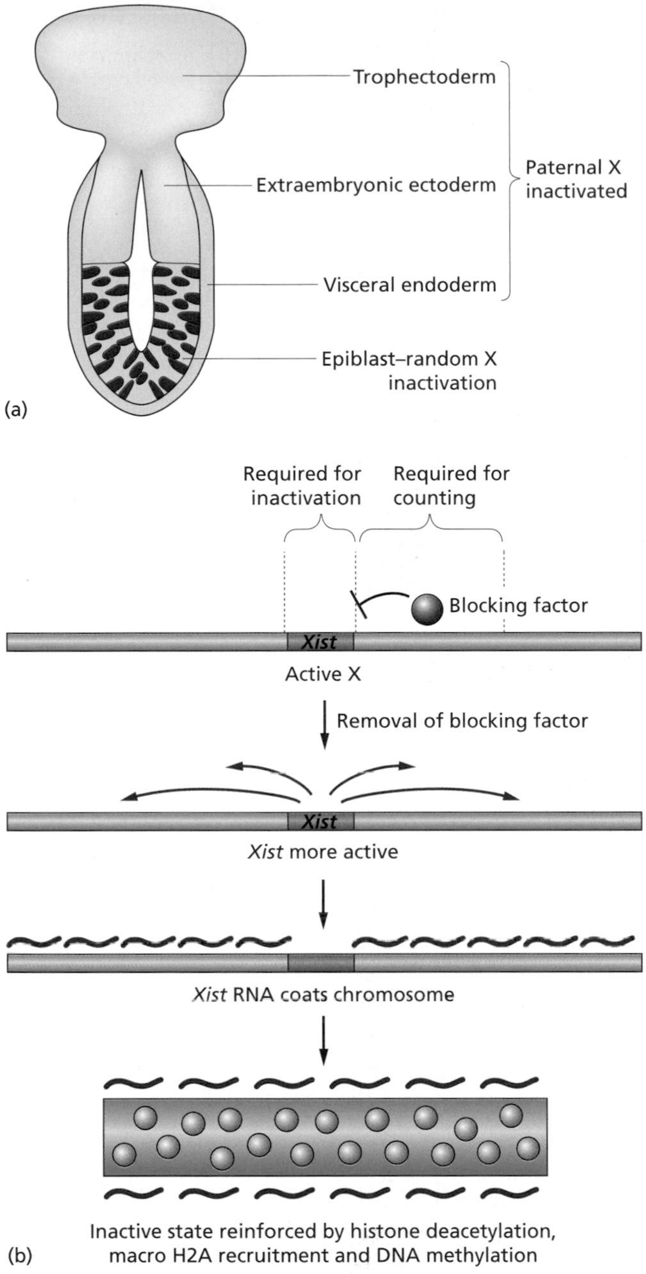

(a)

(b)

Fig. 10.25 Inactivation of the second X-chromosome in females. (a) Expression of beta-galactosidase in a heterozygous H253 mouse embryo which has *lacZ* on the paternal X. (b) Events of X-inactivation.

mammals are naturally mosaic for all X-linked heterozygous loci. X-inactivation is accompanied by late replication of the DNA relative to the rest of the genome, and by hypoacetylation of the histones on the inactive chromosome. In mouse, X-inactivation occurs earlier in the extraembryonic tissues than in the epiblast itself, and in these tissues the paternal X-chromosome is inactivated (Fig. 10.25a). This is known as imprinted X-inactivation and shows some differences from the random X-inactivation found in the embryonic tissues. X-inactivation can be easily visualized in the H253 strain of mice,

which carry a number of copies of *lacZ* on the X-chromosome. In female heterozygotes the chromosome carrying the *lacZ* is inactivated in 50% of cells, so when a specimen is stained with X-Gal individual cells are either blue or colorless. This enables some types of **clonal analysis**, as any structure formed from a single cell must become either all blue or all colorless but cannot be of mixed composition.

X-inactivation depends on the presence in the chromosome of an X-inactivation center (Xic). This contains a gene called *Xist* (pronounced "exist") which encodes a nontranslated RNA. *Xist* is critically important for X-inactivation and works only on its own chromosome, not on others within the same nucleus. It is active at a low level in all X-chromosomes of the early embryo. In the chromosomes that become inactivated the activity increases and the *Xist* RNA coats the whole of the chromosome. If the *Xist* locus is deleted then the chromosome cannot become inactivated. Conversely, a transgene for *Xist* can cause inactivation of the chromosome where it lies, even if this is an autosome. Experiments with an inducible *Xist* transgene in ES cells have shown that it will cause inactivation of its own chromosome when induced and that its presence is required for 48 hours for maintenance of inactivation. After this it is no longer required, as the inactivation becomes stabilized by other mechanisms, such as recruitment of a special histone, macroH2A, and histone hypoacetylation (Fig. 10.25b). The promoter for *Xist* is differentially methylated during gametogenesis, such that it is demethylated in sperm and methylated in oocytes. This pattern is maintained in the extraembryonic tissues, correlating with the preferential inactivation of the paternal X. In the epiblast, the imprint is erased and replaced by random methylation of the promoter. There is an antisense gene called *Tsix* that overlaps the *Xist* gene on the other DNA strand. This is active in maternally derived X-chromosomes and helps make the maternal X resistant to early inactivation.

X-inactivation must involve a counting mechanism that senses the ratio between the number of X-chromosomes and the number of autosomes. This is because individuals that contain multiple X-chromosomes inactivate all but one. The region of the Xic that "counts" lies at the 3′ end of the *Xist* locus. Deletion of this region causes preferential inactivation. A model consistent with these data involves a blocking factor, produced by autosomes, which is produced in limiting amount so that it can only block *Xist* expression on one X-chromosome. The binding sites for the blocking factor would lie in the "counting" region (Fig. 10.25b).

Teratocarcinoma

Most of the thoughts about the nature of ES cells and the uses to which they might be put were anticipated by earlier work on **teratocarcinomas**. These are malignant and transplantable tumors which consist of several types of tissue and also contain undifferentiated cells. The undifferentiated cells, which will grow in tissue culture, are considered to be the stem cells for the tumor and are often called embryonal carcinoma (EC) cells. Three types of teratocarcinoma can be distinguished: spontaneous testicular, spontaneous ovarian, and embryo derived. Spontaneous testicular teratocarcinomas arise in the testes of fetal male mice of strain 129. They are thought to arise from primordial germ cells and the well-known F9 cell line is of this type. Spontaneous ovarian teratocarcinomas arise in females of LT mice at about 3 months of age. They are derived from oocytes that have completed the first meiotic division and undergo approximately normal embryonic development as far as the egg cylinder stage. Embryo-derived teratocarcinomas can be produced by grafting early mouse embryos to extrauterine sites in immunologically compatible hosts, usually the kidney capsule or testis. The embryos become disorganized and produce various adult tissue types together with proliferating EC cells.

Different teratocarcinoma cell lines differ greatly in their properties: some need feeder cells to grow and others do not. Some will grow as dispersed cells in the peritoneal cavity (called ascites tumors) and others will not. Some will differentiate *in vivo* or *in vitro* while others will not. Most lines have an abnormal karyotype, although a few are normal or nearly normal. Most of the differences have probably arisen by selection of variants during establishment of the tumor and during culture, although it may to some extent also reflect a heterogeneity of the parent cell type. Chimeric mice can be produced by injection of some types of teratocarcinoma cells into blastocysts, but this does not work nearly as well as with ES cells. It does, however, have some theoretical interest since it shows that at least one sort of tumor can be made to revert to normal behavior if it is placed in the appropriate biological environment.

Key Points to Remember

- Fertilization comprises several distinct steps. Firstly, recognition of the zona pellucida by the sperm. Secondly, the acrosome reaction leading to release of hydrolytic enzymes that digest a path through the zona. Thirdly, the fusion of the sperm and oocyte plasma membranes with introduction of a phospholipase C into the oocyte cytoplasm. Fourthly, the elevation of intracellular calcium, which provokes the release of cortical granules, the completion of the second meiotic division of the oocyte, and the resumption of DNA synthesis ready for the first cleavage division.
- During the first 4 days of development the mouse embryo is in the preimplantation phase. It lies free in the oviduct or uterus and develops up to the blastocyst stage. Preimplantation embryos can easily be cultured *in vitro* and manipulated. Subsequent postimplantation development occurs anchored to the uterus by the placenta.
- The overall course of development resembles the chick, but there are distinct features associated with the egg cylinder arrangement found in rodents. Fate mapping shows that the animal pole of the early embryo becomes the distal tip of the egg cylinder.
- Transgenic mice are those containing an extra gene. This can be introduced by injection of the DNA into a pronucleus of a fertilized egg. The injected eggs are implanted into the

reproductive tract of a "foster mother" and reared to term. Then they are bred to create a transgenic line of mice. The specificity of transgene expression can be controlled by the promoter to which it is attached.
- Knockout mice are those in which a gene has been inactivated. They are made by homologous recombination of a targeting construct containing the defective gene into ES cells. The cells are injected into mouse blastocysts to create chimeric embryos. They are reimplanted into foster mothers and reared to term. Offspring that transmit the mutation through their gametes are used to breed a knockout line.
- The embryo body plan is established in the egg cylinder epiblast by inductive signals from the extraembryonic ectoderm and the anterior visceral endoderm. Nodal is essential for formation of the mesoderm and the node region has similar properties to the organizer in *Xenopus*. FGF and Wnt signaling are required for formation of posterior body parts.
- The breakage of symmetry that leads to left-right differences depends on the nodal cilia that drive fluid to the left.
- Some genes are expressed only from the maternal or the paternal chromosome (imprinting).
- In early embryos of females, one copy of the X-chromosome becomes inactivated.

Further reading

Website for gene knockouts

http://www.bioscience.org/knockout/knochome.htm

General

Theiler, K. (1989) *The House Mouse. Development and normal stages from fertilization to four weeks of age*, 2nd edn. Berlin: Springer-Verlag.

Kaufman, M.H. (1992) *The Atlas of Mouse Development*. London: Academic Press.

Alexandre, H. (2001) A history of mammalian embryological research. *International Journal of Developmental Biology* **45**, 457–467.

Nagy, M., Gertsenstein, M, Vintersten, K. & Behringer, R. (2002) *Manipulating the Mouse Embryo*. Cold Spring Harbor, NY: Cold Spring Harbor Laboratory Press.

Fertilization

Alberio, R., Zakhartchenko, V., Motlik, J. & Wolf, E. (2001) Mammalian oocyte activation: lessons from the sperm and implications for nuclear transfer. *International Journal of Developmental Biology* **45**, 797–809.

Wasserman, P.M., Jovine, L. & Litscher, E.S. (2001) A profile of fertilization in mammals. *Nature Cell Biology* **3**, E59–E64.

Runft, L.L., Jaffe, L.A. & Mehlmann, L.M. (2002) Egg activation at fertilization: where it all begins. *Developmental Biology* **245**, 237–254.

Saunders, C.M., Larman, M.G., Parrington, J. et al. (2002) PLCζ: a sperm-specific trigger of Ca^{2+} oscillations in eggs and embryo development. *Development* **129**, 3533–544.

Jungnickel, M.K., Sutton, K.A. & Florman, H.M. (2003) In the beginning: lessons from fertilization in mice and worms. *Cell* **114**, 401–404.

Talbot, P., Shur, B.D. & Myles, D.G. (2003) Cell adhesion and fertilization: steps in oocyte transport, sperm-zona pellucida interactions, and sperm-egg fusion. *Biology of Reproduction* **68**, 1–9.

Dean, J. (2004) Reassessing the molecular biology of sperm-egg recognition with mouse genetics. *Bioessays* **26**, 29–38.

Morphology, gene expression, fate map

Davidson, D., Bard, J.B.L., Brune, R., et al. (1997) The mouse atlas and graphical gene expression database. *Seminars in Cell and Developmental Biology* **8**, 509–517.

Tam, P.P.L. & Behringer, R.R. (1997) Mouse gastrulation: the formation of a mammalian body plan. *Mechanisms of Development* **68**, 3–25.

Brune, R.M., Bard, J.B.L., Dubreuil, C., et al. (1999) A three dimensional model of the mouse at embryonic day 9. *Developmental Biology* **216**, 457–468.

Davidson, B.P. & Tam, P.P.L. (2000) The node of the mouse embryo. *Current Biology* **10**, R617–R619.

Piotrowska, K. & Zernicka-Goetz, M. (2001) Role for sperm in spatial patterning of the early mouse embryo. *Nature* **409**, 517–521.

Tam, P.P.L., Gad, J.M., Kinder, S.J., Tsang, T.S. & Behringer, R.R. (2001) Morphogenetic tissue movement and the establishment of body plan during development from blastocyst to gastrula in the mouse. *Bioessays* **23**, 508–517.

Zernicka-Goetz, M. (2002) Patterning of the embryo: the first spatial decisions in the life of a mouse. *Development* **129**, 815–829.

Technology

Melton, D.W. (1994) Gene targeting in the mouse. *Bioessays* **16**, 633–638.

Müller, U. (1999) Ten years of gene targeting: targeted mouse mutants from vector design to phenotype analysis. *Mechanisms of Development* **82**, 3–21.

Smith, A.G. (2001) Embryo-derived stem cells. Of mice and men. *Annual Reviews of Cell and Developmental Biology* **17**, 435–462.

Nagy, A. & Rossant, J. (2001) Chimaeras and mosaics for dissecting complex mutant phenotypes. *International Journal of Developmental Biology* **45**, 577–582.

Gossen, M. & Bujard, H. (2002) Studying gene function in eukaryotes by conditional gene inactivation. *Annual Reviews of Genetics* **36**, 153–173.

Loebel, D.A.F., Watson, C.M., De Young, R.A. & Tam, P.P.L. (2003) Lineage choice and differentiation in mouse embryos and embryonic stem cells. *Developmental Biology* **264**, 1–14.

Ying, Q.L., Nichols, J., Chambers, I. & Smith, A. (2003) BMP induction of Id proteins suppresses differentiation and sustains embryonic stem cell self-renewal in collaboration with STAT3. *Cell* **115**, 281–292.

Early inductive events

Rossant, J. (1995) Development of extraembryonic lineages. *Seminars in Developmental Biology* **6**, 237–246.

Beddington, R.S.P. & Robertson, E.J. (1999) Axis development and early asymmetry in mammals. *Cell* **96**, 195–209.

Bielinska, M., Narita, N. & Wilson, D.B. (1999) Distinct roles for visceral endoderm during embryonic mouse development. *International Journal of Developmental Biology* **43**, 183–205.

Liu, P.T., Wakamiya, M., Shea, M.J., Albrecht, U., Behringer, R.R. & Bradley, A. (1999) Requirement for Wnt3 in vertebrate axis formation. *Nature Genetics* **22**, 361–365.

McMahon, A.P., Harland, R.M., Rossant, J., et al. (2000) The organizer factors Chordin and Noggin are required for mouse forebrain development. *Nature* **403**, 658–661.

Brennan, J., Lu, C.C., Norris, D.P., Rodriguez, T.A., Beddington, R.S.P. & Robertson, E.J. (2001) Nodal signaling in the epiblast patterns in the early mouse embryo. *Nature* **411**, 965–969.

Bachiller, D., Klingensmith, J., Kemp, C., et al. (2003) The Cdx1 homeodomain protein: an integrator of posterior signaling in the mouse. *Bioessays* **25**, 971–980.

Sutherland, A. (2003) Mechanisms of implantation in the mouse: differentiation and functional importance of trophoblast giant cell behavior. *Developmental Biology* **258**, 241–251.

Left–right asymmetry

Mercola, M & Levin, M. (2001) Left–right asymmetry determination in vertebrates. *Annual Reviews of Cell and Developmental Biology* **17**, 779–805.

Hamada, H., Meno, C., Watanabe, D. & Saijoh, Y. (2002) Establishment of vertebrate left–right asymmetry. *Nature Reviews Genetics* **3**, 103–113.

McGrath, J. & Brueckner, M. (2003) Cilia are at the heart of vertebrate left–right asymmetry. *Current Opinion in Genetics and Development*. **13**, 385–392.

Hox genes

Hunt, P. & Krumlauf, R. (1992) Hox codes and positional specification in vertebrate embryonic axes. *Annual Review of Cell and Developmental Biology* **8**, 227–256.

Burke, A.C., Nelson, A.C., Morgan, B.A. & Tabin, C. (1995) Hox genes and the evolution of vertebrate axial morphology. *Development* **121**, 333–346.

van den Akker, E., Fromental-Ramain, C., de Graaff, W., et al. (2001) Axial skeletal patterning in mice lacking all paralogous group 8 Hox genes. *Development* **128**, 1911–1921.

Wellik, D.M. & Capecchi, M.R. (2003) Hox10 and Hox11 genes are required to globally pattern the mammalian skeleton. *Science* **301**, 363–367.

Imprinting and X-inactivation

Moore, T. & Haig, D. (1991) Genomic imprinting in mammalian development: a parental tug of war. *Trends in Genetics* **7**, 45–49.

Tilghman, S.M. (1999) The sins of the fathers and mothers: genomic imprinting in mammalian development. *Cell* **96**, 185–193.

Avner, P. & Heard, E. (2001) X-chromosome inactivation: counting, choice, and initiation. *Nature Reviews Genetics* **2**, 59–67.

Cheng, M.K. & Disteche, C.M. (2004) Silence of the fathers: early X-inactivation. *Bioessays* **26**, 821–824.

Chapter 11

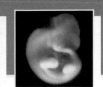

Drosophila

The first organism whose development was understood in molecular detail was the fruit fly *Drosophila melanogaster*. *Drosophila* is highly suited to genetic experimentation because of its small size and its short life cycle of 2 weeks. A number of very sophisticated techniques have been developed for constructing stocks and carrying out mutagenesis screens. Work on *Drosophila* has tended to start with mutagenesis to produce mutants with interesting-looking phenotypes. Then the genes are cloned and the expression patterns studied by *in situ* hybridization. The interactions between genes are then deduced by examining the expression of one gene in embryos in which there is loss of expression, or ectopic expression, of other genes. Finally more detail about the molecular biology may be obtained by *in vitro* studies of interactions between transcription factors and regulatory regions in the DNA.

Insects

All adult and larval insects are built up of an anteroposterior sequence of segments which fall into the three principal body regions of head, thorax, and abdomen. The prototype body plan is most clearly seen at the embryonic stage which is called the **extended germ band** and is the phylotypic stage at which all insect species display their maximum morphological similarity. The head may consist of as many as six segments: three **procephalic** and three **gnathal** (pronounced "naythal"), the gnathal segments bearing leg bud-like appendages which later become the mouthparts. The middle part of the body is the thorax which always consists of three segments: the **prothorax (T1)**, **mesothorax (T2)**, and **metathorax (T3)**. All of these develop a pair of legs on the ventral side and the meso- and metathorax also produce a pair of wings on the dorsal side. The number of abdominal segments varies with species but is usually in the range 8–11.

Drosophila, like other Diptera (two winged flies), follows the general insect pattern except that it does not display the procephalic head segments even in the embryo, the three gnathal segments appear only transiently, and only the mesothorax bears wings, the metathoracic wings being represented by small balancing structures called **halteres**. *Drosophila* has the usual three thoracic segments (T1–3) and has eight abdominal segments (A1–8). Although the conventional segments are

Classic Experiments

THE BREAKTHROUGH FROM MUTANTS TO GENE FUNCTION

Homeotic mutants had been known in *Drosophila* since the 1920s but few investigators were interested in them. One was Ed Lewis who published a classic study of the bithorax complex in 1978 and proposed a model for body-plan development. The full range of developmental genes was revealed by the mutagenesis screens of Christiane Nüsslein-Volhard and Eric Wieschaus which provided the raw material for the work of most labs in the 1980s. These three workers were awarded the Nobel Prize for Physiology in 1995.

The newly invented techniques of molecular biology made it possible to clone the genes that gave rise to developmental mutants and to study their expression patterns. This showed that the homeotic genes were indeed expressed in spatial domains, and that genes with periodic phenotypes, like the pair-rule genes, were expressed in periodic patterns.

Lewis, E.B. (1978) A gene complex controlling segmentation in *Drosophila*. *Nature* **276**, 565–570.

Nüsslein-Volhard, C. & Wieschaus, E. (1980) Mutations affecting segment number and polarity in *Drosophila*. *Nature* **287**, 795–801.

Akam, M.E. (1983) The location of Ultrabithorax transcripts in *Drosophila* tissue sections. *EMBO Journal* **2**, 2075–2084.

Hafen, E., Kuroiwa, A. & Gehring, W.J. (1984) Spatial distribution of transcripts from the segmentation gene fushi-tarazu during *Drosophila* embryonic development. *Cell* **37**, 833–841.

dominant in the larval and adult body plan, during early development the most important repeating units are the **parasegments**, which have the same period but are out of phase with the later segments (see below).

Inside the insect body the principal nerve cord is on the ventral side. The heart is on the dorsal side, and its action moves the **hemolymph** around the body cavity, there being no specialized vascular system as in vertebrates. Oxygen is brought to the tissues by diffusion through **tracheae**, which are long, branched ingrowths of the epidermis.

Drosophila is a holometabolous insect, meaning that it undergoes an abrupt and complete metamorphosis, so the egg hatches into a larva which is quite different in structure from the adult. The larva grows and passes through two molts before becoming a resting stage called a **pupa** in which the body is remolded to form the adult. Much of the adult body is formed from the **imaginal discs** and the abdominal histoblasts which are present as undifferentiated buds in the larva. Imaginal disc development is described in Chapter 17. Some other insect orders are **hemimetabolous**, meaning that the larva resembles the adult and acquires the final adult form via a series of nymphal stages separated by molts.

Normal development

Oogenesis

Events during **oogenesis** are very important for regional specification in the *Drosophila* embryo, and the egg is laid with a considerable amount of its pattern already specified. At the start of oogenesis, one germ cell divides four times to produce 16 cells, one of which becomes the **oocyte** and the other 15 all become **nurse cells**. Interestingly, the *par1* gene, important for cytoplasmic asymmetry in *C. elegans* (see Chapter 12), is also essential for oocyte formation in *Drosophila*, and without it all 16 cells become nurse cells. The whole cluster of oocyte and nurse cells is surrounded by ovarian follicle cells to form the **egg chamber** (Fig. 11.1). The follicle cells are derived from the gonads and are thus of **somatic** rather than **germ-line** origin. As the egg chamber enlarges, the follicle cells become divided into three populations. Those over the nurse cells are **squamous** in form and those over the oocyte are columnar. At both ends of the oocyte there is a special group of follicle cells called **border cells**, which are important in the determination of anteroposterior pattern. The nurse cells become polyploid and export large amounts of RNA and protein into the oocyte, contributing to its increase in size. In the later stages the oocyte becomes visibly polarized in both dorsoventral and anteroposterior axes and a granular **pole plasm** forms at the posterior end. The follicle cells secrete both the **vitelline membrane** and the **chorion** which is a tough outer coat surrounding the egg. The major production site for the **yolk** of the egg is the fat body of the female fly, the yolk proteins being carried to the ovary through the hemolymph.

Embryogenesis

Fertilization occurs in the uterus, the sperm entering at the anterior end through a hole in the chorion called the micropyle. Development of *Drosophila* is very fast compared with most other insects and the larvae hatch after less than 24 hours at normal laboratory temperatures. The early embryonic stages are depicted in Fig. 11.2. The initial period is called "cleavage" but as in most insects it is actually a period of rapid synchronous nuclear divisions without cellular cleavage. This is sometimes called **superficial cleavage** (see Chapter 2) and at this early stage the whole embryo forms a **syncytium**, in which all the nuclei lie in a common cytoplasm. After the first eight divisions, the **pole cells** are formed at the posterior end, incorporating those nuclei lying within the pole plasm. These later become the **germ cells**. After nine divisions most of the nuclei migrate to the periphery to form the syncytial blastoderm, those remaining internally are later incorporated into vitellophages which end up in the gut lumen. After four more nuclear divisions, during the third hour, cell membranes grow inwards from the plasma membrane to separate the nuclei and the **cellular blastoderm** is formed. At this stage there are about 5000 surface cells, 1000 yolk nuclei, and 16–32 pole cells. The division rate of the blastoderm cells slows dramatically, and the pole cells divide just once more

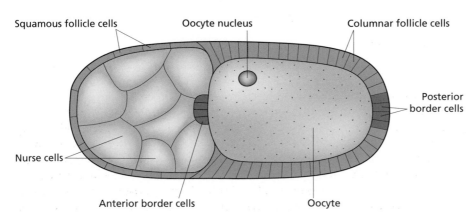

Squamous follicle cells   Oocyte nucleus   Columnar follicle cells

Posterior
border cells

Nurse cells

Anterior border cells   Oocyte

Fig. 11.1 The *Drosophila* egg chamber.

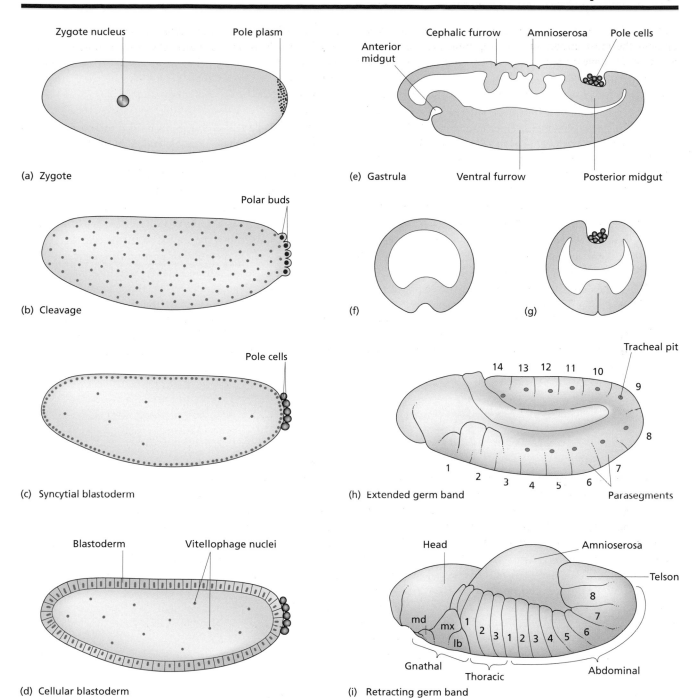

Fig. 11.2 Early development of *Drosophila*. (f) and (g) are, respectively, middle and posterior transverse sections of the gastrula.

before gastrulation. In insect embryology, positions along the anteroposterior axis of the early stages are often expressed as percentage egg length (%EL) from the posterior pole. So, for example, the region designated 100–50%EL is the anterior half, and the region 10–0%EL is the posterior tenth of the embryo.

The columnar epithelium of the blastoderm is quite thick and most of it is destined to become part of the embryo, only a thin dorsal strip becoming the extraembryonic amnioserosa.

Gastrulation commences at about 3 hours with the formation of a deep **ventral furrow** along much of the embryo length (Fig. 11.2e–g). This consists of a mesodermal invagination along the ventral midline, joined shortly later by invaginations of **anterior** and **posterior midgut** at the respective ends. The cephalic furrow appears laterally at 65%EL. Concurrent with gastrulation the germ band begins to elongate, driving the posterior end with the pole cells round to the dorsal side of the egg. By about

4 hours the first **neuroblasts** appear in the neurogenic ectoderm which is now mid-ventral, having closed over the ventral furrow.

Segmentation initially appears at the extended germ band stage. The initial repeating pattern is of **parasegments**. The definitive segments each form from the posterior two-thirds of one parasegment combined with the anterior third of the next. Although their morphological appearance is transient, parasegments are important because they are fundamental units for the construction of the body plan (see below).

At about 7.5 hours the germ band retracts and as it does so the epidermal grooves rearrange themselves so that they separate the definitive segments. The anterior and posterior midgut fuse in the middle and the ventral nerve cord becomes segregated. **Dorsal closure** of the epidermis occurs at 10–11 hours, displacing the amnioserosa into the interior. At about this time the head "involutes" into the interior and is therefore scarcely represented on the outer surface of the larva. This happens also in other Diptera but is unusual for insects in general. In later stages the Malpighian tubules (excretory organs) are formed at the junction of posterior midgut and hindgut; muscles, the fat body,

and the gonads arise from the mesoderm, and the central nervous system is formed by the ganglia of the ventral nerve cord.

Larval stages

The *Drosophila* larva has no legs, its head is tucked away in the interior, and it has three thoracic and eight visible abdominal segments. The specializations of the epidermal cuticle which are formed during late development are very important as they are the features used to assess the phenotypes of late embryo lethal mutations. On the dorsal side, the region from T2 to A8 is covered with fine hairs. On the ventral side are **denticle** belts on each of the thoracic and abdominal segments (Fig. 11.3). Each belt occupies mainly the anterior part of a segment but it also straddles the segment boundary and extends slightly into the posterior of the next segment. The thoracic and abdominal segments can be distinguished by the shapes and sizes of the denticle belts. Segment A8 bears the posterior spiracles, which are openings of the tracheal system. The extreme posterior is

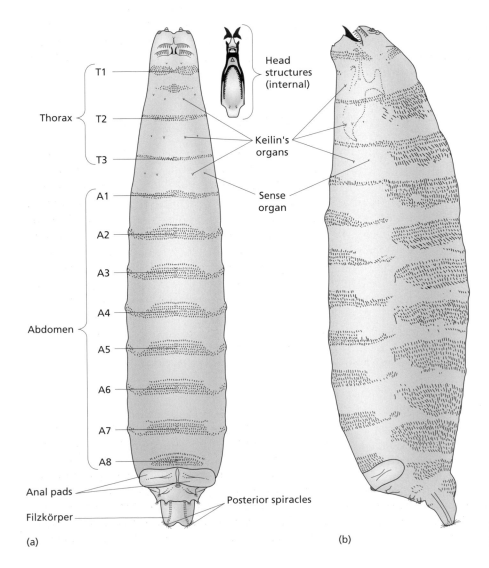

(a)

(b)

Fig. 11.3 *Drosophila* larva. (a) Ventral view. (b) Lateral view.

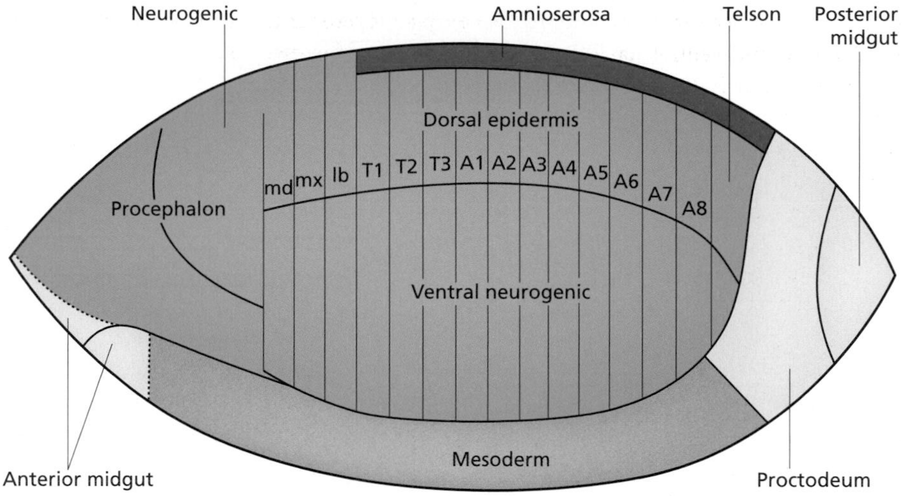

Fig. 11.4 Fate map of the cellular blastoderm stage. (After Hartenstein 1993. *Atlas of* Drosophila *Development*. Cold Spring Harbor Laboratory Press, p. 5.)

unsegmented and is called the **telson**. The structure of the larval head is very complex, but its most prominent component is a horny cephalopharyngeal skeleton secreted by the stomodeal part of the alimentary tract.

The **imaginal discs** are present as small nests of cells in the first-instar larva. They expand slightly during embryonic development, but most of their growth occurs during the larval stages. During pupation they differentiate and form the main epidermal structures of the adult body. The abdominal segments are formed from abdominal histoblasts which grow only in the pupal stage.

Fate map

Fate maps for the cellular blastoderm-stage *Drosophila* embryo have been constructed by localized UV irradiation to produce small defects, and by injection of cells labeled with **horseradish peroxidase** (Fig. 11.4). The principal features of the fate map are as follows: prospective regions exist for all the larval segments, there being no regions of indeterminacy representing later cell mixing or later growth from a small bud. The prospective regions for the recognizable gnathal, thoracic, and abdominal segments are arranged in anteroposterior sequence from 75% to 15% of the egg length. The prospective procephalic head structures occupy the anterior 25%. The prospective anterior midgut also maps to the anterior, while the posterior midgut, Malpighian tubules, proctodeum, and germ cells map to the posterior, as one would expect from the descriptive embryology.

Pole plasm

The pole plasm is formed during oogenesis and contains a **determinant** for germ cells, important components of which are the products of the genes *oskar* and *vasa*. Direct evidence that the pole plasm can program the nuclei to become germ cells was

obtained from transplants of pole plasm to the anterior end of host eggs at cleavage stages. Ectopic pole cells arise from those nuclei that are surrounded by the grafted pole plasm. If these ectopic pole cells are then grafted back to the posterior of a second host, they can be incorporated into the gonad and form functional germ cells. A number of maternal-effect mutations prevent the formation of the pole plasm and result in sterility of the resulting offspring.

Drosophila developmental genetics

The spectacular level of understanding of *Drosophila* development depends largely on the sophistication of the genetic methods available. The genome size is small in comparison with vertebrates, both in gene number and in DNA content, and is fully sequenced. The generation time is short and the animals are small, so experiments involving complicated breeding protocols and large numbers of individuals can be completed in a few weeks. The genetic maps are very detailed as are the cytological maps of the giant **polytene** chromosomes from the larval salivary glands. All these features have been vital for the rapid transition from the discovery of a potentially interesting mutant phenotype to the molecular identification and characterization of the gene responsible.

P-element

A transposable element known as the **P-element** is very important both for the creation of transgenic lines and for insertional mutagenesis. For transgenesis, the gene to be inserted is cloned into a P-element that also contains a marker gene to enable transformants to be identified. This is injected into the posterior pole of the egg and if it is incorporated into the genome of one or more pole cells then the resulting flies will produce some offspring containing the integrated P-element, and a stable line can

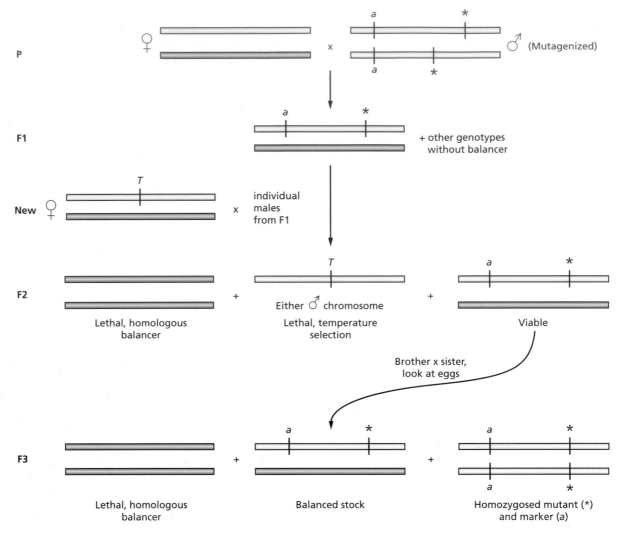

Fig. 11.5 A mutagenesis screen for zygotic autosomal recessive mutants affecting early development. The darker blue chromosome is a **balancer** which suppresses recombination, carries a visible dominant marker gene, and is lethal in homozygous form. "a" is a different visible marker gene, which is recessive. "T" is a dominant temperature-sensitive lethal. "*" represents a new mutation. The matings are carried out as shown and the F3 eggs are examined for abnormalities.

then be bred. The method has been used for several different purposes: to introduce enhancer traps (see also Chapter 10), to create strains carrying reporter constructs, to rescue endogenous mutations, and to express genes in an ectopic manner. For insertional mutagenesis, a line carrying a P-element is crossed to a line carrying a transposase enzyme. In the offspring the transposase can move the P-elements around in the genome, so the gametes contain a range of new insertions.

Identification of relevant genes

The *Drosophila* genome contains about 13,000 genes altogether, including about 5000 genes that can be mutated to lethality. By mutagenesis screening it has been possible to assemble a fairly complete collection of genes which, when mutated, give rise to pattern alterations in the embryo. There are several types of protocol for doing this, of which a simple example is shown in Fig. 11.5. Although the basic principle is the same as the zebrafish screen outlined in Chapter 3, the special tricks available for *Drosophila* select against the unwanted classes of recombinant and greatly reduce the labor involved. The screen illustrated in Fig. 11.5 is for autosomal recessive mutations and like most screens depends on the use of special **balancer chromosomes** which have three important features: they suppress recombination, they carry some visible marker gene, and they are lethal in the homozygous condition. Males carrying a visible marker (a – distinct from the balancer marker) are mutagenized and are then mated to females carrying the balancer. Each *individual* offspring fly represents one mutagenized gamete, so individual males from the F1 generation are isolated and crossed again. For creation of the F2, females are used which carry a dominant temperature-sensitive lethal mutation on the chromosome opposite the balancer and this enables selection against offspring not carrying the mutant chromosome. In the F2 generation each *tube* of flies represents one of the original mutagenized gametes. The viable F2 flies are mated with each other to produce the F3 generation. As the F2s are heterozygous for the mutagenized chromosome and the balancer, 25% of the F3 will have a homozygosed mutant chromosome. If these homozygous mutants are viable, they will display the recessive marker phenotype (*a/a*) from the mutagenized males. If this marker is not visible, then the dead embryos or larvae are examined to see whether they have any significant pattern abnormality. In the case where the mutants are lethal the stock is maintained by breeding from the heterozygotes that have one mutant chromosome and one balancer. These will breed true because both the possible homozygous chromosome combinations that can arise from mating are lethal. The original mutagenesis will induce mutations on all the chromosomes, but those that do not lie on the balanced chromosome are mostly lost in the outcrosses, so the method enables the screen to be carried out for one chromosome at a time. Once obtained, mutations are sorted into groups of alleles of the same gene by **complementation tests**, and are genetically mapped.

In *Drosophila*, much pattern information is laid down during oogenesis and mutations in genes required during oogenesis manifest themselves as **maternal effects**, meaning that the structure of the embryo corresponds not to its own genotype but to the genotype of the mother (see also Chapter 3). Strictly, when discussing maternal-effect mutations one should always say "eggs from mutant mothers" rather than "mutant eggs" but this convention is rarely adhered to. Maternal-effect genes are often identified as female sterile mutations in screens similar in principle to those for zygotic lethals. But some important maternal-effect mutations are also zygotic lethals. In other words the gene is needed for embryonic development as well as for oogenesis. In such a case the homozygous individuals do not grow up to become flies whose fertility can be tested, so maternal screens are usually incomplete. It is, however, possible to test individual zygotic lethals for maternal effects by making mosaic embryos in which the germ line is mutant but the somatic tissues are wild type (see below).

Types of mutation

Most mutations are reduced function or **loss of function**, representing the production of a smaller amount of gene product, or a gene product of reduced effectiveness. These are usually recessive. The alleles are called **hypomorphs** and the phenotypes are called **hypomorphic**. In *Drosophila* genetics much effort goes into the creation of several alleles for each locus to obtain at least one that has lost all function, the so-called **null** alleles giving amorphic phenotypes. Frequently the hypomorphic alleles can be arranged in a series of increasing severity with the amorphic phenotype as the limit form. Such **allelic series** can often give useful information about function, particularly where the weaker alleles give something recognizable and the stronger ones do not.

Some alleles may be **temperature sensitive**, usually because the mutation affects the thermal stability of the protein product. In general the protein is active at a low temperature (the **permissive** temperature) but inactive at a high temperature (the **nonpermissive** temperature). Temperature-sensitive mutants are useful because they can help to establish the developmental stages at which a gene product is required. This is done by shifting between the temperatures, and if the shift to the nonpermissive temperature is made before the time of gene function then most cases will be mutant, while if it is made afterwards, most cases will be normal.

Dominant mutations are sometimes loss of function, in those cases where the locus is **haploinsufficient**. This means that a reduction in the level of the product to 50% of the wild-type level is sufficient to cause a mutant phenotype. In such cases the homozygous phenotype will be more severe than the heterozygous one. More often dominant mutations are **gain of function**, resulting in the production of active gene product in positions or at times when it is not normally found, or **dominant negative** (= **antimorphic**), where the mutant version of the gene product interferes with the function of the wild-type version.

The names of *Drosophila* developmental genes usually derive from the appearance of the mutant phenotype. This means that they may indicate the opposite to the normal function. For example the *dorsal* gene is so called because null mutants are dorsalized, but this is because the normal function of the gene is to form ventral parts. Some genes are named after the adult phenotypes of viable alleles that were discovered before the big screens for embryonic lethal mutations. For example *Antennapedia* is so named because of a gain-of-function allele that converts antenna to leg, although the null phenotype of *Antennapedia* is a conversion of parasegments 4 and 5 to parasegment 3 and is lethal at embryonic stages.

Cloning of genes

Historically, most of the *Drosophila* developmental genes were cloned by **P-element mutagenesis**. When a P-element integrates into or near a gene it may mutate it to inactivity and it is possible to screen for a particular mutation by methods similar to those of Fig. 11.5. A mutation caused by P-element insertion can then be used as a starting point for cloning the gene. A genomic library from this strain is screened with the P-element probe and the resulting clones are tested by *in situ* hybridization to polytene chromosomes to find which particular P-element they represent. The one lying nearest to the genetic map position of the mutation can be used as the starting point for a chromosomal "walk" to obtain the whole of the required gene. Proof that the cloned candidate really *is* the required gene is obtained by three criteria:

1 that several known mutants have identifiable sequence changes in the candidate gene;

2 that the candidate gene is not expressed in null mutants (although sometimes null mutants can produce an inactive product);

3 by rescuing the phenotype of the null mutant by P-element-mediated transgenesis with the candidate gene.

Once the gene is cloned and sequenced the next step is to determine the normal expression pattern using *in situ* hybridization to different stage embryos. The next step is to make antibodies to the protein product, usually of a fusion protein expressed in bacteria, and use this to determine the protein expression pattern.

Gene function is studied by finding how the mutation of one gene affects the expression of others. If A turns on B, then removal of A will cause loss of B, and overexpression of A will cause corresponding ectopic expression of B. On the other hand if A normally represses B, then removal of A will cause ectopic expression of B, and overexpression of A will cause repression of B in its normal domains. But such results do not reveal whether the effects are direct, meaning that the protein product of A is actually interacting with gene B to regulate it. It could equally well be indirect, with gene A turning on something else that regulates gene B. If gene A codes for an inducing factor or receptor then the effect is necessarily indirect. If gene A codes for a transcription factor, then there are three methods of establishing directness:

1 It may be possible to demonstrate interactions between gene A–product and gene B–regulatory region, using **band shift assays**.

2 Genes A and B can be transfected together into *Drosophila* tissue culture cells to see whether B is turned on in the absence of other developmental machinery.

3 It may be possible to do a domain swap such that a new DNA recognition domain is put onto A and the corresponding DNA sequence is inserted into the regulatory region of B. If the effects of A on B are maintained under these circumstances then it must be due to a direct molecular interaction.

When the effect of A upon B is examined, it is quite common to look not at the endogenous B gene product but at a construct composed of the regulatory region of B fused to a reporter gene, usually *lacZ*. The reasons for this may be simply improved sensitivity if the endogenous product is hard to detect. Also in many constructs the β-galactosidase protein is quite stable and so its concentration effectively "integrates" the cumulative gene activity up to the time of examination. Most often it is because the regulatory region of B has been dissected into a number of parts to find which DNA sequences are responsible for each component of the expression pattern. When a reporter construct is used it may include just a single enhancer from the original regulatory region of gene B.

Classic Experiments

DISCOVERY OF THE HOMEOBOX

The homeobox was discovered as a DNA sequence that was present in all of the genes of the Bithorax and Antennapedia complexes. It was called the homeobox because of its location in homeotic genes. It was soon found to exist also in the DNA of other animals, including vertebrates. Excitement grew as it was thought that it might label all homeotic genes, or perhaps all genes concerned with segmentation, and be a real "Rosetta stone" for developmental genetics. Things are not quite so simple however, as it turned out that transcription factors containing the homeodomain DNA binding region are not confined to animals and do not have any single biological function. However within the animal kingdom a high proportion of them are indeed concerned with some aspect of development, and the Hox clusters

homologous to Antennapedia/Bithorax are homeotic genes concerned with anteroposterior patterning.

Carrasco, A.E., McGinnis, W., Gehring, W.J. & Derobertis, E. M. (1984) Cloning of an x-laevis gene expressed during early embryogenesis coding for a peptide region homologous to *Drosophila* homeotic genes. *Cell* **37**, 409–414.

McGinnis, W., Garber, R.L., Wirz, J., Kuroiwa, A. & Gehring, W.J. (1984) A homologous protein-coding sequence in *Drosophila* homeotic genes and its conservation in other metazoans. *Cell* **37**, 403–408.

McGinnis, W., Levine, M.S., Hafen, E., Kuroiwa, A. & Gehring, W.J. (1984) A conserved DNA sequence in homeotic genes of the *Drosophila* Antennapedia and Bithorax complexes. *Nature* **308**, 428–433.

Scott, M.P. & Weiner, A.J. (1984) Structural relationships among genes that control development – sequence homology between the *Antennapedia*, *Ultrabithorax* and *fushi tarazu* loci of *Drosophila*. *Proceedings of the National Academiy of Sciences USA* **81**, 4115–4119.

Hox genes

Although the homeobox and the Hox genes were first discovered in *Drosophila*, we have already met them in Chapters 4, 7, and 10. Like other animals, *Drosophila* contains a Hox cluster and also many other non-Hox homeobox genes that are involved in development, but are not part of the Hox cluster. The Hox cluster is on a single chromosome, but is split into two gene groups: the Antennapedia complex and the Bithorax complex. This split is probably quite recent in evolutionary history, as other insects that have been examined maintain a single cluster.

Overview of the developmental program

Because most of the genes involved are known, the genetic program of

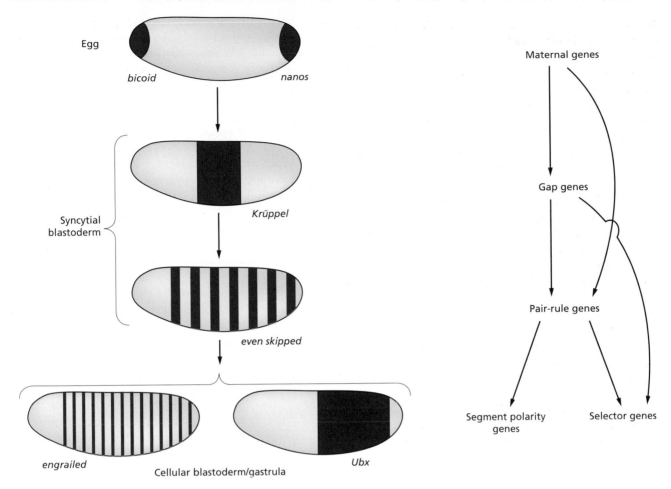

Fig. 11.6 Hierarchy of steps in the development of anteroposterior pattern.

Drosophila appears very complex, although in reality it is simpler than that of vertebrates. The detail can be understood if it is appreciated that the overall program can be regarded as a set of subprograms. One system operates in the **dorsoventral axis** and is responsible for the formation of mesoderm, neurogenic region, and epidermis from ventral to dorsal. As a result of a series of events in the egg chamber of the mother, the product of a maternal gene *dorsal* becomes distributed in the nuclei of the blastoderm in a ventral–dorsal gradient. It regulates a set of zygotic genes including *twist*, *rhomboid*, and *zerknüllt* which control formation of the various bands of tissue from ventral to dorsal.

An independent system operates in the **anteroposterior axis** and is more complex. It is shown in Fig. 11.6, with the expression patterns of a few of the key genes. The first phase of specification occurs in the egg chamber with the establishment of three maternal systems. The anterior system is concerned with the production of a gradient of the bicoid protein from *bicoid* mRNA localized in the anterior. The posterior system, of which the pole plasm is an integral part, deposits the mRNA of a gene called *nanos* in the posterior. There is also a **terminal system**, not shown in Fig. 11.6. The products of the maternal systems divide the embryo into several zones depending on their relative concentrations or activities. Before cellularization, one nucleus can affect the gene activity of nearby nuclei simply by producing a transcription factor; no receptors or signal transduction mechanisms are required. Because of the overlaps between the domains of activity, and because different concentrations of the same substance can have different effects, the maternal systems can activate a spatial pattern of zygotic gene activity which is more complex than their own. The genes activated at this stage belong to the gap class, e.g. *Krüppel*, and to the pair-rule class, e.g. *even skipped*. The gap genes are expressed in one or a few domains while the pair-rule genes are expressed in stripy patterns with a periodicity of two segment widths. Their periodicity arises because many different combinations of maternal and gap genes can activate the same pair-rule gene. The overlapping periodic patterns of pair-rule genes lead to the activation of a repeating pattern of **segment polarity genes**, e.g. *engrailed*, which have single segment periodicity and cause the subdivision of the axis into parasegments. Simultaneously the combined maternal, gap and pair-rule gene product combinations activate the Hox genes, e.g. *Ubx*, which control the character of each parasegment and thus its subsequent pathway of differentiation.

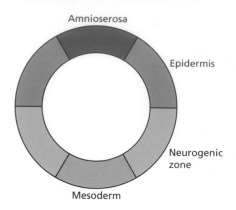

Fig. 11.7 Territories formed during dorsoventral specification of the early *Drosophila* embryo.

The dorsoventral pattern

The pattern along the dorsoventral axis is relatively simple. It consists at the cellular blastoderm stage of four strips committed to become, from ventral to dorsal, the mesoderm, the ventral neurogenic region, the dorsal epidermis, and the amnioserosa (Fig. 11.7). The disposition of these four territories is controlled by a ventral–dorsal nuclear gradient of the *dorsal* gene product, which is a transcription factor. To avoid confusion between dorsal position and the presence of the product of the *dorsal* gene, the gene product will be referred to here as "dorsal protein." The high point of the dorsal protein gradient depends on signaling from the ventral follicle cells (spätzle protein), and this signal exists because it is repressed on the dorsal side by a previous inhibitory signal from the oocyte (gurken protein). The dorsal protein gradient itself works by regulating various zygotic genes such that each ventral to dorsal strip of cells has a different combination of transcription factors active.

Maternal control of dorsoventral patterning

To understand how the dorsoventral system worked, a group of maternal-effect genes affecting dorsoventral pattern were identified from mutagenesis screens. For most of them the loss-of-function phenotype consists of a folded tube of larval cuticle bearing the fine hairs typical of the dorsal epidermis, but lacking structures normally derived from the lateral or ventral territories of the blastoderm. The prototype mutation of this **dorsalizing** class was called *dorsal*. In addition there are three genes, *gurken*, *torpedo*, and *cactus*, whose loss-of-function phenotype is ventralizing, with denticle bands extending all round the embryo. One gene, called *Toll*, has loss-of-function dorsalizing alleles and also has a gain-of-function allele with a ventralizing effect.

Several types of genetic and embryological experiment were important in understanding how these genes worked. Firstly, several of the genes have dorsalizing alleles with different degrees of function and these can be arranged in allelic series in which

structures are progressively lost from the ventral side until, in the amorphic embryos, only the symmetrical tubes of dorsal cuticle remain. This shows that the normal function of these genes is to promote ventral development.

Next, the genes can be arranged in a temporal series by making double mutants of which one has a dorsalizing and the other a ventralizing effect. Whichever predominates is assumed to act later in the developmental program (see also Chapter 3). Some idea of the time of action can also be gained by using temperature-sensitive mutants and shifting between the permissive and nonpermissive temperature at different times.

Next, it can also be asked whether a particular gene is required in the **germ line** (i.e. the oocyte itself and the nurse cells) or in the **soma** (i.e. the ovarian follicle cells). To answer this, embryos were created by grafting mutant pole cells into normal embryos. When these are grown up the females will produce offspring that are mutant if the gene was required in the germ line, or normal if it was required in the soma. Studies of this sort showed that some genes are required in the germ line (e.g. *gurken*, *Toll*, *dorsal*) and some in the soma (e.g. *torpedo*, *pipe*). Nowadays this type of experiment is usually done not by transplantation but by using the **FLP system** (see Chapter 17).

Finally, even before the genes were cloned and the gene products were identified, it was possible to find out something about their function by performing cytoplasmic **transplantations**. Most of the dorsal group mutants can be "rescued" towards a normal appearance by the injection of small amounts of cytoplasm from wild-type embryos before the pole cell stage.

These various types of experiment have led to the following account, in which the term "mutant" will indicate loss of function unless otherwise stated (Fig. 11.8a). The first known gene to act is called *K10*. The mutant has a dorsalizing effect and the function of its product is to sequester the mRNA for *gurken* in the vicinity of the oocyte nucleus. *gurken* and *torpedo* are two genes with similar ventralizing mutant phenotypes. *gurken* codes for a growth factor related to vertebrate TGFα, and is required in the oocyte. *torpedo* codes for a TGFα receptor and is required in the follicle cells. *gurken* mRNA is present just in the vicinity of the nucleus, which is positioned on the dorsal side of the oocyte, while *torpedo* is expressed all over the follicle cells. The protein product of *gurken* is secreted only on the dorsal side, it stimulates the *torpedo* product in the dorsal follicle cells thereby activating the ERK signaling pathway and causing changes of gene expression, in particular the repression of *pipe*, which would otherwise be activated. Mutations in *gurken* and *torpedo* are ventralizing because if the genes are inactive, then *pipe* becomes activated all over.

pipe is one of a group of genes whose function is to create an active extracellular ligand localized on the ventral side of the oocyte. *pipe* itself codes for an enzyme responsible for adding sulfate groups to heparan sulfate, an extracellular glycosaminoglycan. In normal development, *pipe* is expressed just in the ventral follicle cells, since expression on the dorsal side is repressed by the *gurken–torpedo* system. The local sulfation of heparan

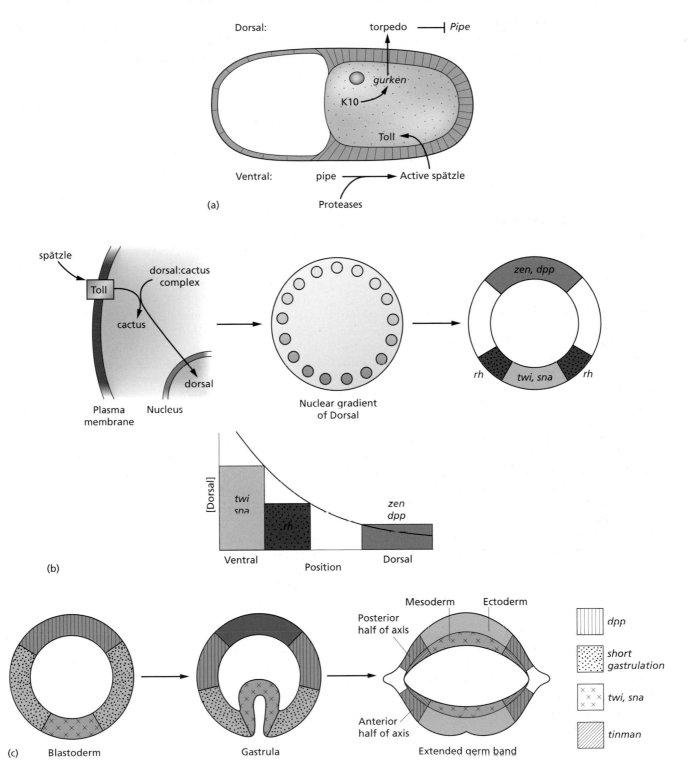

Fig. 11.8 Operation of the dorsoventral system. (a) Dorsally, the Gurken signal causes repression of *pipe*. Ventrally *pipe* is active and enables the activation of Spätzle (actually produced by the oocyte or nurse cells). (b) Spätzle activates Toll, which causes nuclear translocation of Dorsal, and Dorsal regulates zygotic genes. (c) The territories are later refined by means of the gradient of Dpp. *tinman* is maintained in the lateral mesoderm. Abbreviations: *twi, twist; sna, snail; rh, rhomboid; zen, zerknüllt; dpp, decapentaplegic; sog, short gastrulation.*

sulfate sequestrates a group of proteins on the ventral side which include proteases produced by the genes *snake* and *easter* which can activate the actual ventral signal. This signal is the protein product of the *spätzle* gene, active in the oocyte, and is synthesized as an inactive precursor requiring proteolytic cleavage for activation. The function of spätzle is to activate a receptor coded by the *Toll* gene which is present all over the surface of the oocyte. This has some homology to the vertebrate interleukin 1 receptor, although spätzle does not itself resemble interleukin 1. *Toll* has two types of mutant, recessive dorsalizing mutants in which receptor function is lost, and dominant ventralizing mutants in which the receptor is signaling continuously even in the absence of ligand.

The activation of *Toll* on the ventral side commences after fertilization and the subsequent events occur during the embryonic cleavage stages (Fig. 11.8b). *dorsal* is the final gene in the maternal dorsoventral pathway. The dorsal protein is a transcription factor homologous to the vertebrate factor NFκB. Its mRNA is uniformly distributed in the oocyte and the protein is synthesized after fertilization. The distribution is initially uniform but during the syncytial blastoderm stage it enters the nuclei preferentially on the ventral side, forming a ventral–dorsal gradient of nuclear protein. The entry to the nuclei depends on the proximity of activated Toll which causes dissociation of dorsal protein from a complex formed with another protein, IκB. This releases dorsal protein and allows it to enter the nuclei and regulate its target genes. *Drosophila* IκB is encoded by the gene *cactus*. *cactus* mutants are ventralizing because the normal role of cactus is to inhibit the action of dorsal protein, so in the absence of cactus, dorsal protein can enter the nuclei all over the embryo and make it ventral in character all over.

Zygotic control of dorsoventral patterning

The gradient of dorsal protein works by activating or repressing various transcription factors that are encoded by zygotic genes (Fig. 11.8c).

The **mesoderm** is defined by two transcription factors encoded by the genes *twist* and *snail*. These both code for transcription factors and are activated by dorsal protein at high nuclear concentration. *twist* encodes a bHLH protein and is required for correct mesodermal differentiation. *snail* encodes a zinc-finger transcription factor and is needed for invagination of the mesoderm.

Other genes are turned on at lower concentrations of dorsal protein so they become expressed laterally as well as ventrally. Some of these are repressed by snail, leading to a lateral stripe in the prospective neuroectodermal region, for example *rhomboid*, coding for a putative transmembrane receptor, is activated by dorsal protein and repressed by snail, such that it is expressed as a stripe in the neurogenic region.

The **dorsal ectoderm** is defined by a homeoprotein encoded by *zerknüllt*, normally expressed in about 40% of the embryo circumference. It is repressed by dorsal protein, and therefore becomes expressed all round in dorsalizing mutants. Another gene repressed by dorsal protein and normally expressed in the dorsal 40% zone is *decapentaplegic (dpp)*. This encodes a signaling molecule which brings about the patterning of the dorsal half of the embryo circumference and is a homolog of the vertebrate BMP4. Injection of *dpp* mRNA can induce amnioserosa at high dose and dorsal hairs at lower dose. Although this suggests gradient-like behavior, there is not a gradient of production of dpp protein itself. Instead the induction of dorsal genes is conducted by dpp acting together with another BMP homolog called screw, which is expressed ubiquitously. The graded effect of these factors arises because of the action of a inhibitor encoded by the *short gastrulation (sog)* gene. This is expressed in a lateral belt because it is repressed by *snail*, and it inhibits screw, leading to a dorsal–ventral gradient of the overall BMP-like activity. Sog is the homolog of the vertebrate chordin, also an inhibitor of BMP4 (see Chapter 7).

The patterning of the mesoderm also depends on dpp, whose expression later resolves into a pair of broad longitudinal stripes on the dorsal side. *tinman* encodes a homeodomain transcription factor necessary for heart formation. It is initially activated in the whole mesoderm by *twist* but then is turned off except in the lateral region that is adjacent to the *dpp*-expressing epidermis. This lateral region forms the heart while the ventral part forms the body wall muscles.

Thus, the full dorsoventral pattern arises from a maternal gradient of dorsal protein controlling the ventral half, and the zygotic gradient of dpp controlling the dorsal half. Interestingly the dorsal dpp–ventral sog pattern is homologous to the vertebrate ventral BMP4–dorsal chordin pattern, providing evidence that one of the groups must have turned upside down during evolution. Dpp and BMP4 have similar biological activity, as do chordin and sog. In this context it is significant that *tinman* has a vertebrate homolog *Nkx2.5*, and this is also needed for heart development (see Chapter 15).

The dorsoventral system provides examples of several key developmental processes. The gurken and spätzle proteins are both **inducing factors** that cause a regional patterning of their competent tissue. The localized region of activation of Toll is an example of a **cytoplasmic determinant** in the egg, in this case a determinant which is not made up of localized mRNA. The nuclear gradient of dorsal protein, although intracellular, exemplifies the conversion of a simple pattern into a more complex one through the formation of a **gradient**. The various dorsoventral tissue types arise because each is encoded by a combination of transcription factors turned on, directly or indirectly, by the dorsal protein gradient.

The anteroposterior system

The specification of structures along the anteroposterior axis is controlled by maternal systems which are largely, although not entirely, independent from that controlling the dorsoventral pattern. The basic mechanism is summarized in Fig. 11.9. In the anterior, mRNA for *bicoid* is deposited in the egg and a gradient of bicoid protein activates various genes to generate regional subdivisions (Fig. 11.9a). In the posterior, mRNA for *nanos* is deposited, and its protein product lifts the inhibition on gene activation in the future abdomen (Fig. 11.9b).

Anterior system

bicoid is a maternal effect gene coding for a homeodomain transcription factor. Loss-of-function mutations cause a deletion of the head and thorax. Transcription of *bicoid* occurs during oogenesis in both oocyte and nurse cells, and the mRNA becomes localized at the anterior of the oocyte. Study of the bicoid protein by antibody staining showed that it was synthesized during the syncytial stage at 1–3 hours of embryonic development and forms an exponential concentration gradient from anterior to posterior. The localized mRNA is the source and, as the protein appears to have a short half-life, the remainder of the embryo is the sink, and such a system will generate an exponential gradient as explained in Chapter 4.

The level of protein can be manipulated by changing the number of active copies of the gene in the female and this displaces in the expected directions features such as the cephalic furrow or stripes of pair-rule gene expression whose formation depends on the gradient.

A number of microsurgical experiments have been done on this system which show exactly the properties expected for a morphogen source at the anterior end of the egg. An effect similar to the *bicoid* mutation can be produced by pricking the egg at the anterior end and extruding about 5% of the cytoplasm, which contains most of the bicoid mRNA. Conversely, eggs

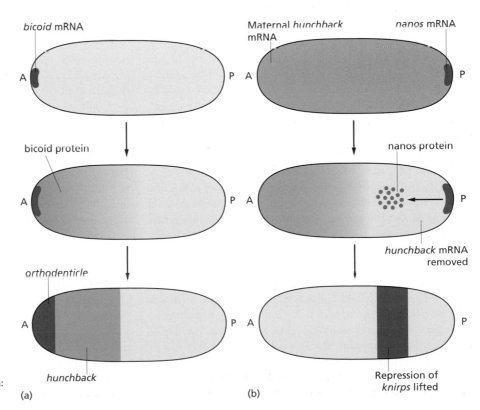

Fig. 11.9 Maternal anteroposterior system: (a) anterior system; (b) posterior system.

lacking *bicoid* can be rescued toward a normal phenotype by injection of *bicoid* mRNA. The position at which the injection is made determines the position of the induced anterior end, so, for example, injection of mRNA to a central position of a *bicoid*⁻ egg produces a head at the injection site, flanked by two thoraxes in mirror symmetrical arrangement.

The function of the bicoid protein gradient is to regulate the zygotic expression of several of the gap genes, such as *orthodenticle*, *hunchback*, and *Krüppel*. *orthodenticle* and *hunchback* both encode homeodomain transcription factors, and *Krüppel* encodes a zinc-finger transcription factor. Each of these genes has a promoter of different sensitivity to bicoid, so they become activated at different anteroposterior levels.

Posterior system

The posterior system depends on the deposition of mRNA for *nanos* in the posterior and the action of its protein product in lifting inhibition on the transcription of gap genes in the future abdomen (Fig. 11.9b). The mechanism of this system has been deduced from a combination of genetic and embryological experiments.

Pricking of the posterior pole and extrusion of a small volume of cytoplasm causes defects in the larva; not in the telson as might be expected, but in the abdomen whose prospective region lies around 50–20%EL. This indicates that something present at the posterior pole is needed for development of the abdomen. There are also a number of mutations in maternal-effect genes that cause loss of the abdomen. Among these are mutations in *nanos*, *oskar*, and *pumilio*. Mutant embryos for all these genes can be rescued toward normality by injection of cytoplasm taken from the pole plasm region of a wild-type embryo and injected into the abdominal region of the mutant, confirming the localization of an "abdomen-forming substance" at the posterior pole of normal embryos. Cytoplasm taken from the nurse cells of mutant egg chambers has a similar rescue activity, except in the case of the *nanos* mutant, from which the cytoplasm has no rescue activity. This shows that *nanos* must encode the functional end product of the pathway. *oskar* is in fact required for the formation of the pole plasm, and sequestration of the *nanos* mRNA, and *pumilio* is required for relaying nanos activity from pole plasm to prospective abdomen.

nanos codes for an RNA binding protein. Its mRNA is normally localized in the pole plasm, but the protein is found in the prospective abdomen. The function is to allow the transcription of the zygotic gap gene *knirps*. The mechanism of action involves a double repression with *knirps* being repressed by hunchback, and translation of *hunchback* being inhibited by nanos. As we have seen, *hunchback* is another zygotic gap gene and is activated by bicoid in the anterior. But *hunchback* is also active during oogenesis such that the egg is normally filled with a uniform concentration of maternally derived *hunchback* mRNA. Translation of this mRNA commences in early cleavage and is normally inhibited in the posterior half of the embryo by the nanos protein. *knirps* becomes turned on in the posterior but not in the anterior because hunchback protein is absent from the posterior. If nanos is missing then the hunchback protein is made all over and *knirps* cannot be turned on anywhere. Since activation of *knirps* depends on a double inhibition it follows that the nanos protein would not be necessary in the absence of the maternal *hunchback* message, and indeed this is the case as embryos from double mutant *hunchback*⁻/*nanos*⁻ mothers are near normal. Nanos is required again at a later stage for germ cell development, and for this it is produced from the zygotic gene.

Initial establishment of anteroposterior and dorsoventral polarity in the oocyte

The initial establishment of the anteroposterior polarity occurs early in oogenesis when the oocyte begins to elongate. At this stage an array of microtubules is formed and is orientated such that the minus end is anterior and the plus end posterior. The mRNAs for *bicoid* and *oskar* are translocated along these microtubules by **motor proteins** so that they end up anterior or posterior, respectively (Fig. 11.10). This tubule array can be demonstrated in flies transgenic for a gene for a fusion protein of kinesin with β-galactosidase. In the egg chambers of these flies, the fusion protein could be located by staining for β-galactosidase, and was found at the posterior end of the oocyte, along with *oskar* mRNA. Kinesin normally migrates to the plus end of microtubules and this experiment shows that the tubule array can serve as a polarized substrate for intracellular localization.

The direction of polarization of the tubules depends on the position of the oocyte in the egg chamber. In mutants of certain genes, such as *spindle-C*, the oocyte is located in the center rather than the posterior of the egg chamber. The oocytes then develop as double anterior in character, with the *oskar* mRNA central, and *bicoid* mRNA at both ends. The key genes required for the tubule polarization process have already been encountered as genes required for dorsoventral patterning, namely *gurken* and *torpedo*. In females lacking either of these, the specialized group of follicle cells called **anterior border cells** develop not just at the anterior but instead form at both ends of the oocyte.

From the behavior of this type of mutant, the course of events in normal development has been deduced. Early in oogenesis the oocyte lies at the posterior of the egg chamber. At this stage the border cells at both ends are committed to develop with an anterior character. *gurken* is expressed uniformly in the oocyte, but, because of the posterior location of the oocyte, gurken protein can only affect the neighboring border cells which thereby become switched to be posterior in character. The posterior state requires activation of torpedo at the cell surface and consequently the ERK signal transduction pathway internally. Following this, the posterior border cells emit another, as yet unknown, signal which brings about the polarization of the

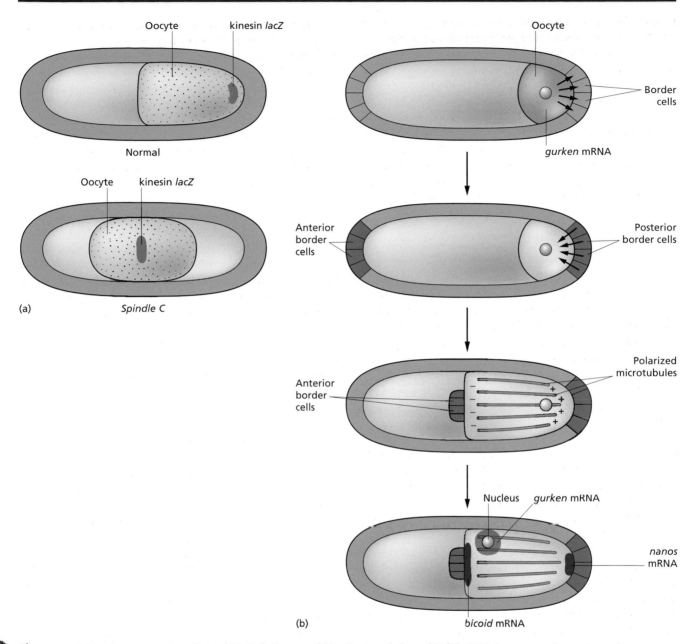

Fig. 11.10 Origin of anteroposterior polarity: (a) *Spindle* C mutant; (b) localization of *oskar* and *bicoid* mRNAs by transport along orientated microtubules.

microtubules in the oocyte such that the plus ends are directed to the posterior. The orientation of the tubules means that plus- or minus-directed motor proteins can, respectively, localize *oskar* mRNA to the posterior and *bicoid* mRNA to the anterior.

As the oocyte grows larger, the nucleus moves back towards the anterior along the microtubule array. The tubules are arranged around the exterior of the oocyte, so the nucleus has to follow a track around the exterior rather than travel straight down the central axis. This movement breaks the former radial symmetry of the oocyte and causes the nucleus to lie closer on one side than the other. Now the *gurken* mRNA, still near the nucleus, causes gurken protein to be secreted just from the side on which the nucleus is present. The adjacent follicle cells become dorsal in character because the gurken signal stimulates torpedo and thereby represses *pipe*, as described above.

Thus, the same genes that are responsible for the dorsoventral patterning of the follicle cells are also, at an earlier stage of oogenesis, responsible for the anteroposterior patterning of the oocyte and border cells. The reason that the same signal can be involved in polarization along two anatomically orthogonal axes is threefold: it acts at different times; the responding populations of follicle cells have different competence; and the intervening growth of the oocyte has changed the effective location of the signal.

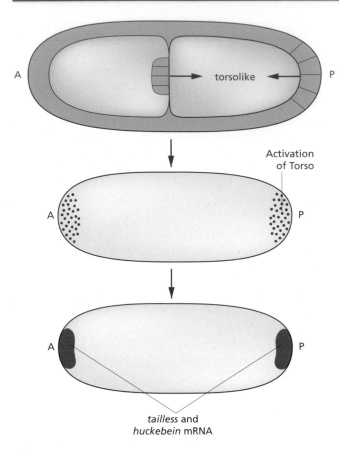

Activation
of Torso

tailless and
huckebein mRNA

Fig. 11.11 The terminal system.

Terminal system

A third maternal system is concerned with the formation of the embryo termini (Fig. 11.11). It involves a signal from the follicle cells activating a receptor at both termini and thereby turning on zygotic genes.

Embryos resulting from loss-of-function mutations of the terminal system show defects at both ends with a normal pattern in between. The key gene in the pathway is *torso*, which also has a gain-of-function allele causing substantial suppression of segmentation in the thorax and abdomen. Cloning and sequencing of *torso* revealed that it encoded a cell-surface receptor of the tyrosine kinase class. It stimulates the ERK signal transduction pathway and the gain-of-function phenotype arises from a constitutively active form of the receptor. As with *Toll*, expression is uniform in the oocyte and in normal development the receptor becomes activated at the termini. The ligand is encoded by *trunk*, which is active in the oocyte and belongs to the same gene family as *spätzle*. It is activated at the termini by the product of *torsolike*, encoding a novel protein, which is present in the anterior and posterior border cells.

Activation of *torso* leads, via the ERK pathway, to activation of two zygotic gap genes: *tailless* and *huckebein*. *tailless* encodes a transcription factor of the nuclear receptor family and *huckebein* encodes a zinc-finger transcription factor.

Although different molecules are involved, the terminal system has a mechanism remarkably similar to the dorsoventral system, and once again the activated receptor protein can be regarded as a type of **cytoplasmic determinant**.

Gap genes

The system of morphogens and determinants bequeathed by the mother becomes elaborated into increasingly complex patterns of gene activity by successive levels of the developmental hierarchy. The first zygotic level is made up by the **gap genes**, so called because mutant embryos have patterns bearing gaps of up to eight contiguous segments. All the gap genes code for transcription factors, and because the early embryo is a syncytium, these can diffuse from one nucleus to another and exert their effects directly, with no need for cell–cell signaling. Some important members of this group are *orthodenticle*, *hunchback*, *Krüppel*, *knirps*, and *giant*. The regulatory relationships have been deduced mainly by two types of experiment:

1 Examining the expression pattern of one gene in the absence of another. Expansion of a domain indicates repression and reduction of a domain indicates activation in normal development.
2 Examining the effect of uniform overexpression of another gene.

The type of result predicted from these two protocols is shown in Fig. 11.12.

Along with bicoid protein, another important early regulator of gap gene expression is the product of the *caudal* gene. This encodes a homeodomain transcription factor and is homologous to the *cdx* gene family in vertebrates. It is expressed maternally to produce a uniform distribution of mRNA in the oocyte. At the syncytial blastoderm stage, mRNA is differentially lost resulting in a posterior-to-anterior gradient of mRNA and protein. Embryos lacking *caudal* have severe posterior defects. Some gap genes are activated by both bicoid and caudal proteins, and since these two factors form inverse gradients in the early embryo, it means that the activation appears to be autonomous and not spatially regulated. *caudal* also has a posterior zygotic domain in the prospective proctodeum.

orthodenticle mutants have defects in the head. The gene encodes a homeodomain transcription factor. Expression is in the head and is activated by high levels of bicoid and by torso. An important vertebrate homolog of *orthodenticle* is *otx2*, which is required for the formation of the forebrain and midbrain.

Embryos homozygous for null alleles of *hunchback* have a large anterior gap which removes the labium and thorax. The gene codes for a zinc-finger transcription factor. Transcription during oogenesis leaves mRNA uniformly distributed in the egg, but as we have seen its stability and translation is antagonized in the posterior half by the nanos protein and so an anterior to posterior gradient of hunchback protein arises during cleavage.

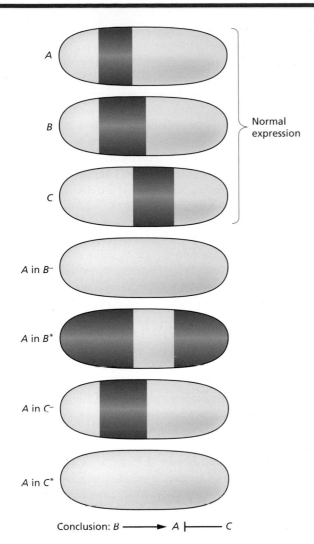

Fig. 11.12 How to work out regulatory relationships between genes. The top three drawings show the normal expression of three genes. The bottom four show expression of gene *A* in embryos mutant (−) or with ubiquitous overexpression (*) of *B* or *C*. The results show that *B* activates *A* and that *C* represses *A*.

Zygotic transcription commences in the syncytial blastoderm in the anterior half of the embryo, and at cellular blastoderm also in a posterior stripe. The anterior, but not the posterior, zygotic domain is activated directly by the bicoid protein at a lower concentration than that required to activate *orthodenticle*. The posterior zygotic domain is activated by torso.

Krüppel is an entirely zygotic gene whose null mutants have a large deletion of the central part of the body, comprising the thorax and abdominal segments 1–5. A second copy of abdominal 6 is often found in inverted orientation. The gene codes for a zinc-finger transcription factor. Expression commences in the syncytial blastoderm as a central band. *Krüppel* is activated by bicoid and hunchback, and repressed by knirps and giant. This ensures its activation as a broad band from about 60 to 50%EL.

Null mutants of *knirps* are similar to the maternal posterior group, having abdominal segments 1–7 replaced by a single large abdominal segment of uncertain identity. The gene codes for a transcription factor belonging to the nuclear receptor family. The expression pattern shows a band between about 45 and 30%EL, coming on at syncytial blastoderm. Activation is constitutive and the position of the main band in normal development is regulated by repression due to hunchback and tailless.

giant mutants have defects in the anterior thorax and in the abdomen at A5–A7. *giant* encodes a leucine zipper transcription factor and expression starts in the syncytial blastoderm in two zones, an anterior zone about 80–60%EL and a posterior zone about 33–0%EL. By the cellular blastoderm the posterior band is fading and the anterior one has resolved into three stripes. *giant* is activated by bicoid and caudal, and repressed in the anterior by hunchback.

tailless and *huckebein* are the zygotic genes activated by torso. The mutants each have terminal defects, which together add up to the phenotype produced by mutants of the maternal terminal genes.

The main regulatory relationships that have been mentioned are summarized in Fig. 11.13.

Pair-rule system

The **pair-rule genes** function as a layer in the developmental hierarchy between the gap genes and the segment polarity genes. They also have a role, along with the gap genes, in controlling the expression of Hox genes and aligning their domains with

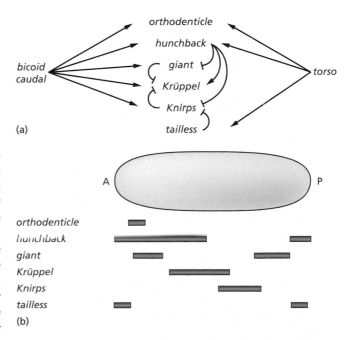

Fig. 11.13 Gap genes: (a) some regulatory relationships; (b) simplified expression domains of six gap genes.

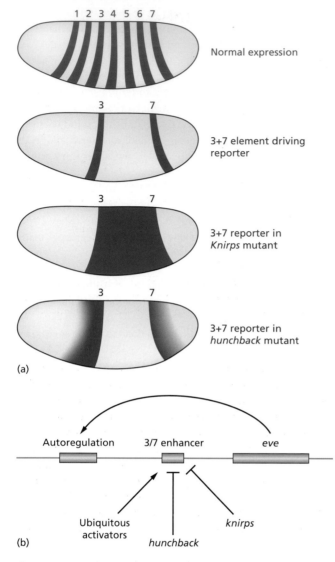

Fig. 11.14 Regulation of *even-skipped*: (a) normal expression pattern, reporter driven by 3+7 enhancer, effects of mutations on pattern of reporter expression; (b) two enhancers controlling *even-skipped* expression.

the segmental repeating pattern. The activation of the pair-rule genes represents the first formation of a reiterated pattern in the embryo.

All the pair-rule genes code for transcription factors and the expression patterns consist of seven stripes corresponding to two segment-wide bands of the syncytial blastoderm (Fig. 11.14). The mutant phenotypes typically show deletions with a periodicity of two segments, although more complex and more severe phenotypes can be found. For example *even-skipped* gets its name from the fact that the even numbered segments are reduced or lost in the original mutant. But this is now known to be a hypomorphic mutant, and in the null mutant all the segments are lost.

Those pair-rule genes designated primary are regulated mainly by the maternal and gap genes, while those designated secondary are regulated mainly by the primary genes. Primary pair-rule genes are *hairy*, *even-skipped (eve)*, and *runt*, and they have complex regulatory regions containing several enhancers. Among the secondary pair-rule genes are *paired*, *odd-paired (odd)*, *sloppy-paired (slp)*, and *fushi-tarazu (ftz)*. *ftz* is regulated to a large extent by repression due to *hairy* and so the *ftz* stripes appear in between the *hairy* stripes.

In general the rule is "one enhancer, one stripe." Each enhancer contains overlapping binding sites for activators and repressors such that it will be on if the activators prevail and off if the repressors prevail. To form a stripe it is necessary that the gene be turned on at one anteroposterior level and turned off again at a slightly different level. The *even-skipped* gene has been subject to very detailed analysis. It encodes a homeodomain protein which acts as a transcriptional repressor and the gene has large regulatory regions containing 12 enhancers. As an example, the stripe 2 element is activated by bicoid and hunchback and is repressed by giant and Krüppel, so in normal development stripe 2 is formed in the thin strip in between the *giant* and *Krüppel* domains. Two of the enhancers control pairs of stripes. The 4+6 and the 3+7 enhancers are both activated by ubiquitous components and repressed by hunchback and knirps. hunchback and knirps repress each others' transcription, ensuring that the *knirps* domain lies in between the two zygotic *hunchback* domains. The two enhancers show different levels of sensitivity to repression so each enhancer controls two expression stripes and they are nested such that stripes 3 and 7 form outside stripes 4 and 6 (Fig. 11.14). In addition to the stripe enhancers, there is an autocatalytic element which stabilizes the seven-stripe pattern once it is formed, and there are other elements active at later stages driving expression in the mesoderm or in subsets of neurons.

The way that the pair-rule transcription factors activate the 14 stripes of the segment polarity genes is very complex indeed. But in

New Directions in Research

As the development of *Drosophila* is rather well understood, its main role in the future is likely to be in the area of cell biology. The genetic tools for ablating genes in specific regions or labeling specific cell types are extremely powerful. These are increasingly enabling the investigation of phenomena such as cell polarity, cell movement, and intracellular trafficking of materials, in a way which is complementary to the continued use of mammalian tissue culture cells.

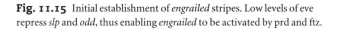

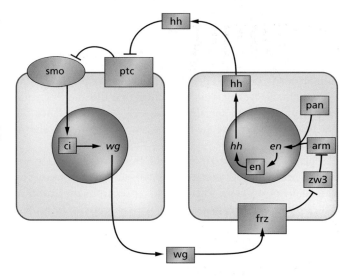

Fig. 11.15 Initial establishment of *engrailed* stripes. Low levels of eve repress *slp* and *odd*, thus enabling *engrailed* to be activated by prd and ftz.

Fig. 11.16 Maintenance of the pattern by the action of Hedgehog and Wingless systems. Abbreviations: *en, engrailed*; hh, hedgehog; ptc, patched; smo, smoothened; ci, cubitus interruptus; *wg, wingless*; frz, frizzled; zw3, zeste white-3; arm, armadillo; pan, pangolin.

principle it is a matter of creating 14 narrow bands in which activators predominate over inhibitors. For example the *engrailed* stripes of the odd-numbered parasegments are activated by *paired*, and *eve* enables this to occur by repressing the expression of *sloppy-paired*, such that there is a domain in which *paired* predominates over *sloppy-paired*. The *engrailed* stripes of the even-numbered parasegments are activated by *ftz* in the region where there is a slight overlap with *eve* (Fig. 11.15). This is because here *eve* represses expression of *odd-paired* and enables *ftz* to prevail. The fact that the odd- and even-engrailed stripes are regulated by different mechanisms explains the origin of the "pair-rule" phenotypes when expression of one pair-rule gene is altered. Note that the gene names *even-skipped* and *odd-skipped* relate to visible segment defects. Because the initially formed parasegments are out of phase with the final segments the *eve* mutants give odd parasegmental defects and *odd* mutants give even parasegmental defects!

Segment polarity system

The **segment polarity genes** function to create the **parasegment** boundaries of the early embryo. Once activated, they maintain their repeating pattern through a positive-feedback loop between the cells on either side of each boundary, defined by activity of the transcription factor genes *engrailed* and *cubitus interruptus (ci)*. The *engrailed* cells emit an inducing factor called hedgehog, which maintains activity of *ci* in the neighboring cells, and the *ci* cells emit a factor called wingless which maintains activity of *engrailed* in the neighboring cells (Fig. 11.16).

After cellularization of the blastoderm, it is no longer possible for one nucleus to influence another simply by producing a

transcription factor and allowing it to diffuse into the surroundings. In a multicellular embryo communication necessarily involves the secretion of inducing factors and the activation of cell-surface receptors. Whereas all the gap and pair-rule genes code for transcription factors, the segment polarity genes, which start to function after cellularization, may code either for transcription factors or for components of the signaling machinery. Most of the segment polarity genes have mutant phenotypes in which the segmental pattern of denticle bands is replaced by a continuous lawn of denticles. This spiky appearance is underlined by some of the gene names, such as *hedgehog, armadillo,* or *pangolin.*

engrailed codes for a homeodomain transcription factor and comes on at the cellular blastoderm stage, forming a 14-stripe pattern by the extended germ band stage. Each band characterizes the anterior quarter of a parasegment. Expression of *engrailed* is initially activated by the pair-rule genes as described above (Fig. 11.15). *ci* encodes a Gli-type zinc-finger transcription factor and is repressed by engrailed such that its expression pattern consists of 14 stripes in between the *engrailed* stripes.

The maintenance of the pattern depends on mutual interactions. This may be shown by the fact that if a gene required for one of the states is absent, the pattern is initially set up correctly, but it then rapidly decays. The key genes for maintenance make up the components of two intercellular signaling systems. In the *ci* cells, a factor called wingless is produced, homologous to the vertebrate Wnt factors. This stimulates a wingless receptor coded by two *frizzled* genes. The signal represses a kinase coded by the *zeste-white-3 (= gsk3)* gene, which in turn represses a protein coded by the *armadillo (= β-catenin)* gene. As two repressions equal one activation this means that Wingless activates armadillo and causes it to move into the nucleus, together with the product of the *pangolin* gene (= *Lef/Tcf*), to activate target genes. Among the targets is *hedgehog*, so hedgehog protein is secreted by the *engrailed* cells and binds to the receptor encoded by *patched*. Patched is a constitutively active repressor of smoothened, another cell-surface protein, which activates the Ci protein. Hedgehog inhibits patched, thus lifts the repression of smoothened and thereby enables Ci protein to enter the nucleus and to activate the *wingless* gene. The system is shown in Fig. 11.16.

There is an essential polarity to this system, because the signals hedgehog and wingless only activate their targets on one side and not on both. This polarity arises from the action of other pair-rule genes that restrict the competence to express *wingless* to the posterior half of the parasegment. The way the system works is well illustrated by examining the mutant phenotype of *patched*. Unlike many of the other segment polarity genes, mutants of *patched* do not show a lawn of denticles. Instead they show a remarkable pattern in which there are extra segment boundaries in between the normal ones, so the pattern approaches a 32-segment one. The reason is that patched is a repressor, so inactivation causes activation of the patched targets, including *wingless*. But *wingless* can only be turned on in its competence domain, roughly the posterior half of the parasegment, so its domain is enlarged but it does not become ubiquitous. Now, the anterior edge of the enlarged *wingless* domain abuts the competence domain of *engrailed*, leading to activation of an ectopic stripe of *engrailed*. Since a parasegment border appears at junctions between *engrailed* and *ci* regions, this causes the appearance of an ectopic border (Fig. 11.17). Exactly the same phenotype is produced by overexpression of *hedgehog*, driven by a ubiquitous promoter. This represses the activity of patched all over and hence also causes the enlargement of the *wingless* domain and the formation of an ectopic *engrailed* stripe.

The segment polarity genes all have vertebrate homologs that are important in development and have been mentioned in previous chapters (Table 11.1). Although these homologs exist and the biochemical pathways are substantially the same, this does not mean that the developmental functions are necessarily the same. The *engrailed–cubitus* loop is the motor of segmentation in insects and other arthropods, but it is probably not involved in vertebrate segmentation. Likewise, the Wnt pathway is an essential feature of dorsoventral polarity determination in *Xenopus*, but has no comparable function in *Drosophila*.

Hox genes

The *Drosophila* Hox complex is split into two regions called the Antennapedia and the Bithorax complexes. The Antennapedia complex corresponds to vertebrate paralog groups 1–6 and contains the genes *labial*, *proboscipedia*, *Deformed*, *Sex combs reduced*, and *Antennapedia*. The Bithorax complex corresponds to vertebrate paralog groups 7–10 and contains the genes *Ultrabithorax (Ubx)*, *abdominal-A*, and *Abdominal-B*. The

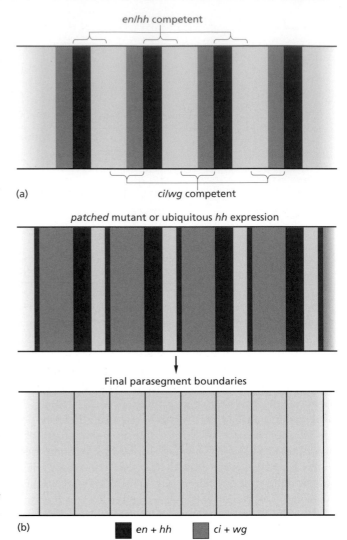

Fig. 11.17 (a) Normal segmentation domains. (b) The effect of ubiquitous expression of *hedgehog* is the same as that of removal of *patched*. *cubitus* becomes activated over its entire region of competence and the resulting Wingless signal induces a second file of *engrailed*-expressing cells from the region of *engrailed* competence. A parasegment border forms wherever *cubitus* and *engrailed* cells are juxtaposed, so the end result is to double the number of segments.

Table 11.1 Vertebrate homologs of segment polarity genes.

Drosophila	Vertebrate
engrailed	*en1* and *en2*
wingless	*wnt1–wnt12*
hedgehog	*sonic, Indian, desert, banded hedgehog*
cubitus interruptus	*gli-3*
frizzled 1,2	*Many frizzleds*
zeste-white-3	*glycogen synthase kinase 3*
armadillo	*beta-catenin*
pangolin	*lef/tcf*
patched	*patched*

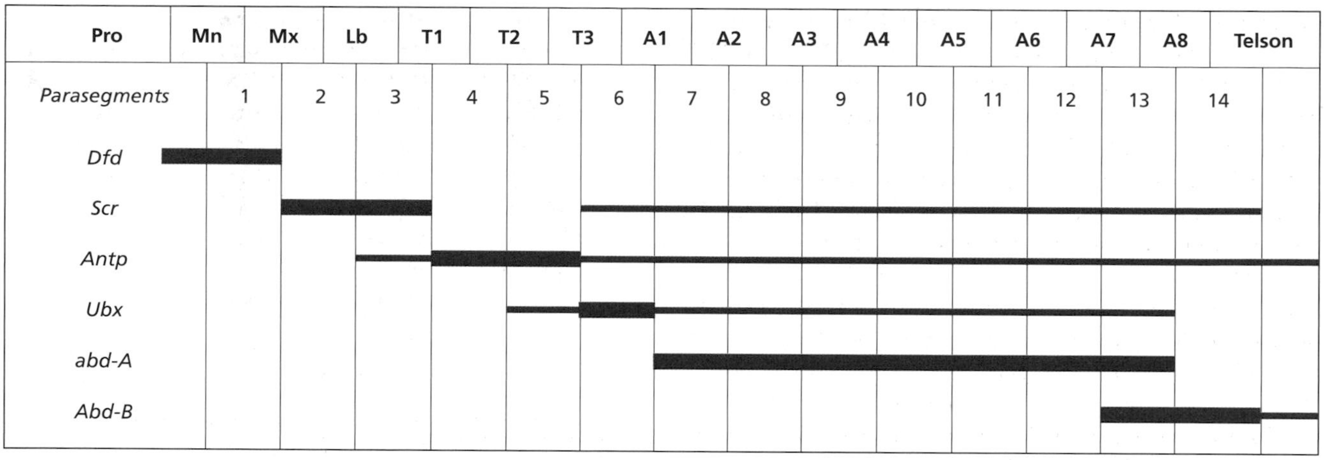

Pro	Mn	Mx	Lb	T1	T2	T3	A1	A2	A3	A4	A5	A6	A7	A8	Telson
Parasegments	1	2	3	4	5	6	7	8	9	10	11	12	13	14	

Rows below the parasegment header (Dfd, Scr, Antp, Ubx, abd-A, Abd-B) show bars indicating Hox gene expression domains.

Fig. 11.18 Schematic expression of Hox genes at the extended germ-band stage.

Antennapedia complex also contains *bicoid* and *zerknüllt*, which do not function as Hox genes in *Drosophila*, but have Hox-like homologs in other insects. Except for *Deformed* where a substantial protein concentration is achieved by the cellular blastoderm stage, the Hox genes are turned on slightly later than the gap and pair-rule classes, and their protein products have generally built up to an effective level by the extended germ-band stage. As in other animals, the order in which the genes are expressed in the anteroposterior axis is also the order of their arrangement on the chromosome.

The approximate expression domains of the Hox genes at the extended germ-band stage are shown in Fig. 11.18 (*proboscipedia* is expressed only in the larva). The function of these genes is to impart different characters to the different segments. In general if the expression domains are altered by mutation or by overexpression experiments, then the identity of the segments is altered in a predictable way. Loss-of-function mutations generally produce anteriorizations and gain-of-function mutations generally produce posteriorizations. However, their action does not depend simply on the combination of gene products present in a region, because individual structures within a segment may also be specified by a peak of expression of one of these genes. For example, expression of *Antennapedia* in parasegment 5 excludes the tracheal pits, which are the only cells of this parasegment to express *Ubx*.

The properties of the system are illustrated by a consideration of the *Ubx* gene, which is the homolog of the Hox7 paralog group in vertebrates. The expression of *Ubx* starts around cellularization and shows an initial peak in parasegment 6 and subsequent lesser expression from parasegments 5 to 13. It is also expressed in the mesoderm. Later expression is prominent in the ventral ganglia of these segments and in the metathoracic (T3) imaginal discs of the larva. Null mutants of *Ubx* show a transformation of parasegments 5 and 6 to parasegment 4.

Initial control of *Ubx* expression is by hunchback which acts as a repressor. There is an activation by fushi tarazu, which gives a transient pair-rule character to the *Ubx* pattern. This pair-rule control is important for aligning the parasegmental register of segment polarity and homeotic genes. Once it is established, the posterior boundary of *Ubx* expression is maintained by repression from abdominal-A. Ubx itself represses *Antp*, so maintaining its posterior boundary. Maintenance, at least in the visceral mesoderm, also depends on a positive feedback loop involving wingless and dpp. To ensure more permanent regulation there are groups of genes that are responsible for Hox maintenance in the long term, such as *Polycomb* and *extra sex combs*. How they work is poorly understood, but they are concerned with chromatin structure, and the demarcation of active and inactive domains within the genome. These genes tend to be ubiquitously active and repressive, so loss of function mutants give ectopic Hox activity (e.g. the *extra sex combs* mutant derepresses *Sex combs reduced* in T2–3, leading to the formation of male sex combs on the second and third legs as well as on the first leg where they are normally found). Homologs of the *Polycomb* group are found in all eukaryotes and are called **chromobox genes.**

The anteroposterior body pattern

The long and complex series of interactions described above leads to a fully specified body plan by the extended germ band stage. This specification has two essential components. There is the repeating pattern composed of parasegments whose boundaries are defined by the juxtaposition of bands of cells in which the *engrailed* (anterior) and *ci* (posterior) systems are active. There is also the nonrepeating sequence of Hox gene expression zones. Although the Hox domains overlap, there is a clear sequence from anterior to posterior in which a single gene predominates.

Each element of this pattern appears to be initiated locally by combinations of concentrations of the products of the maternal systems, the gap genes, and the pair-rule genes. The pair-rule genes have a particularly important role in that they must control the register between the segment polarity and the homeotic selector genes so that each segment acquires the correct identity. By the extended germ-band stage the products of the controlling systems have decayed or are decaying and so the maintenance of the pattern is ensured by separate means, such as: positive-feedback loops, for example of *Ubx*; mutual reinforcement of neighboring states, for example of *engrailed* and *ci*; or inhibition between neighboring states, for example of *Antp* by *Ubx*.

Up to the formation of the general body plan the regional specification of *Drosophila* is quite well understood. But the ultimate role of the developmental genes is to activate appropriate combinations of genes to carry out the differentiated functions of the relevant cells. How this is done is still poorly understood, and this is partly because the cell biology, histology, and general biochemistry of *Drosophila* is not nearly as well studied as that of vertebrates.

Key Points to Remember

- The mutagenesis screens for genes affecting early *Drosophila* development led to the discovery of most of the classes of gene that control development not just in *Drosophila* but in all animals.
- *Drosophila* and other insect embryos initially develop as a syncytium. This means that during the first 3 hours transcription factors can diffuse from one nucleus to another to control gene expression.
- The dorsoventral and anteroposterior patterns are specified largely independently. The dorsoventral pattern is set up by a series of interactions between oocyte and follicle cells, starting with the eccentric position of the oocyte nucleus and culminating in a gradient of dorsal protein in the nuclei of the syncytial blastoderm.
- The cytoplasmic determinants that control anteroposterior pattern are also laid down in the oocyte before fertilization. They are: *bicoid* mRNA in the anterior; *nanos* and *oskar* mRNA in the posterior; and activation of torso at the termini.

- Development proceeds in a stepwise fashion with each domain of cells being defined by the expression of a combination of transcription factors encoded by the gap and pair-rule genes. These regulate the expression of the segmentation genes, defining the 14 repeating units of the embryo; and the Hox genes that define the anteroposterior character of each parasegment.
- Segmentation is controlled by a cross-activation between stripes of engrailed-expressing cells that secrete hedgehog, and cubitus interruptus-expressing cells that secrete wingless.
- Anteroposterior pattern is controlled by transcription factors encoded by Hox genes which are activated in a nested pattern such that all are on at the posterior end with each gene having a specific anterior expression limit. Loss-of-function mutants generally cause anteriorization, while gain-of-function mutants generally cause posteriorization.

Further reading

See also the general textbooks referenced in Chapter 2.

Website for data on *Drosophila* genes
The Interactive Fly: http://www.sdbonline.org/fly/aimain/1aahome.htm

General
Lawrence, P.A. (1992) *The Making of a Fly*. Oxford: Backwell Science.
Bate, M. & Martinez Arias, A., eds (1993) *The Development of Drosophila melanogaster*, vols 1, 2. Cold Spring Harbor, NY: Cold Spring Harbor Laboratory Press.
Campos-Ortega, J.A. & Hartenstein, V. (1997) *The Embryonic Development of* Drosophila melanogaster. Berlin/Heidelberg: Springer-Verlag.

Gehring, W.H. (1998) *Master Control Genes in Development and Evolution. The homeobox story*. New Haven: Yale University Press.

Genetic screen
Nüsslein-Volhard, C. & Wieschaus, E. (1980) Mutations affecting segment number and polarity in *Drosophila*. *Nature* **287**, 795–801.

Dorsoventral patterning
Leptin, M.(1995) *Drosophila* gastrulation: from pattern formation to morphogenesis. *Annual Review of Cell and Developmental Biology* **11**, 189–212.
Morisato, D. & Anderson, K.V. (1995) Signaling pathways that establish the dorso-ventral pattern of the *Drosophila* embryo. *Annual Review of Genetics* **29**, 371–399.

Rusch, J. & Levine, M. (1996) Threshold responses to the dorsal regulatory gradient and the subdivision of primary tissue territories in the *Drosophila* embryo. *Current Opinion in Genetics and Development* **6**, 416–423.

Anderson, K.V. (1998) Pinning down positional information: dorsoventral polarity in the *Drosophila* embryo. *Cell* **95**, 439–442.

Stathopoulos, A. & Levine, M (2002) Dorsal gradient networks in the *Drosophila* embryo. *Developmental Biology* **246**, 57–67.

Maternal anteroposterior systems

St Johnston, D. & Nüsslein-Volhard, C. (1992) The origin of pattern and polarity in the *Drosophila* embryo. *Cell* **68**, 201–219.

Munn, K. & Steward, R. (1995) The anteroposterior and dorsoventral axes have a common origin in *Drosophila melanogaster*. *Bioessays* **17**, 920–922.

López-Schier, H. (2003) The polarisation of the anteroposterior axis in *Drosophila*. *Bioessays* **25**, 781–791.

Zygotic anteroposterior systems

Pankratz, M.J. & Jäckle, H. (1990) Making stripes in the *Drosophila* embryo. *Trends in Genetics* **6**, 287–292.

Finkelstein, R. & Perrimon, N. (1992) The molecular genetics of head development in *Drosophila melanogaster*. *Development* **112**, 899–912.

Forbes, A.J., Nakano, Y., Taylor, A.M. & Ingham, P.W. (1993) Genetic analysis of hedgehog signaling in the *Drosophila* embryo. *Development* (Suppl.) 115–124.

Rivera-Pomar, R. & Jäckle, H. (1996) From gradients to stripes in *Drosophila* embryogenesis: filling in the gaps. *Trends in Genetics* **12**, 478–483.

Small, S., Blair, A. & Levine, M. (1996) Regulation of two pair-rule stripes by a single enhancer in the *Drosophila* embryo. *Developmental Biology* **175**, 314–324.

Hatini, V. & DiNardo, S. (2001) Divide and conquer: pattern formation in *Drosophila* embryonic epidermis. *Trends in Genetics* **17**, 574–579.

González-Gaitán, M. (2003) Endocytic trafficking during *Drosophila* development. *Mechanisms of Development* **120**, 1265–1282.

Evolution

Patel, N.H. (1994) The evolution of arthropod segmentation: insights from comparisons of gene expression patterns. *Development* (suppl.), 201–207.

Gellon, G. & McGinnis, W. (1998) Shaping animal body plans in development and evolution by modulation of Hox expression patterns. *Bioessays* **20**, 116–125.

Hox genes

Mann, R.S. & Morata, G. (2000) The developmental and molecular biology of genes that subdivide the body of *Drosophila*. *Annual Review of Cell and Developmental Biology* **16**, 243–271.

Chapter 12

Caenorhabditis elegans

Caenorhabditis elegans is a small, free-living soil nematode and has been used for developmental biology research since the 1960s. Among developmental biologists it is usually known as "the worm." In one sense it is the best known animal on Earth since the location and lineage of every cell in embryo, larva, and adult is known. Also its genome was the first of any animal to be completely sequenced. The genome contains 19,099 genes, of which about 2000 are mutatable to lethality.

Caenorhabditis elegans is kept on petri plates and feeds on bacteria. Genetic screening is easy because it is possible to examine large numbers of worms, and the generation time is only 3 days. The worms are self-fertilized **hermaphrodites**, so recessive mutations will automatically segregate as homozygotes in two generations without the need to set up any crosses (Fig. 12.1). Genetic stocks can be preserved in liquid nitrogen. This ease of genetic analysis means that large numbers of mutants are available, and, as in *Drosophila*, investigation of a biological problem often starts with a mutant affecting the process in question. As

an alternative to the isolation of mutants it is now also easy to inhibit gene action by **RNA interference** (see Chapter 3). If double-stranded RNA complementary to an endogenous message is introduced this results in the production of embryos resembling the corresponding maternal effect mutation (**phenocopies**). The dsRNA can be administered by injection into the somatic tissues of the worm, or by feeding. A convenient method is to express the required dsRNA from a plasmid in *E. coli* and then use these bacteria as the food for the worms: sufficient dsRNA is absorbed intact to exert its biological effect.

It is also easy to make **transgenics** by injection of the required DNA into the gonad, where it is incorporated as an extra-chromosomal element into the germ cells. Such transgenics are not stable because the element can be lost at meiosis or mitosis. However, transient transgenesis is often sufficient for experimental purposes.

It is possible to make **genetic mosaics** for some parts of the genome. Mosaics arise by the spontaneous loss of small free

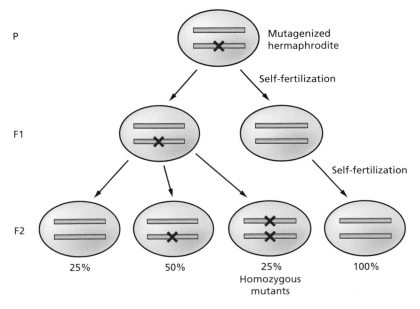

Fig. 12.1 Segregation of homozygotes after mutagenesis and two generations of self-fertilization. In the P generation a particular mutation will be found in just a few gametes, having occurred in a single germ cell.

chromosome fragments which are duplications of normal chromosomal regions. If the main chromosome carries a mutant allele and the free chromosome fragment the wild-type allele, then when this is lost from a single cell its descendant cells will all be mutant but the rest of the animal will be wild type. Mosaics can be very useful for establishing in which region of the embryo the function of a gene is required.

The negative features of *C. elegans* are the small size of the eggs and the tough egg case, both of which make microsurgical experiments difficult.

In *C. elegans*, it is conventional to capitalize the names of proteins. So for example the *glp-1* gene encodes the GLP-1 protein. As usual, many mutations affecting early development are maternal-effect and in such cases it is the genotype of the mother and not the zygote that determines the embryo phenotype. In *C. elegans* a gene that is lacking maternally will also usually be lacking in the embryo. This is because the normal mode of propagation from a self-fertilized hermaphrodite means that a −/− parent will produce −/− zygotes, so both will lack the gene. But it is still necessary to remember that the gene products that control early development are deposited in the egg during oogenesis.

Normal development

Adult anatomy

The adult is highly elongated ("worm shaped"; Fig. 12.2). The outer layer is the **hypodermis** which is one cell thick, largely syncytial, and secretes a thick cuticle. Beneath the hypodermis are four longitudinal bands of mononucleate muscle cells. There is a through gut with a muscular pharynx and an intestine. There is a nerve ring surrounding the pharynx, a ventral nerve cord, and tail ganglia. The main body cavity of nematodes is described as a **pseudocoelom** rather than a coelom because it is not lined all round with mesoderm. The **gonad** opens into a mid-ventral vulva. In the hermaphrodite this has two arms which are both bent back on themselves. Within the gonad, the cells nearest the vulva mature as sperm, while the more distant ones divide continuously as a syncytium and then become cellularized as **oocytes**. These mature into eggs, become fertilized as they encounter the sperm on their way out, and are laid as cleavage stage embryos.

Although most worms are hermaphrodite, there are also occasional males whose gonad has just one arm and opens posteriorly at the cloaca. Hermaphrodites have an XX-chromosome constitution while males are XO. They arise when an X-chromosome is lost by disjunction during meiosis. If a male and hermaphrodite mate, then the male sperm outcompete those of the hermaphrodite, resulting in an outcross.

Even in the adult stage nematodes have rather few cells, and during embryonic and larval development *C. elegans* shows almost complete invariance of cell lineage, meaning that every individual embryo shows exactly the same sequence and orientation for every cell division. Embryos are laid at about 30 cells, hatch at about 14 hours with 558 cells The larva feeds and grows, undergoing four molts before reaching the adult stage with 959 somatic cells plus about 2000 germ cells. The first stage larva also has the option of entering a dormant **dauer larva** phase if nutrients are in short supply (see Chapter 18). After the last molt the adult worm shows no further cell division of somatic tissues and can grow only by cell enlargement.

C. elegans does possess a HOX cluster containing six Hox genes, although as there are some intervening genes it is not a true cluster. The genes obey the rule of colinearity of chromosomal position and anterior expression limit. They are called: *lin39, ceh13,* [gap], *mab5, egl5,* [gap], *php3, nob1.* The last three belong to the Abdominal B, or posterior class. Only *ceh13, php3,* and *nob1* have embryonic phenotypes in loss-of-function mutations.

Embryonic development

Fertilization in *C. elegans* is somewhat unusual. The sperm are amoeboid, with no flagellum or acrosome. Oocytes are fertilized before the first meiotic division. The sperm can enter at any position and the point of sperm entry defines the future **posterior** of the zygote. Following fertilization and the completion of meiosis, there is a cytoplasmic rearrangement associated with a "pseudocleavage" or formation of a furrow which does not progress to a full cleavage. The early cleavages are asymmetrical (Fig. 12.3). The first forms an anterior AB and posterior P cell. AB then forms ABa and ABp while P behaves in a **stem cell**-like manner, keeping a P daughter while successively cutting off EMS, C and D blastomeres. The residual P cell (P4) is the **germ**

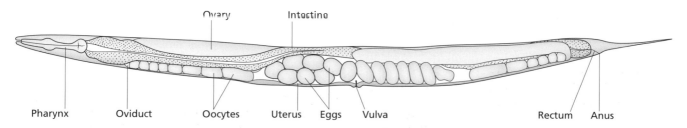

Fig. 12.2 Adult anatomy of *C. elegans.* (After Sulston. In: Bard 1994. *Embryos.* Wolfe, figure 4.4, p. 56.)

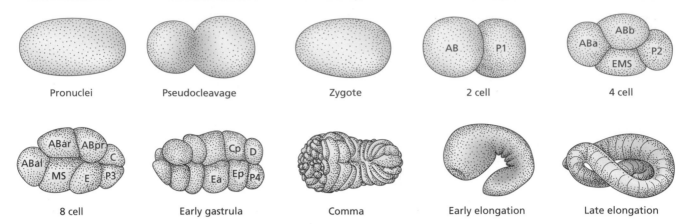

Fig. 12.3 Embryonic development of *C. elegans*.

cell precursor, dividing only once more in embryonic life. Maternal components are sufficient to direct development up to about 26 cells, as this is the earliest stage that defects are apparent if embryos are raised in α-amanitin, an inhibitor of RNA polymerase II. In the germ line zygotic transcription of RNA polymerase II genes remains repressed until about the 100-cell stage.

The egg contains RNA-rich **P-granules** which are initially randomly dispersed but which concentrate in the posterior during the cytoplasmic rearrangement period. During each successive division these granules concentrate in the region that will become the new P cell. Of the original founder cells, AB, MS, and C all produce a variety of cell types while the others generate a single cell type: P4 becoming the germ line, E becoming the gut, and D becoming muscle. "Gastrulation" in *C. elegans* is rather prolonged but can be considered as starting at the 26-cell stage when the two E cells move into the interior. These are followed by the myoblasts derived from C and D, and the pharyngeal cells derived from ABa. The ventral cleft, which resembles a blastopore, closes at about the 300-cell stage.

Because of the relatively small cell number and the invariance such that all individuals undergo exactly the same sequence of cell divisions, the complete **cell lineage** of embryo, larva, and

adult has been determined by direct observation, the first part of which is shown in Fig. 12.4. Although in one respect setting a high standard of precision, the lineage falls short of a complete fate map as it shows only the "family tree" of the cells but not their spatial relationships at the different stages. Development was originally thought to be entirely mosaic in character, because in almost all cases when a cell is removed by laser microbeam irradiation all of its descendants are lost and there is no consequence for the development of neighboring cells. However, a number of inductive interactions are now known, so *C. elegans* does not really differ greatly from the other model species in this regard.

The precision of the cell lineage makes it less useful to define which parts of the embryo belong to the different germ layers than it is for the other animal types. The "official" germ layers are:
ectoderm: AB, Caa, Cpa;
mesoderm: MS, Cap, Cpp, D;
endoderm: E.

However, AB produces the pharyngeal muscles which would normally be considered a mesodermal type and MS produces some pharyngeal neurons which would normally be considered ectodermal.

Classic Experiments

MECHANISM OF UNEQUAL CELL DIVISION

The breakthrough depended on isolation of mutants of maternal-effect genes in which the normal polarization of the zygote was lost. This showed that PAR protein complexes became positioned within the cell by mutual repulsion and controlled degradation. Homologs of the *par* genes are

now known to be involved in asymmetrical cell divisions in many other animals.

Kemphues, K.J., Priess, J.R., Morton, D.G. & Cheng, N. (1988) Identification of genes required for cytoplasmic localization in early *C. elegans* embryos. *Cell* **52**, 311–320.

Guo, S. & Kemphues, K.J. (1995) *Par-1*, a gene required for establishing polarity in *C. elegans* embryos, encodes a putative Ser/Thr kinase that is asymmetrically distributed. *Cell* **81**, 611–620.

Regional specification in the embryo

Asymmetrical cleavages

Asymmetric division is important in numerous cases of tissue differentiation and stem cell behavior in higher animals, and some of the basic mechanism were discovered by studying *C. elegans*. Asymmetrical division involves two processes, the establishment of cytoplasmic

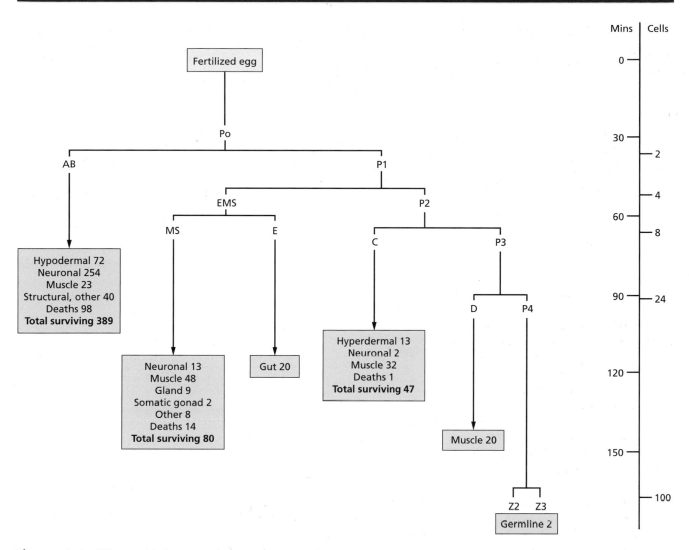

Fig. 12.4 Early cell lineage of *C. elegans*.

polarity and the correct orientation of the mitotic apparatus (Fig. 12.5).

Normally the early blastomeres will divide in a direction at right angles to their last cleavage. The AB cell follows this rule as it divides orthogonal to the first cleavage but the P1 cell does not do so, instead dividing parallel to the first cleavage. It does this because of rotational alignment, which is a 90 degrees rotation of the centrosomes and nucleus, driven by the positioning of microtubule attachments on the cell cortex.

A series of maternal-effect genes affecting the asymmetry of cell divisions have been isolated by mutagenizing worms that are themselves unable to lay eggs. Such worms can still reproduce, because the larvae arising from self-fertilization simply eat their way out of the body of the hermaphrodite. To do the screen, one F1 larva is put in each dish. If it carries a mutation on one chromosome, then 25% of its (F2) offspring will be homozygous for that mutation. If the mutation is zygotic lethal, then the affected embryos will simply fail to develop. However, if

the mutation is a maternal-effect lethal, then the F2 generation will develop into worms but they will then fill up with inviable F3 embryos that cannot develop and so do not eat their way out (Fig. 12.6).

Some of these maternal-effect lethals affect the cleavage planes and character of early blastomeres and are called *par* genes ("partitioning defective"). Embryos produced by homozygous mothers have symmetrical early cleavages and arrest as amorphous cell masses:

par-1 codes for a Ser/Thr kinase which binds nonmuscle myosin. After the cytoplasmic rearrangement it is found in the posterior cortex of the zygote.

par-2 codes for a cytoplasmic protein with adenosine triphosphate (ATP)-binding and zinc-binding (RING) domains. It is also localized to the posterior of the zygote.

par-3 codes for a cytoplasmic protein containing a PDZ (protein-protein recognition) domain. It forms a complex with PAR-6 (another PDZ domain protein) and an atypical

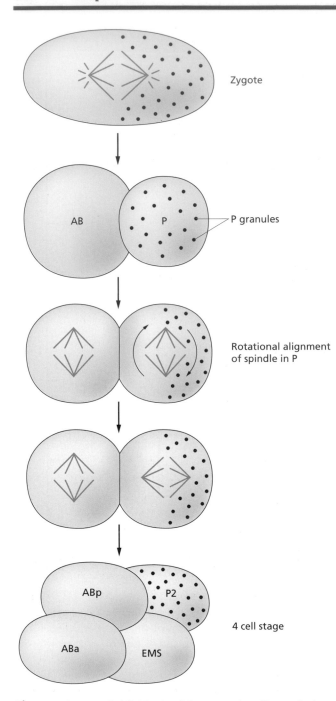

Zygote

AB

P

P granules

Rotational alignment
of spindle in P

ABp

P2

ABa

EMS

4 cell stage

Fig. 12.5 Asymmetrical division involving segregation of P-granules into P cells and rotational alignment of the P1 spindle.

protein kinase C (aPKC3), and the complex becomes associated with the plasma membrane in the anterior of the zygote.

In the unfertilized egg the PAR proteins are uniformly distributed. Following fertilization the sperm aster repels the PAR-3 complex from the posterior. The PAR-3 complex then repels PAR-1 and -2 such that they become concentrated in the posterior cortex (Fig. 12.7). This mutual repulsion seems to be the key element of the cell polarization and effectively serves to amplify

the small change brought about by sperm entry into a big change affecting the overall structure of the zygote. Evidence for the process comes from observing the distribution of one PAR protein in the absence of another. In the absence of PAR-2 there is no movement of the PAR-3 complex to the anterior, and in the absence of PAR-3 there is no movement of PAR-1 and -2 to the posterior. It remains uncertain to what extent the localizations are achieved by actual movement and to what extend by differential degradation, but the mechanism is known to involve the phosphorylation of PAR-3 by PAR-1, which allows binding of other cytoplasmic proteins of the 14-3-3 class and resulting destablization of the PAR-3 complex.

In terms of specification of commitment the function of the PAR proteins is to control the disposition of cytoplasmic determinants in the zygote and early blastomeres as described below. As far as the orientation of mitotic spindles is concerned, embryos lacking PAR-2 show rotational alignment in neither AB nor P. Embryos lacking PAR-3 show rotational alignment of both AB and P cells. The double mutant, *par-2⁻ par-3⁻*, produces embryos showing rotational alignment in both cells, like *par-3⁻*. This means that something other than the *par* genes must be causing the rotational alignment and that the PAR-3 complex normally suppresses it in the AB blastomere and its absence allows it in P1.

There are mammalian and *Drosophila* homologs of the *par* genes, and these are thought also to be involved in the acquisition of cell polarity and in the control of asymmetrical cell division. In mammalian epithelia the PAR-3 complex is found in junctional complexes (see Chapter 13). It has been shown that overexpression of a PAR-1 homolog can alter cell polarity, for example converting a normally columnar epithelial cell type to a liver type with intercellular lumens resembling bile canaliculi. In *Drosophila* neuroblasts the PAR-1 homolog is found in the cortical crescent (see Chapter 14).

Determinants

Several cytoplasmic determinants responsible for regional specification have been identified (Figs 12.7, 12.8). Their mode of action has been deduced from the maternal-effect mutant phenotype, and from the effect on their localization of mutating other genes, including the *par* genes. Localization may be studied by immunostaining for the protein, or by observing the intracellular position of a transgenic GFP fusion protein by fluorescence microscopy.

SKN-1 (pronounced "skin-1") is a transcription factor of the bZIP type and confers an EMS type of development on its nuclei. The mRNA is present maternally and is not localized, but the protein accumulates only in the P1 nucleus, and later in the descendant P2 and EMS nuclei, then becomes lost after the 12-cell stage. Embryos without SKN-1 lack pharynx and intestine because the E and MS blastomeres develop like the C blastomere (hence too much "skin"). Although in the normal embryo

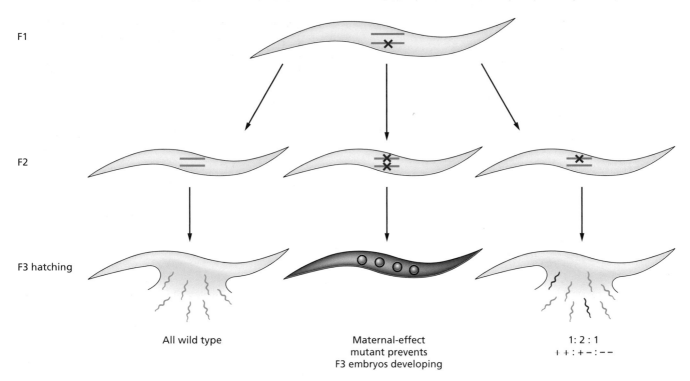

F1

F2

F3 hatching

All wild type

Maternal-effect
mutant prevents
F3 embryos developing

1 : 2 : 1
+ + : + − : − −

Fig. 12.6 Maternal screen. The hermaphrodites are vulvaless and cannot lay eggs, so the larvae eat their way out, destroying the parents. But those F2 worms carrying arrested embryos due to a maternal-effect mutation will persist.

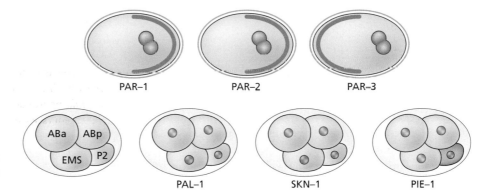

Fig. 12.7 Cytoplasmic determinants in *C. elegans*. The PAR-1 and PAR-2 proteins become localized to the posterior and the PAR-3/PAR-6/aPKC3 complex to the anterior following fertilization. Localization of SKN-1, PIE-1, and PAL-1 proteins occurs as shown by the four-cell stage.

PAR–1   PAR–2   PAR–3

ABa ABp
EMS P2

PAL–1   SKN–1   PIE–1

SKN-1 is present both in EMS and in P2, its transcription factor activity is repressed in P2 by the PIE-1 protein. PIE-1 is responsible for repression of all RNA polymerase II-mediated transcription in the early germ line. Embryos lacking PIE-1 still have a normal distribution of SKN-1 protein, but the P2 cell now develops like EMS, because SKN-1 is active in both cells.

Mex-1 mRNA is initially ubiquitous but becomes lost from cells other than the P lineage. The protein also concentrates in the posterior of the zygote. MEX-1 appears to prevent SKN-1 from entering AB. Embryos lacking MEX-1 have SKN-1 in the nuclei of the two AB cells as well as in P2 and EMS. As a consequence they have AB descendants developing like the normal MS descendants, leading to too much muscle. Embryos lacking both MEX-1 and SKN-1 have a similar phenotype to that caused by lack of SKN-1 alone, with AB normal but EMS developing like C. These results confirm that normal AB behavior depends on the absence of SKN-1.

PAL-1 is a homolog of the *Drosophila caudal* gene and the vertebrate cdx family. Like these genes, it is needed for posterior development. The mRNA is present all over the early embryo, but is normally only translated in EMS and P2. Translation is repressed during the early stages, and in AB cells, by MEX-3, which acts on the 3′UTR of the *pal-3* mRNA. The *mex-3* mRNA and MEX-3 protein are initially uniform then become more abundant in AB cells and are lost after the four-cell stage. Embryos lacking MEX-3 express PAL-1 protein all over and are posteriorized in morphology with the AB descendants resembling the normal descendants of blastomere C.

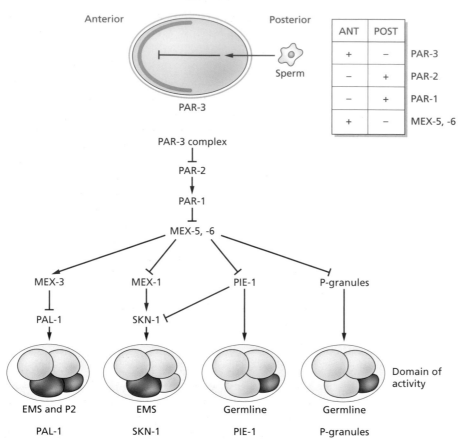

ANT	POST	
+	−	PAR-3
−	+	PAR-2
−	+	PAR-1
+	−	MEX-5, -6

Fig. 12.8 Mechanistic connections between the sperm entry point and the segregation of cytoplasmic determinants. The sperm aster repels the PAR-3 complex and this unleashes a cascade of localization events depending on protein movement, differential degradation, and translational control. Note that SKN-1 is present in P2 but not active because of the presence of PIE-1.

These results show how the character of each of the early blastomeres is specified by the particular combination of determinants which it inherits. The spatial disposition of the determinants is controlled by the PAR system and PAR-1 seems to be the main effector, acting on the cytoplasmic proteins MEX-5 and MEX-6 such that they become localized to the anterior (Fig. 12.8). These in turn act on the MEX-1 and PIE-1 proteins, and the P-granules, to localize them all to the posterior where they direct the formation of the P lineage of blastomeres. The evidence for this is that in the absence of PAR-1, all of MEX-5, -6 and -1 proteins, PIE-1 protein, and the P-granules are uniformly distributed. In the absence of the MEX-5 and -6 proteins, PAR-1 distribution is normal, but MEX-1, PIE-1, and P-granules are uniform, showing that PAR-1 normally regulates the localization of MEX-5 and -6 and that they in turn control the disposition of the other components. These effects are exerted mostly through differential protein degradation.

MEX-5 and -6 also concentrate MEX-3 in the anterior. In the absence of PAR-1, MEX-3 is present all over the embryo. Normally, MEX-3 in the anterior inhibits the production of PAL-1 by translational control, confining the activity of PAL-1 to the posterior. But in the absence of PAR-1, leading to uniform MEX-3, there is no expression of PAL-1 and posterior development is defective.

Inductive interactions in C. elegans

It was initially thought that *C. elegans* functioned entirely on the basis of cytoplasmic determinants, because of the invariant fate map and the **mosaic** behavior of most cells after laser ablation of their neighbors. However, it is now known that there are many inductive interactions as well.

A structure that depends on induction for its formation is the pharynx. There are two successive signals of which the first is repressive from the P2 cell, and the second is positive from the descendants of MS (Fig. 12.9). Both signals operate through the Notch pathway, but using different ligands of Notch.

Normally the anterior part of the muscular pharynx is produced by the ABa cell. If ABa and ABp are interchanged then ABp will form the anterior pharynx instead of ABa, showing that position of the cell rather than its lineage is important. However, if P2 is prevented from touching ABp then ABp forms pharynx as well as ABa, showing that there must normally be a signal from P2 to ABp that suppresses pharynx formation.

This repressive signal is encoded by the maternal-effect *apx-1* (anterior pharynx excess) gene. Embryos lacking APX-1 show formation of anterior pharynx from both AB blastomeres, instead of just ABa. *apx-1* codes for a Delta-like ligand. The receptor is encoded by another maternal-effect gene, *glp-1* (the

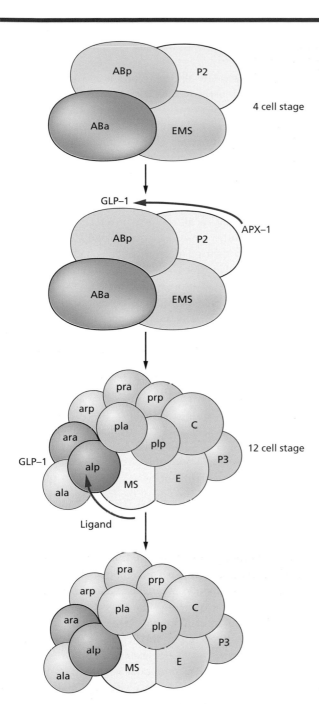

Fig. 12.9 Two inductive interactions leading to formation of the pharynx.

name refers to an effect on **g**erm-**l**ine **p**roliferation) whose product is a Notch-type receptor. Embryos lacking GLP-1 also have ABp developing as ABa in most respects, but unlike *apx-1⁻*, *glp-1⁻* mutants do not actually go on to form pharynx from the two equivalent ABa-like blastomeres. This is because formation of pharynx is not, in fact, a default for ABa, but depends on a subsequent positive inductive interaction. This may be shown by the fact that an isolated AB cell does not produce any pharynx.

Also, laser ablation of the MS cell between eight and 12 cells prevents ABa forming pharynx, showing that its presence must be necessary during this time interval. At the 12-cell stage the MS blastomere touches the two ABa grand-daughters (ABalp and ABara), and emits the second signal responsible for inducing the pharynx. Moreover in *apx-1⁻* embryos the descendants of the ABp cell that produce pharynx are ABpra, ABprp, and ABplp, all of which also contact MS at the 12-cell stage. Remarkably, it seems that the receptor for this second signal is also GLP-1, since, as we have seen, the mutant embryos do not form a pharynx, even though in other regards the two AB lineages behave the same. The *glp-1* mRNA is uniformly distributed in the embryo up to the eight-cell stage but the protein is found only in the AB descendants. This is because of differential translation regulated by a sequence in the 3′-UTR of the message, with translation in the posterior being repressed by a cascade of factors ultimately controlled by PAR-1. The GLP-1 protein disappears at the 28-cell stage (when there are 16 AB descendants). The ligand for GLP-1 expressed by MS is distinct from APX-1.

Hence there is a double requirement for GLP-1, first as the receptor mediating repression of pharynx formation by the action of P1 on ABp, and then as the receptor mediating positive induction of pharynx formation by the action of MS on ABa. These two separate requirements are clearly shown by the phenotypes of temperature-sensitive mutants of *glp-1*, kept for different time periods at the nonpermissive temperature. If the nonpermissive temperature is given only around the four-cell stage then the phenotype is just like the maternal effect phenotype of *apx-1⁻*, with ABp as well as ABa forming pharynx. If the nonpermissive temperature is maintained until the 12-cell stage then the phenotype is like the maternal effect *glp-1* null mutant, with equivalent cell divisions of ABp and ABa but no subsequent pharynx formation.

Ultimately, formation of the entire pharynx is dependent on the zygotically expressed gene *pha-4*, encoding a winged helix transcription factor homologous to the FoxA genes important in vertebrate gut development (see Chapter 16). This is responsible for activating a "pharyngeal enhancer" controlling expression of pharyngeal genes in all the component cell types of the organ. The loss-of-function mutant lacks the entire pharynx, both the part formed from ABa and the part formed from MS. Use of a temperature-sensitive allele shows that there is a requirement for *pha-4* throughout development, for both early and late differentiation events.

The intestine is composed of 20 cells derived from the E blastomere. These cells polarize, intercalate with each other, and become arranged around a gut lumen and joined with junctional complexes. The developmental specification of the E blastomere depends on a signal from the P2 cell. This emits a Wnt-type signal that causes the nearer part of EMS to become E and the further part to adopt the default specification of MS. This may be shown by removing the P2 cell, which causes both progeny of EMS to resemble MS. A series of *mom* (more mesoderm) mutants have a similar effect to loss of P2. These turned out to encode members of the Wnt pathway, and it was shown by mosaic analysis that the signaling components including the Wnt homolog itself (*mom2*) were required in the P2 cell while the receptor homolog (*mom5*) was required in EMS. Loss-of-function maternal-effect mutants of these Wnt pathway components will convert E into a second MS. However the reverse phenotype results from loss of function of *pop-1*, which is an HMG domain transcription factor comparable to the Tcf and Lef factors in vertebrates. This converts MS into a second E, suggesting that formation of E depends on inhibition of POP-1 activity by the Wnt signal, whereas in vertebrates the Wnt signal will normally activate the POP-1 homologs.

A zygotically active gene fulfilling the description of a master regulator for the intestine is *end-1*. This encodes a transcription factor of the GATA class and is expressed only in the E cell and its progeny. It activates a set of intestine-specific target genes. Transcription of *end-1* is activated by SKN-1 and repressed by POP-1, meaning that the formation of intestine requires both the correct early placement of SKN-1 by the PAR-1/MEX-1 system, and the later Wnt signal from the P2 blastomere.

In summary, the molecular basis for regional specification of the blastomeres in *C. elegans* is now fairly well understood. It depends both on the correct placement of cytoplasmic determinants, and on the occurrence of inductive signals between adjacent blastomeres. The asymmetrical localization of both the determinants and the components of the signaling systems depend on the operation of the PAR system.

Analysis of postembryonic development

The vulva

The vulva is the epidermal structure that is formed in larval life around the mid-ventral opening of the gonad (Fig. 12.10). Its formation is controlled by an EGF-like signal from an internal cell called the **anchor cell**. It arises from the cells called P5p, P6p, and P7p, which are the three posterior daughters of embryonically generated ectodermal cells P5, P6, and P7:

P5p makes seven vulval descendants;
P6p makes eight vulval descendants;
P7p makes seven vulval descendants.

In the fourth larval stage these 22 cells undergo various movements and fusions to make the vulva itself. In addition, the surrounding cells, called P3p, P4p, and P8p, are competent to make vulva, but in normal development they each divide just once to make two cells that later enter the syncytial hypoderm. In discussions of the vulva the following convention is used:

formation of eight cells = primary fate 1°; normally followed by P6p;

formation of seven cells = secondary fate 2°; normally followed by P5p and P7p;

formation of two cells = tertiary fate 3°; normally followed by P3p, P4p, P8p.

Hermaphrodites are able to reproduce without a vulva, because the larvae just chew their way out of the body. Therefore it is possible to screen worms for viable mutations with various vulval defects. These viable mutations are often hypomorphic, with the corresponding null alleles of the same genes being lethal. The main classes of mutant are *vulvaless* and *multivulva*, the latter forming supernumerary vulvas from the same P3p–P8p cell group.

The six P3p–P8p cells are said to make up an **equivalence group**, because they are all competent to form vulva and they can replace each other in various experimental situations. This is clearly shown by their relations with the gonadal **anchor cell**, which lies internally adjacent to P6p. If the anchor cell is ablated by laser microbeam radiation, then all the P3p–P8p cells follow the tertiary fate 3° and no vulva is produced. If one of the P3p–P8p cells is removed by laser microbeam, then its neighbor will take its place and a normal vulva will result. If the anchor cell is moved relative to the P3p–P8p cells, as in various *displaced gonad* mutants, then whichever three of the P3p–P8p cells are nearest will produce the vulva.

There are several *vulvaless* mutants, giving a similar phenotype to anchor cell ablation. The anchor cell ligand is encoded by *lin-3*, and is a homolog of EGF. The receptor is an EGF receptor homolog encoded by *let-23*. *let-60* is a gene with several different alleles. Loss-of-function mutants give a vulvaless phenotype while gain-of-function mutants give a multivulva phenotype. *let-60* in fact encodes a homolog of the Ras protein, familiar as an intermediate in the ERK signal transduction pathway activated by EGF or FGF signaling (see Appendix). Constitutively active Ras will produce multivulva, while inactive Ras will produce vulvaless. Double mutant combinations work in a predictable way, for example the combination of *let-23⁻* and *let-60 gof* gives a multivulva phenotype, confirming that the Ras requirement lies downstream of the receptor. It is possible to visualize a gradient of EGF response centered on P6p in worms transgenic for an EGF reporter, which is a *lacZ* gene whose transcription is activated by EGF signaling. The EGF signal also works partly by lifting of a continuous inhibition from the syncytial hypoderm, which is in contact with the P3p–P8p cells. In the *lin-15* mutation a multiple vulva is formed from all six cells with or without the anchor cell. Genetic mosaic experiments show that the P3p–P8p cells themselves need not be *lin-15⁻*, and so it is thought that the mutation must prevent the formation of an inhibitor by the syncytial hypoderm which normally represses vulva formation, and which is overcome by the anchor cell signal.

Although the gradient of EGF signaling should theoretically be enough to generate the three cell fates, there is also a secondary signal emitted by P6p that activates the Notch pathway in P5p and P7p. In normal development the combination of both these signals serves to control the formation of the vulva.

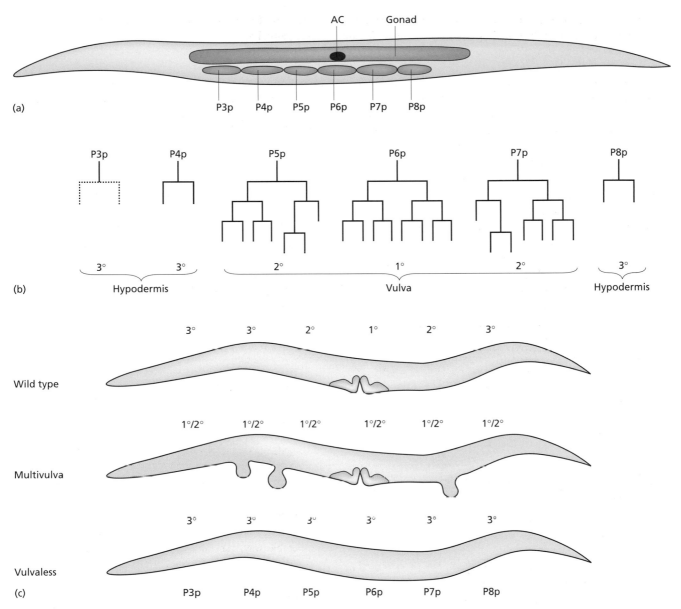

Fig. 12.10 Development of the *C. elegans* vulva. (a) Relationship of the equivalence group cells to the gonad and anchor cell. (b) Normal lineage of each cell. (c) Mutant phenotypes. The fate of each of the cells is indicated for each phenotype.

The germ line

An important aspect of postembryonic development is the maturation of the **germ line**. This derives from the P lineage which inherits various determinants, some associated with the P granules. As mentioned above, the PIE-1 protein causes a general repression of genes transcribed by RNA polymerase II during the early stages. In the newly hatched larva, the germ line consists of just two cells descended from the P lineage: Z2 and Z3. These express the *cgh-1* gene, which encodes an RNA helicase homologous to *vasa* in *Drosophila*. The *cgh-1* gene is active in the

germ line thereafter, and the CGH-1 protein is one of the components of the P-granules found in the egg and segregated to the germ-line lineage in early development. Treatment of worms with RNAi directed against *cgh-1* causes death of the oocytes and formation of nonfunctional sperm, suggesting a key function in the later stages of germ cell development.

In the larval and adult worm the germ cells lie within the gonad. The most mature cells lie near the vulva and the most immature cells, which are still mitotic, at the blind ends of the gonad. These mitotic germ cells form a syncytium and become cellularized as they enter meiosis. In the hermaphrodite, the

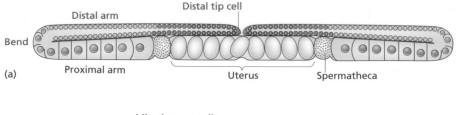

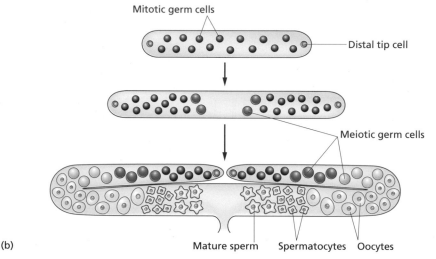

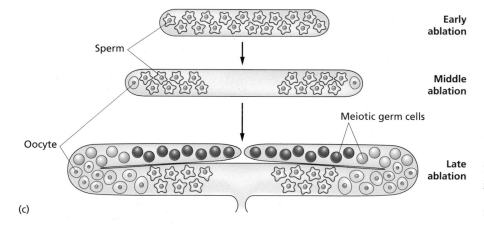

Fig. 12.11 Development of the *C. elegans* gonad. (a) Adult hermaphrodite gonad. (b) Normal development. (c) Effect of removing the distal tip cell at different stages.

early cohorts of germ cells differentiate as sperm and the later ones as oocytes. The tip of each branch of the gonad contains an important somatic cell called the **distal tip cell** whose function is to maintain the neighboring germ-cell nuclei in mitosis. As the gonad grows during larval life, the germ-cell syncytium elongates and cells progressively leave the range of influence of the distal tip cell, whereupon the cells stop mitotic division and enter meiosis. Ablation of the distal tip cell at any stage by laser irradiation causes all the remaining mitotic nuclei to enter meiosis. The composition of the gonad following this is then appropriate to the stage of maturation reached, so an early ablation would create all sperm while a late ablation would

result in a nearly normal arrangement of oocytes and sperm (Fig. 12.11). The distal tip cell acts by expression of *lag-2*, which is another homolog of *Delta*. The receptor, encoded by *glp-1* (the same gene required for pharynx induction) is present on the germ-cell syncytial membrane. Zygotic loss-of-function mutants of *glp-1* or of *lag-2* have the same effect as ablation of the distal tip cell.

The GLP-1 signal inhibits the activity of a pair of proteins GLD-1 and -2, respectively an RNA binding protein and a polyA polymerase, that are present in the germ line and are needed for progression to meiosis. It also activates expression of *daz-1*, encoding another RNA binding protein required for oocyte maturation. *Daz-1* is a homolog of the human gene *daz* (**d**eleted in **az**oospermia), although this is required for spermatogenesis rather than oogenesis. The importance in *C. elegans* of genes such as *cgh-1* and *daz-1* that encode homologs of proteins required for germ cell development in higher animals is another example of the remarkable similarity of developmental mechanisms across the animal kingdom.

Programmed cell death

Cell death, or **apoptosis**, is important in many developmental contexts, and is now known to depend on the action of proteases called caspases. As with asymmetrical cell division, the discovery of the mechanism of programmed cell death is an area in which *C. elegans* genetics has made an important contribution to general cell biology.

During normal development of *C. elegans* about 1 in 8 cells die. They are mainly small cells and collectively represent only 1% of the biomass. Most deaths are autonomous and occur shortly after the cell was born. A few depend on signals from neighboring cells, shown by the fact that the cell will survive when its neighbor has been ablated by laser radiation. The sequence of events is the same as mammalian apoptosis, with a condensation of the nucleus, a shrinkage of the cell to a membrane-bound body, and engulfment by neighboring cells.

Cell-death-defective (*ced*) mutants affect all the cell deaths in the organism, while some other mutants affect the decisions of particular cells to die. Most of the *ced* mutants interfere with the engulfment of the dead cells, but three of them are components of the actual death program itself. In loss-of-function mutants of *ced-3* and *ced-4*, all of the cells normally destined to die now survive. The *ced-9* gene has both loss-of-function and gain-of-function alleles. In loss-of-function mutants there is excessive cell death, while in gain-of-function mutants there is some survival of cells that normally die. Excess cell survival is also shown on overexpression of wild-type *ced-9*. Double mutants of the type *ced-9⁻/ced-3⁻* or *ced-9⁻/ced-4⁻* also show target cell survival, so it follows that the normal function of *ced-9* must be to repress *ced-3/4*, which are themselves downstream of *ced-9* and necessary to execute the death program (Fig. 12.12). Genetic mosaic experiments show that the *ced-3/4* wild-type gene has to be present in the target cell itself in order for it to die.

ced-9 codes for the homolog of the mammalian protein BCL2. Discovered originally as an **oncogene** product, it is a cytoplasmic protein which is an inhibitor of apoptosis. CED-9 and BCL2 are interchangeable, therefore worms transgenic for mammalian BCL2 show inhibition of cell death, and *ced-9⁻* mutants can be rescued by BCL2. *ced-3* codes for a homolog of the interleukin 1β-converting enzyme (ICE). This is a cysteine protease that cleaves at Asp-X sequences, and is the prototype member of the family of caspases, of which many are now known. The caspases are the enzymes that actually bring about cell death, and the targets for the CED-3 protease include polyADPR polymerase

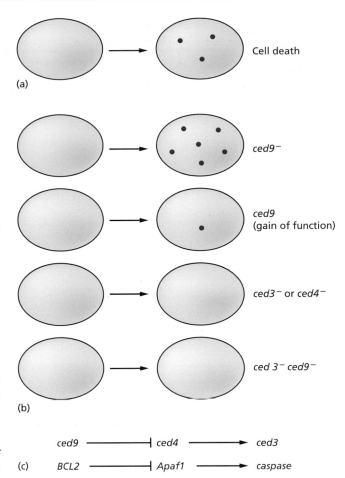

Fig. 12.12 The cell-death pathway. (a) Normal cell death. (b) Phenotypes of mutants. (c) The pathway in *C. elegans* and in mammals.

(involved in DNA repair), lamins (nuclear membrane proteins) and other nuclear proteins. CED-3 itself will cause apoptosis if introduced into mammalian cells. *ced-4* codes for a protein whose mammalian homolog is called Apaf1 and which activates procaspase 9. CED-4 itself similarly activates CED-3 and is inhibited by CED-9.

The final stage of the cell death process is the engulfment of the apoptotic cells by neighboring cells. *C. elegans* does not possess specialized phagocytes and the engulfment is carried out by nonspecialized cells. But as in mammalian phagocytes a critical step is the recognition of a particular phospholipid, phosphatidyl serine, on the surface of the cell to be engulfed. A number of the *ced* genes are defective in engulfment and their homologs are thought also to be important for the properties of mammalian phagocytes. Examples are *ced-1*, encoding a homolog of the mammalian SREC (scavenger receptor from endothelial cells), and *ced-7*, encoding an ABC (ATP binding cassette) transporter protein, one of a large class of proteins responsible for transport of small molecules and ions across cell membranes.

Classic Experiments

CELL LINEAGE AND CELL DEATH

The first two references are the papers describing the complete cell lineage of *C. elegans*, derived from painstaking observation by interference microscopy. The third is a detailed anatomical study. In the course of this work a number of programmed cell deaths were described. Later analysis of the cell death (*ced*) mutants showed that the biochemistry of the process was common to higher animals and allowed the elucidation of the pathway.

Sydney Brenner, Robert Horvitz, and John Sulson were awarded the Nobel Prize for Physiology for this work in 2002.

Sulston, J.E. & Horvitz, H.R. (1977) Postembryonic cell lineages of the nematode

Caenorhabditis elegans. Developmental Biology **56**, 110–156.

Sulston, J.E., Schierenberg, E., White, J.G. & Thomson, J.N. (1983) The embryonic-cell lineage of the nematode *Caenorhabditis elegans. Developmental Biology* **100**, 64–119.

White, J.G., Southgate, E., Thomson, J.N. & Brenner, S. (1986) The structure of the nervous-system of the nematode *Caenorhabditis elegans. Philosophical Transactions of the Royal Society of London Series B: Biological Sciences* **314**, 1–340.

Yuan, J.Y., Shaham, S., Ledoux, S., Ellis, H.M. & Horvitz, H.R. (1993) The *C. elegans* cell-death gene *ced-3* encodes a protein similar to mammalian interleukin-1-beta-converting enzyme. *Cell* **75**, 641–652.

Hengartner, M.O. & Horvitz, H.R. (1994) *C. elegans* cell-survival gene *ced-9* encodes a functional homolog of the mammalian protooncogene *bcl-2. Cell* **76**, 665–676.

Key Points to Remember

• *C. elegans* is very favorable for genetic experimentation. It is a self-fertilized hermaphrodite with a short generation time.

• The precise cell lineage of all cells in the embryo and adult has been described. This serves as a resource for all kinds of experimental work.

• The early regional pattern of the embryo arises through the action of the PAR proteins, which become segregated along the anteroposterior axis after fertilization and control the distribution of cytoplasmic determinants between the early blastomeres.

• Like other types of embryo, various steps of *C. elegans* development depend on inductive interactions. These utilize the same molecular pathways as other animals. For example, the formation of the pharynx from the AB cells depends on a Delta-like signal from the P2 and MS cells. The formation of the E lineage depends on a Wnt signal from the MS cell.

• Postembryonic development also involves inductive interactions. The EGF/Ras pathway is important for vulval development and the Notch pathway for the maintenance of the mitotic germ line.

• *C. elegans* has made important contributions to cell biology by helping to elucidate the mechanisms of cell polarization and of apoptotic cell death.

Further reading

Websites

Caenorhabditis elegans server: http://elegans.swmed.edu/
Introduction:
http://www.biotech.missouri.edu/Dauer-World/Wormintro.html
"Wormbase": http://www.wormbase.org/

General

Riddle, D.L., ed. (1997) *C. elegans* II. Cold Spring Harbor, NY: Cold Spring Harbor Laboratory Press.

Hope, I.A. (ed.) (1999) *C. elegans: a practical approach.* Oxford: Oxford University Press.

Singson, A. (2001) Every sperm is sacred: fertilization in *Caenorhabditis elegans. Developmental Biology* **230**, 101–109.

Maduro, M.F. & Rothman, J.H. (2002) Making worm guts: the gene regulatory network of the *Caenorhabditis elegans. Developmental Biology* **246**, 68–85.

Genetics

Kuwabara, P.E. & Kimble, J. (1992) Molecular genetics of sex determination in *C. elegans. Trends in Genetics* **8**, 164–168.

Salser, S.J. & Kenyon, C. (1994) Patterning in *C. elegans*: homeotic cluster genes, cell fates and cell migrations. *Trends in Genetics* **10**, 159–164.

Hunter, C.P. (1999) A touch of elegance with RNAi. *Current Biology* **9**, R440–R442.

Blumenthal, T. & Seggerson-Gleason, K. (2003) *Caenorhabditis elegans* operons: form and function. *Nature Reviews Genetics* **4**, 112–120.

Yochem, J. & Herman, R.K. (2003) Investigating *C. elegans* genetics through mosaic analysis. *Development* **130**, 4761–4768.

Asymmetric division and determinants

Rose, L.S. & Kemphues, K.J. (1998) Early patterning of the *C. elegans* embryo. *Annual Reviews of Genetics* **32**, 521–545.

Kemphues, K. (2000) PARsing embryonic polarity. *Cell* **101**, 345–348.

Lyczak, R., Gomes, J.E. & Bowerman, B. (2002) Heads or tails: cell polarity and axis function in the early *Caenorhabditis elegans* embryo. *Developmental Cell* **3**, 157–166.

Nance, J. & Priess, J.R. (2002) Cell polarity and gastrulation in *C. elegans*. *Development* **129**, 387–397.

Wodarz, W. (2002) Establishing cell polarity in development. *Nature Cell Biology* **4**, E39–E44

Schneider, S.Q. & Bowerman, B. (2003) Cell polarity and the cytoskeleton in the *Caenorhabditis elegans* zygote. *Annual Review of Genetics* **37**, 221–249.

Macara, I.G. (2004) Parsing the polarity code. *Nature Reviews Molecular Cell Biology* **5**, 220–231.

Induction, the germ line, the vulva, cell death

Priess, J.R. & Thomson, J.N. (1987) Cellular interactions in early *C. elegans* embryos. *Cell* **48**, 241–250.

Horvitz, H.R. & Sternberg, P.W. (1991) Multiple intercellular signaling systems control the development of the *C. elegans* vulva. *Nature* **351**, 535–541.

Sundaram, M. & Han, M. (1996) Control and integration of cell signaling pathways during *C. elegans* vulval development. *Bioessays* **18**, 473–480.

Kornfeld, K.(1997) Vulval development in *C. elegans*. *Trends in Genetics* **13**, 55–61.

Cryns, V. & Yuan, J. (1998) Proteases to die for. *Genes and Development* **12**, 1551–1570.

Ikenishi, K. (1998) Germ plasm in *Caenorhabditis elegans*, *Drosophila* and *Xenopus*. *Development, Growth and Differentiation* **40**, 1–10.

Metzstein, M.K., Stanfield, G.M. & Horvitz, H.R. (1998) Genetics of programmed cell death in *C. elegans*: past, present and future. *Trends in Genetics* **14**, 410–417.

Labouesse, M. & Mango, S.E. (1999) Patterning the *C. elegans* embryo. *Trends in Genetics* **15**, 307–313.

Seydoux, G. & Strome, S. (1999) Launching the germ line in *Caenorhabditis elegans*: regulation of gene expression in early germ cells. *Development* **126**, 3275–3283.

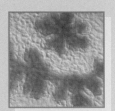

Section 3
Organogenesis

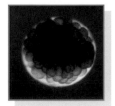

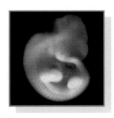

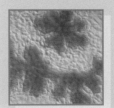

Chapter 13

Tissue organization and stem cells

The chapters of this section will describe the development of various selected organ systems, mostly in higher vertebrates, although certain aspects are illuminated by studies on the lower vertebrate and invertebrate models.

The techniques used in organogenesis research are essentially similar to those used in early development. In addition, much use is made of *in vitro* **organ cultures** of rudiments from mouse and chick embryos as these are more accessible than the same stage organ *in vivo*. **Knockout** mouse strains provide most of the **loss-of-function** data, and it is often possible to grow an organ culture from a knockout embryo beyond the stages at which the whole embryo remains viable. Expression studies are done by *in situ* **hybridization** and **immunostaining**. Biological activity data are acquired from addition of proteins to the organ cultures, or introduction of the genes encoding them by **electroporation** or viral infection.

This chapter deals with the chief tissue types found in the vertebrate body with special attention to their cellular renewal. On the basis of light microscopy there are about 200 types of differentiated cell, although *in situ* hybridization and immunostaining reveal many more. They are arranged in tissues, each of which contains several different cell types. An organ or body part contains several tissue types arranged to fulfill a common function, and they are usually derived from more than one embryonic cell lineage.

Types of tissue

Epithelia

An **epithelium** (plural **epithelia**) is a sheet of cells, arranged on a **basement membrane** with each cell joined to its neighbors by specialized junctions. About 60% of visible cell types are constituents of epithelia. The cells show a distinct apical–basal polarity, where the basal surface is that next to the basement membrane

and the apical surface is on the opposite side, often facing a fluid-filled lumen. The basement membrane consists of a basal lamina secreted by the epithelium itself, together with some additional extracellular material from the underlying connective tissue. It is composed of laminin, type IV collagen, entactin, and heparan sulfate proteoglycan. The junctional complexes consist of three components: **tight junctions**, adherens junctions, and desmosomes (Fig. 13.1). The tight junction belt prevents liquid leaking between the cells and also isolates the components of the apical and basolateral membranes; adherens junctions are attachment points joining the microfilament networks of the cells; and desmosomes are point contacts joining bundles of cytokeratin filaments. Cell–cell contacts through adherens junctions and desmosomes are made by cadherins, with their homophilic calcium-dependent binding. The cells are anchored to the basement membrane by cell–matrix adherens junctions and by hemidesmosomes. These are similar to the cell–cell junctions but utilize integrins for attaching the cell to the matrix components. Apical surfaces often bear cilia, and may bear **microvilli** if the epithelium has an absorptive function (see Appendix for some further notes about these various cell components).

Epithelia may be simple, with one layer of cells, stratified, with many layers of cells, or pseudostratified, meaning that they look stratified but in fact all cells contact the apical and basal surfaces. They may be **squamous**, with flattened cells, cuboidal, or columnar. Many epithelia are glandular and secrete materials into their surroundings. Glands may be simple or branched, and tubular or acinar (Fig. 13.2). The duct of a gland represents its original site of invagination during development. **Exocrine** glands secrete into the duct, **endocrine** glands have lost their ducts and secrete into the bloodstream. **Myoepithelial cells** are often found surrounding the **acini**, and their contraction helps drive the secretion down the duct. The terms mucous membrane or **mucosa** are often used to refer to a moist internal epithelium together with the immediately underlying connective tissue layer.

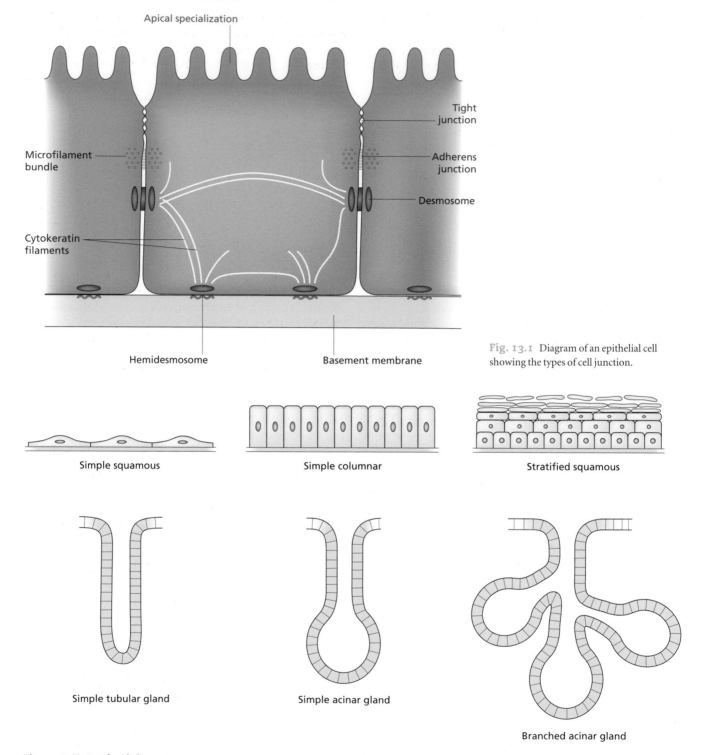

Fig. 13.1 Diagram of an epithelial cell showing the types of cell junction.

Apical specialization

Tight junction

Microfilament bundle

Adherens junction

Desmosome

Cytokeratin filaments

Hemidesmosome

Basement membrane

Simple squamous

Simple columnar

Stratified squamous

Simple tubular gland

Simple acinar gland

Branched acinar gland

Fig. 13.2 Types of epithelium.

Although commonly thought to be ectodermal in origin, epithelia are, in fact, derived from all three of the germ layers of the embryo. The organization and cell renewal in epidermis and intestinal epithelium are described below, and of neuroepithelium in Chapter 14.

Connective tissues

The term **connective tissue** refers to those tissues dominated by fibroblasts, such as the dermis of the skin and the fibrous capsules surrounding most organs. In some histology or biology textbooks it may be used in a wider sense to include the skeletal tissues, muscle, and even the cells of the blood.

Much of the connective tissue is derived from the mesoderm of the embryo, although some is also formed by the **neural crest**. Mature connective tissue consists of **fibroblasts** embedded in an extracellular matrix. Fibroblasts are cells specialized to secrete the matrix components which include hyaluronan, proteoglycans, fibronectin, type I collagen, type III collagen (reticulin), and elastin. Also found in connective tissue are histiocytes, which are macrophages resident in the tissue, and mast cells, which are histamine-secreting cells similar to the basophils of the blood but also resident in the tissues. Both these types originate from the bone marrow. **Adipose tissue** is closely related to loose connective tissue, as fibroblasts can become adipocytes under appropriate conditions.

The skeletal tissues are composed of cartilage and bone and arise both from embryonic mesoderm and neural crest. Much of the skeleton is formed initially as cartilage which is then gradually replaced by bone. Skeletal parts formed in cartilage are known as **cartilage models**. Some parts, particularly in the skull, differentiate directly from mesenchyme into bone, and these are known as **membrane bones**. Skeletal tissues are discussed further in Chapter 18.

A term causing much confusion is **mesenchyme**. This is not a synonym for connective tissue nor for mesoderm. It is a descriptive term for scattered stellate cells embedded in a loose **extracellular matrix** (see also Chapter 2). Mesenchyme, derived either from mesoderm or from neural crest, fills up much of the embryo and forms fibroblasts, adipose tissue, smooth muscle, and skeletal tissues, however these tissues should not be referred to as "mesenchymal" once they are differentiated.

Muscle

There are three main types of muscle (Fig. 13.3): skeletal muscle is composed of elongated multinucleate cells called **myofibers**. Bundles of myofibers are gathered together in fascicles surrounded by a fibrous sheath, the perimysium, and the whole muscle is surrounded by another sheath called the epimysium. Skeletal muscle is derived from the **myotomes** of the **somites** and its development is further described in Chapter 15. Smooth (= visceral) muscle exists as bundles of individual spindle-shaped mononuclear cells. These contain a similar contractile apparatus to skeletal muscle but it is not arranged as visible sarcomeres. Smooth muscle is derived from the lateral plate of the embryo

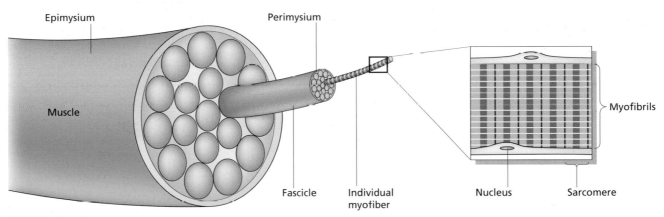

(a) Striated muscle

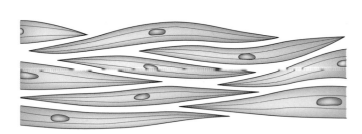

(b) Smooth muscle

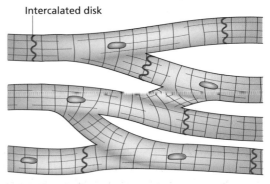

(c) Cardiac muscle

Fig. 13.3 Types of muscle.

and is found mainly around the gut, blood vessels, and the ducts of glands, where inherent rhythmic contraction is required. Smooth muscle is usually mitotically quiescent but can be stimulated to grow following tissue damage. Cardiac muscle occurs only in the heart. It derives from the anteroventral margin of the lateral plate mesoderm of the embryo. Like skeletal muscle it has visible myofibrils, but like smooth muscle it remains as individual cells. The cells are joined end to end by intercalated discs which contain structural junctions (adherens and desmosomes), together with gap junctions that allow rapid spread of electrical signals through the myocardium. Cardiac muscle, like skeletal muscle, is postmitotic, although some growth can occur by cell enlargement. Development of the skeletal muscle and heart are described in Chapter 15.

Neural tissues

Neural tissues comprise those cell types formed from the **neural tube** and some of those formed from the **neural crest**. The neural tube is composed of a specialized epithelium, the **neuroepithelium**, and produces both central neurons and glial cells, while the neural crest produces autonomic neurons of the peripheral nervous system, together with **Schwann cells** and pigment cells. Development of neural tissues is described in Chapter 14.

Blood and blood vessels

Blood contains a variety of cell types. In addition to the red cells, there are **granulocytes**, **monocytes**, and lymphocytes. All these, as well as other cells such as histiocytes, osteoclasts, and Langerhans cells of the skin, arise from **hematopoietic** tissue in the bone marrow. The hematopoietic system is a state of continuous cell production and renewal throughout life, and is further described below.

The circulatory system consists of arteries taking blood from the heart to the tissues, capillaries supplying the tissues, and veins returning blood to the heart (Fig. 13.4). All blood vessels have three layers. The inner layer is composed of a single layer of **endothelial cells** sometimes with a little underlying connective tissue. The middle layer is composed of smooth muscle, which may be very thick in arteries and thinner in veins, and the outer layer is composed of fibrous connective tissue.

The capillaries consist of a single layer of endothelial cells with a basal lamina on the exterior surface. There is no smooth muscle but there may be some associated contractile cells called pericytes. Usually the capillary wall is continuous but sometimes, as in the sinusoids of the liver, it contains gaps. Endothelial cells can divide throughout life and there is usually a low level of growth associated with tissue remodeling. The formation of new capillaries by endothelial cell division and cell movement is known as **angiogenesis**. A number of growth factors are active in promoting angiogenesis, particularly vascular endothelial cell

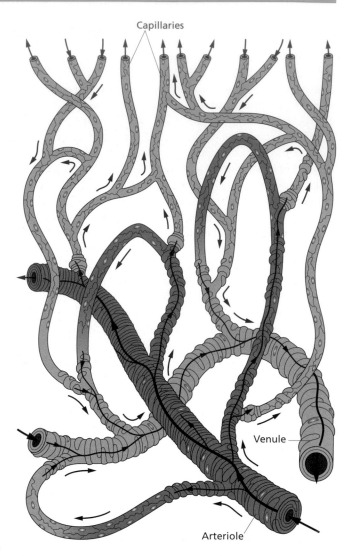

Capillaries

Venule

Arteriole

Fig. 13.4 A microcirculatory unit, showing joining of terminal arteriole and venule by capillaries. (After *Gray's Anatomy*, 35th edn, 1973. Longman, figure 6.10, p. 595.)

growth factor (VEGF) and the fibroblast growth factors (FGFs). The embryonic development of blood vessels is considered in Chapter 15.

Tissue renewal

Measurement of cell turnover

The brief sketch above focused on the visible appearance of tissues. But overall morphology tells us little about the cell turnover which is so critical to their maintenance. Although tissue culture cells in optimal medium may grow exponentially, this is rarely true of cells within the body. Usually cell turnover is slow, and particularly in epithelia it is often compartmentalized with separate proliferative and differentiating zones.

The multiplication of cell populations can be estimated by counting the proportion of visible mitoses to obtain a **mitotic index**. However, mitosis usually occupies a short period within the cell cycle, and a population has to be growing fast to show a significant mitotic index. More sensitive are methods that identify cells in **S phase**, which represents a longer fraction of the cell cycle and hence enables more cycling cells to be observed. One simple method is immunostaining for a protein associated with DNA replication: proliferating cell nuclear antigen (PCNA). This will give an estimate of the proportion of cells in S phase at the time of fixation. Alternatively, cells, tissues, or whole animals can be labeled by administration of a DNA precursor, usually **bromodeoxyuridine (BrdU)**, a thymidine analog that is incorporated into DNA and can be detected with a specific antibody (see

Classic Experiments

THE HEMATOPOIETIC STEM CELL AND ITS CELL LINEAGE

In the 1940s it was known that irradiated mice could be rescued by a bone marrow graft. But it was thought that some substance or hormone present in the marrow was responsible. The paper by Ford et al. showed by identification of a chromosomal marker that the rescue activity of the marrow graft was actually due to colonization by blood-forming cells.

The second paper describes the ability of hematopoietic cells to form monoclonal colonies in the spleen of irradiated animals. This provided a method for quantifying numbers of particular types of progenitor and showed the existence of multipotent cells forming clones of mixed cell type.

The third paper established an *in vitro* assay for colony formation which resulted in the characterization of further multipotent cell types and was used to purify the hematopoietic growth factors.

Finally, the paper of Spangrude et al. describes the isolation of pure HSCs from mouse bone marrow by cell sorting, using the criterion of $Sca1^+Lin^-Thy1^{lo}$.

Ford, C.E., Hamerton, J.L., Barnes, D.W.H. & Loutit, J.F. (1956) Cytological identification of radiation chimaeras. *Nature* **177**, 452–454.

Till, J.E. & McCulloch, E.A. (1961) A direct measurement of the radiation sensitivity of normal mouse bone marrow cells. *Radiation Research* **14**, 213–222.

Bradley, T.R. & Metcalf, D. (1966) The growth of mouse bone marrow cells in vitro. *Australian Journal of Experimental Biology and Medical Science* **44**, 287–300.

Spangrude, G.J., Heimfeld, S. & Weissman, I.L. (1988) Purification and characterization of mouse hematopoietic stem-cells. *Science* **241**, 58–62.

Chapter 5). This will reveal which cells were in S phase at the time that the label was administered. It makes it possible to label at one time and then to trace the subsequent position and differentiation class of the cells that were labeled. If the cells continue dividing, then the incorporated BrdU will be diluted out after a few rounds of division and will no longer be detectable, therefore long term retention of the label is taken to indicate that the cells underwent their final S phase at the time of administration, and divided only once before becoming postmitotic. This final division is sometimes called the cell's "birthday." Before BrdU became available, many similar studies were conducted using **³H-thymidine (³HTdR)**, which is also incorporated into DNA in S phase and can subsequently be localized by **autoradiography** (see Chapter 5).

Tissue types in the body can be classified on the basis of their proliferative behavior, visualized with BrdU or ³H-thymidine:

1 Postmitotic, such as neurons and skeletal or cardiac muscle. Once formed these cells do not divide again, although it is now known that their numbers can be replaced to a small extent from undifferentiated progenitors. There is a limited new formation of neurons of the olfactory bulbs and the hippocampus from neuronal stem cells in the ependyma. There is also some new formation of myofibers from satellite cells.

2 "Expanding." These tissues grow while the animal is growing and stop when adult size is attained. In the adult such tissues are mostly quiescent although there may be a slow turnover of cells. In addition they remain capable of growth to a greater or lesser degree when stimulated by wounding. In this category are connective tissue, smooth muscle, and liver.

3 Renewal. Here the tissue is in a constant state of cell turnover. There is an active proliferative zone containing stem cells (see below) and this feeds a population of differentiated cells, which itself has a finite lifetime and is constantly dying and being repopulated. Examples are the hematopoietic system, the epidermis of the skin, and the epithelium of the gut.

Much of the developmental biology of postnatal life concerns the behavior of the **renewal tissues**, some of which are described below. Of course renewal involves cell death as well as cell birth. The index of apoptotic cell death can be measured by several methods. The most popular are immunostaining for the presence of one of the apoptosis-associated proteins such as the caspase enzymes, and the detection of DNA breaks by a method called TUNEL (TdT-mediated dUTP nick end labeling). Here the enzyme terminal nucleotidyl transferase is used to add a modified nucleotide, usually biotin-labeled, to the fragemented DNA of the dying cell. This is then detected with a fluorescent or enzyme-linked streptavidin.

Although most measurements of cell turnover look at the proportion of cells in cycle, or the proportion of cells in apoptosis, what is really required to understand the situation is a measure of cell production rate and cell removal rate. To obtain a cell production rate it is necessary to know, as well as the S-phase labeling index, the duration of the cell cycle and the proportion of the cycle spent in S phase. For example if cells divide on average once per 24 hours and the S phase lasts 6 hours, then a short pulse of BrdU would enable the observation of about 6/24, or one quarter, of the cells in cycle. So in this case the cell production rate is about 4× the cell labeling index. The cell removal rate is rarely calculated, and often underestimated. Because the duration of apotosis is quite short (1–4 hours) the flux to cell death per 24 hours is a high multiple of the apoptotic labeling index. For example, if the apoptotic index is 1% and the dying cells are observable for only 2 hours, then the flux to cell death is actually about 1× (24/2) or 12% per day.

Stem cells

Stem cells (Fig. 13.5) are defined as cells:

1 that can divide without limit;
2 that are visibly undifferentiated;
3 whose progeny include both further stem cells and cells destined to differentiate.

In addition, there are other properties characteristic of some but not all types of stem cell:

1 some can give rise to more than one type of differentiated progeny (**pluripotent**);
2 some undergo obligatory **asymmetrical division** to yield one stem cell daughter and one daughter destined to differentiate.

Embryonic stem cells (**ES cells**, see Chapter 10) are usually considered to be similar to the cells of the mammalian embryonic epiblast or inner cell mass and are capable of forming any cell type in the body if reimplanted into an embryo. **Tissue stem cells**, sometimes called **adult stem cells,** are found in renewal tissues of the postnatal animal and are normally thought to be committed to form one particular tissue type. Thus, an intestinal stem cell can form only intestinal types and an epidermal stem cell can form only keratinocytes. This corresponds to the idea that embryonic development is hierarchical, with the cells of the early blastula or epiblast being able to form anything, and then being progressively restricted in their potency by a succession of inductive signals. For example, in the course of development the precursors of an intestinal stem cell would have been committed to endoderm and then to intestine, while the embryonic precursors of a hematopoietic stem cell would have been committed to mesoderm and then to the blood-forming tissue. According to this view, the stem cell for each tissue type may be quite similar to the cells of the appropriate organ rudiment at the **phylotypic stage**. This conception of stem cells resembling embryonic tissue rudiments has recently been challenged by experiments showing repopulation of many tissues by a single type of stem cell, but this area remains controversial (see below).

By definition, stem cells are capable of unlimited division. However, they are by no means the only dividing cells in the tissue because their direct progeny are usually **transit amplifying cells** capable of dividing only a few times but whose division can be regulated to correspond to the demand for new differentiated cells. The stem cells are usually a minority among the dividing population and generally grow more slowly than the transit amplifying cells. Although they must repopulate the stem-cell compartment as well as feed cells to the transit amplifying and differentiated compartments, this does not necessarily mean that every individual cell division need be an asymmetrical one. It is simply required that, on average, the progeny of the stem cell consists of 50% stem cells and 50% cells destined to differentiate (Fig. 13.6). This is important, for example, when considering the acquisition of monoclonality in intestinal crypts (see below).

In the adult body, cell growth is mainly confined to the regions of the renewal tissues containing the stem cells and the transit amplifying cells, and elsewhere most cells are quiescent. It is to be expected that cell division should be under strict inhibitory control, as otherwise the unregulated growth of even one single cell could easily become a macroscopic cancer and destroy the organism in a few months.

Stem cell niches

Many tissues have a histological substructure such that they consist of many small repeating modules or units, for example glandular acini or intestinal crypts. These units are not only the functional units of the tissue, but are also often the units within which cell proliferation and turnover are organized. The places where stem cells are found contain a specific microenvironment known as the **stem cell niche** suitable for their persistence and growth. If the stem cells are removed from this microenvironment they will grow no more. Conversely if they are reintroduced to the niche then they will grow again.

The molecular identity of stem cell niches is now becoming known. One well characterized example lies in the germarium regions of the *Drosophila* ovary (Fig. 13.7). The ovary comprises a set of ovarioles, each of which contains a string of egg chambers with the germarium at the proximal end. The egg chambers consist of one oocyte and 15 nurse cells surrounded by follicle cells, as previously described in Chapter 11. The oocyte and nurse cells of each egg chamber are formed from four divisions of a single germ cell in the germarium region. This cell is known

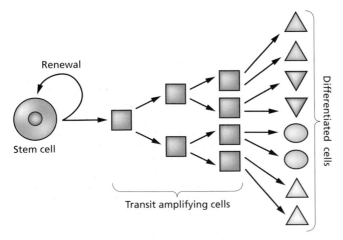

Fig. 13.5 Cell lineage in a renewal tissue, showing stem cells, transit amplifying cells, and differentiated cells.

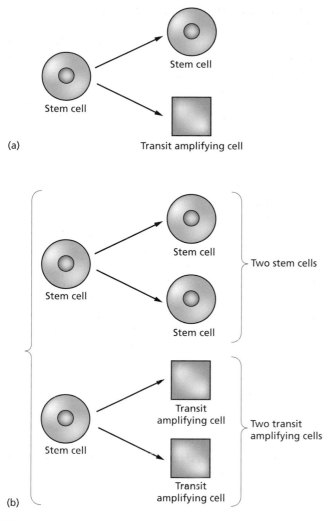

(a)

(b)

Fig. 13.6 Stem cells can maintain themselves either (a) by repeated asymmetrical division or (b) by generating stem cell and transit amplifying daughters with equal frequency but at separate divisions.

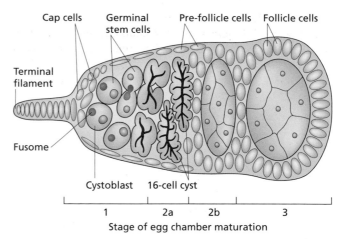

Fig. 13.7 The stem cell niche in the *Drosophila* ovary germarium.

absence of *bam* the the germinal stem cells do not differentiate but continue to proliferate, forming a germ cell tumor. Conversely the overexpression of *bam* under control of a heat shock promoter causes premature cystoblast formation in all the germinal stem cells. Evidence that the repression of *bam* depends on the dpp signal from the somatic cap cells comes from experiments showing that overexpression of *dpp* has a similar effect to loss of function of *bam*, and loss of function of *dpp* has the same effect as overexpression of *bam*. Normally the germinal stem cell divisions are asymmetrical, because the daughter not contacting the cap cells turns on *bam* and becomes a cystoblast. But one of the properties of the niche is that if some stem cells are removed by genetic means then this creates some space adjacent to the cap cells. This space can become occupied by progeny from the surviving stem cell(s) that would normally have become cystoblasts, and thus the production of cystoblasts is suspended while the normal number of stem cells is restored. This behavior resembles the repopulation of mammalian bone marrow, following irradiation and grafting of marrow from a healthy donor (see below).

as a **cystoblast**, and its precursors are the germinal stem cells (= **oogonia**) which divide mitotically about once per day. Each germarium contains two or three germinal stem cells, adjacent to five or six somatic cap cells and it is these somatic cap cells that define the niche. They do so by secretion of the decapentaplegic (dpp) protein. This represses expression of the *bag of marbles* (*bam*) gene in the adjacent cells. *bam* encodes a cytoplasmic protein which interacts with a germ cell-specific organelle called the fusome and is needed for the cystoblast maturation. In the

Classic Experiments

Cell turnover in tissues

The first paper is a study of mitoses in the epithelium of the small intestine and arrives at the conclusion that cells must be being continuously produced in the crypts and shed from the villi. The second paper uses the incorporation of radioactive debris from ^{3}H-thymidine-labeled cells that have died as a cell marker for neighboring cells. From studying the subsequent distribution of

these radioactive debris it is postulated that a single type of stem cell produces all four cell types of the small intestinal epithelium.

Leblond, C.P. & Stevens, C.E. (1948) The constant renewal of the intestinal epithelium in the albino rat. *Anatomical Record* **100**, 357–377.

Cheng, H. & Leblond, C.P. (1974) Origin, differentiation and renewal of the four main epithelial cell types in the mouse small intestine. V. Unitarian theory of the origin of the four epithelial cell types. *American Journal of Anatomy* **141**, 537–562.

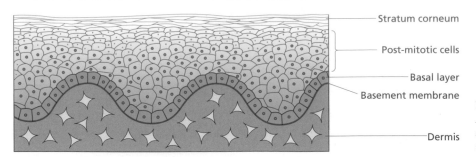

Stratum corneum

Post-mitotic cells

Basal layer
Basement membrane

Dermis

Fig. 13.8 Organization of the epidermis. All keratinocytes are born in the basal layer and differentiate progressively as they move up to the surface.

Skin

Skin consists of a squamous stratified epithelium, the **epidermis**, on top of a connective tissue, the **dermis** (Fig. 13.8). The main cell type in the epidermis is the **keratinocyte**. Cell division is confined to the basal layer. This contains **stem cells**, which can both renew themselves and generate keratinocytes, and also **transit amplifying cells**, which are formed from the stem cells and have a finite division potential before they differentiate. In humans the epidermis is renewed from the basal layer about every 2 weeks. The entire basal layer of the epidermis, and of other squamous epithelia such as those of the esophagus or vagina, depends on the activity of a transcription factor called p63. This is expressed throughout the basal layer and is switched off when cells migrate upwards. The knockout mouse lacking p63 is unable to form any squamous epithelia and dies soon after birth.

The dermis is a dense fibroelastic connective tissue derived from the dermatome and neural crest of the embryo. At deeper levels it is largely adipose tissue. The junction between dermis and epidermis is marked by a basement membrane and, in humans, this is wrinkled with epidermal ridges and corresponding dermal papillae (note that the specialized dermal core of the hair bulb is also called a dermal papilla) . The dermis contains the usual nerves and blood vessels, and pressure receptors called Pacinian corpuscles, as well as the epidermis-derived specializations.

The factors required for proliferation of the epidermal basal layer include both factors produced by the epidermis itself, such as TGFα, and factors secreted by the underlying dermis, including keratinocyte growth factor (KGF), a member of the FGF family. Keratinocyte cultures can be grown *in vitro* and form the same stratified arrangement as the natural skin. There is no need for the dermis in such cultures as the growth factors which it normally supplies are present in the medium. Labeling studies show that about 60% of basal layer cells are in cycle, but only a fraction of these are stem cells. The functional test for a stem cell is that it can form a large self-sustaining epidermal colony either in culture, or after grafting to a **nude mouse** (a type of mouse that cannot reject tissue grafts). Transit amplifying cells, by contrast, can only form small colonies of a few cells and then stop growing. By this criterion about 10% of basal layer cells are thought to be stem cells. A similar proportion, presumably the same stem cells, are capable of forming large colonies that can repopulate the epidermis following severe radiation damage.

The stem cells defined by these criteria are characterized by a higher level of β-1 integrin than the transit amplifying cells. This is a cell adhesion molecule involved in the recognition of collagen, laminin, and fibronectin. *In vivo*, in human foreskin, the high-integrin cell clusters are found at the tips of the dermal papillae, suggesting that this may define the stem cell niche for the epidermis. The stem cells also show an enhanced level of nuclear β-catenin, indicating a possible role for Wnt signaling in the maintenance of the niche. If either β-1 integrin or β-catenin are introduced into keratinocyte cultures by retroviral infection, then the cells receiving the gene will acquire stem cell properties. The stem cells, both *in vivo* and *in vitro*, also contain an elevated level of the Notch ligand delta-1. If delta-1 is introduced into cells of a keratinocyte culture by retroviral infection, then the high delta-expressing cells tend to cluster and to be inhibited from differentiation, while their neighbors are stimulated to differentiate. This suggests that the distinction between stem cell clusters and the surrounding transit cells may be maintained by a **lateral inhibition** mechanism similar to that involved in neurogenesis and pancreatic differentiation (see Chapters 4, 14, and 16). Although this picture is incomplete, it seems likely that the dermal papillae emit a Wnt signal, and that the adjacent basal layer cells respond by increasing synthesis of β-1 integrin. This helps maintain the stem cells as a cluster and also has additional intracellular signaling effects leading to the increase of delta-1 expression and the spatial segregation of the basal layer into stem cell and transit amplifying cell zones (Fig. 13.9).

Once cells leave the basal layer they stop dividing and enter a program of further differentiation. The progress of maturation is reflected by the names given to successive layers of the epidermis: stratum germinativum (the basal layer), stratum spinosum (the "prickle" layer – the apparent prickles are abundant desmosomes), stratum granulosum (with granules), and stratum corneum (cells have lost nuclei and have become flat sacs of keratin). Keratin is a generic name for the large family of fibrous proteins forming the cytokeratin **intermediate filament** family and found in all epithelial cells. There are many different keratins coded by different genes, and the repertoire expressed changes as cells move up from the basal layer. In the granular and cornified layers the cells also contain a tough internal sheath of an insoluble protein called involucrin.

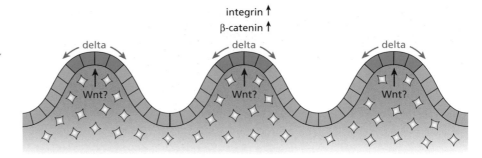

Fig. 13.9 The stem cell niche in the epidermis. Signals from the dermis, probably including Wnt, maintain groups of stem cells. The signals increase the level of β1-integrin, which causes the stem cells to remain as a small cluster. The stem cells display delta-1 on their surfaces, thereby repressing the surrounding cells from stem cell behavior.

Hair follicles

The structure of the hair follicle is shown in Fig. 13.10a. The hair shaft is composed of dead keratinocytes, which are produced in the epidermal **matrix** region at the base. This lies in close proximity to the **dermal papilla**, a projecting bud of fibroblastic cells, and also contains **melanocytes** which transfer pigment granules to the keratinocytes of the hair. Surrounding the whole is a layer of cells continuous with the surface epidermis called the outer root sheath. High up, near the junction with the surface epidermis, lies a sebaceous gland. The entire region at the base of the follicle comprising the dermal papilla, the proliferative epidermal zone, and the outer layers, is known as the hair bulb. Hair does not grow continuously but in a cycle. The active growth phase is known as anagen, which lasts about 3 weeks in mice but can be much longer in humans. The period of regression of the follicle is called catagen, and the period of quiescence is called telogen (Fig. 13.10b).

Hair follicles start life in the late mammalian embryo as epidermal invaginations. Their formation requires both Wnt signaling and also inhibition of BMP signaling from the dermis. BMPs are produced by the epidermis, and noggin by the dermis. A combination of Wnt3A and noggin will induce new follicle buds. The BMP inhibition causes transcription of Lef1 while the Wnt signal stabilizes β-catenin and the combination carries the Lef1 into the nucleus to control target genes (see Appendix). One target is E-cadherin, whose expression becomes repressed, thus reducing the mutual adhesion of the cells and leading to the invagination behavior to form the bud. Evidence for this mechanism is based on the the following:

1 Invaginating buds are produced by adding Wnt3A + noggin to keratinocyte cultures, or in transfecting in *Lef1* plus a constitutive form of β-catenin.

2 Knockout mice lacking either *noggin* or *Lef1* have few hair follicles.

3 The Wnt signaling pathway is active in the early buds. This may be seen in a reporter strain of transgenic mice which contains *lacZ* driven by a Wnt-sensitive promoter.

4 Ectopic hair follicles are formed in mice transgenic for constitutive β-catenin, driven by the *keratin 14* promoter which is active only in the basal layer of the epidermis.

5 Bud formation is suppressed in mice transgenic for production of the Wnt-inhibitor Dickkopf, also driven in the basal layer by the *keratin 14* promoter.

The understanding of hair follicle initiation makes it possible to contemplate a possible "cure" for human baldness. However the overexpression of Wnt pathway components in human patients would probably not be acceptable because of the risk of inducing cancers.

Although the initial signal for bud formation comes from the dermis, the formation of the **dermal papilla** depends on a second signal from the invaginating bud to the dermis (Fig. 13.11). The dermal papilla secretes growth factors needed by the proliferative zone of the epidermal matrix region. Isolated papillae will induce new epidermal proliferative zones from the upper halves of follicles, and in some situations can induce complete new follicles from epidermis. Examination of mouse **aggregation chimeras** in which one component is labeled and the other unlabeled suggests that each follicle contains about four epidermal stem cells, all of which contribute to all the layers of the hair shaft. In aggregation chimeras the cells derived from the two embryos are intimately mixed and so, allowing for similar adjacent clones, the number of labeled and unlabeled patches in a small structure like a hair approximate to the number of clones, and therefore the number of stem cells (see also Chapter 10).

As far as stem cells are concerned, attention in the past concentrated on the epidermal matrix region and the dermal papilla because of their obvious role in hair shaft formation. However it is now thought that the real epidermal stem cell population of the hair follicle lies not in the bulb but in a lateral bulge half way up the outer root sheath. The evidence for this is as follows:

1 A 3-day label with BrdU will label many cells in the epidermis of the hair follicle. But a long chase period shows that the cells that retain the label, and are thus very slowly dividing, are concentrated in the bulge region rather than in the bulb. In the next active growth cycle (anagen) some of these labeled cells are seen to have migrated into the bulb.

2 If a bulge region from a *lacZ* positive (**Rosa26**) mouse is grafted to a large vibrissal follicle and then cultured under the kidney capsule of a **nude mouse**, the β-galactoside expressing cells are seen to populate the entire follicle.

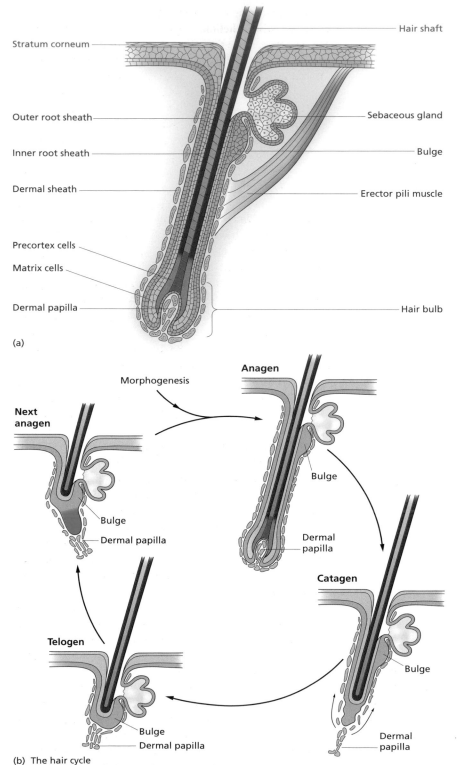

Fig. 13.10 (a) Structure of the hair follicle. (b) The hair growth cycle.

3 Individual *lacZ* positive bulge regions grafted to the back of unlabeled late embryos will generate labeled surface epidermis, hair follicles, and sebaceous glands (Fig. 13.12).
4 If the surface epidermis is wounded, it can be repopulated by label-retaining cells from the bulge.

The hair follicle bulge is certainly a stem cell niche. Although it seems that the stem cells of the bulge are "more primitive" than those of the basal layer of the surface epidermis, it is not clear whether there is a slow continuous repopulation of the surface epidermis from the bulge, or whether this occurs only as

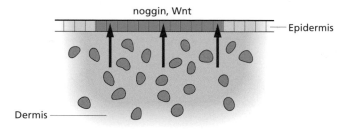

noggin, Wnt

Epidermis

Dermis

First mesenchymal signal

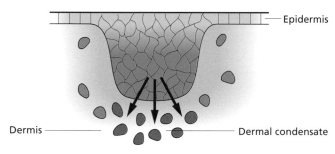

Epidermis

Dermis

Dermal condensate

Epidermal signal

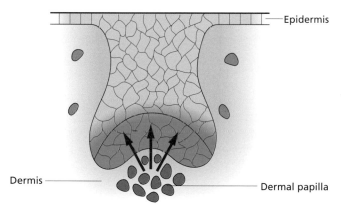

Epidermis

Dermis

Dermal papilla

Second mesenchymal signal

Fig. 13.11 Initial formation of hair follicles. The first phase involves induction by noggin and Wnt from the dermis. This is followed by a signal from the epidermal bud inducing a specialized dermal papilla.

a specific response to wounding. The bulge region does, in a sense, represent the most secluded part of the epidermis as it is continuous with the surface epidermis and is the lowest level of the follicle that persists throughout the hair cycle. In the human there is no visible bulge, but the stem cells also reside in the lowest permanent part of the outer root sheath.

Intestine

The gastrointestinal tract of vertebrates consists of a muscular tube running from the mouth to the anus. It is lined by a number of different epithelia: pharyngeal, esophageal, gastric, small intestinal, and colonic, separated by abrupt discontinuities of cell type. This **epithelium** is derived from the **endoderm** of the early embryo, while the other cell layers of the gut are derived from the **splanchnic mesoderm**. The epithelium, together with underlying connective tissue called the lamina propria and a thin muscle layer called the muscularis mucosa, is often known as the **mucosa**. Outside the mucosa lie further thick layers of connective tissue and smooth muscle.

The program of cell renewal is best understood for the small intestine. This contains regions called the duodenum, jejunum, and ileum, although the difference of cell type between them is not great. On a microscopic scale the intestinal epithelium is arranged on finger-like villi projecting into the lumen, and between the villi lie crypts of Lieberkuhn sunk below the surface (Fig. 13.13a,b). Cell proliferation takes place only in the crypts, and differentiated cells are continuously moving out from the crypts, moving up the villi, then dropping off into the gut lumen. The main cell types are **enterocytes** (absorptive cells) and **goblet cells**. The absorptive cells are characterized by a brush border at their apical surface, consisting of numerous close-packed microvilli. Goblet cells contain a large vesicle filled with mucins. In addition there are **Paneth cells** located at the base of the crypts, which secrete antibacterial substances, and several types of **enteroendocrine cells** each secreting a particular peptide hormone. The crypts themselves are set up shortly before birth by folding of the endodermal epithelium which before this stage is a simple columnar epithelium. During the growth of the animal they can divide by budding, starting at the base (Fig. 13.13c). The signal for crypt division is not known but may be an increase in the number of stem cells. Some experimental studies have also been performed on the colon because of its importance in terms of colonic cancer. Its structure is similar to that of the small intestine, but without villi and without Paneth cells.

Cell division in the crypts is rapid. The stem cells are located near the crypt base, above the Paneth cells, with several layers of transit amplifying cells lying above them. A mouse crypt contains about 250 cells of which roughly 160 are dividing, with a cycle time of about 13 hours. The progeny move up and out of the crypts and the tissue is arranged such that each crypt feeds more than one villus, and each villus draws cells from several crypts. A small proportion of cells also move down the crypt to replenish the Paneth cells at the crypt base.

It has been possible to examine the clonal composition of the crypts by making aggregation chimeras between mouse strains that differ in the expression of a marker, *Dolichos* lectin receptor (Fig. 13.14a). This carbohydrate is expressed by intestinal cells in some mouse strains and is absent in others. In an aggregation chimera the cells of the two donor embryos are intimately mixed, so at the time of crypt formation most crypts will receive cells of both donor types. However, over the first 1–2 weeks of postnatal life the crypts lose one of the two types, such that all cells in each crypt are either one type or the other. In other words, the crypts become **monoclonal**. Initially it was thought that this meant there was only one stem cell per crypt. However,

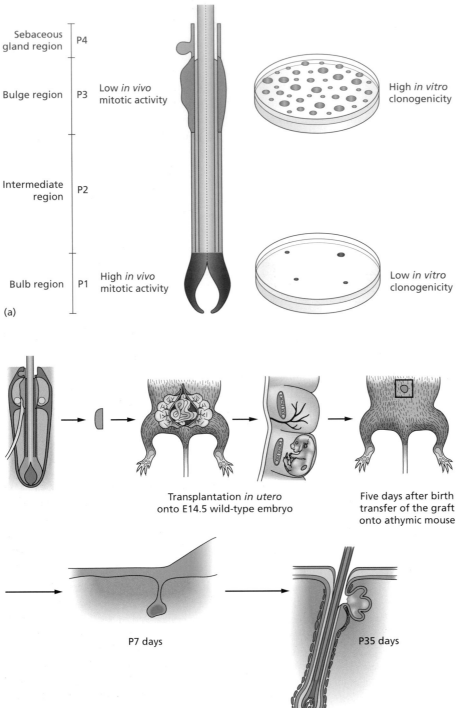

(a)

(b)

Fig. 13.12 Evidence that the bulge region of the hair follicle contains the stem cells. (a) The bulb region contains more cells in cycle but the bulge region contains more clonogenic cells. (b) A β-galactosidease labeled graft from the bulge region can repopulate the surface epidermis and other epidermal sturctures as well as the entire hair follicle.

it now seems that there are about six stem cells per crypt and that the monoclonality arises by a random cell selection process. This probably arises because the stem cells, on division, are able to produce either two stem cells or one stem and one transit amplifying, or two transit cells. Over several cell cycles the genetic diversity of the stem cells within a crypt will become progressively reduced because any stem cell that produces two transit amplifying cells will be lost to the stem-cell pool. Eventually this random loss of stem-cell diversity will result in monoclonality. An additional reason for the acquisition of monoclonality is the fact that the crypts themselves multiply by budding, sharing their stem cells between the two daughter crypts. By this process it is possible to lose all the stem cells of one genotype in one or both of the daughter crypts.

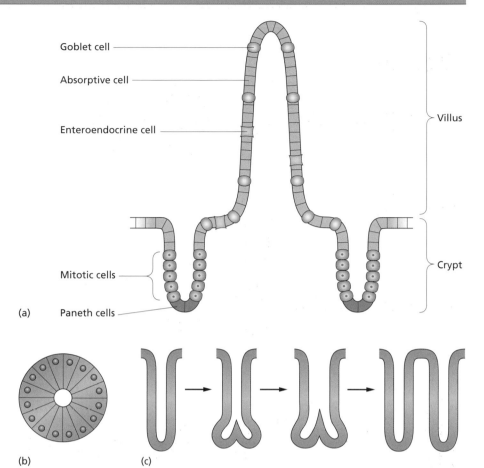

Fig. 13.13 Diagrammatic organization of the small intestinal epithelium. (a) Longitudinal section of two crypts and a villus. (b) Transverse section of a crypt. (c) Multiplication of crypts by budding from the base.

Estimates of the proportion of stem cells among the dividing cells of the crypts have been made by several methods including modeling of the cell-cycle kinetics of the whole crypt. The estimate of six is based mainly on mutagenesis studies (Fig. 13.14b). If a mouse of a suitable strain is mutagenized, then a proportion of cells will acquire reactivity for *Dolichos* lectin. Shortly after the mutagenesis, many crypts containing mutant clones are seen. Most of the clones arise in transit amplifying cells and are soon lost. But if they arise in stem cells they are retained in the long term. In principle, if there are n stem cells per crypt, then a mutation in a stem cell should appear as a labeled sector within the crypt occupying about $1/n$ of its circumference. In fact, after a few weeks many mutant crypts are uniformly composed of mutant cells, showing the same drift to monoclonality as aggregation chimeras, probably for the same reasons. Mutagenesis also enables the study of the contribution of each crypt to the villi, as a labeled crypt will emit a stream of labeled cells that form a strip running up each of the villi to which it contributes (Fig. 13.14b).

A different type of estimate of stem-cell number can be made based on radiation toxicity (Fig. 13.14c). A given dose of X- or gamma rays will sterilize a proportion of cells in the epithelium. Cells that are capable of growing and repopulating the tissue are known as **clonogenic** cells. It is assumed that a crypt can only regenerate if it includes at least one clonogenic cell that survived the radiation. Measurements of crypt survival following various dose regimes suggest that there are about 80 clonogenic cells per crypt. This is substantially greater than the likely number of stem cells per crypt calculated from mutagenesis, and the difference in the two estimates suggests that a proportion of the transit amplifying cells are capable of becoming stem cells again under conditions of severe tissue damage. This makes sense if it assumed that a transit amplifying cell will become promoted to stem cell status if it can enter the stem cell niche. The mutagenesis and radiation experiments also provide data to support the idea that the stem cells of the small intestine are **pluripotent**, being able to generate all four of the usual cell types: columnar, goblet, Paneth, and enteroendocrine. Crypts entirely populated by one mutant clone contain all four cell types, so one cell must have generated them all. Similarly, after high doses of radiation from which only a minority of crypts survive, these will mostly have regenerated from a single clonogenic cell, but will nevertheless all acquire the four cell types.

Knowledge of the molecular characteristics of intestinal stem cells remains limited. An RNA-binding protein called Musashi-1 is expressed in a few cells between and just above the Paneth cells, and may be a stem cell marker. There is an increase in the number of Musashi-1 positive cells during the recovery

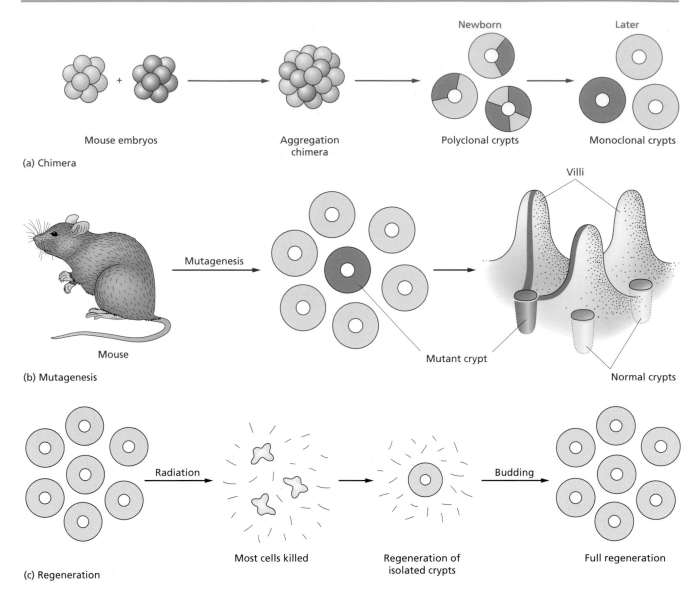

(a) Chimera

Mouse embryos   Aggregation chimera   Newborn   Later

Polyclonal crypts   Monoclonal crypts

(b) Mutagenesis

Mouse   Mutagenesis   Villi   Mutant crypt   Normal crypts

(c) Regeneration

Radiation   Most cells killed   Regeneration of isolated crypts   Budding   Full regeneration

Fig. 13.14 Methods for studying intestinal crypt organization. Crypts are shown diagrammatically in transverse section. (a) Aggregation chimeras. Early crypts are polyclonal but later become monoclonal. (b) Mutagenesis produces occasional cells that can be visualized by binding of Dolichos lectin. One mutant stem cell may often come to populate an entire crypt, and its progeny form streams up to the tips of the adjacent villi. (c) A dose of X-radiation that destroys most cells leads to regeneration of whole crypts from individual clonogenic survivors.

from radiation damage, which is consistent with the increase in the number of clonogenic cells in this situation. As in the epidermis, there is good evidence that the intestinal proliferative compartment depends on the Wnt signal transduction pathway (Fig. 13.15). It is known that the mouse knockout of the transcription factor gene *tcf4* fails to form any proliferative compartment. TCF4 is one of the HMG-type transcription factors that is activated by β-catenin and nuclear β-catenin is normally found in the cells of the bottom third of the crypt, which represents the proliferative compartment. So it is likely that Wnt signaling from the lamina propria is needed to maintain proliferation in the epithelium, and something additional is needed to define the much smaller stem cell niche.

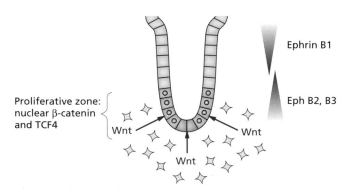

Ephrin B1

Eph B2, B3

Proliferative zone: nuclear β-catenin and TCF4

Wnt   Wnt   Wnt

Fig. 13.15 Role of the Wnt pathway in controlling the proliferative structure of the intestine.

There is also an intimate connection between the control of proliferation and the actual structure of the epithelium, because the crypts are characterized by expression of the adhesion molecules Eph B2 and B3, while the villi are characterized by expression of their ligand ephrin B1. Transcription of these molecules is, respectively, activated and repressed by β-catenin. If both Eph B2 and B3 are removed by targeted mutagenesis then the intestine loses its structure and both dividing and differentiated cells are found mixed together. If only EphB3 is removed then the structure is normal but Paneth cells are found all over the epithelium instead of being confined to the crypt bases. APC, the product of the *adenomatous polyposis coli* gene, is a cytoplasmic protein required to enable the phosphorylation of β-catenin by GSK3. In loss-of-function mutants β-catenin is not inactivated, and is therefore constitutively active. This leads to the inability to shut off Eph B2 and B3, and to the formation of **polyps**, which are projections into the lumen of differentiated but abnormally organized intestinal tissue. Human patients suffering from adenomatous polyposis coli have many such polyps and a high risk of a polyp developing to cancer (see below). The disease is hereditary and due to loss of one copy of the *APC* gene. When the other, good, copy is lost from an individual cell due to occasional somatic mutations, then that cell will acquire constitutively active β-catenin and develop into a polyp.

In the intestine, as is generally the case for renewal tissues, the stem cells are responsible for producing several types of differentiated cell. It now seems that this is achieved using the Delta-Notch **lateral inhibition** mechanism (see also Chapters 4, 14, and 16). There is a "master switch" at the level of the decision whether to become an ordinary absorptive cell or one of the three specialized cell types (goblet, enteroendocrine, or Paneth), and this is controlled by a bHLH type transcription factor called Math1, which also promotes the formation of Delta (Fig. 13.16). All the cells express Notch and initially have a similar level of Math1. Cells which by chance have a slightly higher level of Math1 produce a little more Delta and signal to surrounding cells. Notch is stimulated in these surrounding cells leading to inhibition of expression of Math1, and hence reduction of Delta. The process will run on until there are a few high Math1-high Delta cells surrounded by a larger number of low Math1-low Delta cells. The cells with low Math1 become absorptive cells, while those with high Math1 become either goblet or enteroendocrine or Paneth, the subsequent decisions depending on further unknown mechanisms. The main evidence for this process is that the knockout of *math1* has an intestinal epithelium which

is normal in overall structure, and which contains normal absorptive cells, but totally lacks all the types of specialized cell. Another knockout, of the transcription factor gene *hes1*, shows an opposite phenotype with an elevation of Math1-expressing cells and of the proportion of specialized cell types in the epithelium. Hes1 is on the Notch signaling pathway and its loss will reduce the inhibition of Math1 expression by Notch signaling.

New Directions in Research

Stem cell research is thought to be rich in practical applications, mostly having the aim of repairing damaged tissues and organs by grafts of stem cells. In recent years many biotech companies have been founded on the basis of these opportunities.

In terms of basic knowledge it is important to establish exactly how tissue stem cells arise during development and how similar they really are to the whole embryonic rudiment from which the tissue develops.

We need to know if Wnt signaling really controls stem cell behavior, or just proliferative behavior generally.

We also need to understand whether the Notch lateral inhibition system controls the differentiation of multiple cell types in all cases, or whether there are other similar systems.

Ultimately, understanding the signals that control the self-renewal and differentiation of the stem cells should enable the design of culture conditions for growing tissue stem cells without limit *in vitro*.

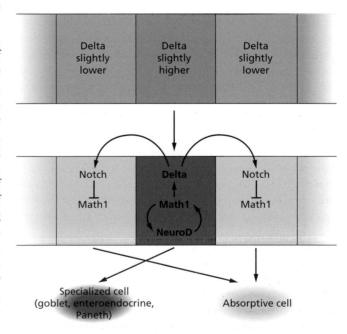

Fig. 13.16 Control of cell differentiation in the intestinal epithelium by lateral inhibition.

Hematopoietic system

In the adult mammal, the **hematopoietic** system resides in the bone marrow within the larger bones of the skeleton. In the embryo it is found at various other sites. Initially it is extra-embryonic, in the yolk sac, then in the **AGM (aorta-gonad-mesonephros) region** of the mesoderm, then in the liver, then in the spleen and lymph nodes, and finally in the bone marrow. The first two of these sites of production arise *in situ* while the later ones are colonized by cell migration from the earlier sites. This process of cell migration has been established by the grafting of marked cells in amphibian and avian embryos and localizing the progeny of the grafted cells at a later stage.

Like the skin and gut epithelium, the hematopoietic system is a state of continuous cell production and renewal throughout life. There exists a **hematopoietic stem cell (HSC)** that can both renew itself and also differentiate into a wide variety of cell types. These include all cells of the blood and immune system together with histiocytes, osteoclasts, and Langerhans cells of the skin.

The cellular components of the blood are as follows:

1 **Erythrocytes** (red cells).
2 **Granulocytes**, comprising neutrophils (phagocytes), eosinophils, and basophils (similar to mast cells).
3 **Monocytes** (similar to **macrophages**/histiocytes).
4 **Megakaryocytes**, giant cells that break up to become platelets. The above four cell types are collectively known as **myeloid cells**.
5 **Lymphocytes** (T and B cells).

Much of the evidence for the existence of HSCs comes from reconstitution experiments. The bone marrow is the most sensitive tissue in the body to irradiation, and so there is a dose range that will kill by bone marrow failure while other tissues are still potentially able to recover. If a mouse, lethally irradiated with such a dose, is given a graft of marrow cells by injection into the bloodstream, then the graft will colonize the marrow of the host, proliferate extensively, and enable survival of the host. After a few weeks the counts of the various cell types mentioned above have returned to normal and the observation of genetic markers shows that they are all derived from the graft. The ability permanently to rescue lethally irradiated mice is often taken to be the defining feature of the HSC. The use of reconstitution assays to examine the hematopoietic populations in the mouse embryo shows that the earliest HSCs are found in the AGM region at 10–11 days of gestation and in the yolk sac and liver from about 12 days of gestation. This suggests that the hematopoietic population seen in the yolk sac at earlier stages does not contain HSCs capable of long term repopulation.

It is now possible to isolate HSCs directly from bone marrow using the technique of fluorescence-activated cell sorting (**FACS**, see Chapter 5). Mouse HSCs are characterized by high levels of Sca1, low but finite levels of Thy1, and the absence of all other differentiation markers (Sca1$^+$, Thy1^{lo}, Lin1$^-$). Sca1 (stem cell antigen 1) and Thy1 (thymus 1) are both cell surface glycoproteins, Thy1 being abundant on mature T cells. The study of mice transgenic for GFP driven by the *Sca1* promoter shows

that the very first HSCs arise in the endothelium of the dorsal aorta. Mouse HSCs can also be isolated because they preferentially exclude certain fluorescent dyes such as Hoechst 33324 or Rhodamine 123, so after exposure to these dyes give a lower fluorescence signal than all other cells in the marrow. The self-renewing properties of the HSC seem to depend on the proto-oncogene *bmi-1*, which encodes a *polycomb* type transcriptional repressor. Knockouts for *bmi-1* develop HSCs but the numbers are greatly reduced postnatally and they have very little reconstitution activity. HSCs are difficult to grow in culture but the numbers, measured by the mouse reconstitution assay, can be increased by introduction of certain genes using retroviruses. These include stabilized (constitutive) β-catenin, suggesting that Wnt signaling may be necessary for HSC self-renewal, as it is for the stem cells of the skin and the intestine.

In addition to the HSC, the marrow contains other cells that can be isolated by different combinations of cell surface markers. In the reconstitution assay they give rise to only a subset of the complete HSC repertoire. This shown the existence of various multipotent progenitors including the common lymphoid progenitor, the common myeloid progenitor, the granulocyte-macrophage progenitor, and the megakaryocyte-erythrocyte progenitor. There are also pluripotent stem cells that have only a temporary repopulating ability, which are believed to represent the next step of maturation after the permanent HSC. A current consensus model for the cell lineage of the hematopoietic system is shown in Fig. 13.17.

A second line of evidence for this model comes from *in vitro* colony assays. It is possible to obtain clones of hematopoietic cells *in vitro* by plating marrow cells in soft agar or methyl cellulose in the presence of the appropriate growth factors. Since most single colonies will be clones derived from a single cell, the production of multiple cell types by a single colony indicates the existence of a multipotent progenitor, and the same potency classes are recovered in these assays as in the whole mouse reconstitution assay. Although the model is generally accepted, it has not yet been confirmed by prospective labeling, which would require the insertion of a permanent genetic marker into HSCs *in vivo*, followed by identification of each type of progeny cell as it is formed.

The *in vitro* colony formation assay has been used to isolate a number of colony-stimulating factors (CSFs), otherwise known as hematopoietic growth factors. Interleukin 3 and stem cell factor (steel factor) can stimulate proliferation of the HSCs, while granulocyte–macrophage CSF (GM-CSF), granulocyte CSF (G-CSF), macrophage CSF (M-CSF), and erythropoietin work in combinations to stimulate the division of the various transit amplifying lineages. Most of the feedback control over the production of the various cell types is exerted by varying the growth rate at the transit amplifying cell level. This is because there are large numbers of such cells and a rapid response can be obtained to changing demand. By contrast, it would take several weeks to alter the rate of production by regulation at the level of the HSC. Several of the hematopoietic growth factors have been prepared

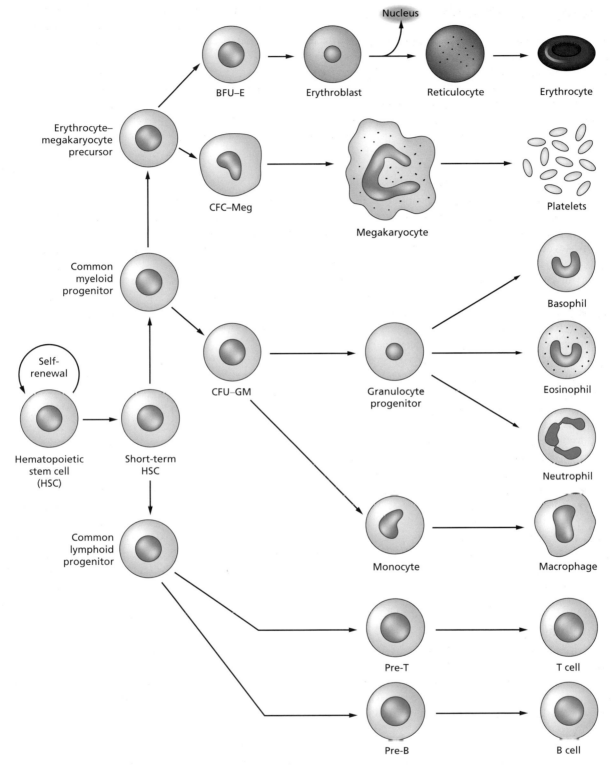

Fig. 13.17 A consensus model for the cell lineage of the cells of the blood and immune system.

in therapeutic quantities by recombinant DNA methods and are now very useful in clinical practice, particularly for the treatment of various types of anemia and for enabling people to rebuild their marrow after cancer therapy.

The transplantable properties of the HSC have also made marrow grafting one of the earliest types of cell therapy to be adopted in human medicine. Indeed the first human bone marrow grafts between identical twins were performed in the 1950s.

Grafting of the patient's own marrow (**autologous graft**) is now routinely used in cancer therapy. Some of the patient's bone marrow is extracted before treatment, the patient is then given a lethal dose of radio- or chemotherapy to kill the cancer, and then the marrow is re-infused to rescue the patient from the radiation. One of the limitations of this method is that there are often cancer cells in the graft which escape the treatment and are reintroduced into the patient. In principle the effectiveness of such grafts could be increased by using pure HSCs instead of whole marrow. It is also possible to graft bone marrow between individuals (**allograft**). However, because the bone marrow is also a factory for the production of immunoactive lymphocytes it is necessary to have a match of major histocompatibility (HLA) antigens in donor and host to avoid both rejection of the graft by the host, and also graft-versus-host disease caused by reaction of lymphocytes in the graft against the host. Even matched grafts are likely to need immunosuppressive treatment to limit the reaction to minor histocompatibility antigens.

Various types of **gene therapy** have been proposed which depend on the possibility of inserting a missing gene into HSCs and then introducing them into the patient. This is likely to be appropriate in cases where the missing gene has a metabolic function, and so would be effective regardless of the tissue in which it is expressed. These methods are still experimental because of the difficult of obtaining enough HSCs, of efficiently introducing genes, and the safety problems associated with random gene insertion events which can sometimes produce cancer-causing mutations.

The reason that the reconstitution assay for HSCs works is that there are a number of stem cell niches for the HSC within the marrow. These are now known to be formed by the **osteocytes** lining the trabecular bone surfaces in the marrow cavity, to which the HSCs attach via N-cadherin (Fig. 13.18). Various treatments that increase the number of trabecular osteoblasts, including downregulation of BMP receptor 1A or injection of parathyroid hormone, also increase the number of HSCs.

Mesenchymal stem cells and "transdifferentiation"

In addition to the HSCs, the bone marrow contains another type of stem cell. These are called mesenchymal stem cells, or marrow stromal cells, both conveniently abbreviating to **MSC**. It is not yet possible to purify them by cell sorting but they adhere to plastic and long-term cultures can be grown in which the other cell types of the marrow are selected out and disappear. MSCs will form adipocytes, chondrocytes, or osteocytes *in vitro* when cultured in appropriate media. The normal function of MSCs is to produce the various nonhematopoietic cell types found in the bone marrow.

A number of recent studies have suggested that bone marrow cells are capable of colonizing a wide variety of other tissue types

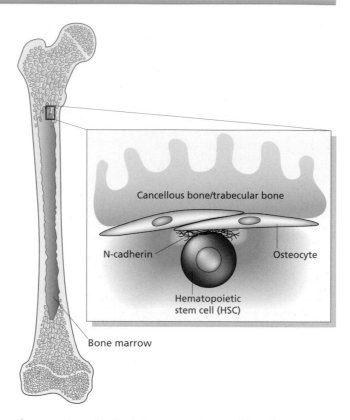

Fig. 13.18 The niche for the hematopoietic stem cell in the bone marrow.

when transplanted into irradiated hosts. Some of these are performed with unfractionated marrow, some with enriched or purified HSCs or MSCs. The tissues colonized can include virtually everything including epithelia, muscle, and neurons. This work has generated considerable controversy because it suggests a very different model of development from the conventional one. Instead of cell populations undergoing a series of decisions during embryonic development, in each of which their competence is restricted, the idea is that the whole body is continuously being renewed by highly pluripotent cells from the bone marrow. The phenomenon is known as "transdifferentiation" although this is unfortunate as the term was previously used in a more restricted sense to refer to the rare but well-established cases of direct transformation between differentiated cell types. It now appears that some of the results were due to cell fusion, whereby the genetic markers from donor cells became incorporated into host cells by direct fusion. Other results appear to indicate genuine reprogramming of marrow-derived cells to various other tissue types, but only at very low frequency. Because the hosts are nearly always irradiated, and therefore have considerable tissue damage and tissue regeneration all over the body, it is thought that this situation allows favorable circumstances for the occasional reprogramming event. It is not, however, likely that reprogramming can occur on a large scale, or that the bone marrow is a repository for cells that can regenerate the rest of the body.

Key Points to Remember

- The organs of the body are mostly composed of several tissue layers. Each tissue layer contains multiple cell types.
- Tissues may be classified as postmitotic, quiescent but capable of growth, and renewal. In renewal tissues there is a balance of cell production and cell death.
- Stem cells are cells that can both renew themselves and produce offspring destined to differentiate. They are normally found in "niches" defined by signals from surrounding cells.
- The epidermis is a stratified squamous epithelium. It has proliferative cells only in the basal layer, and the successive layers represent degrees of postmitotic cell maturation. The stem cells are found in integrin-rich clusters.

- Hair follicles contain stem cells in the bulge of the outer root sheath that can populate the entire epidermis and its specializations.
- The small intestine has its proliferative zone in the lower part of the crypts. The stem cells are found just above the Paneth cells at the crypt base. Four types of differentiated cells are produced which move continuously to the upper crypt and villi.
- The bone marrow contains hematopoietic stem cells (HSCs) and mesenchymal stem cells (MSCs). The HSCs continuously renew the cells of the blood and immune system. Their niche is in association with osteocytes on the trabecular bone of the marrow cavity.

Further reading

Histology

Le Gros Clark, W.E. (1971) *The Tissues of the Body*. Oxford: Clarendon Press.

Cormack, D.H. (1987) *Ham's Histology*. Philadelphia: Lippincott Williams and Wilkins.

Wheatear P.R. & Burkitt, H.G., eds (2000) *Wheater's Functional Histology: a text and colour atlas*. Edinburgh: Churchill Livingstone.

Alberts, B., Johnson, A., Lewis, J., et al. (2002) Histology: the lives and death of cells in tissues. In: *The Molecular Biology of the Cell*, 4th edn, chapter 22. New York: Garland.

Kaye, G.I., Ross, M.H. & Pawlina, W. (2002) *Histology: a text and atlas*. Philadelphia: Lippincott Williams and Wilkins.

Stem cells and their niches

Potten, C.S. & Loeffler, M. (1990) Stem cells: attributes, cycles, spirals, pitfalls and uncertainties. Lessons for and from the crypt. *Development* **110**, 1001–1020.

Slack, J.M.W. (2000) Stem cells in epithelial tissues. *Science* **287**, 1431–1433.

Marshak, D.R., Gardner, R.L. & Gottlieb, D., eds (2001) *Stem Cell Biology*. Cold Spring Harbor, NY: Cold Spring Harbor Laboratory Press.

Spradling, A., Drummond-Barbosa, D. & Kai, T. (2001) Stem cells find their niche. *Nature* **414**, 98–104.

Raff, M. (2003) Adult stem cell plasticity: fact or artifact. *Annual Reviews of Cell and Developmental Biology* **19**, 1–22.

Fuchs, E., Tumbar, T. & Gunsch, G. (2004) Socializing with the neighbors: stem cells and their niche. *Cell* **116**, 769–778.

Wagers, A.J. (2004) Plasticity of adult stem cells. *Cell* **116**, 639–648.

Epidermis

Watt, F.M. (1998) Epidermal stem cells. Markers, patterning and the control of stem cell fate. *Philosophical Transactions of the Royal Society of London Series B: Biological Sciences* **353**, 831–837.

Gat, U., Dasgupta, R., Degenstein, L. & Fuchs, E. (1998) *De novo* hair follicle morphogenesis and hair tumors in mice expressing a truncated β-catenin in skin. *Cell* **95**, 605 614.

Alonso, L. & Fuchs, E. (2003) Stem cells in the skin: waste not Wnt not. *Genes and Development* **17**, 1189–2000.

McKeon, F. (2004) p63 and the epithelial stem cell: more than status quo. *Genes and Development* **18**, 465–469.

Intestine

Winton, D.J. & Ponder, B.A.J. (1990) Stem cell organization in mouse small intestine. *Proceedings of the Royal Society of London Series B: Biological Sciences* **241**, 13–18.

Gordon, J.I., Schmidt, G.H. & Roth, K.A. (1992) Studies of intestinal stem cells using normal, chimeric and transgenic mice. *FASEB Journal* **6**, 3039–3050.

Wright, N.A. (2000) Epithelial stem cell repertoire in the gut: clues to the origin of cell lineages, proliferative units and cancer. *International Journal of Experimental Pathology* **81**, 117–143.

Marshman, E., Booth, C. & Potten, C.S. (2002) The intestinal epithelial stem cell. *Bioessays* **24**, 91–98.

Vidrich, A., Buzan, J.M. & Cohn, S.M. (2003) Intestinal stem cells and mucosal gut development. *Current Opinion in Gastroenterology* **19**, 583–590.

Bone marrow stem cells

Prockop, D.J. (1997) Marrow stromal cells as stem cells for non-hematopoietic tissues. *Science* **276**, 71–74.

Weissman, I.L. (2000) Translating stem and progenitor cell biology to the clinic: barriers and opportunities. *Science* **287**, 1442–1446.

Orkin, S.H. (2001) Hematopoietic stem cells: molecular diversification and developmental interrelationships. In: Marshak, D.R., Gardner, R.L. & Gottlieb, D, eds. *Stem Cell Biology*, pp. 289–306. Cold Spring Harbor, NY: Cold Spring Harbor Laboratory Press.

Pittenger, M.F. & Marshak, D.R. (2001) Mesenchymal stem cells of adult human bone marrow. In: Marshak, D.R., Gardner, R.L. & Gottlieb, D., eds. *Stem Cell Biology*, pp. 349–373. Cold Spring Harbor, NY: Cold Spring Harbor Laboratory Press.

Zhu, J. & Emerson, S.G. (2004) A new bone to pick: osteoblasts and the hematopoietic stem cell niche. *Bioessays* **26**, 595–599.

Chapter 14

Development of the nervous system

Overall structure and cell types

The vertebrate central nervous system (CNS) consists of the brain and spinal cord and is formed from the neural plate of the early embryo. The peripheral nervous system (PNS) consists of the cranial, spinal and autonomic nerves and their associated ganglia. It is formed from the neural crest, from epidermal placodes, and from axons growing from cell bodies in the CNS. The nerve cells, or neurons, of the CNS generally have numerous dendrites receiving input from other neurons, and one axon emitting signals from the cell body and connecting to the dendritic trees of numerous other neurons (Fig. 14.1a). In the CNS, "gray matter" means mainly cells and "white matter" means mainly nerve axons, the whiteness coming from the high lipid content of the myelin sheaths that enwrap the axons. Condensations of neuronal cell bodies in the CNS are often called nuclei, not to be confused with cell nuclei.

The connections between neurons, or between neurons and non-neuronal target cells, are called synapses. They work by the release of a neurotransmitter from the presynaptic terminal and the opening of a ligand-gated ion channel in the postsynaptic terminal which initiates an action potential in the receiving cell. The overall complexity of the vertebrate nervous system is enormous: the human brain containing around 10^{11} neurons, and each of these forming synapses with around 10^3 others. The capabilities of the system depend largely on the placement and connectivity of the neurons which is achieved during embryonic development, but the vast complexity of the end product is built up by the sequential operation of a number of quite simple processes.

There are many types of neuron (Fig. 14.1a). They can be classified by the morphology of the axon and the dendritic tree, by the neurotransmitter used, for example acetylcholine, noradrenaline (= norepinephrine), or dopamine, and by whether their target is another neuron, a muscle (motor neurons), a gland, or a sense organ (sensory neurons). In addition to the neurons there are about 10 times as many **glial cells** in the nervous system (Fig. 14.1b). **Astrocytes** are the most numerous cell type in the CNS and perform many structural and metabolic functions. Type 1 (= fibrous) astrocytes are more numerous in white matter, and type 2 (= protoplasmic) astrocytes are more numerous in gray matter. **Oligodendrocytes** are the cells of the CNS that form the myelin sheaths around axons, and the **Schwann cells** are their equivalents in the PNS. **Radial glial cells** are present only during development and are now known to be a type of neuronal stem cell. They have cell bodies near the lumenal surface of the neural tube and extend fine processes to the outer surface, which guide the radial migration of neurons. Microglia are not derived from the neural plate at all but are phagocytic cells derived from the bone marrow. Finally, the **ependyma** is the simple cuboidal epithelial lining of the ventricles and spinal canal, and also functions as a repository of neuronal stem cells in the early developmental stages.

Neurons carry out most of their protein synthesis in the cell body. Materials are transported along the axon by slow axonal transport, or, in the case of neurotransmitter vesicles, by fast axonal transport. This flow of materials is balanced in a mature neuron by retrograde transport of materials from the axon terminal to the cell body. This feature of neuronal behavior is very useful to the neuroanatomist as it is possible to trace the pathways of axons through the complexity of the CNS by injecting at the periphery a marker such as the enzyme **horseradish peroxidase (HRP)**. This will be taken up by the axons and transported back to the cell bodies, which can then be located by histochemical staining for peroxidase. For example if HRP is injected into a single muscle, then it will be transported back into the motor neurons of the spinal cord that are innervating the muscle and enable them to be visualized.

The brain

The main parts of the brain (Fig. 14.2) appear from the neural tube during early development. The anterior forebrain becomes the **telencephalon**, which forms the paired cerebral hemispheres and serves, particularly in lower vertebrates, as the receptive area for the olfactory organs. The olfactory sensory cells themselves

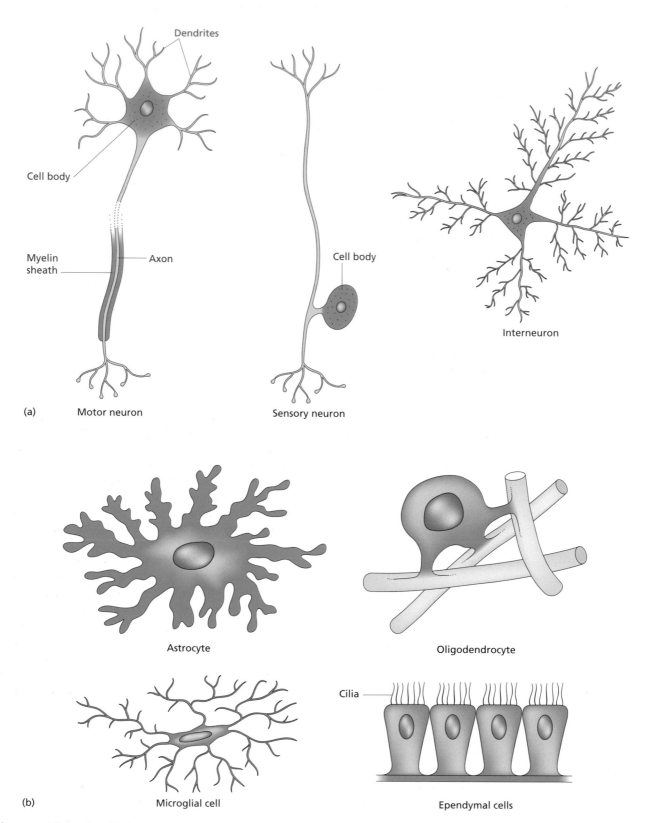

Fig. 14.1 Cell types found in the central nervous syste: (a) neurons; (b) glia. Note that some cells with glial morphology are now considered to be neural stem cells.

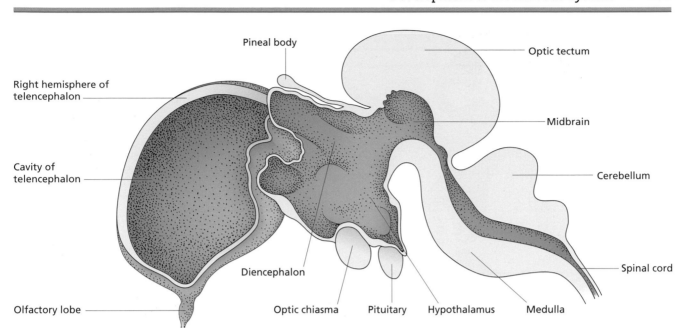

Fig. 14.2 Structure of the brain of an 8-day chick embryo. The mid- and hindbrain are sectioned in the median plane, while the right telencephalon and cerebellar swelling are shown in parasagittal section. (After Lillie.)

arise from the nasal placodes, which are thickenings of the epidermis adjacent to the telencephalon. The cerebral hemispheres are very large in higher vertebrates where more complex associative and control behaviors are apparent, and in mammals the multilayered cerebral cortex is often called the **neocortex**. The posterior forebrain becomes the **diencephalon**, which produces the optic vesicles. These grow out and invaginate to form the **optic cups** (Fig. 14.3). The lens of the eye arises from the epidermis where it is contacted by the optic cup. This forms a thickened lens placode which then rounds up and invaginates to be incorporated into the eye. The inner layer of each cup becomes the sensory retina, the outer layer becomes the **retinal pigment epithelium**, and the tube leading from the brain to the eyecup later carries the optic nerve. When it differentiates, the retina forms a multilayered structure containing the sensory photoreceptors, and various other neurons. These send processes back to the brain down the optic stalks. The optic nerve fibers terminate not in the diencephalon but in the **midbrain**, whose dorsal region becomes enlarged into paired optic lobes, called **optic tecta** (singular **tectum**) in nonmammals, which are the primary centers for reception of visual information. The midbrain also forms the auditory lobes.

Apart from the eyes, the diencephalon also forms the thalamus, the pineal body (= epiphysis) dorsally, and the pituitary (= hypophysis) ventrally. The thalamus occupies most of the side walls and is an important center of integration. The pineal is thought to be concerned with diurnal rhythms, and is later a source of melatonin. The pituitary, a key endocrine organ secreting a wide variety of hormones, has a dual origin. The anterior part arises from a placode initially just anterior to the neural folds, which invaginates through the oral aperture to become Rathke's pouch in the dorsal pharynx. The posterior part arises from the hypothalamus, the floor of the diencephalon, to which it remains permanently connected.

The hindbrain contains the cerebellum, important in the control of movement, dorsoanteriorly, and the medulla, containing a series of cranial nerve nuclei, more posteriorly. The early embryonic hindbrain becomes segmented to form seven **rhombomeres** of which the first forms the cerebellum. Opposite rhombomeres 5 and 6 are the **otic placodes**, which invaginate and develop to form the inner ear, consisting of the cochlea, which is the organ of hearing, and the semicircular canals, which are concerned with perception of motion and control of balance.

The ventricles of the brain and the central canal of the spinal cord represent the persistent original lumen of the neural tube. There is a basic set of 10 **cranial nerves** in all vertebrates: I is the olfactory nerve; II the optic nerve; III, IV, and VI control the eye muscles; V is the trigeminal; VII the facial; VIII the auditory; IX the glossopharyngeal; and X the vagus.

Spinal cord

The structure of the spinal cord is of an inner ependymal layer, a central region of gray matter, containing most of the cell bodies, and an outer layer of white matter containing mainly myelinated fibers (Fig. 14.4). In the gray matter, the motor neurons are located ventrally, while the interneurons, which receive sensory input, and the **commissural neurons**, which connect one side of the spinal cord with the other, lie dorsally.

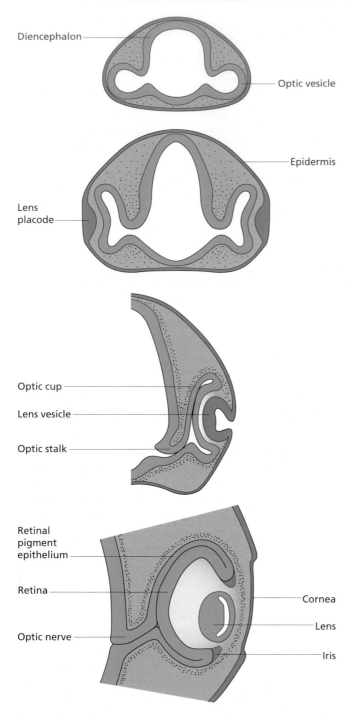

Fig. 14.3 Development of the eye. The optic vesicle grows out of the diencephalon while the lens develops from the epidermis. The inner layer of the cup becomes the retina, while the outer layer becomes a pigmented epithelium. (After Hildebrand 1995. *Analysis of Vertebrate Structure.* Wiley, figure 19.15.)

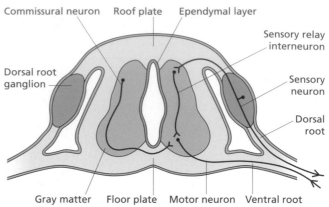

Fig. 14.4 Main structures of the developing spinal cord.

muscles. The dorsal roots consist of sensory fibers which connect the peripheral sense organs with the interneurons of the dorsal spinal cord. The sensory neurons are located in the **dorsal root ganglia**, which are formed from the neural crest (Fig. 14.4). At the levels of the fore- and hindlimb buds the spinal nerves join to form networks, respectively, called the **brachial plexus** and the **lumbosacral plexus**, from which emerge the nerves supplying the limbs.

The autonomic system with its numerous small ganglia and CNS connections is also formed from the neural crest. It is divided into a distinct sympathetic and parasympathetic system, also producing an extensive system of enteric neurons in the gut wall.

Anteroposterior patterning of the neural plate

The initial induction and regionalization of the neural plate has already been covered in Chapter 7. Neural induction is due to inhibition of bone morphogenetic proteins (BMPs) by substances such as chordin released from the **organizer**. This induces a **neuroepithelium** with an anterior specification such that, in the absence of further signals, it will express transcription factors such as Otx and form forebrain structures. The determination of more posterior structures starts with the action of various posteriorizing factors. Important among these are the fibroblast growth factors (FGFs) which induce Cdx transcription factors, which in turn induce Hox genes of the trunk and tail (see Chapter 7). It is thought that Wnt factors and retinoic acid are also important components of the posterior signal.

Forebrain

These early inducing factors produce a CNS that is divided into just a few large anteroposterior domains, and this then becomes refined by a number of local interactions. Patterning of the forebrain works in the same way as other embryological interactions:

One pair of **spinal nerves** originates from each trunk segment and is formed by the combination of nerves called the ventral and dorsal roots. The ventral roots consist of motor fibers from the spinal motor neurons which are destined to innervate the inducing factors cause activation of different combinations of transcription factors in different regions and these control the subsequent pattern of differentiation (Fig. 14.5). It was formerly thought that patterning of the cerebral cortex depended on innervation from the thalamus. But mouse knockouts in which this innervation is lost nonetheless show a normal initial pattern of transcription factor domains in the cortex. One inducing signal required for forebrain regionalization is FGF from the anterior neural ridge. This was first discovered in zebrafish as a region at the extreme anterior of the neural plate that caused expansion of the diencephalon when removed. When grafted posteriorly it could induce ectopic telencephalon. A sim-

Classic Experiments

PATTERNING OF THE CNS

The "real" neural inducer was eventually established to consist of factors inhibiting BMP signaling. The first paper is the culmination of this work, described in Chapter 7.
Wilson, P.A. & Hemmatibrivanlou, A. (1995) Induction of epidermis and inhibition of neural fate by BMP-4. *Nature* **376**, 331–333.
The next two papers describe the hindbrain in terms of developmental units characterized by expression of particular combinations of transcription factors. This helped to unify the traditional neuroanatomical description with the principles of modern developmental biology.
Lumsden, A. & Keynes, R. (1989) Segmental patterns of neuronal development in the chick hindbrain. *Nature* **337**, 424–428.
Wilkinson, D.G., Bhatt, S., Cook, M., Boncinelli, E. & Krumlauf, R. (1989) Segmental expression of Hox-2 homeobox-containing genes in the developing mouse hindbrain. *Nature* **341**, 405–409.
The last paper describes the signaling activity of the floor plate that controls the dorsoventral pattern in the neural tube.
Yamada, T., Placzek, M., Tanaka, H., Dodd, J. & Jessell, T.M. (1991) Control of cell pattern in the developing nervous-system – polarizing activity of the floor plate and notochord. *Cell* **64**, 635–647.

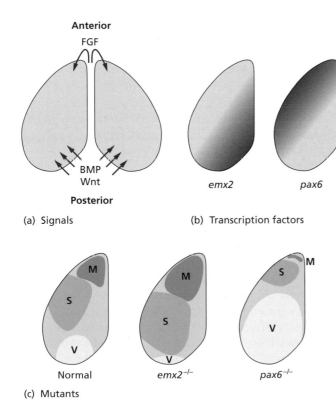

(a) Signals

(b) Transcription factors

(c) Mutants

Fig. 14.5 Regionalization of the cerebral cortex of the mouse. (a) Source regions of inducing factors are the anterior rim and the cortical hem. (b) Various transcription factors show early regionally specific expression. (c) The phenotype of loss-of-function mutations suggests that these transcription factors control subsequent development. M, motor area; S, sensory area; V, visual area. For simplicity the diagrams represent the 11.5-day mouse embryo, but the signals operate at earlier stages.

ilar source of FGF8 and FGF15 exists in the mouse. At the posterior edge of the developing cerebral cortex, in the cortical hem region, a number of Wnts and BMPs are expressed. Mouse knockouts for *wnt3A* or *Lef1* show defects in the posterior telencephalon, such as lack of the hippocampus, suggesting that a Wnt signal is required there. Some of the transcription factors that are expressed in the forebrain show gap phenotypes in loss of function mutants. This suggests that they have to be present in their normal expression domain for the later structures to form. For example this is true of *emx2*, which is expressed in the posteromedial cortex, and *pax6* expressed in the anterolateral cortex.

Isthmic organizer

An important anteroposterior signaling center is the isthmic organizer. The constriction between midbrain and hindbrain vesicles is known as the **isthmus** (Fig. 14.6). At the open neural plate stage, transcription factor genes such as *otx2* and *lim1* are expressed in the prospective fore- and midbrain. At their posterior boundary, just anterior to the future isthmus, a strip of cells starts to produce the signaling factors FGF8 and Wnt1. This organizing region controls the regional specification of the neural tissue on either side. To the anterior it induces the **midbrain**, and also establishes a gradient of expression of the transcription factors engrailed 1 and 2. To the posterior it induces rhombomere 1 and thence the **cerebellum**. These effects may be shown by transplantation of the isthmic region, which can induce midbrain from the prospective diencephalon and cerebellum from the anterior part of the hindbrain. These activities are shared by pure FGF8 protein, applied on a bead. FGF8 will also induce expression of *engrailed* and *wnt1* and of its own gene,

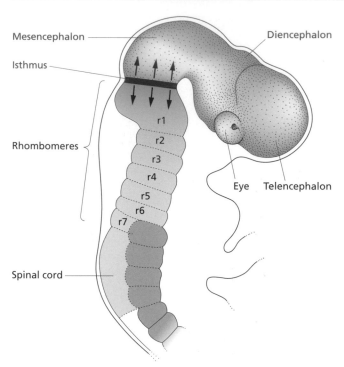

Fig. 14.6 Position of the isthmic organizer, in relation to main parts of the brain of the chick embryo.

showing that the organizer can maintain itself after its initial establishment. The mouse knockouts of both *wnt1* and *engrailed1* lack not just the isthmus region where they are expressed, but also large parts of the midbrain and cerebellum, confirming that these factors are essential to the signaling function of the isthmic organizer. The mouse knockout of *fgf8* is an early lethal because it has a large posterior defect, but there is a hypomorphic mutant of *fgf8* that has enough residual function to survive early development, and this mutant also displays a midbrain–cerebellum defect. A similar phenotype arises in the zebrafish mutant *acerebellar*, which has a loss-of-function mutation of *fgf8* but is viable for a longer period than the corresponding mouse mutant.

A consequence of the activity of the isthmic organizer is the establishment of a "Pax code", such that expression of different combinations of pax family transcription factor genes characterize different parts of the brain. The *pax6* gene is expressed in forebrain and hindbrain, while *pax2* and *pax5* are expressed in the midbrain and rhombomere 1 region. A mouse double knockout of *pax2* and *–5* lacks the midbrain and cerebellum, showing that these genes are necessary for development of the corresponding structures.

Segmentation of the hindbrain

The pattern is more obviously segmental in the hindbrain than in the more anterior parts of the brain. A series of seven (some-

times considered to be eight) bulges becomes visible after neural tube closure, called **rhombomeres** (Figs 14.6, 14.7). The first rhombomere gives rise to the **cerebellum**. Each subsequent pair of rhombomeres contains a reiterated set of motor **nuclei** and contributes fibers to one **cranial nerve**. Thus, the trigeminal (V) nerve comes from rhom-bomere 2–3, the facial (VII) nerve from rhombomere 4–5, and the glossopharyngeal (IX) nerve from rhombomere 6–7. Every other rhombomere also contributes **neural crest** cells to one of the **branchial arches**. Therefore rhombomere 2 contributes cells to the first arch, rhombomere 4 to the second arch and rhombomere 6 to the third arch. The segmental character of the rhombomeres is also evident from **clonal analysis**. If a single cell is labeled, its progeny will cross between rhombomeres at an early but not at a late stage. This clonal restriction arises because of the aggregation of cells in the individual rhombomeres. This aggregation is at least partly due to the expression by alternating rhombomeres of ephrins and Eph-type receptors. Ephrin B-type ligands are expressed on rhombomeres 1, 2, 4, and 6, while Ephs are found on rhombomeres 3 and 5. A repulsive interaction between the cell groups bearing Ephs and ephrins is thought to create the boundaries, as the boundaries are not formed if a dominant negative Eph is overexpressed in the region. In the open neural plate stage the identity of the rhombomeres is still labile and so prospective hindbrain regions can be exchanged in position without affecting the final pattern. However, once the rhombomere boundaries begin to appear, their character can no longer be re-specified by transplantation. Actual formation of the boundary cells depends on the Notch system. In the zebrafish it has been shown that cells overexpressing constitutive Notch are driven into the boundary regions while those over-expressing a dominant negative Su(H) are excluded from the boundaries.

Hox genes are not expressed in the fore- and midbrain but are important for anteroposterior patterning from the hindbrain posteriorly. The most anterior Hox genes have their anterior boundaries in the hindbrain, and the combinations of activity are such that each rhombomere carries a unique code (Fig. 14.7). The Hox genes are thought to be activated by other transcription factors which are in turn activated by the early gradients of FGF and retinoic acid. An example of one such intermediate transcription factor is Krox 20 which is a zinc-finger protein expressed in rhombomeres 3 and 5. This is the rhombomere pair that expresses Eph-type receptors and that does not emit a neural crest stream. If *krox20* is knocked out in the mouse, then rhombomeres 3 and 5 fail to form and 2/4/6 are all fused together.

Manipulations of retinoic acid levels show particular effects on the hindbrain. Retinoic acid is produced from dietary vitamin A and the enzyme that makes it (retinaldehyde dehydrogenase, RALDH2) is found in the somites and lateral plate of the trunk region. The enzyme that destroys it (Cyp26, a cytochrome P450) is found in the fore- and midbrain. Since the source lies to the posterior and the sink lies to the anterior it is thought that the retinoic acid forms a concentration gradient falling from posterior to anterior across the hindbrain region. Quail embryos can

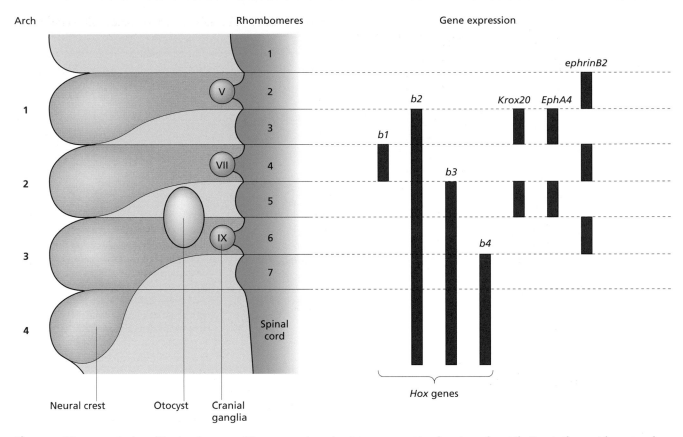

Fig. 14.7 Diagrammatic view of the rhombomeres of the mouse embryo showing gene expression domains and contributions to the cranial nerves and neural crest.

be made almost entirely retinoid-free by depriving the mothers of vitamin A. Such embryos show small hindbrains completely lacking rhombomeres 5–7, but the missing parts do develop if the embryos are given vitamin A before the early somite stage. There is also a loss of ventral neuronal types in the anterior spinal cord. Mouse embryos with RALDH2 knocked out die at mid-gestation and also lack rhombomeres 5–7. Similar effects are shown in *Xenopus* by overexpression of a dominant negative retinoic acid receptor. Mouse embryos can be treated with an excess of retinoic acid by dosing the mothers. Treatment at 7.5 days causes *hoxb1* to be expressed in rhombomere 2 as well as the normal expression in rhombomere 4. This alters the character of the neural crest cells emitted by rhombomere 2 and so converts the first arch into a copy of the second arch. This is a good example of a **teratogenic** effect that is now well understood at the molecular level. Retinoic acid has in fact presented a teratogenic risk to humans as it can be used in oral form for acne therapy. Before the risk was understood, this led to many abortions and to a number of birth defects with brain damage.

Spinal cord

Within the spinal cord the anteroposterior pattern is not so evident as in the brain. However, there are Hox boundaries at various levels and, in addition to their characteristic dorsoventral pattern, there is an anteroposterior arrangement of certain groups of motor neurons. For example, the lateral motor-neuron columns supplying the limbs are present only at forelimb and hindlimb levels, and the column supplying the axial muscles extends from the cervical to the beginning of the lumbar region.

Dorsoventral patterning of the neural tube

The spinal cord has the same basic dorsoventral arrangement along its length, and in fact many of the features extend through the hindbrain and midbrain as well. Mid-ventrally there is a **floor plate**, composed of columnar cells in close proximity to the notochord. Flanking this is the region containing the motor neurons. There are various concentrations of these, the medial column supplying the axial and body wall muscles and the lateral columns the limb muscles. The motor neurons in each column express a particular combination of Lim-type transcription factors and this is associated with the group of muscles to which they later project. If the combination of *lim* genes expressed is experimentally altered, then the motor neuron connectivity pattern is changed accordingly. In the dorsal half of the spinal cord are the dorsal sensory relay interneurons and the

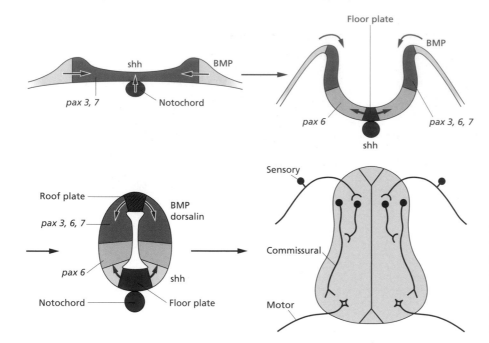

Fig. 14.8 Dorsoventral patterning of the neural tube. Sonic hedgehog from the notochord induces expression of its own gene in the floor plate. BMP from the epidermis induces the roof plate. The signals from the two centers induce a nested expression of *pax* genes that later brings about the pattern of neuronal differentiation.

commissural neurons which project to the opposite side of the spinal cord. In the dorsal midline is the **roof plate**.

This dorsoventral pattern is controlled by signals, initially from the tissues neighboring the neural plate, and later from the floor plate and the roof plate (Fig. 14.8). The **notochord** emits a signal which induces the floor plate and, at lower concentration, the motor neurons. This can be shown by recombination experiments in which the notochord is grafted adjacent to different dorsoventral levels of the neural tube. An ectopic floor plate will form in close proximity and motorneurons a bit further away. The signal is believed to be sonic hedgehog (Shh) for three reasons:

1 *shh* is expressed in the notochord;
2 Shh will induce floor plate and motor neurons when applied to the neural tube;
3 the floor plate and motor neurons are lost in the mouse knockout of *shh*.

In fact fate-mapping studies in the chick have shown that in normal development the medial part of the floor plate is directly derived from Hensen's node, while the lateral parts are induced from the surrounding neuroepithelium.

The floor plate itself also makes Shh, so once it has been formed it takes over the signaling role of the notochord. Shh represses the expression of various transcription factor genes such as *pax3* and *pax7*. These are initially expressed all over the neural tube but become repressed ventrally by the action of the signal. Misexpression of these factors will prevent the ventral cell types from forming. Later specific transcription factors become expressed at each dorsoventral level. For example the bHLH transcription factors olig1 and olig2 are expressed in the premotor-neuron region and are necessary for the formation of the motor neurons and of oligodendrocytes.

The dorsal midline of the neural tube arises from the lateral margin of the neural plate, where it contacts the epidermis. The epidermis expresses *bmp4* and *bmp7*, and these BMP proteins can elevate expression of *pax3* and *pax7* when applied to the neural plate. After neural tube closure, the roof plate expresses *bmp4* and *bmp7* and also the gene for another TGF-β superfamily member called dorsalin. These factors all have the activity of inducing dorsal cell types such as sensory relay neurons. If formation of the roof plate is prevented, as for example in the mouse knockout of the *lim1a* gene, then the dorsal neuronal types fail to form. For these reasons it is thought that the overall dorsoventral pattern arises from the combined effects of the gradients of Shh from the floor plate and of BMPs and dorsalin from the roof plate.

Neurogenesis and gliogenesis

Drosophila

Some of the principles of neurogenesis have been learned from studies on *Drosophila*. In *Drosophila* the ventral nerve cord arises from a neurogenic region which originates lateral to the mesoderm and becomes mid-ventral once the mesoderm has invaginated. The neurogenic region contains a segmentally reiterated pattern of **proneural clusters** defined by expression of the bHLH transcription factors produced by the achaete–scute complex (AS-C). The positions of the clusters depend on the anteroposterior and dorsoventral patterning systems described in Chapter 11. Each cluster produces only one **neuroblast** together with a number of epidermal cells and this is controlled by the Notch **lateral inhibition** system. AS-C factors activate

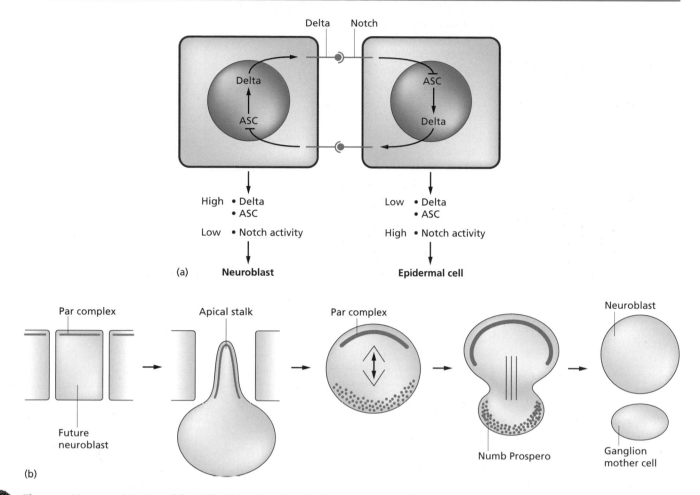

Fig. 14.9 Neurogenesis in *Drosophila*. (a) The Delta-Notch lateral inhibition system amplifies initial small differences between cells of a proneural cluster ensuring that only one ends up with high levels of products of the achaete–scute complex (AS-C) and becomes the neuroblast. (b) Asymmetrical division of neuroblast to generate another neuroblast and a ganglion mother cell.

production of Delta and this stimulates Notch on the adjacent cells. The Notch signal represses expression of AS-C and so reduces Delta production in the adjacent cells (Fig. 14.9). This positive feedback system amplifies initially small differences such that eventually there is just one cell with a high level of AS-C, which becomes the neuroblast, surrounded by cells with a low level of AS-C that become epidermal. This mechanism explains why loss of function of *Notch* leads to an overproduction of neuroblasts. The mechanism is an example of a **symmetry-breaking** system. The initial uniform state is unstable because there is slight fluctuation in the levels of AS-C or Delta between cells, and this will automatically become amplified by the feedback inherent in the interaction.

Once formed the neuroblasts detach from the neighboring cells and sink into the interior, remaining temporarily connected by an apical stalk in the plane of the epithelium. This apical specialization contains the Par complex described in Chapter 12 (i.e. the complex of Par3/Par6/aPKC; the *Drosophila* homolog of *par6* being called *bazooka*). The Par complex phosphorylates components of the intracellular trafficking system so

as to repel other proteins to the basal region, including Numb which is a tyrosine phosphatase and Prospero, which is a homeo-domain transcription factor. The next division is asymmetrical with a large neuroblast inheriting the apical stalk, and a smaller ganglion mother cell enriched in Numb and Prospero, on the basal side. Numb further inhibits Notch signaling by causing endocytosis of the Notch protein from the cell surface, while Prospero activates ganglion mother cell-specific genes and represses those needed for further cell division. Evidence for this mechanism is largely derived from the behavior of mutants, for example embryos lacking components of the Par complex do not show localization of Numb and Prospero, and their neuroblasts divide symmetrically.

Vertebrate primary neurogenesis

In lower vertebrates the first neurons arise directly from the neuroepithelium of the open neural plate, which gives a good opportunity to study the process. There are several neurogenic transcription factors that drive neuronal differentiation. Among these are neurogenin and neuroD, both bHLH-type factors belonging to the same biochemical family as AS-C in *Drosophila*. *neurogenin* is expressed in the areas of neural plate that give rise to the first primary neurons and it will cause activation of *neuroD* on overexpression. *neuroD* is also expressed in primary neurons from open neural plate stages. Overexpression of either factor in *Xenopus* will increase the proportion of neurons formed from the neuroepithelium. The spacing of the primary neurons is determined by a lateral inhibition mechanism very similar to that in *Drosophila* (Fig. 14.10a). All of the cells in a *neurogenin*-expressing patch are competent to become neurons. But this tendency is inhibited by Notch signaling, and Notch on one cell is stimulated by Delta on the neighboring cells. Overexpression of Delta, or of a constitutive form of Notch, in the *Xenopus* neural plate reduces the formation of primary neurons. Conversely, overexpression of a dominant negative Delta generates a higher density of primary neurons.

Later neurogenesis

Neurogenesis continues throughout embryonic development and into postnatal life. The neuroepithelium is only one cell thick but appears stratified because the nuclei are present at many different levels. In terms of epithelial morphology, the **apical** side is that facing the lumen of the neural tube while the **basal** side is that facing the exterior (this means that apical is usually at the *bottom* of a diagram and basal at the *top* – opposite to the convention for other epithelia). At early stages all of the cells are mitotic and there is a migration of nuclei such that they approach the exterior surface during S phase and the lumenal surface during mitosis. Once a neuron is formed it ceases to divide and this means that a pulse of BrdU or ³HTdR (see Chapter 13) can reveal the time of formation of each class of neuron. If the tissue is examined some time after a pulse label, all the cells that have continued to divide will have lost the label by dilution with unlabeled precursors, whereas those that ceased dividing just after incorporating the label will still retain it. The time of the last S phase before neuronal differentiation is often called the "birthday" of the neuron.

This final division is a tangential one separating the newly formed neuron from the lumenal surface of the neural tube. It is associated with the localization to the lumenal side of the Numb protein. As in *Drosophila*, this is controlled by the Par complex

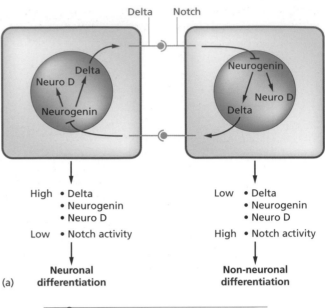

(a)

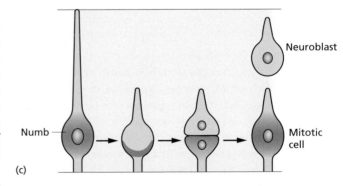

(b)

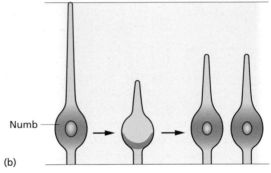

(c)

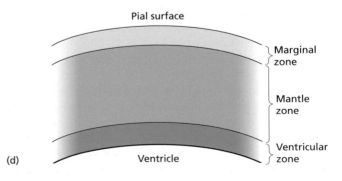

(d)

Fig. 14.10 (*right*) (a) Lateral inhibition mediated by the Delta–Notch system in vertebrates ensures that only some cells from the neurogenic regions become neurons. (b,c) Symmetrical and asymmetrical divisions of cells at the lumenal surface (= ventricular zone; apical surface of neuroepithelium). In a symmetrical division Numb protein is partitioned equally. In an asymmetrical division the lumenal daughter keeps the Numb, and the internal daughter, which in early neurogenesis will become a neuroblast, loses it. (d) Three-layered structure of the later CNS.

which is associated with the intercellular junctions at the lumenal ends of the cells. However, unlike *Drosophila*, in vertebrates the Par complex seems to attract Numb to the same region of the cell rather than repelling it to the opposite side. After division, the mitotic lumenal cell retains the Numb protein and the postmitotic, internalized neuron is depleted in Numb (Fig. 14.10b,c). This cytoplasmic localization is thought to represent an important aspect of neuronal determination, however the fate of the Numb-containing cell seems to change with stage. In early mouse embryos Numb represses neurogenesis and maintains proliferation, while at later stages it may segregate to the neuronal daughter, promote neurogenesis, and repress proliferation. Knockout mice for *numb* (and the similar gene *numblike*) show increased neurogenesis, which is consistent with the early role of *numb* being to promote proliferation. In the chick, the Numb protein localizes to the basal rather than the apical side, which also suggests that the presence of Numb may have different effects in different circumstances.

The occurrence of the tangential divisions means that the neural tube ceases to be one cell thick (Fig. 14.10d). The mitotic lumenal layer then becomes known as the **ventricular zone** (and eventually as the **ependymal** layer). The outer surface becomes known as the **pial surface**, after the enveloping connective tissue layer called the pia mater (Latin for "tender mother"). The layered structure of the CNS emerges through an "inside-out" pattern of morphogenesis. Cells moving inwards from the ventricular layer form a **mantle layer**, later known as the gray matter, containing both neurons and glia. As the neurons produce their axons, these build up to form a cell-poor **marginal zone** under the pial surface. When the axons are myelinated this becomes the white matter. As the neural tube thickens, the neurons need to migrate from the site of their formation in the ventricular zone to their final positions. This migration is guided by the **radial glial cells** (see below) which extend processes from the inner to the outer surface of the tube. The neurons with early birthdays tend to migrate shorter distances and those with late birthdays migrate further.

The initial three layered structure of ventricular, mantle, and marginal zones is maintained in the spinal cord, but in some parts of the brain it later gets much more complex. In the mammalian cerebral cortex, neurons from the mantle zone invade the marginal zone to establish six further layers of **neocortex**. Most cortical neurons of mouse embryos are formed during the period 12.5–17.5 days of gestation. In the cerebellum some mitotic cells migrate to the outer layer to establish an external germinal layer, with subsequent "outside-in" neurogenesis.

Classic Experiments

NEUROGENESIS AND NERVE CONNECTIONS

Le Douarin, N.M. (1986) Cell line segregation during peripheral nervous system ontogeny. *Science* **231**, 1515–1522.
This is actually a review, but nicely summarizes a series of papers published in French in the 1970s. It establishes the fate and potency of regions of the neural crest using quail to chick grafts.

Kennedy, T.E., Serafini, T., Delatorre, J.R. & Tessier-Lavigne, M. (1994) Netrins are diffusible chemotropic factors for commissural axons in the embryonic spinal-cord. *Cell* **78**, 425–435.
This paper marked the first discovery of a guidance molecule that controlled the direction of growth zone growth.

Lee, J.E., Hollenberg, S.M., Snider, L., Turner, D.L., Lipnick, N. & Weintraub, H. (1995) Conversion of *Xenopus* ectoderm into neurons by NeuroD, a basic Helix-Loop-Helix protein. *Science* **268**, 836–844.
This paper provides evidence that overexpression of a single proneural transcription factor could drive the formation of neurons in a vertebrate.

Also, the mantle zone becomes subdivided into several layers including one for the Purkinje neurons, which are large cells, found only in the cerebellum, contacting as many as 10^5 other neurons. In the eye, the outer region of the optic cup becomes **pigment epithelium** and the inner layer becomes the retina, consisting of six cell layers with the photoreceptors on the original ventricular surface and the fibers running to the optic nerve on the original pial surface. In addition to the ubiquitous radial migration, there is also some lateral migration of neurons in various parts of the brain.

Neural stem cells

The mitotic cells of the ventricular zone can both maintain themselves and generate progeny that differentiate into neurons and glia. Later, when this layer becomes known as the **ependyma**, it is still mitotic although cell divisions are then few and far between. It is now thought that the CNS is organized rather like other cell-renewal systems (see Chapter 13) with a small number of **stem cells** and a larger number of **transit amplifying cells**. In lower vertebrates there is cell renewal throughout life, and it is now known that this is also true for two regions of the brain in mammals: the hippocampus and the regions adjacent to the lateral ventricles. The new neurons from the lateral ventricles migrate anteriorly to end up in the olfactory bulbs.

The potency of individual CNS mitotic cells has been examined by two methods. One is *in vivo* **clonal analysis** by infection with a replication-incompetent retrovirus expressing β-**galactosidase** or **GFP**. If this is done at low multiplicity, the infected cells are well separated and each positive cell group can be considered as a single clone. This method confirms the predominant radial migration of the progeny and shows that most clones contain either neurons or glia, indicating infection of a transit amplifying cell. However, a minority of clones contain both neurons and glia, showing that the originally labeled cell was **pluripotent**. The other method is *in vitro* culture of individual cells. In cultures from the lateral ventricular zone of the rat embryo cerebral cortex, the majority of clones form small numbers of neurons, a few form small numbers of glia, and about 7% form large clones containing neurons, astrocytes, and oligodendrocytes. In both types of experiment the large mixed clones are presumed to arise from **neural stem cells** and the small single-type clones from transit amplifying cells. During the period of peak neurogenesis it now seems that the neural stem cells are actually the cells called **radial glia**. Radial glia have nuclei near the ventricular surface but extend a fine process all the way to the pial surface. Labeling by retroviral infection shows that they divide repeatedly and that the progeny are neurons or neuronal precursors. In order to reach their final cell layer these neurons migrate towards the pial surface along the elongated radial glial mother cell. In mammals, after neurogenesis is completed, the radial glia differentiate into astrocytes, but in other vertebrate classes they may persist into adult life.

Various markers have been proposed to characterize the CNS stem cells, although it is not clear to what extent they identify transit amplifying cells rather than true stem cells. They include an intermediate filament protein called nestin, a GTP binding protein, nucleostemin, and a complex carbohydrate SSEA-1 found also on embryonic stem cells. It is possible to culture cells from the stem cell regions of the CNS in the presence of FGF and epidermal growth factor (EGF). If grown in suspension they form small aggregates called **neurospheres** which contain a mixture of stem cells and transit amplifying cells. When plated on a laminin substrate the neurospheres will differentiate into neurons, astrocytes, and oligodendrocytes.

The stem cells in the regions of adult neurogenesis exist in a subventricular zone, which is immediately adjacent to the ependyma. The **stem cell niche** (Fig. 14.11) is thought to be formed by the proximity of blood vessels, with their basal laminae sequestering growth factors. Evidence that the stem cells are astrocyte-like cells is as follows:

1 The subventricular astrocytes are the only cells that show long-term retention of **BrdU**.
2 They survive treatment with the antimitotic drug ara-C, which kills the rapidly dividing transit amplifying cells. Since neurogenesis resumes after the cessation of treatment it probably originates from these survivors.
3 If glial cells are specifically labeled with GFP their progeny can be seen to include dividing transit amplifying cells and neurons migrating to the olfactory bulbs. This specific labeling is done by viral infection of transgenic mice that express a receptor for avian leukosis virus under the control of a glial-specific (GFAP) promoter. This virus cannot normally infect mice, but in this transgenic line it will specifically infect and label the glial cells.
4 Neurospheres can be grown which contain some of these labeled cells. But neurospheres will not grow if all astrocytes are killed, which can be done using the drug gancyclovir on cells from a transgenic mouse carrying the GFAP-TK gene (see Chapter 10).

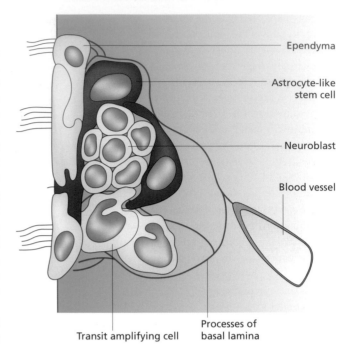

Fig. 14.11 A model for the stem cell niche in the subventricular zone of the mammalian lateral ventricle. (After Doetsch 2003. *Current Opinion in Genetics and Development* **13**, 543–50.)

Evidence for the role of blood vessels in maintaining the niche is that endothelial cells can maintain self-renewal of neural stem cells *in vitro*, even in a **transfilter** culture situation.

Glial cell clusters arise from the ventral part of the ventricular zone, near the floor plate. These cells migrate dorsally and can reach all parts of the CNS. Unlike the neurons, they are still proliferative during migration. Among these cells is an oligodendrocyte precursor cell (formerly called the O2A cell), which has a bipolar morphology. Their normal fate is to form oligodendrocytes. But in culture they can also form type 2 astrocytes on exposure to ciliary neurotrophic factor (CNTF). If treated with high serum followed by FGF they will become neuronal stem cells, generating neurons, oligodendrocytes, or type 1 astrocytes. These recently discovered properties of the subventricular astrocytes, the radial glia, and the oligodendrocyte precursors indicate that some cell types classified as glia may have much more developmental potential than formerly realized.

The neural crest

The neural crest is a population of cells derived from the neural folds. They are present in the dorsal part of the neural tube at the time of closure but soon migrate away from the neural tube into the surrounding tissues. They form a variety of different cell types of which the following are the most significant:

1 neurons and glia of the sensory and autonomic systems;
2 adrenal medulla and calcitonin cells of thyroid;

3 pigment cells (excluding those of the pigmented retina);
4 skeletal tissues of the head;
5 part of cardiac outflow tract.

The normal fate of the crest has been established by two major techniques. One is the grafting of segments of neural tube from quail to chick embryos, followed by localization of the quail cells at a later stage (see Chapter 9). The second method is *in vivo* labeling of neural crest by localized injection of dyes. Extracellular injection of **DiI** would be used to label a patch of cells, or intracellular injection of a fluorescent dextran would be used to label an individual cell. Following application of the label, the embryo is allowed to develop for a while and then the position and cell types of the labeled cells identified at a later stage.

Experiments of these types show that the crest is divided into distinct anteroposterior domains as far as normal fate is concerned (Fig. 14.12). The cephalic crest gives rise to a remarkable range of structures. Initially it is the source of most of the head mesenchyme. As well as general connective tissue, this later becomes cartilage, bone, cranial ganglia, the pericytes and smooth muscle of blood vessels, and the odontoblasts (dentine-forming cells) of the teeth. The trunk crest shows two distinct pathways of migration (Fig. 14.13). The dorsolateral pathway passes between the epidermis and the somites and give rises to melanocytes. The ventral pathway passes through the sclerotome to become the dorsal root ganglia, the sympathetic ganglia, the adrenal medulla, and also makes a contribution to the outflow tract of the heart (cardiac crest). This migration occurs only through the anterior half of each sclerotome segment (see Chapter 15), and each dorsal root ganglion includes a contribution from the crest adjacent to the posterior sclerotome segments on either side. Finally there are two regions of crest known as the vagal and sacral, which contribute cells to the network of enteric ganglia.

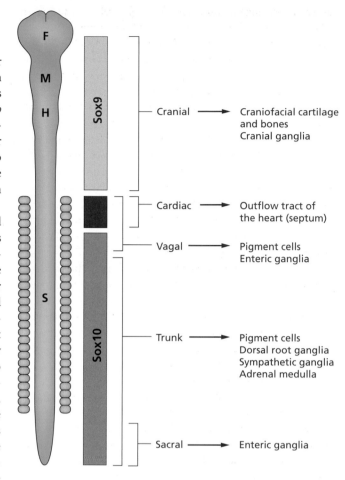

Fig. 14.12 Fate map of neural crest compiled from chick–quail orthotopic grafts. F, forebrain; M, midbrain; H, hindbrain; S, spinal cord.

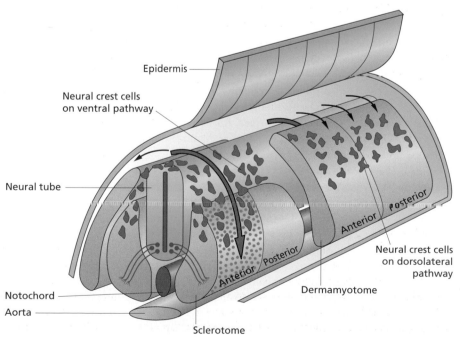

Fig. 14.13 Dorsolateral (black arrows) and ventral (red arrows) pathways of trunk neural crest migration.

Progressive commitment of crest cells

If individual cells within the neural folds are labeled, it is found that their clonal descendants can include epidermal, neural, or neural crest cells, or any combination of the three. This suggests that there is no specific neural crest rudiment, but that there is an environment in this region that promotes the formation of neural crest cells. When the neural folds are removed and the neuroepithelium joined directly to the epidermis, neural crest can be generated from cells on both sides of the junction, indicating that an interaction between epidermis and neuroepithelium can lead to its formation. BMP4 from the epidermis is an important ingredient in this signal, but may not be sufficient on its own to induce neural crest formation. Experiments in *Xenopus* and chick suggest that FGFs, Wnts, and retinoic acid are all likely to be required as well.

It seems that the molecular pathway for neural crest determination has some properties of a linear sequence, but also with some feedback effects. The initial response to BMP signaling is the expression of the Msx1 and 2 homoedomain transcriptional repressors. Following this are two zinc-finger transcriptional repressors called Snail and Slug, and further transcription factors including Sox9 and 10. Evidence for the pathway is derived from overexpression, inhibition, and epistasis experiments. The *Xenopus* results are described here but evidence from mouse knockouts and chick retroviral infection experiments are broadly consistent. In *Xenopus* it is found that overexpression of the Msx factors will induce expression of *snail*, overexpression of Snail will induce expression of *sox10*, and overexpression of Sox10 will induce expression of *slug*. All these factors can drive neural crest formation and injection of the appropriate antisense **morpholinos** or dominant negative **domain-swap** constructs of any of them will inhibit neural crest formation. The sequence based on the overexpression data is therefore Msx→Snail→Sox10→Slug. However, it is probably not strictly linear. Slug will induce Sox10, and Sox10 will induce Slug, suggesting some autocatalytic feedback in the pathway. Also a **dominant negative** inhibitor to Slug will reduce expression of Snail and Sox10. A glucocorticoid-inducible version of the dominant negative Slug will block crest formation if induced early, and if induced later, after the crest has formed, it will block the cell migration. This shows that Slug function is required for an extended time to control the properties of the crest cells. Sox10 is initially expressed in the whole crest, but is soon replaced by Sox9 in the cranial crest (Fig. 14.12). Sox9 is known to be important for cartilage differentation from the mesoderm (see Chapter 15). The importance of Sox9 for cartilage formation from the neural crest is shown by a tissue-specific knockout in the mouse. In embryos of the constitution *wnt1-cre; floxed sox9* the Cre recombinase is expressed in the neural tube but not in the mesoderm. The resulting ablation of *sox9* in the neural crest inhibits formation of the cranial skeleton without affecting the formation of the trunk skeleton.

The migration of crest cells depends on a wide variety of extracellular matrix components. Neural tube cells express the adhesion molecule N-cadherin. This is lost by migratory crest cells but re-expressed when they reach their destinations. Migration of crest cells is blocked by **neutralizing antibodies** to several extracellular matrix components such as fibronectin or laminin. *In vivo* the migration of the trunk crest of the ventral pathway is specifically inhibited by the **posterior sclerotome**, leading to the preferred migration route through **anterior sclerotome**. The reason is that the crest cells express EphB3 and the posterior sclerotome expresses ephrin B1. These make up a repulsive combination such that when the crest cell processes touch the ephrin B1-containing surface they collapse due to actin depolymerization mediated by the RhoGTPase system (see Appendix). Addition of an ephrin inhibitor to chick embryos enables the crest cells to migrate through the posterior as well as anterior sclerotome. In rhombomeres 3 and 5 of the hindbrain the outward migration of the crest is also inhibited by the Eph system.

The overall anteroposterior commitment of the neural crest has been investigated by interchanging segments of neural tube and examining the fate of the crest cells that emerge from the graft. At least in higher vertebrates, this type of experiment has shown that the distinction between cranial and trunk crest arises early, as only cranial crest can generate skeletal tissues. Trunk crest will not generate cartilage even when grafted to neural tube of the head, whereas cranial crest will contribute to skeletal structures of the body after grafting to the trunk, in addition to forming the various neuronal types of the trunk. However trunk crest cells can form cartilage if they are actually grafted into the facial primordia, rather than into the hindbrain, so the distinction may be more to do with migration capacity than actual competence to form cartilage. In general, any region of the crest will form parasympathetic, sympathetic or sensory ganglia if grafted to the appropriate position along the axis. For example, the vagal crest normally produces cholinergic (parasympathetic) neurons and the thoracic crest adrenergic (sympathetic/adrenal medulla). But if these regions are interchanged the cell types formed are appropriate to the new position of the graft.

Labeling of individual crest cells *in vivo* shows that those leaving the neural tube early may be **pluripotent**, as some labeled clones include several cell types such as sensory neurons, pigment, adrenal medulla, and glial cells. Those leaving the neural tube late are more likely to populate just one terminal cell type, for example melanocytes. This shows that at least some of the cells in the crest are pluripotent, and in recent years these have started to be called neural crest stem cells (NCSC) to indicate their affinity with other types of stem cell.

The final differentiation of the cells does depend to a large extent on the environment through which they migrate (Fig. 14.14). For example the formation of dorsal root ganglia depends on exposure to brain-derived neurotrophic factor (BDNF) from the neural tube. If a barrier is inserted between the neural tube and the crest cells, then they fail to form the dorsal root ganglia; however, they will do so if supplied with BDNF protein. Mice lacking the *bdnf* gene do not form dorsal root ganglia. The decision whether to form autonomic neurons, glia, or smooth

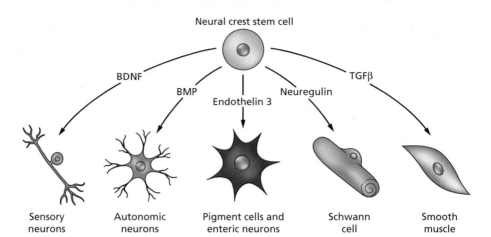

Fig. 14.14 Preferred differentiation pathways of trunk neural crest stem cells exposed to different inducing factors.

muscle also depends on environmental signals. Cultured trunk crest cells will preferentially form **Schwann cells** when exposed to neuregulin (= glial growth factor), smooth muscle when exposed to TGFβ, and autonomic neurons when exposed to BMP2 or −4. These effects are primarily instructive although the factors may also contribute to differential survival of their own cell type. These factors are expressed in the appropriate regions of the embryo through which the cells migrate, neuregulin in the nerve sheaths, TGFβ in the major outflow vessels of the heart, and BMP2/4 in the dorsal aorta and gut. Mouse knockouts have the predicted effects, for example a knockout of *neuregulin* reduces the number of Schwann cells, and knockouts of the *TGFβ* genes interfere with heart development. The autonomic neurons express the transcription factor gene *mash1*, encoding a bHLH transcription factor which is a homolog of the *Drosophila* Achaete-Scute family and the knockout of *mash1* lacks most autonomic neurons. Commitment to form the pigment cells and enteric ganglia depends on the inducing factor endothelin 3. Mice lacking this factor or its receptor are deficient in both cell types. The requirement for activity is in the time interval 9.5–12.5 days of gestation as may be shown by making transgenic mice for an inducible form of the receptor, and inducing its activity at different times.

In addition to these data showing control of differentiation of NCSC by environmental signals, there are also various experiments in which crest cells show differentiation into just a single cell type even when they are exposed to a wide range of environmental conditions, suggesting that there is also an element of autonomous commitment to particular fates that occurs progressively with time.

Development of neuronal connectivity

The growth cone

The axons of neurons grow out and innervate their targets. Motor neurons in the ventral spinal cord innervate muscles or glands, and sensory neurons in the dorsal root ganglia innervate a variety of peripheral receptors and sense organs. The final pattern of connections is one of exquisite complexity, so how do they know where to go? There is in fact a whole hierarchy of different mechanisms controlling the specificity of connections. Each of them is of quite low specificity but together they yield a very precise final result.

Each developing axon has at its tip a **growth cone** (Fig. 14.15). This is a structure somewhat resembling a migrating fibroblast. It is constantly extending and retracting **filopodia** and it crawls actively across the substrate. The cytoskeleton of the growth cone is in a dynamic steady state. Its **lamellipodia** and filopodia are built around microfilaments that are constantly elongating at the plus (exterior) ends. At the same time the filaments are being drawn back into a central depolarized domain by myosin-type **motor proteins**. In this region terminate the microtubule bundles that maintain the structure of the axon. Stimuli that cause growth cone advance will increase the rate of actin polymerization, decrease the activity of the myosin motors, and increase the elongation of the microtubules. As the growth cone advances new material becomes laid down to elongate the axon. The cell body and proximal axon do not move, but they are necessary to the process as they carry out much of the synthesis of the new materials required for elongation. Because of the relatively large size of the growth cone, it is possible for different parts of it to be exposed to different signals, and a unilateral stimulus will bring about a change in the direction of growth. The coupling between receptors for signals and the cytoskeleton depends on the small GTP exchange proteins. RhoA causes growth cone collapse in its active form, but active Rac and Cdc42 promote elongation.

The factors known to be most potent at promoting axonal extension are the neurotrophins. These comprise nerve growth factor (NGF), brain-derived neurotrophic factor (BDNF), neurotrophin 3 (NT3), and neurotrophin 4/5 (NT4/5). The receptors are called Trk (pronounced "track") proteins, which are receptor tyrosine kinases. NGF binds TrkA, BDNF and NT4/5 bind TrkB, and NT3 binds predominantly TrkC, providing some specificity to the system. The neurotrophins also all bind to a lower affinity receptor called p75. This has a tonic function of activating RhoA,

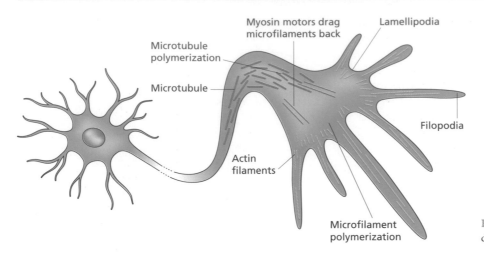

Fig. 14.15 A growth cone at the tip of a developing axon.

and in the presence of the neurotrophin the activation is stopped, thus enabling growth cone elongation. In addition extracellular matrix proteins such as fibronectin and laminin often provide a permissive environment for elongation. The neurotrophins can maintain survival if applied only to the cell body, but for elongation they must be applied to the growth cone. It is thought that endocytotic vesicles containing active receptors are transmitted to the cell body by retrograde transport in order to bring about the changes in gene expression.

Guidance molecules

The guidance of growth cones is facilitated by some factors and antagonized by others. Four categories of guidance factors can be recognized, those responsible for contact attraction, contact repulsion, long-range attraction, and long-range repulsion. Their identification depends mainly on the use of *in vitro* assays (Fig. 14.16). The contact factors can simply be coated onto a culture dish to see whether they will support outgrowth of axons from an explant of neural tube. However, a more discriminating assay is provided by offering a choice between alternative substrates, since growth cones *in vivo* will be constantly choosing between alternative environments. For example, ephrins with growth-cone-repelling activity were isolated from the optic tectum using this sort of assay. For long-range factors it is necessary to show that they are diffusible and that growth cones can sense a concentration gradient. This is done by setting up two explants in a collagen gel. If axons grow directly from one to the other this shows that there must be a long-range chemo-attractant secreted by the target explant. A conclusive proof of chemoattraction requires that the source be moved and produce a change of direction of the growth of the axons.

The contact attractive factors include extracellular matrix components such as laminin and fibronectin, and also adhesion molecules on other neurons such as N-CAM, NgCAM, or N-cadherin. The contact repulsion factors are ephrins, which bind receptors of the Eph group, and some extracellular matrix components such as tenascin. The long-range guidance molecules

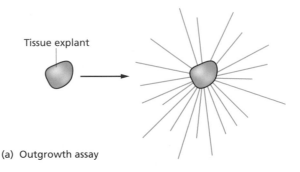

(a) Outgrowth assay

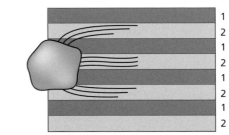

(b) Competition assay

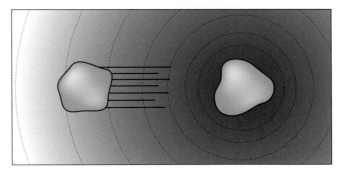

(c) Chemoattraction assay

Fig. 14.16 Assays for attraction and repulsion. (a) Simple assay for outgrowth of fibers by plating a tissue explant on a coated surface, or by treating it with factors. (b) Competition assay. Two substrates, 1 and 2, are coated onto the thin alternating strips so that the axons growing out of the explant can choose between them. (c) Chemoattraction assay. The fibers from one explant grow towards the second explant which is emitting a chemoattractant.

are netrins, which are diffusible extracellular factors related to laminin, semaphorins, and slits. In general, netrins are attractive while semaphorins and slits are repellant to axon growth, but this does depend on the competence of individual cell populations. Receptors for the netrins are called DCC ("deleted in colon carcinoma"), receptors for semaphorins are called neuropilins and plexins, and receptors for slits are called robos (encoded by homologs of the *Drosophila roundabout* gene). Although some semaphorins are secreted and act as long-range factors, others are transmembrane molecules and act at short range through cell contacts.

Pathways

Studies of the initial pathways have been made mainly in the limb regions where the spinal nerves fuse into the **brachial** or **lumbosacral plexus** and then branch again into a few principal nerves supplying the limb. The spinal nerves innervating the leg first grow through the dorsoanterior sclerotome surrounding the neural tube. As is the case for neural crest cells, the ventral and the posterior sclerotome carries a repulsive factor and prevents growth. The mesenchyme of the plexus region forms a permissive environment for growth, but the nerves are obliged to pass through two holes in the forming pelvic girdle. If more holes are created, then the nerves will grow through. If a segment of neural tube is rotated or displaced by a small amount, the nerves will still find their correct normal destinations. But if it is displaced by a large amount the nerves from a region that does not normally innervate the limb can grow into the limb and form an approximately normal-looking pattern. These experiments show that the crudest level of control must be the presence of pathways through the tissues that can be followed by any axons. These pathways may simply be spaces, or regions rich in suitable substrates such as laminin, or regions free from repulsive factors.

The most discriminating studies on this scale of connectivity have been made in insects. However, it is important to remember an important difference between insect and vertebrate neural development. In vertebrates, most neurons arise in the CNS and send axons out to peripheral organs, while in insects many neurons arise in peripheral organs and send axons into the CNS. Studies on the development of these pathways have shown that the complete pathway is divided into short segments of about 100 μm. Over this length, growth cones will grow along the pathway. When they get to a choice point, which may be a pre-existing neuron, or a change in the character of the substrate, they will stop, and extend processes in all directions to explore the surroundings and locate the most favorable pathway for the next segment of growth. These initial connections are made by "pioneer neurons" when the whole embryo is quite small so the overall length of a pathway is also small. Later pathways mainly depend on growth along pre-existing axonal tracts, often called **fascicles**. These provide a "highway" along which subsequent axons will grow. Whether axons remain bundled into a fascicle

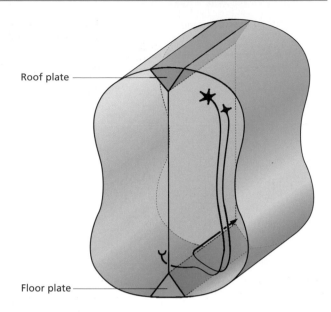

Fig. 14.17 Pathway of the axons from the commissural neurons.

or not depends on a local balance of attraction and repulsion. If axons attract each other more than the substrate then the fascicle will persist; if they attract the substrate more than each other then the fascicle will break up. For example defasciculation in vertebrates may arise because of addition of polysialic acid to N-CAM. This introduces a high surface negative charge and so makes the axons repel each other.

In the CNS there is a population of **commissural neurons** in the dorsal part of the spinal cord. Their axons grow down to the ventral midline, then most cross to the other side, and either contact a neuron immediately, or turn and grow longitudinally along the midline to make eventual synapse at another anteroposterior level (Fig. 14.17). The ventral direction of growth depends on the secretion of netrins from the **floor plate**. Evidence for this is that a floor plate explant will attract the growth of commissural axons *in vitro*, as will cells transfected with *netrin1*. Furthermore, the mouse knockout of *netrin1*, or its receptor *DCC*, shows a deranged pattern of commissural axon growth. When the axons have reached the midline, those that cross do so because they are expressing an NgCAM capable of binding a similar molecule on the floor plate cells. Addition of a neutralizing antibody to NgCAM will cause most of the axons to grow longitudinally without crossing. In addition there is an active repulsive mechanism within the floor plate region consisting of the presence of the repulsive factor called slit, secreted by midline glia. The receptor for slit, robo, becomes expressed once the growth cones have entered the midline region. Activation of robo by slit tends to repel the growth cone from the floor plate and it also inactivates the guidance (but not the growth promoting) function of DCC. This ensures that once the axons have exited the floor plate the netrin signal continues to promote axonal elongation but does not attract the axons back towards the midline.

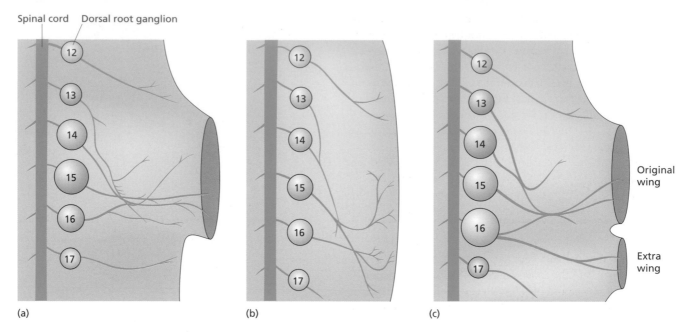

Spinal cord   Dorsal root ganglion

Original wing

Extra wing

(a)   (b)   (c)

Fig. 14.18 Effects of neurotrophins on the dorsal root ganglia. (a) Normal brachial plexus in a chick embryo. (b) Reduced ganglion size after removal of the wing bud. (c) Increased ganglion size after grafting of extra wing bud. (After Hamburger. In Alberts et al. 1989. *Molecular Biology of the Cell,* 2nd edn. Garland, figure 19.72.)

Neurotrophins and cell survival

The final selection of peripheral targets by vertebrate neurons depends on the secretion of **neurotrophins** by those targets. For example the pineal gland and the external ear normally have a sympathetic innervation. This is lost in the mouse knockout of *neurotrophin 3*, but can be restored in organ culture experiments by application of the factor.

Neurotrophins are responsible for regulating the survival of pools of neurons in the CNS and dorsal root ganglia. Both sensory and motor neurons die in large numbers after the initial connections to the target organ have been made. This is a normal aspect of neuronal development, particularly in higher verte- brates. Several growth factors can function as survival factors for neurons in culture. They include the NGF group (NGF, BDNF, NT3, NT4/5), some FGFs, hepatocyte growth factor, and some other factors, including glial-derived neurotrophic factor (GDNF) and ciliary neurotrophic factor (CNTF). The survival of neurons depends on absorbing enough of the relevant neuro- trophin from the target organ. If the target organ is removed, then most of the neurons that would normally form connections to it die off. For example, initially all of the dorsal root ganglia are the same size. But there is more cell death in those that do not innervate the limbs, and so they become smaller. If a limb bud is removed, then its ganglia will suffer more cell death and will shrink. If an extra limb bud is grafted in, then more cells will survive in the ganglia that supply it, and they will remain larger (Fig. 14.18). This mechanism is known as the "neuro- trophic hypothesis."

NGF and the other neurotrophins have two distinct actions on sensory and sympathetic neurons in culture. They will promote the outgrowth of axons and support the survival of the cell bodies. If an experiment is set up so that only the growth cones and apical part of the axons are exposed to the factor, the cell bodies still sur- vive, showing that the factor can be absorbed at the growth cone and its effects are transmitted down to the cell body by retrograde transport.

New Directions in Research

In this area the agenda of developmental biology intersects with that of neuroscience generally so there are a myriad of interesting issues. But sticking to the specifically developmental questions some of the most exciting areas are those that relate to possible recovery from traumatic injury or neurodegeneration:

1 Is it possible to reactivate ependymal cells in adult life so that they revert to stem cell behavior?

2 Can transplanted neural stem cells really repair damaged parts of the CNS?

3 Peripheral axons will regenerate following transection but central axons will not. Is this because of the presence of inhibitory substances in the CNS and, if so, can they be removed?

The protective effect on dorsal root ganglia of grafting an extra limb bud into their vicinity can be mimicked by an injection of NGF into the tissue of the flank. Conversely, the mouse knockouts for the neurotrophins, or their receptors, show an excessive loss of sensory and sympathetic neurons, the effect on the dorsal root ganglia being greatest for the knockout of NT3.

Axonal competition

Once the initial synaptic connections have been made and the central neurons have been thinned out by cell death, there is then a reshuffling of connections. For innervation of muscle this means that an initial situation of several neurons projecting to one multinucleate myofiber becomes reorganized so that each myofiber is innervated by only one neuron, but with a greater number of synapses (the specialized synapses on muscle are called neuromuscular junctions). Unlike the initial, rather crude and diffuse, pattern of connections, this later phase occurs postnatally and depends on electrical activity. It is governed by a rule which states that each neuromuscular junction becomes strengthened by muscle excitation if it has recently been active itself, and is weakened by muscle excitation if it has not been recently active. Therefore an impulse from one motor neuron will provide mutual reinforcement for all of its own neuromuscular junctions and will weaken all those from other neurons. This process will automatically proceed until each myofiber has eliminated the neuromuscular junctions from all but one neuron. A similar principle applies to neuron–neuron interactions within the CNS, where a connection strengthened by activity is known as a "Hebb synapse."

Neuronal connections in the visual system

The visual system represents one of the most complex examples of neuronal specificity. As described above, the retina is formed from the inner surface of the **optic cup**. Because of the camera-like optics of the vertebrate eye, each point on the retina receives light from a particular point in the visual field. The retinal neurons send axons down the optic tube, which becomes the optic nerve (the second cranial nerve). The fibers grow back into the midbrain when they form synapses on the **optic tecta** (or, in mammals, the lateral geniculate nuclei). Each point on the retina sends fibers to a particular point on the tectum, so that the surface of the tectum has a one-to-one, or **topographic**, relationship to the retina and therefore also to the external visual field. This projection can be visualized by illuminating a point in the visual field and recording electrical activity from the corresponding point on the tectum. In lower vertebrates and birds there is a complete crossing of optic fibers at the optic chiasma such that the right retina projects to the left tectum and vice versa, while in mammals there is a projection from both retinas to both tecta. The nature of the projection is such that the anterior retina projects to the posterior tectum and the dorsal retina

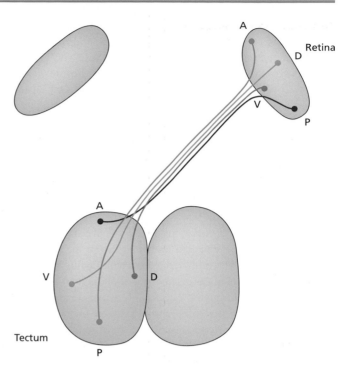

Fig. 14.19 Normal retinotectal projection in a lower vertebrate.

projects to the ventral tectum (Fig. 14.19). Note that in the neuroscience literature, the anteroposterior axis of the retina is called the nasal–temporal axis, but the usage here is harmonized with the rest of developmental biology.

It is clear that the specificity of connections requires some system of cell labeling in both retina and tectum, and some means for matching the two sets of labels. This principle is known as the "chemoaffinity hypothesis" of Sperry. The pattern in the eye is set up at an early embryonic stage. If a *Xenopus* eye is rotated before the tailbud stage, then the projection to the tectum will still grow normally. If it is rotated later, the projection will form but remains inverted. In lower vertebrates the optic nerve will regenerate if cut and it is also possible to examine the topography of the regenerated map. In normal regeneration the connections re-establish the correct projection. If half of the eye is removed at the time that the nerve is cut, then the other half will project across the whole tectum, with normal orientation. Likewise, if half the tectum is removed when the nerve is cut, the whole eye will project to the remaining half with normal orientation. Numerous experiments of this sort showed that the labeling systems are quantitative but with a preferred hierarchy of connectivity. They also showed that the projection must be dynamic and be continually rearranging itself. This is because the retina grows radially, adding new cells from a germinative zone at the margin. By contrast, the tectum grows on the posterior–medial surface, adding cells at one end. These different modes of growth mean that the connections must be continually readjusting themselves in order to maintain the overall topographic projection.

The basis for the topographic projection is now known to be a gradient of adhesivity across both the retina and the tectum. Disaggregated cells from a small region of the retina will adhere to tectal explants with the same specificity as the normal *in vivo* neuronal projection (i.e. anterior retina to posterior tectum and dorsal retina to ventral tectum). The molecular basis is at least partly understood. Using the stripe assay (see Fig. 14.16), it was shown that an extract of posterior tectum could inhibit the growth of posterior retinal axons. Anterior tectum had no repulsive activity and anterior retina was not responsive to the tectal factor. The factors were purified and found to be members of the ephrin family of adhesion molecules, with a gradient of expression from high posterior to low anterior. There is a corresponding gradient of Eph-type receptor on the retinal cells, again with high expression posterior and low anterior. So the molecular basis of the system is a short-range repulsion due to the ephrin system which causes posterior fibers preferentially to adhere to anterior tectum. Although the anterior fibers could also adhere to anterior tectum, the competition between fibers displaces them to the posterior tectum resulting in the observed specificity (Fig. 14.19).

The expression of the ephrins is controlled by the transcription factor engrailed, which itself forms a posterior–anterior gradient across the tectum, initially established by the action of the isthmic organizer on the midbrain (see above). If *engrailed* is misexpressed in chick embryos using a retrovirus, then the regions of ectopic expression on the tecta show an increased expression of ephrins and an exclusion of posterior retinal fibers.

Later refinement

The initial retinotectal projection can be formed without any neuronal activity. This may be shown by treating embryos with tetrodotoxin, which blocks voltage-gated Na channels, to suppress neuronal activity. However, the initial projection is quite crude and it becomes refined later on by processes that do require neuronal activity.

A similar principle underlies the projection from the primary visual centers to the cerebral cortex. In mammals where each retina projects to both sides of the brain, different layers of cells in the lateral geniculate nuclei (the equivalent of the optic tecta) receive inputs from the two retinas, but both project to the same layer of cells in the visual cortex. At birth the receptive fields for the two eyes in the visual cortex are mixed. But with exposure to visual stimuli, they sort out into stripes of about 0.5 mm width called ocular dominance columns. These columns can be visualized by injecting a tracer, such as radioactive proline, into one eye. It is taken up by the retinal cells and transported all the way to the cortex by anterograde transport, first by the retinal cells and then by midbrain cells. Neighboring points on adjacent stripes correspond to the same point in visual space, perceived through the two eyes. These columns do not form if neuronal activity is blocked with tetrodotoxin. Nor do they form if the

young animal is exposed to strobe light, which gives a uniform simultaneous stimulus through both eyes. If one eye is blocked, the input from the neighboring eye will expand to fill the cortical space. The explanation seems to be the same type of principle that refines the neuromuscular connections. If two midbrain cells are connected to the same cortical cell, then the connection will be strengthened if inputs are simultaneous, but weakened if they are not simultaneous. If the midbrain cells respond to adjacent points on the retina of one eye they are likely to fire simultaneously most of the time. But if they are responding to the same visual point viewed through different eyes the input will never be quite simultaneous, and the competitive principle will force the sorting into adjacent files of cortical cells.

Athough this is the conventional explanation, recent work has shown that columns can form in the simple absence of input from one eye so there may be an element of programmed development as well.

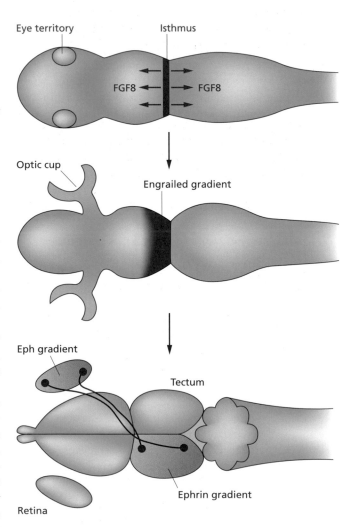

Fig. 14.20 Stages in the acquisition of retinotectal specificity. The axonal projection pattern is controlled by ephrin–Eph interactions, and the initial polarity of the tectal ephrin gradient depends on the gradient of Engrailed set up earlier in response to the signal from the isthmic organizer.

From simplicity to complexity

The development of the visual system is understood, in outline, from the initial neural induction to the final physiological function. It demonstrates at some stage all the principles of neuronal specificity. The ephrin gradients that determine the topographic mapping from retina to tectum depend on a gradient of engrailed, which in turn depends on the FGF8 emitted from the isthmic organizer (Fig. 14.20). The early stages of the map are constructed without the need for neuronal activity or visual experience, while the later details do require an input of experience. There are numerous processes and interactions involved, each of which is intrinsically quite simple. But in combination and in sequence they can build up a structure of quite extraordinary complexity and sophistication.

Key Points to Remember

- Initial neural induction depends on inhibition of BMP activity by factors secreted from the organizer. Subsequent regional patterning in the anteroposterior axis depends on a posterior to anterior gradient of FGF/Wnt/retinoic acid which activates the Hox genes at different anteroposterior levels. Dorsoventral patterning depends on a gradient of sonic hedgehog from the floor plate and of BMPs from the roof plate.
- Subsequent regional specification depends on signals from local centers such as the anterior neural rim, the cortical hem, and the isthmic organizer; these include FGFs, Wnts, and BMPs. Segmental identity of rhombomeres in the hindbrain depends on a gradient of retinoic acid from the posterior mesoderm.
- Neurogenesis in both *Drosophila* and vertebrates depends on a symmetry breaking process of lateral inhibition whereby incipient neurons suppress neuronal differentiation of their neighbors by stimulation of the Notch pathway. The asymmetrical divisions that produce the neuroblasts and neurons depend on the Par complex.
- The CNS contains some neural stem cells that can persist throughout adult life. These can be grown in culture as neurospheres and on differentiation can produce both neurons and glia.
- A variety of tissues arise from the migratory cells of the neural crest. The cephalic crest produces skeletal tissues while the trunk crest does not. Commitment to particular cell types depends partly on the environment through which the cells migrate, but there also seems to be some commitment of cells before migration.
- Growth of axons depends on the elongation of the growth cone. This occurs in response to neurotrophins. The direction of growth depends on permissive components of the environment and also on specific attractants and repellents. A succession of simple decisions can generate a very complex pattern of neuronal connectivity. Some systems of connection, such as that from the retina to the brain, maintain a topographic mapping determined by matching of adhesive gradients on the axons and the target brain regions.
- Initial nerve connections are quite crude. They are refined by differential neuronal survival, depending on the supply of neurotrophins from the target tissues; and axonal competition for innervation of the same target cell. At late stages this may include the effects of sensory experience.

Further reading

Anteroposterior patterning

Bally Cuif, L. & Wassef, M. (1995) Determination events in the nervous system of the vertebrate embryo. *Current Opinion in Genetics and Development* 5, 450–458.

Joyner, A.L. (1996) Engrailed, Wnt and pax genes regulate midbrain-hindbrain development. *Trends in Genetics* 12, 15–20.

Lumsden, A. & Krumlauf, R. (1996) Patterning the vertebrate neuraxis. *Science* 274, 1109–1115.

Marshall, H., Morrison, A., Studer, M., Pöpperl, H. & Krumlauf, R. (1996) Retinoids and Hox genes. *FASEB Journal* 10, 969–978.

Blumberg, B. (1997) An essential role for retinoid signaling in anteroposterior neural specification and in neuronal differentiation. *Seminars in Cell and Developmental Biology* 8, 417–428.

Simeone, A. (2000) Positioning the isthmic organizer. Where otx2 and Gbx2 meet. *Trends in Genetics* 16, 237–240.

Ragsdale, C.W. & Grove, E.A. (2001) Patterning the mammalian cerebral cortex. *Current Opinion in Neurobiology* 11, 50–58.

Wilson, S.I. & Edlund, T. (2001) Neural induction: toward a unifying mechanism. *Nature Neuroscience* 4, 1161–1168.

O'Leary, D.D.M. & Nagakawa, Y. (2002) Patterning centers, regulatory genes and extrinsic mechanisms controlling arealization of the neocortex. *Current Opinion in Neurobiology* 12, 14–25.

Lumsden, A. (2004) Segmentation and compartition in the early avian hindbrain. *Mechanisms of Development* **121**, 1081–1088.

Dorsoventral patterning

Placzek, M. & Furley, A. (1996) Patterning cascades in the neural tube. *Current Biology* **6**, 526–529.

Jessell, T.M. (2000) Neuronal specification in the spinal cord: inductive signals and transcriptional codes. *Nature Reviews Genetics* **1**, 20–29.

Marquandt, T. & Pfaff, S.L. (2001) Cracking the transcriptional code for cell specification in the neural tube. *Cell* **106**, 651–654.

Ruiz i Altaba, A., Nguyen, V. & Palma, V. (2003) The emergent design of the neural tube: prepattern, SHH morphogen and GLI code. *Current Opinion in Genetics and Development* **13**, 513–521.

Strähle, U., Lam, C.S., Ertzere, R. & Rastegar, S. (2004) Vertebrate floor plate specification: variations on common themes. *Trends in Genetics* **20**, 155–162.

Neurogenesis

Sanes, J.R. (1989) Analysing cell lineage with a recombinant retrovirus. *Trends in Neuroscience* **12**, 21–28.

McKay, R.D.G. (1997) Stem cells in the central nervous system. *Science* **276**, 66–71.

Gage, F.H. (2000) Mammalian neural stem cells. *Science* **287**, 1433–1438.

Knust, E. (2001) G protein signaling and asymmetric cell division. *Cell* **107**, 125–128.

Cayouette, M. & Raff, M. (2002) Asymmetric segregation of Numb: a mechanism for neural specification from *Drosophila* to mammals. *Nature Neuroscience* **5**, 1265–1269.

Kintner, C. (2002) Neurogenesis in embryos and in adult neural stem cells. *Journal of Neuroscience* **22**, 639–643.

Doetsch, F. (2003) The glial identity of neural stem cells. *Nature Neuroscience* **6**, 1127–1134.

Henrique, D. & Schweisguth, F. (2003) Cell polarity: the ups and downs of the Par6/aPKC complex. *Current Opinion in Genetics and Development* **13**, 341–350.

Neural crest

Le Douarin, N.M. & Smith, J. (1988) Development of the peripheral nervous system from the neural crest. *Annual Reviews of Cell Biology* **4**, 375–404.

Anderson, D.J. (1997) Cellular and molecular biology of neural crest cell lineage determination. *Trends in Genetics* **13**, 276–280.

LaBonne, C. & Bronner-Fraser, M. (1999) Molecular mechanisms of neural crest formation. *Annual Reviews of Cell and Developmental Biology* **15**, 81–112.

Graham, A. & Smith, A. (2001) Patterning the pharyngeal arches. *Bioessays* **23**, 54–61.

Krull, C.E. (2001) Segmental organization of neural crest migration. *Mechanisms of Development* **105**, 37–45.

Aybar, M.J. & Major, R. (2002) Early induction of neural crest cells: lessons learned from frog fish and chick. *Current Opinion in Genetics and Development* **12**, 452–458.

Le Douarin, N.M. (2004) The avian embryo as a model to study the development of the neural crest: a long and still ongoing story. *Mechanisms of Development* **121**, 1089–1102.

Neuronal growth and connectivity

Colman, H. & Lichtman, J.W. (1993) Interactions between nerve and muscle: synapse elimination at the developing neuromuscular junction. *Developmental Biology* **156**, 1–10.

Müller, B.K., Bonhoeffer, F. & Drescher, U. (1996) Novel gene families involved in neural pathfinding. *Current Opinion in Genetics and Development* **6**, 469–474.

Tessier-Lavigne M. & Goodman, C.S. (1996) The molecular biology of axon guidance. *Science* **274**, 1123–1133.

Bibel, M. & Barde, Y.A. (2000) Neurotrophins: key regulators of cell fate and cell shape in the vertebrate nervous system. *Genes and Development* **14**, 2919–2937.

Tannahill, D., Britto, J.M., Vermeren, M.M., Ohta, K., Cook, G.M.W. & Keynes, R.J. (2000) Orienting axon growth: spinal nerve regeneration and surround-repulsion. *International Journal of Developmental Biology* **44**, 119–127.

Lisman, J.E. & McIntyre, C.C. (2001) Synaptic plasticity: a molecular memory switch. *Current Biology* **11**, R788–R791.

Oster, S.F. & Sretavan, D.W. (2002) Connecting the eye to the brain: the molecular basis of ganglion cell axon guidance. *British Journal of Ophthalmology* **87**, 639–645.

Goldberg, J.L. (2003) How does an axon grow? *Genes and Development* **17**, 941–958.

Schneider, V.A. & Granato, M. (2003) Motor axon migration: a long way to go. *Developmental Biology* **263**, 1–11.

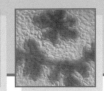

Chapter 15

Development of mesodermal organs

In vertebrates, the mesoderm becomes partitioned at an early stage into four zones from medial to lateral. The **notochord** occupies the midline; next is the **paraxial mesoderm**, which will become the somites; then the **intermediate mesoderm**, which will form the gonads, kidneys, and adrenals; then the **lateral plate** mesoderm. The lateral plate becomes subdivided by the **coelom** into an outer **somatic** mesoderm, later forming the limb buds, and the inner **splanchnic mesoderm**, forming the mesenteries and the heart. The skeleton originates from three regions: most of the skull is formed from the **neural crest**; the vertebrae are formed from the **somites**; and the bones and girdles of the limbs arise from the limb buds and associated lateral plate.

Invertebrate mesoderm does not show the same regional subdivision as vertebrate mesoderm, although some similar tissue types are formed including muscle, gonad, connective tissues, and blood cells.

Somitogenesis and myogenesis

Normal development of the somites

Along with the rhombomeres of the hindbrain, the pattern of somites represents the clearest indication of the segmental arrangement of the body pattern in vertebrates. Somites arise in anteroposterior sequence from the mesoderm flanking the notochord which is called the **paraxial mesoderm** or **segmental plate** (Fig. 15.1). This is distinguished from the intermediate mesoderm by expression of the forkhead transcription factors FoxC1 and C2. The double knockout of these genes in the mouse lacks somites, and overexpression by retroviral infection in the chick increases the width of the somite files. As somite formation is progressive and clearly visible, the number of somites formed is often used as an accurate indication of the stage of a vertebrate embryo. The total number formed over the course of development is a property of the species and can vary widely between vertebrate taxa.

Although all vertebrates form somites, most experimental work has utilized the chick embryo, with some additional information coming from mouse, *Xenopus*, and zebrafish. In the chick, somites first become visible as loose cell associations called somitomeres. These condense into epithelial somites, which are vesicles of cells enclosing a cavity. Expression of fibronectin and N-cadherin increases with the formation of the epithelial somite and may be responsible for the changes in cell adhesion driving the process. The epithelial somite is a transient structure as it soon undergoes an epithelial to mesenchymal transformation. This occurs on the medial side, in the vicinity of the notochord, to form the **sclerotome**, which is a mesenchymal condensation that will form the vertebrae and ribs (Fig. 15.2). The dorsal part of the sclerotome later forms tendons and is now called the syndetome. Like many other parts of the skeleton, the vertebrae are formed initially of cartilage which is later replaced by bone (see Chapter 18). Each vertebra is formed by the sclerotome of two somites, the posterior half of one somite combining with the anterior half of the next, so the vertebrae end up half a segment out of phase with the somite pattern. This is known as resegmentation and is somewhat reminiscent of the relationship between parasegments and definitive segments in *Drosophila*.

The lateral part of the epithelial somite forms a plate called the **dermamyotome**. This in turn becomes subdivided into two parts, the **dermatome**, which forms the dermis of the dorsal midline, and the **myotome**, which forms skeletal muscle. The epaxial myotome, or region near the midline, forms the segmental muscles of the main body axis, and the hypaxial myotome, which is more lateral, forms the muscles of the ventral body wall, limbs, and diaphragm.

Segmentation mechanism

The segmental repeating pattern of the somites arises from the operation of a molecular oscillator, or clock, operating in conjunction with a spatial gradient. One cycle of the clock causes the

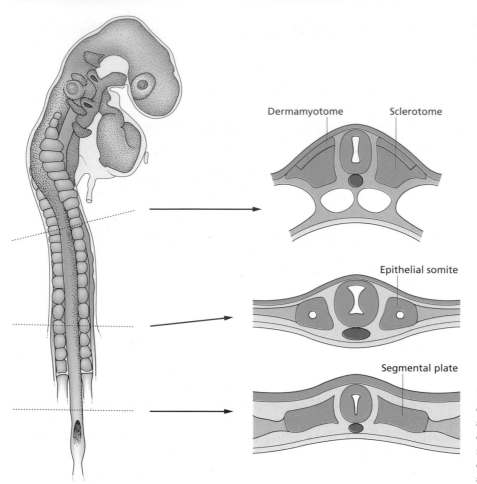

Fig. 15.1 Somitogenesis in the chick embryo. The process starts in the anterior and new segments are added at the posterior end of the file for a considerable time. The morphology is initially epithelial and then changes to an outer dense dermamyotome and an inner loose sclerotome.

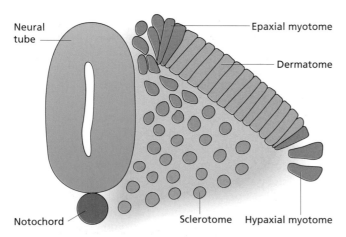

Fig. 15.2 Somite differentiation.

formation of one somite, and the gradient determines that the somites are formed in anterior to posterior sequence. The clock starts operating in the early (stage 3) chick embryo, and the third cycle corresponds to the first somite. The clock can be visualized as periodic expression of *hairy1*, which codes for a bHLH transcription factor. This is a homolog of *Drosophila hairy*, a pair-rule gene also involved in segmentation. The *hairy1* mRNA levels oscillate with a periodicity corresponding to frequency of formation of new somites, about one every 90 minutes in the chick embryo. In most of the presomite plate this oscillation is synchronous, indicating that some cell communication must be occurring to maintain the same phase of the cycle in adjacent cells. In the region adjacent to the last-formed somite, the lower level of the posterior to anterior gradient causes the oscillation to slow such that it lags relative to the rest of the presomite plate. Eventually it stops altogether in the region that will form the next somite. Because the slowdown of the clock in this region is progressive rather than instantaneous, the anterior part of the newly formed somite is left with a low level of *hairy1* and the posterior part with a high level (Fig. 15.3).

hairy1 drives expression of another gene *lunatic fringe* (*lfng*). This is one of a group of genes homologous to *Drosophila fringe* (see Chapter 17) and codes for a glycosyl transferase. It can modify the components of the Delta–Notch signaling system such that Notch activation occurs only along a boundary between a *fringe*-expressing and nonexpressing region. In somitogenesis, *notch1* and *delta1* are both expressed in the presomitic mesoderm and resolve into a spatial periodic pattern as segmentation takes place, both being expressed in the posterior half of each somite. It is likely that the oscillator causes the formation of a steep gradient of fringe activity across the posterior part of each prospective somite. This causes activation of Notch in the posterior part and this in turn drives expression of other components such as ephrins, that contribute to the cell-adhesion changes required to form the epithelial somite.

Evidence for this mechanism comes from study of expression patterns, knockouts, and overexpression experiments. The periodic nature of *hairy1* and *lunatic fringe* expression was discovered because the static *in situ* patterns appeared very variable between specimens until it was eventually realized that they were rapidly changing in time.

Classic Experiments

SOMITOGENESIS AND MYOGENESIS

A clock-based model for somitogenesis was first proposed by Cooke and Zeeman in the first paper, but it was largely ignored and it was not until many years later that the somite oscillator was discovered (second paper).

Meanwhile the understanding of somite cell differentiation got a boost from the discovery of MyoD, the first "master controller" among transcription factors that was able to reprogram other cell types to become muscle (third paper).

Cooke, J. & Zeeman, E.C. (1976) A clock and wave front model for control of the number of repeated structures during animal morphogenesis. *Journal of Theoretical Biology* **58**, 455–476.

Palmeirim, I., Henrique, D., Ish-Horowicz, D. & Pourquie, O. (1997) Avian hairy gene expression identifies a molecular clock linked to vertebrate segmentation and somitogenesis. *Cell* **91**, 639–648.

Davis, R.L., Weintraub, H. & Lassar, A.B. (1987) Expression of a single transfected cDNA converts fibroblasts to myoblasts. *Cell* **51**, 987–1000.

The mouse knockouts of *notch1* or *delta1* or *lunatic fringe* all have similar phenotypes with respect to somitogenesis, such that the segmentation pattern is disrupted, although the subsequent cell differentiation of the somite derivatives is fairly normal. Overexpression of *notch* or *ephrin* in *Xenopus* or zebrafish also disrupts the segmentation pattern. Thus, to obtain segmentation, not only must the components be present, they must also be expressed in a spatially periodic manner. If their activity is the same in the prospective anterior and posterior of each somite, then the segmentation does not occur correctly.

The actual oscillator mechanism probably depends on autoregulation of transcription by the hairy-type transcription factors, which are known as hes in the mouse and her in zebrafish. These proteins will repress their own transcription and are very unstable. Transcription is allowed when the level of the product is low, this causes the protein to build up so transcription becomes repressed, but because the protein is unstable this means its level then falls, and so the transcription increases again (Fig. 15.4a). The main evidence for this negative feedback type mechanism is that the inhibition of hairy synthesis by

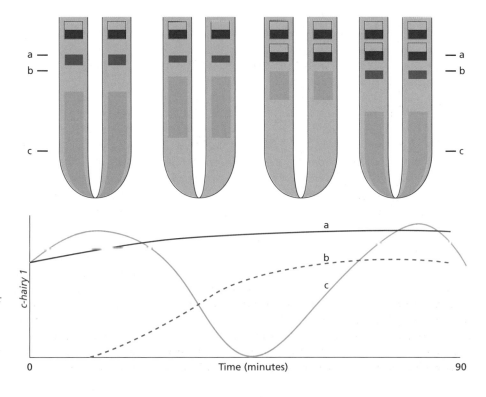

Fig. 15.3 Operation of the somite oscillator over one cycle of somite formation in the chick. The diagrams show the expression pattern of *c-hairy* at four times in the cycle, and the graphs show how the level of transcript varies at the points a, b, and c.

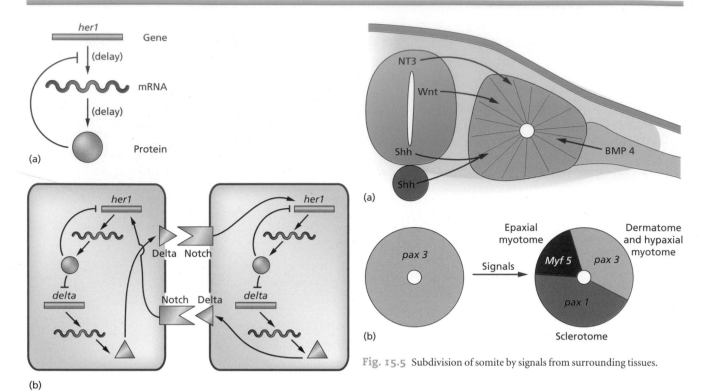

Fig. 15.4 (a) The feedback and delay model for the oscillator. (b) Coupling between cells by the Delta-Notch system.

Fig. 15.5 Subdivision of somite by signals from surrounding tissues.

mutation or **morpholino** treatment will stop the oscillator. In zebrafish the oscillations are rapid, every 30 minutes, and probably arise simply from this mechanism. In mouse the oscillation has a longer period, about 90 minutes, and it probably involves the Delta-Notch system as well. In this more complex cycle, hairy proteins repress *delta* transcription, this will reduce Notch signaling in adjacent cells, but Notch signaling increases *her* transcription, so hairy protein goes down in adjacent cells, this leads to more *delta* transcription so more Notch signaling in the original cell, so *her* transcription goes up again (Fig. 15.4b). Inhibition of Delta in mice suppresses the oscillations, wheras inhibition of Delta in zebrafish permits the oscillations to continue, but abolishes the synchrony of oscillations in adjacent cells. This suggests that in the mouse the Delta-Notch system is part of the oscillator itself whereas in zebrafish it serves simply to maintain the same phase in the population of presomite plate cells.

The gradient part of the system seems to be the posterior to anterior gradient of fibroblast growth factor (FGF) that is responsible for other aspects of posterior body pattern. Evidence for a gradient of FGF controlling somitogenesis is clearest in zebrafish. There is a gradient of phosphorylated extracellular signal regulated kinase (ERK) detectable by immunostaining. The formation of this is blocked by the inhibitor of FGF signaling: SU5402. A pulse of this inhibitor gives large somites, presumably because a larger than normal region has low FGF

signaling and is then incorporated into the somite formed in each cycle. Conversely the administration of FGF protein on a slow-release bead gives rise to smaller than normal somites.

The internal anterior–posterior subdivision of the somite remains important in later development. As we have seen in Chapter 14, the neural crest cells and the spinal nerves are both repelled from the posterior sclerotome because of its high ephrin level, and therefore preferentially grow through the anterior sclerotome of each segment.

Subdivision of the somite

The subdivision of the somite into sclerotome, dermatome, and myotome depends on interactions with the surrounding tissues (Fig. 15.5). This may be shown by the fact that the subdivision will occur normally with respect to the whole body axes, even after the epithelial somites are microsurgically rotated or inverted. Each of the specific inductions was identified by showing that they failed to occur when the presomite tissue was isolated, and that they occurred at the site of contact when the signaling tissue was cultured together with presomite tissue.

The **sclerotome** is induced by the notochord and the ventral part of the neural tube. The epaxial part of the **myotome** is induced by the notochord and dorsal neural tube, and the spread of myotomal induction is limited by the lateral plate mesoderm. The **dermatome** is induced by the neural tube. The signals responsible for each of these interactions have been deduced from the usual criteria for proof described in Chapter 4: namely expression, activity, and inhibition.

The signal for sclerotome induction is Sonic hedgehog (Shh).

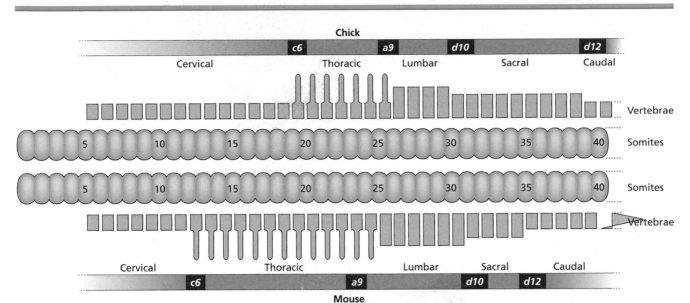

Fig. 15.6 Formation of vertebrae from the somites of the chick (above) and the mouse (below). Each vertebra is formed from two half somites, and the regional character of vertebrae correlates with the expression domains of Hox genes. *c6, a9, d10, d12* show the anterior boundaries of these Hox genes in the two species.

This is expressed in the notochord and floor plate of the neural tube, and is also required for patterning of the neural tube as explained in Chapter 14. Cells expressing Shh will induce cartilage from any part of the epithelial somite. The early events involve upregulation of the transcription factor genes *pax1* and *sox9* and of myogenic inhibitors of the *Id* class (see below). The mouse knockout for *Shh* lacks most derivatives of the sclerotome showing that it is really needed for sclerotome formation *in vivo*. As mentioned above, each vertebra is later formed from the posterior sclerotome of one somite combined with the anterior sclerotome of the next posterior somite. The regional character of vertebrae (cervical, thoracic, lumbar, sacral, and caudal) is controlled by the combination of Hox genes expressed at that level. The mouse and the chick have different numbers of vertebrae in each of these body regions, but the Hox codes are the same (Fig. 15.6). For example, the anterior boundary of *Hoxc6* expression marks the transition from cervical to thoracic vertebrae, the anterior boundary of *Hoxd10* the transition from lumbar to sacral, and the anterior boundary of *Hoxd12* the transition from sacral to caudal. Overexpression and knockout of Hox genes can bring about predictable changes in vertebral character (see Chapter 10).

The induction of the myotome is more complex as it seems to involve an early exposure to Shh at the presegmental stage, followed by a Wnt signal from the dorsal neural tube. This results in the induction of myogenic genes of the bHLH class such as *myf5* (see below) and the repression of *pax3*, initially expressed in the whole epithelial somite, which has an inhibitory effect on myogenesis. A similar effect is shown by removal of the lateral plate mesoderm, which must therefore emit some signal that inhibits myogenesis. The lateral plate expresses BMP4, and this will repress myogenesis if supplied to the premyotomal region by one of the standard overexpression methods (cell pellet, slow-release bead, or retroviral overexpression). The spatial extent of the myotome seems to be determined by a balance between the effects of the Wnt signal from the dorsal neural tube, and the BMP signal from the lateral plate. One function of the Wnt signal is the induction of *noggin* expression in the medial myotome, and Noggin of course inhibits the lateral plate-derived BMP4. The induction of the hypaxial myotome presumably proceeds through a different route, as the primordium is much further from the neural tube. A Wnt signal from the dorsal epidermis is one candidate.

The **dermatome** arises because of a neurotrophin 3 signal from the dorsal neural tube. A neutralizing antibody to this factor will prevent the epithelial to mesenchymal transition which is involved in the formation of the dermis.

Myogenesis

Vertebrate muscle is generally subdivided into skeletal muscle, smooth muscle, and cardiac muscle (see Chapter 13). All skeletal muscle is derived from the myotomes of the somites, or from the corresponding regions of the unsegmented head mesoderm. Smooth muscle is formed from lateral plate mesoderm and cardiac muscle from the myocardium of the early heart.

Skeletal muscle is the familiar voluntary muscle making up a large proportion of the mass of a vertebrate animal. It consists of multinucleate **myofibers** which contain myofibrils composed of the contractile proteins actin and myosin. Differentiation of skeletal muscle involves first the commitment of cells to become

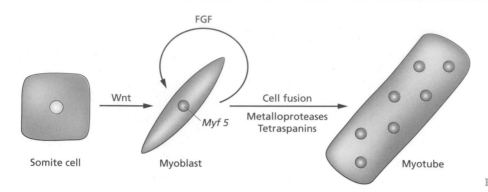

Fig. 15.7 Factors controlling myogenesis.

myoblasts, then a period of multiplication and/or migration, then a cessation of division and fusion of the myoblasts to form the characteristic multinucleate myofibers (Fig. 15.7).

The fact that several myoblasts fuse to form one myotube was proved to occur *in vivo* by the use of **chimeric** mice. If a chimera is made from embryos from strains expressing different electrophoretic variants of the enzyme isocitrate dehydrogenase, then the muscle is found to contain not only the two separate forms but also an intermediate form. This is because the enzyme is dimeric. Single cells can produce only AA or BB dimers, depending on the embryo from which they are descended, but fused cells will also produce the heterodimer AB because both forms are present in the same cytoplasm. The fusion of myoblasts is thought to depend on at least two classes of molecule, as shown by promotion of fusion by transfection with the appropriate genes, and inhibition of fusion by treatment with specific antibodies. They are ADAM class metalloproteases called meltrins, and transmembrane proteins called tetraspanins. Both of these will bind to integrins and both are also involved in mammalian fertilization (see Chapter 10). In tissue culture, myoblasts keep growing while they are supplied with FGF, but when this is removed they stop growing and start fusing into **myotubes**, which are similar to the **myofibers** found *in vivo*. FGF maintains expression of the transcriptional repressor Msx1 which inhibits myogenesis, and to some extent can even cause dedifferentiation if introduced into existing myotubes.

Central to skeletal muscle differentiation is the family of **myogenic** proteins. These are all transcription factors of the bHLH class, and are called MyoD, Myf5, myogenin, and MRF4 (Muscle Regulatory Factor 4). They were identified because, when the genes were transfected into various types of tissue-culture fibroblast, they could cause the cells to become myogenic. The transfected cells would turn on genes for muscle proteins and start to fuse into multinucleate myotubes. MyoD has been shown directly to activate genes for some muscle proteins such as creatine phosphokinase or acetylcholine receptor. It also activates its own transcription, so, once turned on, it stays on (see "bistable switch," Chapter 4). bHLH factors act as dimers, usually as heterodimers with another ubiquitous bHLH protein called an E protein. There also exist inhibitory HLH proteins containing the dimerization but not the transcriptional activation domain, and these work by sequestering the bHLH proteins in unproductive dimers. For example, the Id protein activated in the sclerotome as a response to Shh is an HLH-type inhibitor that forms unproductive dimers with myogenic bHLH proteins and ensures that this part of the somite does not form muscle.

In normal development *myf5* is activated in the epaxial myotome and *myoD* in the hypaxial myotome. Knockouts of either gene have only slight effects on muscle development because if one is removed the other will upregulate and take its place. However, knockout of both together leads to a severely affected mouse lacking both myoblasts and skeletal muscle. *myogenin* is activated later than *myf5* and *myoD*, and is expressed all over the myotome. When it is knocked out there is a serious defect of skeletal muscle formation, showing that the effects of *myoD* and *myf5* must to a large extent operate through myogenin.

Once they have differentiated, myofibers may survive for the entire life of the animal, and muscles can usually increase in size only by enlargement of pre-existing fibers. Growth is controlled by a feedback system involving a circulating inhibitor called myostatin, which is described in Chapter 18. Although the myofibers are postmitotic there are associated with them some persistent progenitor cells called muscle **satellite cells**. These are small mononuclear cells located under the basement membrane of individual myofibers. They are characterized by expression of the transcription factor pax7, and in *pax7* knockout mice, they do not form. They are normally quiescent but can proliferate following muscle damage due to release of hepatocyte growth factor (HGF) from the damaged fibers. The dividing satellite cells show some evidence of asymmetrical division with a segregation of Notch and Numb proteins, rather as seen in neurogenesis (see Chapter 14). Once proliferating they start to express myogenic factors and contribute to tissue repair by differentiating and fusing with each other or with existing myofibers. In addition to the satellite cells some workers have reported finding multipotent stem cells in muscle, which are able to differentiate into osteoblasts or adipocytes in appropriate media. These resemble the mesenchymal stem cells found in bone marrow (see Chapter 13) but their role in muscle development is not yet clear.

The kidney

The kidney is familiar as the principal excretory organ of the vertebrate body and because of the shortage of human kidneys for transplantation it is one of the principal targets for replacement through **tissue engineering** or some other form of applied developmental biology. The embryonic development of the kidney is representative of a large class of organs, as it involves reciprocal inductive interactions between an **epithelium** and a **mesenchyme**. The kidney has a reiterated small-scale pattern and its development can be studied in **organ culture**, so it has been possible to do most of the experimental work on mammals. An explanted rudiment will grow and differentiate for several days in culture and this enables various types of procedure to be carried out which cannot be done on a mammalian embryo *in utero*. In particular, tissues can be separated and recombined, components can be added to or removed from the culture medium, growth factors can be locally applied on slow-release beads, and the expression of individual genes can be inhibited by the addition of specific antisense oligonucleotides or RNAi.

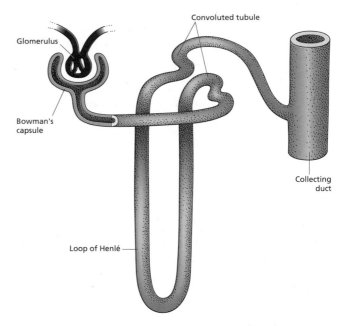

Fig. 15.8 Structure of a nephron.

Normal development of the kidney

Lateral to the paraxial mesoderm that forms the somites lies the **intermediate mesoderm** which gives rise to the kidney and the gonads. Amniotes (reptiles, birds, and mammals) have three kidneys, from anterior to posterior there is the **pronephros**, the **mesonephros**, and the **metanephros**. The pronephros is vestigial and nonfunctional, the mesonephros may function transiently in embryonic life, and the metanephros is the definitive kidney. In lower vertebrates (fish and amphibians) there is no metanephros. The pronephros may function in larval life and the mesonephros forms the definitive kidney.

The basic structural unit of the kidney is the nephron, consisting of a renal corpuscle with associated tubule and collecting duct. The renal corpuscle serves to filter fluid from the blood into the tubule. The fluid is processed into urine as it passes down the tubule and ions are added or removed, with corresponding movements of water. Eventually it passes into the collecting duct system, and thence to the ureter and, in mammals, fish, and amphibians, a urinary bladder. The renal corpuscle itself consists of a convoluted ball of capillaries (the glomerulus) invested with kidney epithelial cells (podocytes), within a Bowman's capsule (Fig. 15.8).

The kidney develops from two components, both arising within the intermediate mesoderm. The **nephric duct** (= **Wolffian duct**) originates in the anterior, at the level of somite 10 in the chick, and grows posteriorly to the cloaca. Labeling with **DiI** or by introduction of a *lacZ* gene shows that it elongates by intrinsic growth not by recruitment of surrounding cells. It requires the neighboring epidermis in order to form, and this requirement can be substituted by a BMP slow-release bead.

A part of the surrounding intermediate mesoderm is already committed to become kidney, and is known as the **nephrogenic** mesoderm. As the duct grows posteriorly, tubules differentiate from this nephrogenic mesoderm and join up with it. The mesonephros in particular shows a segmented arrangement with nephrons arranged in anteroposterior sequence.

The metanephros differs from the other parts of the kidney in that its collecting system is not formed from the main nephric duct, but from an outgrowth called the **ureteric bud**. This grows into the nephrogenic mesenchyme and starts to branch extensively. The mesenchyme differentiates into nephrons which join up with the branches to form a compact, nonsegmental organ. This process is an example of a mesenchymal-to-epithelial transition. First the mesenchyme aggregates and turns on expression of the transcription factor pax2, then it forms epithelial bodies in the shape of a comma, these become an S-shape, and elongate into tubules (Fig. 15.9). During the differentiation process *pax2* expression is lost. Initially the mesenchyme secretes a typical matrix containing fibronectin and collagen I and III. As the transition occurs, these products are replaced by laminin and collagen IV, typical components of an epithelial basement membrane. At the same time, N-CAM on the cell surfaces becomes replaced by E-cadherin.

The process of ureteric branching and nephron formation continues for some time. The remaining metanephrogenic mesenchyme becomes a population of *pax2*-expressing renal **stem cells** at the periphery of the kidney which continue to produce nephrons while also renewing themselves. In the mouse the formation of the metanephric kidney begins at about day 10 of

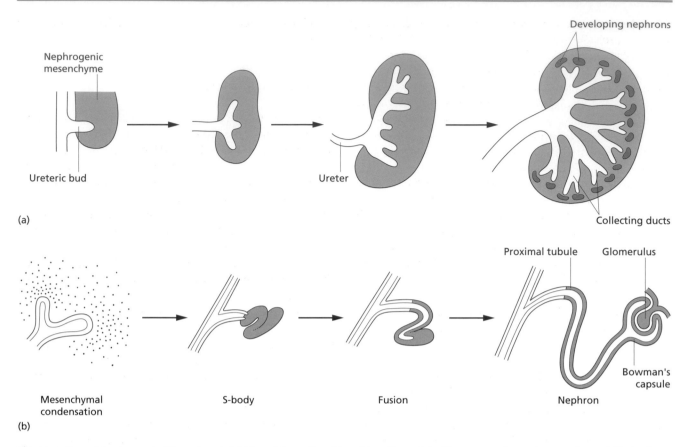

Fig. 15.9 Normal development of the metanephric kidney: (a) growth and branching of ureteric bud into metanephrogenic mesenchyme; (b) induction of a nephron.

embryonic life, and formation of peripheral nephrons continues until about 10 days postnatally.

Tissue interactions in kidney development

If the tip of the nephric duct is destroyed, its further growth is inhibited. Mesonephric tubules develop normally at the level where the duct is present, but they do not develop more posteriorly. This indicates that the duct is needed for formation of the nephrons. The same is true for the metanephros. It is possible to culture the metanephrogenic mesenchyme *in vitro*, but without the ureteric bud there is no tubule formation, showing that the bud emits an inductive signal necessary for tubule formation. This is an example of a **permissive** induction, because the mesenchyme cannot form anything other than kidney. In fact the induction is a reciprocal process because there is also an effect of the mesenchyme on the ureteric bud. Without the presence of the mesenchyme, the bud will not arise from the nephric duct and it will not continue to grow and branch. So there are at least two signaling systems at work, one from the mesenchyme making the bud grow and branch, and one from the bud making the mesenchyme differentiate into tubules.

The signal from the mesenchyme is glial-derived neurotrophic factor (GDNF). This is expressed in the mesenchyme and its receptor, ret, is expressed in the ureteric bud. ret is a tyrosine kinase working through both ERK and PI3K pathways. Mice with either the *gdnf* or *ret* gene knocked out form no kidney. If a GDNF slow-release bead is placed on a culture of nephrogenic mesenchyme from these embryos, then branching of the duct is restored in the *gdnf* knockout, which lacks the factor, but not in the *ret* knockout, which lacks the capacity to respond to it. Conversely, a *ret* gain-of-function mutant with constitutive receptor signaling activity shows an unregulated growth of the ureteric bud. Hence, this system passes all three tests of expression, activity, and inhibition.

Various tissues other than the ureteric bud will induce tubules when cultured in contact with metanephrogenic mesenchyme, and much of the earlier experimental work on this problem used spinal cord as the inducer. This is very active, and probably works through secretion of Wnt factors. A Wnt family member that is expressed in the ureteric bud is Wnt9b, and in the *wnt9b* knockout both meta- and mesonephric tubules are missing suggesting that this factor could be a major component of the signal. This knockout also lacks the epididymal tubules and the Müllerian duct, showing that Wnt9b has a wider role in the urogenital system. Another component of the tubule-inducing signal is probably leukemia inhibitory factor (LIF). LIF is expressed in the ureteric bud and its receptors, LIFR and gp130, are expressed by the metanephrogenic mesenchyme. If the mesenchyme is treated with LIF *in vitro*, this will induce tubule formation. The mouse knockout of *gp130* has a small kidney with a reduced nephrogenic zone, although the fact that the kidney is not lost altogether shows that LIF cannot be solely responsible for tubule induction.

The competence of the mesenchyme to become induced depends on expression of a zinc-finger transcription factor WT1. The mouse knockout for *WT1* shows a complete failure of metanephric development. WT1 also has later functions. Its level increases in the mesenchyme following induction and it is required to turn off *pax2*, which is activated at the condensation stage but whose removal is necessary to enable differentiation of the nephron. The "WT" in WT1 stands for Wilm's tumor, a human pediatric tumor in which renal stem cells continue to proliferate, and of which some cases are caused by somatic loss of the *WT1* gene. This probably reflects loss of the late function, repression of *pax2*, rather than the early function, competence of the mesenchyme.

In the absence of the inducing signals there is massive cell death in the mesenchyme. FGF2 is produced by the ureteric bud and it will both maintain survival of the mesenchyme and also induce aggregation and the expression of *pax2*. But it will not support further development. The *fgf2* knockout shows a fairly normal phenotype, showing that, if there is a function for FGF2, it can be replaced by other FGFs. BMP7 is also expressed in the bud and will maintain survival of isolated mesenchyme and induce tubules. The knockout of *bmp7* has severely dysplastic kidneys containing many nephrons blocked at the comma or S stage of development, suggesting that the main requirement occurs shortly after the initial induction of tubules. Wnt4 is expressed in the developing tubules themselves. It will also induce tubulogenesis *in vitro* and the knockout is blocked at the aggregation stage. To summarize these data: the nephrons are initially induced from the nephrogenic mesenchyme by Wnt9b and LIF from the ureteric bud, and further development depends on Wnt4 expressed in the mesenchyme itself, with FGFs and BMP7 probably also being required as survival or trophic factors.

Germ cell and gonadal development

The **gonads** are both the repositories of the gametes and important endocrine organs. They have a dual embryonic origin. The somatic tissues of the gonads arise from **genital ridges** which are formed from the **intermediate mesoderm**. The germ cells that produce the gametes are derived from **primordial germ cells** (**PGCs**). In most embryo types these form at a distant site in early development and undergo a migration to the genital ridges. Both the mature germ cells and the gonad have a different morphology depending on whether the individual is male or female.

Germ cell development

In the mouse the primordial germ cells originate from a **proximal** part of the **egg cylinder**. This was shown by fate mapping using **horseradish peroxidase (HRP)** injection (see Chapter 10). The PGCs are induced by the effect of the extraembryonic mesoderm since they do not develop in the absence of this tissue, and in recombination experiments it is able to induce PGCs from both proximal and distal epiblast. The signal may be BMP4, as extraextraembryonic ectoderm from the *bmp4* knockout mouse fails to induce PGCs. From 7.5 days the PGCs become visible as large cells containing the enzyme alkaline phosphatase. Their migration can be followed even before this stage in mice transgenic for *gfp* driven by the promoter for *oct4*. The transcription factor oct4 is required for the **pluripotency** of ICM and ES cells (see Chapter 10) and it remains on in the PGCs but not in the somatic tissues. The PGCs become incorporated into the base of the allantois at about 8 days. From there they enter the hindgut at about 9–10 days and migrate up the mesentery, reaching the genital ridges, which form the gonads, between 11 and 13 days (Fig. 15.10).

In addition to oct4, certain gene products discovered in *Drosophila* as being necessary for germ cell development are also required in mice. These include vasa (an RNA helicase) and nanos (an RNA binding protein) (see Chapter 11). It is now known that the migration of the primordial germ cells is controlled by a **chemokine**, SDF1 (stromal cell derived factor 1), which is expressed on the lateral plate mesoderm. The receptor for this chemokine, CXCR4, which is a G-protein-coupled receptor, is expressed on the PGCs. In zebrafish it was shown that the PGCs are attracted to a source of SDF1 *in vitro* and that injection of embryos with a morpholino to either the ligand or the receptor will block germ cell migration. In the mouse the situation is very similar and it has been shown that in the knockout for *cxcr4* the PGCs fail to colonize the genital ridges.

Normal development of the gonads

Male and female gonads are different and so the development of the gonads is intimately associated with sex determination. Unlike most other aspects of development, the mechanism of sex determination is not well conserved between animal groups, so the account that follows applies only to mammals.

The **gonads** develop from the intermediate mesoderm in close proximity to the kidneys. The lateral part of the intermediate mesoderm forms the mesonephros and the medial part the gonad. The **genital ridge** appears at 9.5 days on each side of the trunk, projecting slightly into the coelom (Fig. 15.11 and see also Fig. 9.17 for the layout in the avian embryo). The ridges express

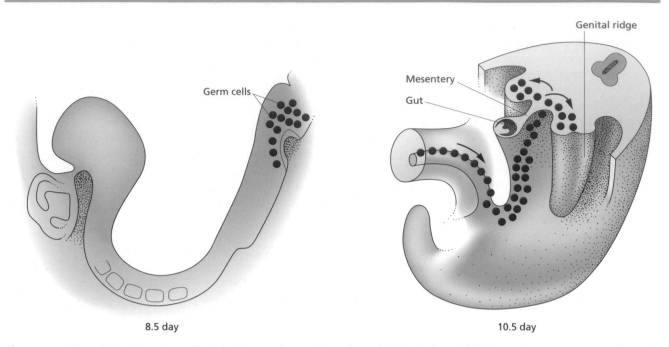

Germ cells

8.5 day

Genital ridge
Mesentery
Gut

10.5 day

Fig. 15.10 Origin and migration of germ cells. (After Hogan et al. 1994. *Manipulating the Mouse Embryo*. Cold Spring Harbor Laboratory Press, figure 3.)

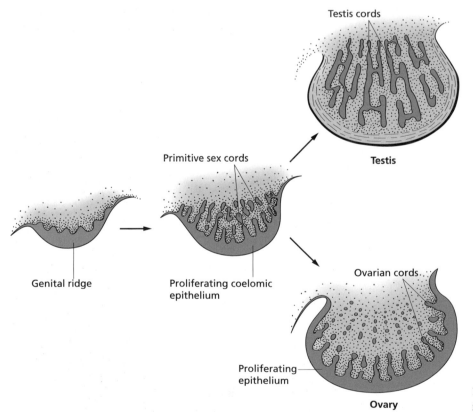

Testis cords

Testis

Primitive sex cords

Genital ridge

Proliferating coelomic epithelium

Ovarian cords

Proliferating epithelium

Ovary

Fig. 15.11 Formation of gonads from the genital ridge.

the Lim-homeodomain transcription factor Lhx9, and the knockout mouse for this factor fails to form any gonads. Slightly after the appearance of the genital ridge, cords of cells begin to form from the coelomic lining epithelium and to grow into the underlying mesenchyme. Two ducts also arise in this period which are important elements in the formation of the sex organs. The **Wolffian duct** is another name for the **nephric duct**. Once the development of the metanephros is underway it has no further excretory function but will eventually form the vas deferens in the male. The **Müllerian (= paramesonephric)**

duct appears alongside the Wolffian duct, and will eventually become the oviduct, uterus, and proximal vagina of the female. Up to this point (12.5 days) there is no difference in the visible differentiation of the two sexes.

In the male, the cords of coelomic lining cells growing into the genital ridge form a complex system of seminiferous tubules composed of **Sertoli** cells, into which the germ cells are integrated. Cells responsible for secretion of testosterone, called **Leydig** cells, differentiate from the mesenchyme between the tubules. The tubules become connected to the Wolffian duct and the developing gonad shortens and becomes encapsulated to form the testis. Meanwhile the Müllerian duct regresses. The tubules of the testis remain solid until after birth, when they start to hollow out and the **spermatogonia**, derived from the germ cells, appear. Spermatogenesis involves the continuous production of sperm from type A spermatogonia attached to the basement membrane of the tubule, via mitotic type B spermatogonia which move toward the lumen and eventually undergo meiosis each to produce four spermatozoa (Fig. 15.12).

In the female, the cords of cells from the coelomic lining do not grow so far into the mesenchyme but stay near the surface as granulosa (= **follicle**) cells, in proximity to the oogonia. **Thecal cells**, responsible for estrogen production, differentiate from the mesenchyme. The gonad becomes encapsulated as an **ovary** and

the Wolffian duct degenerates. The two Müllerian ducts fuse at their posterior ends and form the proximal part of the vagina and the uterus, which is Y shaped in the mouse. At the anterior end they form the **oviducts**, whose collecting funnels are closely apposed to the ovaries. The **oogonia**, derived from the germ cells, become **oocytes** following their final mitosis. It is generally believed that all mammalian oocytes are formed and enter their first meiotic prophase before birth. However this has recently been questioned because estimates of the rate of postnatal oocyte death in the mouse suggest that this must be balanced by some postnatal production of new oocytes. Each primary oocyte becomes invested by granulosa cells to form a primordial follicle. Once sexual maturity has been attained, a number of follicles become activated during each reproductive cycle, the granulosa cells multiply and the follicle with its oocyte expands (Fig. 15.13). In the mouse 8–12 oocytes are likely to be ovulated at one time. Each ruptured follicle becomes a **corpus luteum**, which secretes progesterone, helping to prepare the uterus for implantation.

Sex determination

The gonads can develop normally in various mutants defective in germ-cell migration, so gonadal development does not

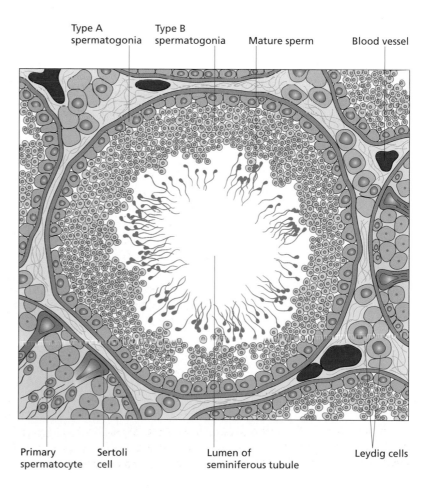

Type A spermatogonia   Type B spermatogonia   Mature sperm   Blood vessel

Primary spermatocyte   Sertoli cell   Lumen of seminiferous tubule   Leydig cells

Fig. 15.12 Organization of a seminiferous tubule. (After Hildebrand 1995. *Analysis of Vertebrate Structure*, 4th edn. Wiley, figure 16.5.)

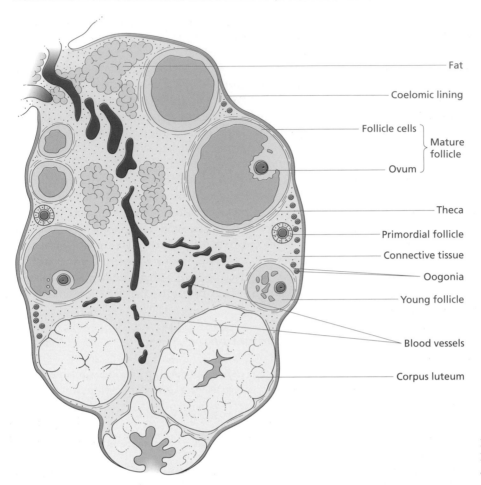

Fat

Coelomic lining

Follicle cells ⎫
⎬ Mature
⎭ follicle
Ovum

Theca

Primordial follicle

Connective tissue

Oogonia

Young follicle

Blood vessels

Corpus luteum

Fig. 15.13 A mammalian ovary. (After Hildebrand 1995. *Analysis of Vertebrate Structure*, 4th edn. Wiley, figure 16.4.)

Classic Experiments

MAMMALIAN SEX DETERMINATION

Since the discovery of sex chromosomes it has been known that possession of a Y-chromosome is necessary to be male, as human individuals of XO constitution are female while those with one Y- and multiple X-chromosomes are male. But there is a very small minority of apparently XX individuals who are male. They have experienced a translocation of the sex-determining gene from the Y- to the X-chromosome. Conversely there is a very small minority of apparent XY females who have lost the sex-determining gene from the Y chromosome. The application of positional cloning techniques to the DNA of these unusual cases resulted in the discovery of the *sry* gene. Studies on the mouse showed that the gene is expressed in the correct location, and

that the human gene could masculinize genetically female mice.

Gubbay, J., Collignon, J., Koopman, P., Capel, B., Economou, A., Munsterberg, A., Vivian, N., Goodfellow, P. & Lovell Badge, R. (1990) A gene mapping to the sex-determining region of the mouse Y-chromosome is a member of a novel family of embryonically expressed genes. *Nature* **346**, 245–250.

Koopman, P., Munsterberg, A., Capel, B., Vivian, N. & Lovell Badge, R. (1990) Expression of a candidate sex-determining gene during mouse testis differentiation. *Nature* **348**, 450–452.

Sinclair, A.H., Berta, P., Palmer, M.S., Hawkins, J.R., Griffiths, B.L., Smith, M.J., Foster, J.W., Frischauf, A.M., Lovell Badge, R. & Goodfellow, P.N. (1990) A gene from the human sex-determining region encodes a protein with homology to a conserved DNA-binding motif. *Nature* **346**, 240–244.

Koopman, P., Gubbay, J., Vivian, N., Goodfellow, P. & Lovell Badge, R. (1991) Male development of chromosomally female mice transgenic for sry. *Nature* **351**, 117–121.

require the presence of germ cells. In the early indifferent phase there is a requirement for at least two gene products, WT1, the zinc-finger transcription factor also required for kidney development, and SF1 (steroidogenic factor 1), a member of the nuclear hormone receptor family. Knockouts of either of these genes prevent gonadal development.

Whether the gonad forms an ovary or a testis ultimately depends on the chromosome constitution: in mammals this is XY for males and XX for females. However, there is an asymmetry to the situation because the Y-chromosome contains a region not homologous to the X-chromosome. Within this lies the gene *SRY* (sex-determining region of Y), which is the critical switch controlling the pathway of sexual development. It codes for a transcription factor of the HMG (High Mobility Group) class and is the prototype of the Sox family of transcription factors. The evidence for its role is as follows. If *SRY* is deleted from the Y chromosome, then an XY mouse will develop as a female, even producing viable oocytes. If *SRY* is introduced as a transgene, then an XX mouse will develop as a male, although it does not produce sperm. Similar although not identical effects are found in naturally occurring chromosomal abnormalities of humans. XXY individuals are male (Klinefelter's syndrome) while XO individuals are female (Turner's syndrome), so it follows that possession of a Y-chromosome must confer maleness. Occasional XX individuals who are phenotypically male are found to have an *SRY* gene on one X-chromosome due to an aberrant recombination at meiosis. Conversely, occasional XY individuals who are female are found to have a mutated *SRY* gene.

In males, *SRY* is expressed in the gonad just before the sexual dimorphism appears. Its immediate function is probably to repress expression of **DAX1**, which encodes a nuclear hormone receptor and is initially expressed at the same time. *DAX1* expression normally declines in males but persists in females. *DAX1* was discovered because, if duplicated on the X-chromosome, it is a cause of female development in XY individuals. This suggests that the critical decision depends on a titration of SRY against DAX1, and if SRY prevails the result is a male, because the function of DAX1 is to repress various male functions. These include the gene for anti-Müllerian hormone (AMH, or Müllerian inhibitory substance, MIS), which is secreted by the Sertoli cells. AMH is a member of the TGFβ superfamily and it has a very specific role in causing regression of the Müllerian duct in the male, as witnessed by the fact that the knockout mouse for *AMH* has a persistent Müllerian duct. Another important male function is the *SF1* gene, which is expressed not only in early gonadal development (see above) but again during testis differentiation. This is also thought to activate *AMH* as well as the genes for the testosterone synthesis pathway in the Leydig cells. Circulating testosterone is the key to the remainder of male development. This, metabolized to the active form dihydrotestosterone, brings about the differentiation of the male external genitalia, and is also responsible for the various secondary sexual characteristics, such as the facial hair and deeper voices of human males.

In the absence of *SRY* activity, *DAX1* expression in the genital ridge persists and the gonad will become female. An early step in the female pathway is the activation of *Wnt4*, also required for development of the kidney. *Wnt4* is normally expressed in the early genital ridges, then is repressed in males but persists in females. The *Wnt4* knockout has an ovary containing few oocytes, with cells secreting male steroids, and also lacks the Müllerian ducts.

Although the germ cells belong to a different cell lineage from the gonads, their pathway of development does depend on the route followed by the gonad. In XY animals any germ cells that fail to find the gonad during their migration will persist for a while in their ectopic location and will develop as oocytes, not spermatogonia. The converse situation can be seen in a chimeric mouse formed from both XX and XY cells. Such mice are male, because the *SRY* gene becomes activated in the XY Sertoli cells and sets off the male program, leading to release of AMH and testosterone. These hormones exert the same effects on cells whatever their chromosome constitution, so the chimeric mouse develops with a male morphology but with both testes and germ cells of mixed XY and XX chromosome constitution. The XX germ cells that enter the testis do become spermatogonia, but are incapable of developing all the way to mature sperm.

Limb development

The vertebrate limb has been subject to intense study over many years and its development is now understood better than most other topics in organogenesis. It is of particular interest to the developmental biologist because their visible nature makes congenital defects affecting the limbs of humans particularly obvious and distressing. The bulk of the experimental work has been done on the chick embryo. At 3–4 days of incubation it is possible to operate on the limb bud *in ovo* through a window cut in the egg shell. The shell can then be sealed and the egg returned to the incubator to allow further development. Many experiments have been performed using classical microsurgical techniques, and in recent years this information has been supplemented by experiments involving gene overexpression or ablation. In the chick, genes may be ectopically expressed by introduction in a replication-competent retrovirus, or by electroporation, or substances can be administered from slow-release beads or from pellets of transfected cells. It is not possible to knock out genes in the chick, but mouse limb development is very similar and therefore data from mouse knockouts is generally incorporated into the account of events.

Normal development of the limb

The amphibians, reptiles, birds, and mammals are known as **tetrapods** because they have four limbs; a pair of forelimbs which may be legs, arms, or wings, and a pair of hindlimbs. The few tetrapods that have no limbs, such as snakes, have lost them secondarily in the course of evolution. The structure of limbs and the course of limb development are very similar in all tetrapods. Following the phylotypic stage of the embryo, limb buds arise in the flank, each consisting of a core of mesenchyme overlaid by epidermis. The buds elongate, flatten, and the various muscles and cartilage elements differentiate in proximal to distal sequence, that is from the body outwards (Fig. 15.14).

The skeleton of the differentiated forelimb consists of a **humerus** in the upper limb, a **radius** and **ulna** in the lower limb, **carpals** in the wrist, and **metacarpals** and **phalanges** in the digits (Fig. 15.15). In the hindlimb the skeleton comprises the **femur** in the upper limb, **tibia** and **fibula** in the lower limb, **tarsals** in the ankle, **metatarsals** and **phalanges** in the toes. The **proximal–distal** axis runs from humerus/femur to phalanges. The **anteroposterior** axis runs from digit I to the highest

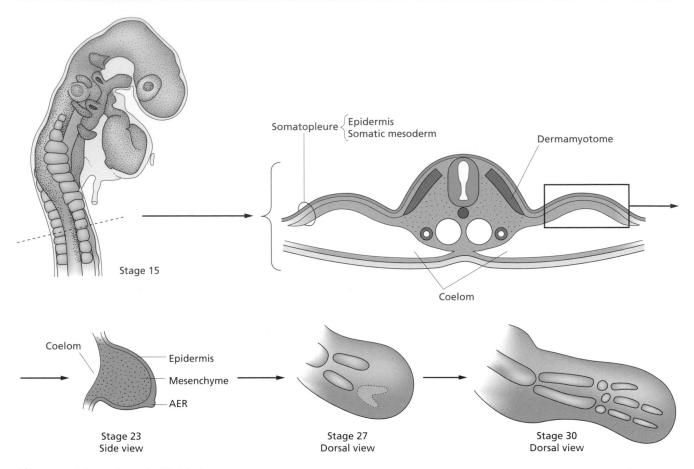

Fig. 15.14 Origin and growth of limb buds.

numbered digit. This is the same as the thumb-to-little finger axis of a human hand and corresponds to the anteroposterior axis of the whole body when the arm is extended with the palm facing ventrally. The vertebrate limb is described as **pentadactyl** because the number of digits rarely exceeds five. However, the number is often reduced, so for example the chick wing only has three digits, which are confusingly numbered II, III, and IV, and the chick leg has four digits. The **dorsal–ventral** axis runs from the upper to the lower surface of the limb (from back of the hand to the palm). The results of many experiments are scored by wholemount staining to reveal the pattern of cartilage elements. This is a useful guide to the proximal–distal and the anterior–posterior pattern, but does not reveal the dorsal–ventral pattern because all the skeletal elements are in one plane. There are, however, various other features that differ along the dorsal–ventral axis and can be used to analyze unusual patterns, in particular the muscle pattern and the types of epidermal specializations such as feathers, scales, claws, and nails.

The region of mesoderm from which the limbs arise lies lateral to the intermediate mesoderm and is known as the **somatic mesoderm**. The somatic mesoderm together with the overlying epidermis is sometimes called the **somatopleure**. The early limb bud consists of an undifferentiated mass of loose mesenchyme surrounded by epidermis. At the distal edge the epidermis is thickened to form the **apical ectodermal ridge (AER)**, which is an essential structure controlling outgrowth (see below). The limb bud mesenchyme forms all of the skeletal structures and connective tissues of the limb: the cartilages, tendons, ligaments, dermis, and the sheaths surrounding the muscles. However, the actual **myofibers** of the muscles are derived from the somites. This may be shown by making grafts of somites from quail to chick at a stage before the limb bud starts to grow out (Fig. 15.16). When the limbs have developed it can be seen that all of the myofiber nuclei are quail type, while all other cell types in the limb are of chick type.

A differentiated limb contains many other cell types in addition to the skeleton and the muscles. All the skin specializations, such as scales and claws, are formed from the epidermis. Blood vessels grow from the capillary network already present in the mesoderm at the time of limb bud outgrowth. Nerves grow in from the spinal cord, arriving as the limb bud is elongating but still undifferentiated. The relevant spinal nerves join together to form the **brachial plexus** to innervate the forelimb and the **lumbosacral plexus** to innervate the hindlimb. As elsewhere in the body, the pigment cells are derived from the neural crest. The skeleton initially differentiates as cartilage and this is later

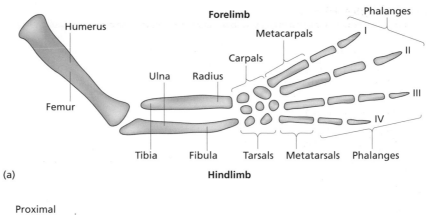

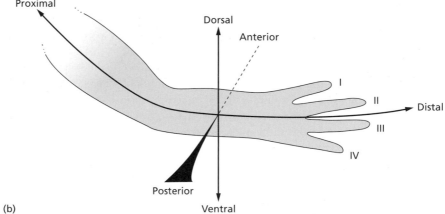

Fig. 15.15 Anatomy of the tetrapod limb: (a) nomenclature of the skeletal elements in the forelimb and the hindlimb; (b) anatomical axes used for description of the limb.

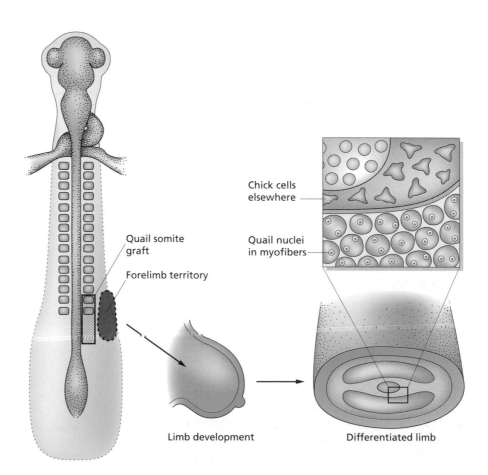

Fig. 15.16 Evidence that the limb muscles derive from the somites was obtained by grafting somite mesoderm from quail to chick and analyzing the resulting limbs.

replaced by bone (see Chapter 18). Unlike the initial phase of limb development, ossification of the cartilage elements does not proceed from proximal to distal, but it does follow a predetermined sequence, generally starting earlier in the larger skeletal elements and later in the small ones.

Experimental analysis of limb development

Limb determination

The limb rudiment is initially specified as a territory in the mesoderm. This may be shown by transplantation of the prospective limb mesoderm to a site on the flank in between the limb buds. In this position, it will provoke formation of a limb bud whereas the prospective limb epidermis will not. However, there is a stage before this in which even the prospective limb mesoderm will not produce a limb unless it is accompanied by somite tissue from the same anteroposterior level, suggesting that an early induction from the somites is necessary for the acquisition of limb commitment.

It is possible to induce a new limb from the region of the flank between the normal limb buds by implantation of a slow-release bead of FGF (Fig. 15.17). Forelimbs arise in the anterior and hindlimbs in the posterior flank in response to such a bead implantation. *fgf10* is expressed in the early presomite plate and subsequently in the prospective limb mesoderm itself. Furthermore, we know that FGF10 is necessary for limb development because the *fgf10* knockout mouse forms no limb buds. FGF10 therefore satisfies the criteria of expression, activity and inhibition for being responsible for the initial induction of the limb-forming territories in the somatic mesoderm. The FGF10 is probably activated by Wnt signaling. The forelimb bud region expresses *wnt2b* and the hindlimb region *wnt8b*. These factors can both induce expression of *fgf10* and outgrowth of limbs. In support of this, the Wnt inhibitor Axin can suppress limb development, and mouse knockouts of *Lef1* or *Tcf1*, components of the Wnt signal transduction pathway, form no AER or limb buds.

The fore- and hindlimb are distinguished by expression of T-box genes, with *Tbx5* expressed in the early forelimb and *Tbx4* in the early hindlimb, and these genes are regulated, directly or indirectly, by the combination of Hox genes active at different body levels. Overexpression of *Tbx4* can bring about a partial conversion of the chick wing bud to a leg bud, suggesting that the T-box genes are themselves responsible for determining the difference between forelimb and hindlimb. Like Wnt and FGF, Tbx4 and 5 will cause ectopic limbs to appear if they are overexpressed in the chick flank. The various components that are capable of inducing limbs may be arranged in a linear pathway, for example: Hox→Tbx→Wnt→FGF→induction of AER, but it is quite likely that some of these components operate in parallel and that there are autocatalytic loops involved.

The migration of myogenic cells from the somites occurs some hours before the commencement of outgrowth of the bud, and any somite can contribute cells if it is grafted to a limb level, not just those which normally supply the limbs. The chemotactic signal for migration is probably hepatocyte growth factor (HGF, = scatter factor). This is expressed in the prelimb bud mesenchyme, and its receptor, called met, is expressed in the myotomes. When *met* or *hgf* is removed by targeted mutagenesis in the mouse there is no migration of myoblasts into the limbs and they develop without muscles.

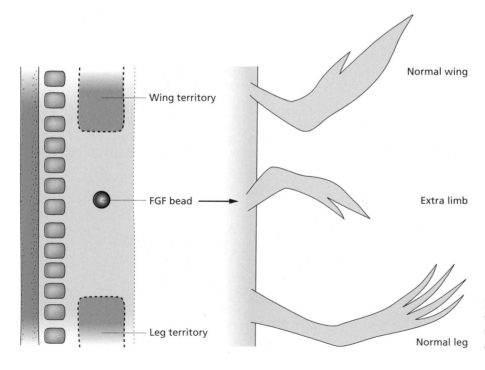

Wing territory

FGF bead

Leg territory

Normal wing

Extra limb

Normal leg

Fig. 15.17 Induction of a limb bud from the flank by an FGF bead. The AP polarity of the ectopic limb is inverted because of the action of the normal forelimb ZPA.

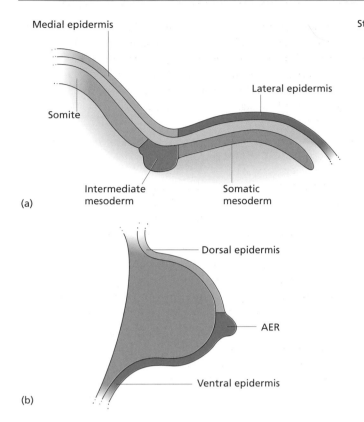

Fig. 15.18 Origin of dorsal and ventral limb-bud epidermis.

Proximal–distal outgrowth and patterning

The earliest visible event in limb-bud development is the formation of the **apical ectodermal ridge** (**AER**) in the epidermis. The AER is induced by the mesoderm, as may be shown by grafting an extra prelimb mesoderm under the flank epidermis. But it also requires for its formation an interaction between two epidermal territories: the medial territory overlies the somites and intermediate mesoderm, while the lateral territory overlies the somatic mesoderm. Fate mapping of the prospective limb epidermis using small marks of **DiI** shows that the medial epidermis becomes the dorsal epidermis of the limb bud, and the lateral epidermis becomes the ventral epidermis of the limb bud, and also forms the AER itself (Fig. 15.18). The AER expresses the homeodomain transcription factors Msx1 and Msx2, which are usually activated by BMP signaling. BMP signaling is in fact necessary for AER formation as it can be prevented in transgenic mice by expressing a dominant negative BMP receptor under control of an epidermis-specific promoter. The lateral epidermis expresses the transcription factor gene *engrailed-1*, which probably functions to repress AER formation, as the mouse knockout of *engrailed-1* has an AER extended onto the ventral side.

The AER persists during the outgrowth of the limb bud. If it is surgically removed at any stage this leads to a distal truncation of the final pattern (Fig. 15.19). The earlier it is removed the more structures are lost and it is presumed that this represents a

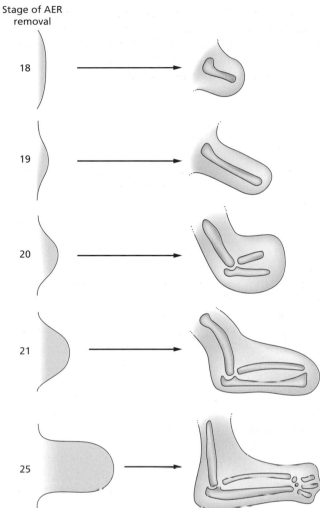

Fig. 15.19 Increasing extent of distal truncation the earlier the AER is removed.

sequence of determination of successive proximal to distal levels of the pattern in the course of limb-bud outgrowth. The region about 300 μm from the AER is known as the **progress zone**. This is the labile region of the limb bud within which the new proximal–distal levels are being laid down. If a progress zone is grafted to a limb-bud stump of a different stage there is relatively little interaction between them. The stump forms what is already determined, and the progress zone lays down what it normally would produce. These observations suggest that the tip mesenchyme is kept in a labile state by the AER, and that structures are specified in sequence just before they leave the tip region (progress zone model). The AER itself needs factors from the mesenchyme for its survival as if it is grafted to a non-limb-bud site it will quickly regress. This factor is called the AER maintenance factor. Thus, the relationship between AER and mesenchyme is a reciprocal one, a signal from each being required for the function of the other. There are now known various molecular markers of the progress zone. One of these is the

Classic Experiments

THE ZONE OF POLARIZING ACTIVITY IN THE LIMB

The ZPA is one of the classic signaling centers of developmental biology. Anteroposterior respecification was first described by J.W. Saunders in the course of tip rotation experiments on the chick limb bud. But the interpretation of this effect as a graded morphogen was made by Tickle et al. in the first of these papers. In the second paper they came up with retinoic acid as a candidate morphogen, but in the third the true morphogen was shown to be Sonic hedgehog.

Tickle, C., Summerbell, D. & Wolpert, L. (1975) Positional signaling and specification of digits in chick limb morphogenesis. *Nature* **254**, 199–202.

Tickle, C., Alberts, B., Wolpert, L. & Lee, J. (1982) Local application of retinoic acid to the limb bond mimics the action of the polarizing region. *Nature* **296**, 564–566.

Riddle, R.D., Johnson, R.L., Laufer, E. & Tabin, C. (1993) Sonic-hedgehog mediates the polarizing activity of the ZPA. *Cell* **75**, 1401–1416.

Lim-homeobox gene *lhx2*. Overexpression of an inhibitory **domain swap** *lhx-engrailed* will inhibit limb outgrowth, suggesting that this gene is necessary for progress zone function. Interestingly, *lhx2* is a homolog of the gene *apterous*, required for wing outgrowth in *Drosophila* (see Chapter 17).

The AER itself expresses various *fgfs* (*fgf2*, *fgf4*, and *fgf8*). If the AER is removed, a slow-release FGF bead can substitute for it and support continued outgrowth and distal patterning. Knockouts of *fgf4* and *fgf8* are both early lethals preventing the embryos developing to the limb-bud stage. However, an AER-specific knockout of *fgf8* using the Cre-lox system does develop and has distal limb defects. So the FGFs satisfy the usual three criteria of expression, activity, and inhibition, as the active factors released by the AER. The maintenance activity from the mesenchyme is partly accounted for by Sonic hedgehog from the **zone of polarizing activity** (**ZPA**, see below). Several BMPs are expressed both in the limb mesenchyme and the AER. If the BMP activity is inhibited by application of Noggin the AER will expand causing increased limb outgrowth, so presumably one of the functions of BMPs is to limit the size of the AER. Later in limb development, BMPs are also required for the cell death that occurs in between the forming digits. If this is prevented in transgenic mice by using the *keratin 14* promoter to drive expression of *noggin*, then webbed feet will result, resembling those of a duck.

Each successive cohort of cells to leave the progress zone forms a different proximal–distal part of the limb. The most proximal territory, the **stylopod**, seems to be encoded by expression of the homeobox genes *Meis1* and *Meis2*, as overexpression of these genes can cause proximalization of the more distal parts. The nested expression in the early limb bud of the *Hoxa10*, *Hoxa11*, and *Hoxa13* genes (Fig. 15.20a) is often thought to control proximal–distal differentiation, but the exact role of Hox genes in limb patterning remains uncertain. The joints that form between the different cartilage elements require Wnt14. This is expressed in early forming joints and ectopic expression will cause the formation of extra joints. Wnt14 seems to work by amplifying its own expression at short range but repressing it at long range, hence causing formation of a periodic pattern of expression prefiguring the arrangement of joints.

Anteroposterior patterning

The **anteroposterior** pattern is controlled by the **zone of polarizing activity (ZPA)** located at the posterior margin of the bud. This behaves like the source of a diffusible morphogen which is held at a constant concentration in the ZPA itself, can diffuse into the surrounding tissues, and is there destroyed or removed. Such a mechanism will establish an exponential concentration gradient from posterior to anterior (see Chapter 4). If a second ZPA is grafted to the anterior margin of the bud, it induces from the anterior tissue the formation of a second set of structures in a **mirror symmetrical** arrangement to the original set (a double

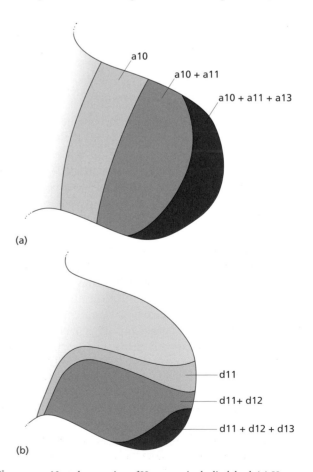

(a)

a10

a10 + a11

a10 + a11 + a13

(b)

d11

d11+ d12

d11 + d12 + d13

Fig. 15.20 Nested expression of Hox genes in the limb bud: (a) *Hoxa* genes; (b) *Hoxd* genes.

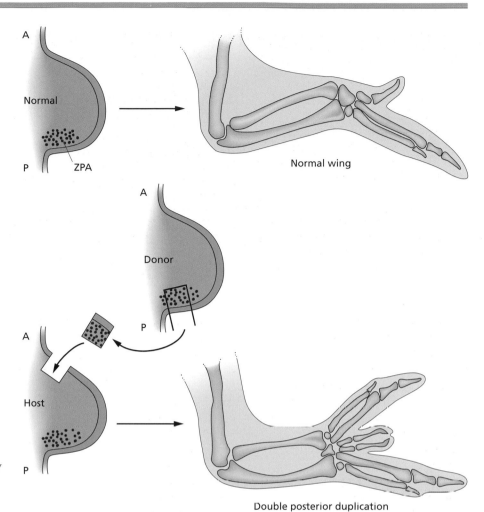

Fig. 15.21 The ZPA graft leads to the formation of a double posterior duplicated limb. Only the distal parts (digits) are usually affected because the more proximal parts are already determined when the graft is carried out.

posterior duplication: Figs 15.21, 15.22). The structure of the duplication can vary from the six-digit IV–III–II–II–III–IV to the three-digit IV–III–IV. The number of structures included in the duplication increases with distance between the two ZPAs. This is because if they are far apart the central minimum of the U is lower and so the thresholds for the formation of more anterior-type structures will be triggered.

The ZPA signal is the same in different vertebrate species, so grafts of lizard or mouse ZPA into the anterior side of the chick limb bud will also provoke the formation of a duplicated limb. All parts of the duplicate limb have a chick type of morphology, rather than that of the donor species, because they are made from chick tissue. Limbs induced in the flank by FGF beads have an anteroposterior polarity that is inverted with regard to the normal limbs. This is because the ZPA of the induced limb bud develops from the same region that forms the ZPA of the normal forelimb, so it is on the anterior rather than the posterior side of the bud.

Implantation of a slow-release bead containing retinoic acid has the same effect as a ZPA graft, inducing an extra set of posterior parts. Retinoic acid and retinoic acid receptors are present in the limb bud, and quail embryos, retinoid deficient because their mothers received a diet lacking vitamin A, show limb bud abnormalities. However it does not seem that retinoic acid is the endogenous ZPA signal because transgenic mice containing a retinoic acid **reporter** (consisting of a retinoic acid response element linked to *lacZ*) do not show a gradient of retinoic acid, or indeed any sign of retinoic acid activity, in the limb bud itself. Retinoic acid beads will in fact induce a functional ZPA from anterior tissue, and it is likely that a requirement for the initial formation of the ZPA is the true role of retinoic acid *in vivo*.

The actual morphogen is the product of the *sonic hedgehog* gene (*shh*). *shh* is expressed in the region thought to be the ZPA on the basis of grafting experiments. Measurement of the amount of active Shh in explants of limb tissue using a tissue culture-based bioassay shows about five times higher concentration on the posterior than the anterior side. A pellet of tissue culture cells producing Shh can induce duplications just like the ZPA. Shh can also support the continued existence and *fgf* expression of the AER, thus behaving like the putative AER maintenance factor. *shh* expression can be activated by retinoic acid, but maintenance of its expression by the ZPA requires continued contact with the AER or with an FGF bead. This shows that the

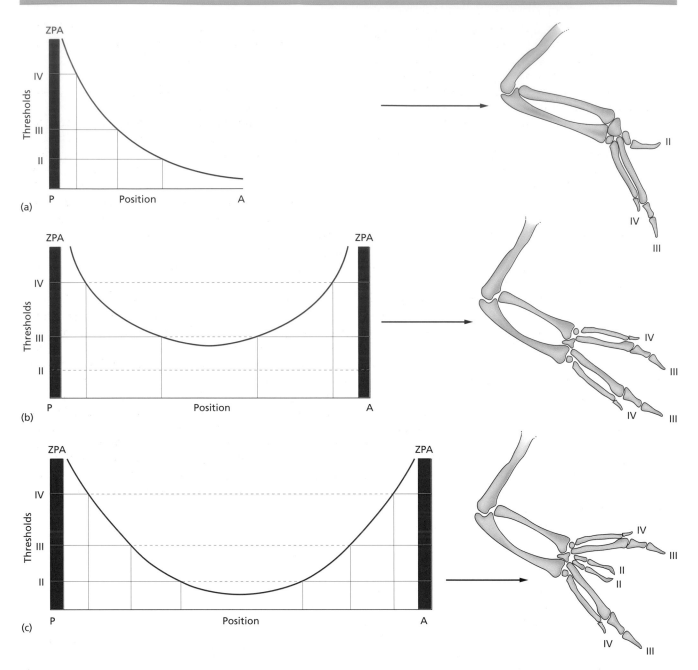

Fig. 15.22 How the gradient from the ZPA controls the pattern of digits. (a) Shows normal development; (b) and (c) show ZPA grafts, because the separation of graft and host ZPAs is greater in (c), the minimum of the U-shaped gradient is lower and the duplication contains more elements than in (b).

genes for Shh and FGF constitute a positive-feedback loop. FGF from the apical ridge maintains the ZPA, and Shh from the ZPA maintains the apical ridge. The *shh*-knockout mouse has abnormal limbs with severe distal defects and lacking anteroposterior polarity. As might be expected, the knockout limb buds contain no ZPA activity as assessed by grafting to the anterior side of normal limb buds. The knockout phenotype is due to the combination of loss of anteroposterior patterning and loss of support for the AER. The initial activation of Shh expression can be brought about by retinoic acid, but the mechanism that

establishes it in normal development seems to be a mutual repression between the transcription factors dHAND and Gli3. dHAND is a bHLH factor which is initially expressed in the whole limb mesenchyme, while Gli3 is on the hedgehog signal transduction pathway. In normal development the two factors achieve an equilibrium with dHAND being expressed in the posterior of the future limb bud and activating expression of Shh, and Gli3 being expressed elsewhere. Evidence for the mutual repression is that if either factor is knocked out in the mouse, the other becomes expressed in a wider domain.

Gli3 is one of three Gli-type transcription factors that are on the Shh signaling pathway. In principle the Shh signalling stabilizes the Gli factors against degradation and enables them to enter the nucleus and regulate target genes. In fact only Gli2 behaves this way. Gli1 is not very active and Gli3 is predominantly repressive, antagonizing the expression of the Shh target genes. This explains why the mouse knockout of *gli3* has too many digits (**polydactyly**). This is partly because the *dHAND*, and therefore also *shh* domain, is bigger than usual, and also because the Shh signal is more effective than usual and induces digits from more of the limb bud. Loss or reduction of Gli3 is also one cause of polydactyly in humans, as seen in Greig syndrome.

The genes *Hoxd9, -10, -11, -12,* and *-13* have expression patterns showing nested territories running across the limb bud from posterior to anterior (Fig. 15.20b). They can be activated ectopically in this nested pattern by a ZPA graft, or by application of retinoic acid, or Shh. It has been widely thought that these genes encode states of determination for different parts of the limb. Retroviral overexpression of *Hoxd11* gives a homeotic transformation of digit I to II in the chick foot, providing some evidence for this view. However the results of mouse knockouts are not consistent with it as loss of subsets of the *Hoxd* cluster tend to give distal deletions rather than homeotic transformations. The *Hox* genes have expression patterns that are somewhat different at different stages, and the nested patterns of Fig. 15.20 are found only at the stage indicated. In such cases the loss-of-function phenotype of a gene generally reflects the earliest expression pattern which is needed for the formation of the subsequent morphology. It remains unclear at which stage in limb development the Hox genes really are required, and whether or not they really code for states of determination.

Dorsoventral patterning

The dorsoventral pattern is apparent mainly in the arrangement of muscles or of epidermal specializations such as feather germs and claws. The dorsoventral pattern of the internal limb tissues can be inverted by inverting the epidermis at an early stage. This is achieved by removing the limb bud from the embryo, separating the epidermis from the mesenchyme by treatment with trypsin, recombining them with the epidermis inverted, then reattaching to the correct position on the embryo with small platinum pins.

The initial polarity of the epidermis comes from the fact that the dorsal epidermis was originally above the somites while the ventral epidermis overlay the lateral plate (see Fig. 15.18). *engrailed-1* is normally expressed in the ventral epidermis. The *engrailed-1* knockout mouse has a partial dorsalization of the ventral paw, suggesting that *engrailed-1* encodes the ventral state, and, as we have seen, the knockout also has a ventral extension of the AER suggesting that engrailed-1 somehow limits the extent of the AER. If the prospective ventral epidermis is separated from the underlying mesoderm by a barrier, then *engrailed-1*

does not become activated, and the resulting limbs are double-dorsal in character. This shows that a signal from the lateral mesoderm is ultimately responsible for the dorsoventral polarity of the limb bud. The signal may be BMP4 which is produced by the lateral mesoderm and is also partly responsible for regionalizing the somites (see somitogenesis above).

A gene for a Wnt factor, *wnt7a*, is normally expressed in the dorsal but not the ventral epidermis. One of the functions of engrailed-1 is to repress expression of *wnt7a*, as may be shown by ectopic expression of *engrailed-1* on the dorsal side. Overexpression of *wnt7a* produces double-dorsal limb and the *wnt7a* knockout mouse has double-ventral paws. Hence, the evidence is good that Wnt7a is the signal from the epidermis conferring dorsal identity on the underlying limb bud mesenchyme. Wnt7a turns on a LIM-homeobox gene, *lmx1*, in dorsal mesenchyme. Wnt7a also contributes to the maintenance of the *shh* expression by the ZPA, and this indirectly helps maintain the AER.

The limb differs from many other organs in that it is asymmetrical in three dimensions, and because it is formed from a solid mass of tissue rather than by morphogenetic movements of cell sheets. The evidence is good that the pattern in each of the three axes is controlled by a separate process. The proximal–distal patterning is still not fully understood but relies on a continuous supply of FGF from the apical ridge. The anteroposterior pattern is controlled by a gradient of Shh from the ZPA. The dorsoventral pattern is controlled by Wnt7a from the dorsal epidermis. Development in the three axes is tied together by the fact that the apical ridge requires both Shh and Wnt7a for its own continued survival and function (Fig. 15.23). This means that the three processes work together to make an integrated organ.

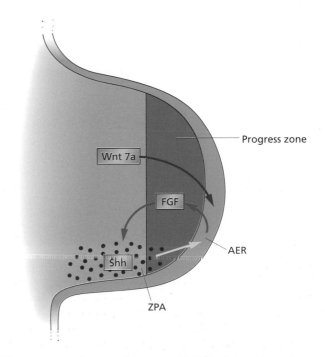

Fig. 15.23 The three signaling systems driving limb elongation and patterning.

Heart and blood vessels

The heart is very important in human embryology because as many as 1% of live births have a congenital defect of heart development. The hearts of the different vertebrate classes all originate in a similar way although the end product is rather different. Fish have a heart consisting of a single muscular tube, divided into sinus venosus, atrium, ventricle, and outflow tract. This drives a single circulation through the gills to the body organs and back to the heart. The same basic regions exist in the hearts of higher vertebrates, although the sinus venosus is reduced to the sinoatrial node in mammals. Amphibians have separate pulmonary and systemic circulations. The atria are separated but there is a single ventricle which seems to be able to keep the two blood flows fairly separate. Birds and mammals have a complete double circulation with the right atrium receiving blood from the organs, the right ventricle sending to the lungs, the left atrium receiving from the lungs, and the left ventricle sending to the organs. Since this arrangement cannot work until the lungs are functional, the system has to be designed for a rapid transition at the time of birth or hatching. In humans the main organ of fetal respiratory exchange is the placenta whose blood returns through the umbilical vein to the right atrium. It then proceeds through the **foramen ovale**, a gap in the interatrial septum, and thence to the left ventricle and the systemic circulation. The output of the right ventricle is also diverted into the systemic circulation through a connection between the pulmonary artery and the aorta called the **ductus arteriosus**. At birth both the foramen ovale and the ductus arteriosus are closed, preventing fluid flow from right to left atrium and making the right ventricular output go through the lungs.

The earliest stages of heart development are best known in the chick, although the situation is similar in other vertebrates. Fate mapping shows that the cardiogenic mesoderm originates in the epiblast lateral to the node (Fig. 15.24). It passes through the anterior third of the primitive streak forming lateral territories by stage 4, which then move anteriorly to form two elongated strips on either side of the embryonic axis. From about stage 5 various transcription factors become expressed in an anterior crescent region which does not entirely correspond to the prospective region of the heart. These include Nkx2.5 (homeodomain), GATA4-6 (zinc finger), MEF2 (MADS box), and Tbx5 (T box). This is also the stage at which interchange of regions or removal of explants starts to cause heart defects so corresponds to the time of determination. By stage 6 there begins to be some segregation in the fate map into regions destined to form the different levels of the heart: sinus venosus, atria, ventricles, and outflow tract.

In effect the heart represents the anteroventral sector of the embryo body, although this is more apparent in a fish or frog embryo than in the flattened chick blastoderm. As we saw in Section 2, anterior character is specified by inhibition of Wnt and Nodal signaling by factors such as Cerberus and Dickkopf, while ventral character is specified by BMP signaling. These factors are all expressed in the anterior endoderm and in recombination experiments the anterior endoderm can induce additional posterior mesoderm to form cardiogenic mesoderm. Cardiogenesis can also be induced by administration of BMP plus a Wnt inhibitor. Finally the inhibition of BMP signaling by the addition of noggin to chick embryos *in ovo* will suppress heart development. The most sensitive period for this is during the period of anterior migration (stages 4–6).

As the head lifts off the blastoderm surface and the foregut begins to form, the heart rudiments move underneath it towards the midline (Fig. 15.25). This migration depends on fibronectin and on FGF8, both from the anterior endoderm. A failure of migration leads to the condition of **cardia bifida** where two separate hearts form side by side. Normally the two rudiments fuse to form a single tube which has four layers: the **endocardium** within, a layer of extracellular matrix called the cardiac jelly, the **myocardium** which forms the actual cardiac muscle, and the pericardium which becomes the thin outer connective tissue sheath. The cardiogenic cells begin to segregate into endocardial and myocardial populations during the migration phase. Cre-lox labeling in mouse embryos, using the

New Directions in Research

Much organogenesis research is motivated by medical objectives. There is an obvious interest in heart development in terms of isolating cardiogenic stem cells, or reprogramming other cell types to be able to repair damaged hearts.

The recent discovery of the molecular differences between arteries and veins should be of importance for tissue engineering. It will not be possible to create real artificial organs without a vascular supply and the principles of molecular recognition currently being uncovered will be critical for this.

Similarly in the kidney more attention will probably focus on the persistent population of renal stem cells that continues to generate new nephrons until after birth.

One particular interest of germ cell development relates to the prospects for therapeutic cloning of human cells. In countries where these techniques are legal the main practical obstacle is the shortage of human oocytes to act as hosts for the transplanted nuclei. A better understanding of germ cell development could, for example, enable the production of oocytes from ES cells.

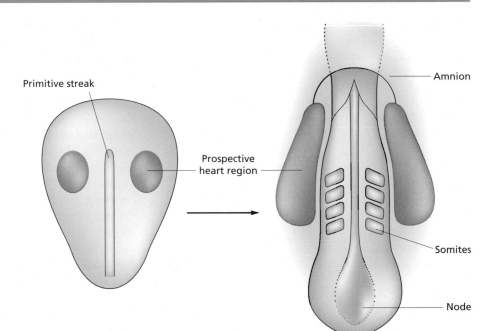

Fig. 15.24 Heart-forming regions of the chick embryo fate mapped with DiI.

Stage 4

Stage 8

Nkx2.5 promoter to drive the *Cre*, shows that all the layers of the heart tube are formed from the cells that formerly had the *Nkx2.5* promoter active. *Nkx2.5*, which is a homolog of the *Drosophila tinman* gene, is the closest thing to a "master regulator" for the heart. In *Xenopus* it has been shown that overexpression of *Nkx2.5* will enlarge the size of the heart, and a dominant negative version will suppress heart development completely. The mouse knockout for *Nkx2.5* does have a heart but its development is arrested at the looping stage.

Shortly after fusion of the two rudiments the heart tube begins to undergo slow pulsations. These are an intrinsic property of the differentiating cardiac muscle and will occur in isolation in tissue or organ culture. *In vivo* the heartbeat will later become controlled by the pacemaker (oscillator) in the sinoatrial node. Also concurrently with fusion, the heart tube starts to become asymmetrical by looping to the right. This depends on the left–right asymmetry system described in Chapter 9, with left-sided expression of nodal as the principal factor. The initial asymmetry leads to the asymmetric expression of various transcription factors affecting the heart. In the mouse knockouts of *dHAND* and *mef2C* both have asymmetrical effects, disrupting the right ventricle specifically. Another important transcription factor for regionalization of the early heart is Tbx5. This is normally graded in expression from posterior to anterior. The mouse knockout lacks the sinoatrial region, and overexpression in the ventricular region disrupts their development. The human condition Holt–Oram syndrome is due to the loss of one copy of the *tbx5* gene, and it involves defects both in the heart and the upper limbs (see above).

As the regionalization and looping of the heart proceeds there is a further recruitment of cells into its anterior end. These come from what is now called the anterior heart field which initially lies within the pharyngeal mesenchyme and later anterior to the region expressing the cardiac transcription factors. It is characterized by expression of the LIM-homeodomain transcription factor islet-1. In addition there is a contribution of **neural crest** cells to the outflow tract of the heart. They are responsible for forming the septum in the outflow tract that divides the pulmonary and aortic circulation (see below).

The aspect of heart development that has been very thoroughly studied in the human embryo is the process that converts a simple tube into the four-chambered heart with separate right- and left-sided circulation (Fig. 15.26). This involves looping, asymmetrical growth, movement of blood vessel insertion sites and, most important, the formation of internal septa to divide up the lumen. First the looping of the heart tube brings the atria to the anterior and the ventricles to the posterior (Fig. 15.26a). Then the venous returns move such that they run into the right side of the atrium, and a new pulmonary vein sprouts from the left side of the atrium. Then two atrial septa grow down towards the junction with the ventricles. These contain openings that will allow blood through until the time of birth. At the same time, four **endocardial cushions** appear at the atrioventricular junction. These arise from the endocardium by epithelial-to-mesenchymal transition, with the cells invading the hyaluronan-rich cardiac jelly. This transition is induced by TGFβ from the myocardium. Addition of TGFβ will cause the transition, and addition of inhibitors of TGFβ will prevent it. Two of the cushions meet to form a **septum intermedium**, dividing the ventricle into right and left sides. The other cushions later contribute to the atrioventricular valves: the biscupid or mitral on the left and the triscupid on the right. Initially the

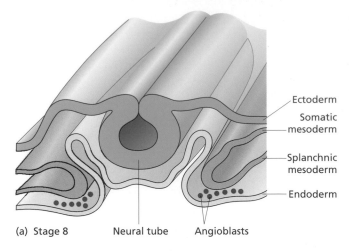

(a) Stage 8   Neural tube   Angioblasts

Ectoderm

Somatic mesoderm

Splanchnic mesoderm

Endoderm

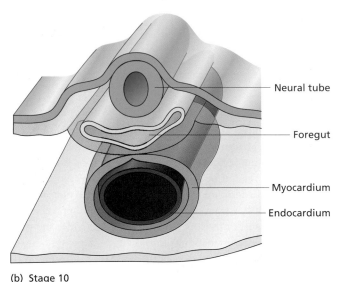

(b) Stage 10

Neural tube

Foregut

Myocardium

Endocardium

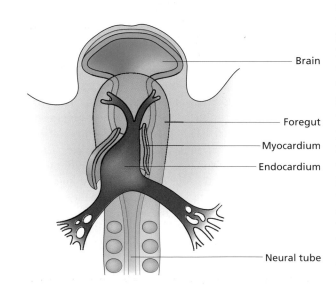

(c) Stage 10 (ventral view)

Brain

Foregut

Myocardium

Endocardium

Neural tube

primitive atrium is connected only to the prospective left ventricle with the prospective right ventricle having the whole outflow tract. During the period of formation of the septum intermedium a remodelling occurs that aligns what are now right and left atria with the corresponding side of the ventricle (Fig. 15.26b) At the same time the outflow tract becomes symmetrically related to the prospective ventricles. Then a muscular **ventricular septum** grows from the ventricular wall to separate the right and left sides. This is controlled by Tbx transcription factors. At the corresponding stage in the chick Tbx5 is expressed in the left ventricle and Tbx20 in the right. If *tbx5* is electroporated into a region of the right side, it will suppress *tbx20* expression and a ventricular septum starts to form at the junction. Simultaneous with these events, longitudinal **truncoconal swellings** develop in the outflow tract and grow towards each other, eventually separating the pulmonary artery from the aorta (Fig. 15.26c). All the various ingrowths, the atrial septa, the outflow tract septa, and the ventricular septum, eventually meet at the septum intermedium and the heart has become fully divided into two. This process is generally similar in the chick except that there is just a single atrial septum and this has many small perforations instead of one large one.

Blood vessels

De novo formation of blood vessels in the embryo is called **vasculogenesis**, while the formation of new capillaries from existing ones by cell division and cell movement is known as **angiogenesis** (Fig. 15.27). Vasculogenesis is intimately associated with the formation of the blood itself. At early stages there is present in the lateral plate mesoderm a cell type called the **hemangioblast** that forms both hematopoietic stem cells and the endothelial cells of the vessels. The evidence for the existence of the hemangioblast is as follows:

1 Early blood-forming regions of *Xenopus* and zebrafish embryos coexpress genes for hematopoiesis and vasculogenesis. In particular these include *SCL* (encodes "stem cell leukemia" transcription factor), *flk1* (encodes receptor for vascular endothelial growth factor, VEGF), and *GATA -1* and *-2*.

2 If a cell surface marker is expressed in transgenic mice under the control of the *SCL* promoter then it is possible to sort SCL-positive cells in viable form from early embryos. These cells are mostly also Flk1 positive, and they will grow to form large clones containing both hematopoietic and vasculogenic progeny.

3 Similar large multipotent clones may be isolated from ES cell-derived embryoid bodies.

4 The mouse knockout for *SCL* loses the blood cells but not the vessels, while the knockout for *flk1* loses both blood cells and vessels.

Fig. 15.25 (*left*) Formation of the heart tube in the chick embryo. (a) At stage 8 there are two separate rudiments on either side of the foregut. (b) By stage 10 these move together to form a single tube with endocardial and myocardial layers. (c) Stage 10 ventral view. (After Fishman 1997. *Development*, **124**, 2099–117, figures 2 and 3.)

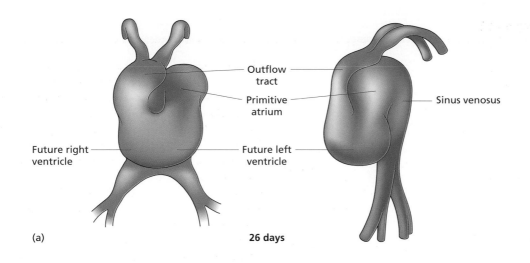

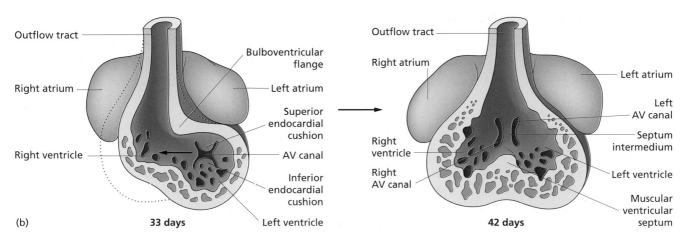

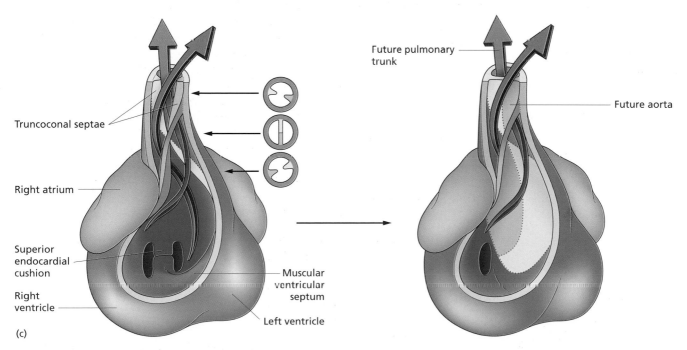

Fig. 15.26 Later development of the human heart. (a) External views of the heart after looping, 26 days. (b) Formation of the septum intermedium and remodelling of the position of the outflow tract relative to the ventricle. (c) Formation of the longitudinal septum in the outflow tract. (After Larsen 1993. *Human Embryology*. Churchill Livingstone, figures 7.9, 7.17 and 7.20.)

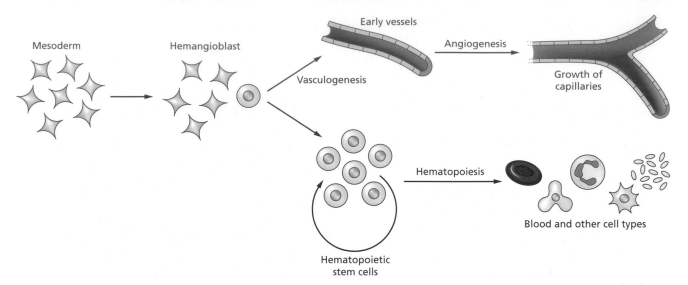

Fig. 15.27 Formation of blood cells and blood vessels.

SCL is a bHLH transcription factor also known as Tal1. VEGF is the most potent of a number of angiogenic factors. Its receptor Flk1 is a tyrosine kinase and is also known as kdr or VEGF2 (*cloche* in the zebrafish). The GATA factors are zinc-finger transcription factors. In zebrafish or *Xenopus*, injection of mRNA for SCL, together with a Lim domain factor Lmo2, will induce blood and vessel formation from most parts of the mesoderm. Inclusion of GATA1 as well will enable the complete development of red blood cells. It seems that the progeny of the hemangioblasts either lose expression of SCL to become endothelial precursors, or they lose expression of Flk1 to become hematopoietic stem cells (HSCs).

Capillaries are composed of a single layer of **endothelial cells** with a basal lamina on the exterior surface. Larger blood vessels have a layer of smooth muscle and a layer of connective tissue outside the endothelium. Endothelial cells can divide throughout life and there is usually a low level of growth associated with tissue remodelling. The formation of new capillaries by cell division and cell movement may occur quite actively in wound healing or in the growth of tumors. A number of growth factors are active in promoting angiogenesis, particularly VEGF, FGFs, and angiopoietins.

The capillary beds of the later embryo arise by fusion of capillaries growing from the arterial and the venous sides. The former carry ephrin B2 and the latter Eph B4, and it is presumed that the molecular complementarity between these molecules enables the two components to fuse. Knockouts of genes for either component leads to severe angiogenic defects. Ephrin B2 expression persists into adult arteries, in the smooth muscle layer as well as the endothelium, and its expression increases during angiogenesis. In zebrafish it has been shown that VEGF promotes arterial development by activation of Notch signaling which in turn activates expression of the ephrin B2.

Anatomists have known for centuries that major nerve tracts tend to be associated with major blood vessels. It now seems that they are bound together by common signaling systems. This is because VEGF binds not only to Flk1 but also to neuropilins, which are receptors for semaphorins. Neuropilin 1 is expressed on arteries and neuropilin 2 on veins. The double knockout of both neuropilins not only has axonal growth defects but also has no blood vessels. VEGF acts as a positive signal and semophorin 3A as a negative one for the growth of both nerves and vessels, ensuring that they will tend to run together. Moreover nerves often secrete VEGF and so can be expected to attract growth of blood vessels directly.

Key Points to Remember

- The repeated somites of the vertebrate body arise in anterior to posterior sequence. The sequence of formation is controlled by the posterior to anterior gradient of FGF. The number of cells in each somite is controlled by an oscillator based on the feedback inhibition of transcription of bHLH transcription factors of the hairy family.
- The somite is divided into regions forming the muscle, the vertebrae, and the dermis in response to inducing signals from the surrounding tissues.
- Muscle differentiation is driven by bHLH transcription factors of the MyoD group. Individual myoblasts fuse to form multinucleate myofibers. Although the myofibers are postmitotic, satellite cells associated with them can divide and contribute to muscle growth.
- The functional kidney in higher verterbates is the metanephros. This is formed by a reciprocal interaction between the ureteric bud from the nephric duct, and the metanephrogenic mesenchyme. The mesenchyme causes branching of the ureteric bud by a GDNF signal, and the ureteric bud induces tubules from the mesenchyme by a signal including Wnt9b and LIF.
- Germ cells in mammals are induced from the egg cylinder epiblast by the extraembryonic ectoderm. They then migrate to the base of the allantois, round the hindgut, and up the mesentery to the genital ridges.
- The gonads develop from the genital ridges. In the male the SRY protein represses expression of DAX1, which therefore fails to repress expression of AMH, and this leads to regression of the Müllerian duct. In the female, DAX1 represses expression of AMH so the Müllerian duct persists and becomes the female reproductive tract.
- The limb buds originate in the somatic mesoderm of the flank. They require a Wnt signal, an FGF signal, and expression of T-box transcription factors (Tbx5 or Tbx4) to become established.
- The limb buds are asymmetrical in three dimensions and need three signals to control the formation of the pattern. One is associated with FGFs from the apical ectodermal ridge. One is sonic hedgehog from the zone of polarizing activity. One is Wnt 7a from the dorsal epidermis. Shh and Wnt7a both help maintain FGF expression in the apical ridge, establishing a reciprocal feedback system.
- The heart arises as bilateral mesodermal primordia that fuse in the midline to form a simple tube. Nkx2.5 is a key transcription factor controlling cardiac development. The four-chambered heart of higher vertebrates arises from the simple tube through the formation of various septa combined with remodelling of the proportions of the heart tube and of the insertion sites of major vessels.
- Blood and blood vessels arise from a common precursor: the hemangioblast. Initial formation of blood vessels occurs from hemangioblasts, and later formation by growth from pre-existing vessels (angiogenesis). Arteries and veins differ in the expression of various receptors and adhesion molecules and this complementarity is essential to creating the capillary beds.

Further reading

Somitogenesis

Burke, A.C., Nelson, C.E., Morgan, B.A. & Tabin, C. (1995) Hox genes and the evolution of vertebrate axial morphology. *Development* **121**, 333–346.

Balling, R., Neubüser, A. & Christ, B. (1996) Pax genes and sclerotome development. *Seminars in Cell and Developmental Biology* **7**, 129–136.

Christ, B., Huang, R. & Wilting, J. (2000) The development of the avian vertebral column. *Anatomy and Embryology* **202**, 179–194.

Pourquié, O. (2001) Vertebrate somitogenesis. *Annual Reviews of Cell and Developmental Biology* **17**, 311–350.

Brent, A.E. & Tabin, C.J. (2002) Developmental regulation of somite derivatives: muscle, cartilage and tendon. *Current Opinion in Genetics and Development* **12**, 548–557.

Giudicelli, F. & Lewis, J. (2004) The vertebrate segmentation clock. *Current Opinion in Genetics and Development* **14**, 407–414.

Rida, P.C.G., Le Minh, N. & Jiang, Y.J. (2004) A Notch feeling of somite segmentation and beyond. *Developmental Biology* **265**, 2–22.

Myogenesis

Rudnicki, M.A. & Jaenisch, R. (1995) The MyoD family of transcription factors and skeletal myogenesis. *Bioessays* **17**, 203–209.

Tajbakhsh, S., Rocancourt, D., Cossu, G. & Buckingham, M. (1997) Redefining the genetic hierarchies controlling skeletal myogenesis: pax3 and myf5 act upstream of MyoD. *Cell* **89**, 127–138.

Seale, P. & Rudniki, M.A. (2000) A new look at the origin, function, and "stem cell" status of muscle satellite cells. *Developmental Biology* **218**, 115–124.

Buckingham, M. (2001) Skeletal muscle formation in vertebrates. *Current Opinion in Genetics and Development* **11**, 440–448.

Parker, M.H., Seale, P. & Rudniki, M.A. (2003) Looking back to the embryo: defining transcriptional networks in adult myogenesis. *Nature Reviews Genetics* **4**, 497–507.

Pownall, M.E., Gustafsson, M.K. & Emerson, C.P. (2003) Myogenic regulatory factors and the specification of muscle progenitors in vertebrate embryos. *Annual Reviews of Cell and Developmental Biology* **18**, 747–783.

Kidney

The kidney development database: http://golgi.ana.ed.ac.uk/kidhome.html

Lechner, M.S. & Dressler, G.R. (1997) The molecular basis of embryonic kidney development. *Mechanisms of Development* **62**, 105–120.

Sariola, H. & Sainio, K. (1997) The tip top branching ureter. *Current Biology.* **9**, 877–884.

Barasch, J., Yang, Y., Ware, C.B. et al. (1999) Mesenchymal to epithelial conversion in rat metanephros is induced by leukaemia inhibitory factor. *Cell* **99**, 377–386.

Kuure, S., Vuolteenaho, R. & Vainio, S. (2000) Kidney morphogenesis: cellular and molecular regulation. *Mechanisms of Development* **92**, 31–45.

Shah, M.M., Sampogna, R.V., Sakurai, H. et al. (2004) Branching morphogenesis and kidney disease. *Development* **131**, 1449–1462.

Germ cells, gonads, and sex determination

Bogan, J.S. & Page, D.C. (1994) Ovary? Testis? – A mammalian dilemma. *Cell* **76**, 603–607.

Jiménez, R. & Burgos, M. (1998) Mammalian sex determination: joining pieces of the genetic puzzle. *Bioessays* **20**, 696–699.

Swain, A. & Lovell-Badge, R. (1999) Mammalian sex determination: a molecular drama. *Genes and Development* **13**, 755–767.

Wylie, C. (1999) Germ cells. *Cell* **96**, 165–174.

Capel, B. (2000) The battle of the sexes. *Mechanisms of Development* **92**, 89–103.

Matova, N. & Cooley, L. (2001) Comparative aspects of animal oogenesis. *Developmental Biology* **231**, 291–320.

Lovell-Badge, R., Canning, C. & Sekido, R. (2002) Sex determining genes in mice: building pathways. In: *The Genetics and Biology of Sex Determination. Novartis Foundation Symposium* **244**, 4–22. Chichester, Wiley.

McLaren, A. (2003) Primordial germ cells in the mouse. *Developmental Biology* **262**, 1–15.

Raz, E. (2003) Primordial germ cell development: the zebrafish perspective. *Nature Reviews Genetics* **4**, 690–700.

Limb development

Johnson, R.L. & Tabin, C.J. (1997) Molecular models for vertebrate limb development. *Cell* **90**, 979–990.

Martin, G.R. (1998) The roles of FGFs in the early development of vertebrate limbs. *Genes and Development* **12**, 1571–1586.

Vogt, T.F. & Duboule, D. (1999) Antagonists go out on a limb. *Cell* **99**, 563–566.

Capdevila, J. & Izpisa-Belmonte, J.C. (2001) Patterning mechanisms controlling vertebrate limb development. *Annual Reviews of Cell and Developmental Biology* **17**, 87–132.

Christ, B. & Brand-Saberi, B. (2002) Limb muscle development. *International Journal of Developmental Biology* **46**, 905–914.

Logan, M. (2003) Finger or toe: the molecular basis of limb identity. *Development* **130**, 6401–6410.

Niswander, L. (2003) Pattern formation: old models go out on a limb. *Nature Reviews Genetics* **4**, 133–143.

The heart

Fishman, M.C. & Chien, K.R. (1997) Fashioning the vertebrate heart: earliest embryonic decisions. *Development* **124**, 2099–2117.

Harvey, R.P. (1999) Seeking a regulatory roadmap for heart morphogenesis. *Seminars in Cell and Developmental Biology* **10**, 99–107.

Srivastava, D. & Olson, E.N. (2000) A genetic blueprint for cardiac development. *Nature* **407**, 221–226.

Harvey, R.P. (2002) Patterning the vertebrate heart. *Nature Reviews Genetics* **3**, 544–556.

Brand, T. (2003) Heart development: molecular insights into cardiac specification and early morphogenesis. *Developmental Biology* **258**, 1–19.

Blood and blood vessels

Risau, W. & Flamme, I. (1995) Vasculogenesis. *Annual Reviews of Cell and Developmental Biology* **11**, 73–91.

Robb, L. & Elefanty, A.G. (1998) The hemangioblast – an elusive cell captured in culture. *Bioessays* **20**, 611–614.

Adams, R.H. (2003) Molecular control of arterial-venous blood vessel identity. *Journal of Anatomy* **202**, 105–112.

Kubo, H. & Alitalo, K. (2003) The bloody fate of endothelial stem cells. *Genes and Development* **17**, 322–329.

Ruhrberg, C. (2003) Growing and shaping the vascular tree: a multiple role for VEGF. *Bioessays* **25**, 1052–1060.

Chapter 16

Development of endodermal organs

The endoderm is the innermost of the three germ layers formed during gastrulation. It forms the epithelial lining of the gut together with its outgrowths that include the liver, pancreas, and respiratory system. The outer coats of these organs, comprising smooth muscle, connective tissue, and blood vessels, is formed from the **splanchnic mesoderm**, which is the inner subdivision of the lateral plate following the opening of the **coelomic** cavity. Just as the combination of somatic mesoderm and epidermis is called the **somatopleure**, the combination of splanchnic mesoderm and endoderm is known as the **splanchnopleure**.

Normal development

Formation of the gut tube in amniotes

As with much organogenesis research, the majority of work has been conducted with the chick embryo, with some assistance from mouse knockouts and experiments on early stage *Xenopus* and zebrafish. In the chick the endoderm is formed during gastrulation as a lower layer of **definitive endoderm** which emerges from the primitive streak and displaces the **hypoblast** cells to the periphery of the blastoderm. The gut lumen originates from the space beneath the blastoderm. The gut tube itself is formed by the folding of the body away from the blastoderm, which can be thought of as a sort of evagination of all three germ layers (Fig. 16.1). The formation of the head fold resembles the situation that would arise if a hypothetical miniature finger were pushed up from below to deform the blastoderm, then pushed sideways to elongate the projection. Thus, the anterior part of the body becomes lifted off the blastoderm, and the endoderm in this region becomes a tube of **foregut**. The junction of foregut and midgut, or **anterior intestinal portal**, is the level where the tube opens out into the flat sheet of endoderm. Somewhat later a similar process happens at the rear end and the **hindgut** is formed, with the junction between midgut and hindgut as the **posterior intestinal portal**. The mouth forms at the start of the

third day by fusion of the ventral foregut with the overlying epidermis, and a **cloaca** forms at the posterior end, although does not become perforate until hatching. Growth and morphogenetic movements increase the size of the extremities of the embryo such that the residual area of open endoderm rapidly shrinks. By 4 days this has become reduced to a **vitellointestinal duct** connecting the midgut to the yolk mass. As the gut closes, the splanchnic mesoderm from the two sides fuses to form dorsal and ventral mesenteries (Fig. 16.2a,b). The dorsal mesentery persists and holds the mature gut while the ventral mesentery is lost except in the vicinity of the heart, liver, and cloaca. The process continues by the epidermis and mesoderm closing around the gut to form the ventral body wall (Fig. 16.2c). This eventually becomes complete except for the **umbilical tube** which is the connection to the extraembryonic tissues enclosing the vitellointestinal duct, the vitelline blood vessels, and the **allantois**. The allantois is a ventral outgrowth from the hindgut (Fig. 16.1b). From the second day it grows out of the embryo into the amniotic cavity and fuses with the chorion to form the **chorio-allantoic membrane** (see Fig. 9.5). This is the richly vascularized membrane which is visible immediately on opening the egg of a late embryo and it serves as an important respiratory organ before hatching. The stalk and blood vessels of the allantois become incorporated into the umbilical tube. By about 4 days of incubation the chick gut is fully formed as a tube and has commenced asymmetrical coiling, controlled by the left–right asymmetry system described in Chapter 9. It is also starting to show some regional pattern in terms of the shape and size of the various organ rudiments (Fig. 16.3).

The course of events in other higher vertebrates is fairly similar. In the mouse the definitive endoderm starts out as a ventral strip of tissue on the outside of the egg cylinder, bordered by visceral endoderm. It becomes incorporated into the fore- and hindgut pockets in a similar way to the chick, assisted by the "**turning**" process described in Chapter 10. The overall structure of the gut is similar, with the allantois contributing to the placenta.

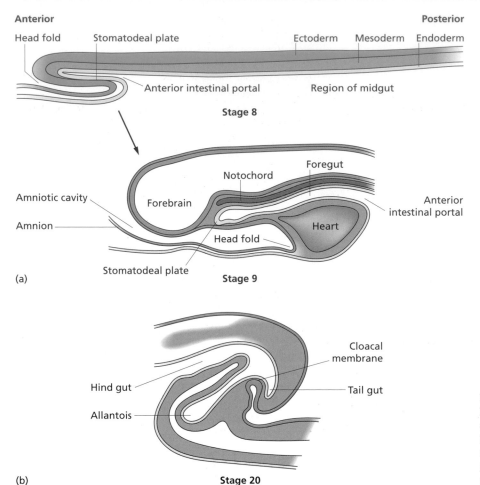

Fig. 16.1 Formation of the fore- and hindgut in the chick embryo. (a) Formation of headfold and foregut. (b) Formation of hindgut (later stage). (After Bellairs and Osmond 1998. *The Atlas of Chick Development*. Academic Press, figures 31, 32 and 33.)

Organization of the gut tube

In the region of the pharynx are formed the **pharyngeal pouches**, which are aligned with the branchial arches formed from alternating rhombomeres (see Chapter 14). There are four major pouches and sometimes one or two more rudimentary ones, which are associated with a segmental arrangement of endodermal outgrowths (Fig. 16.4). Buds from the first pouch form the cavities of the middle ear and the Eustachian tubes. The ventral midline region opposite the second arch forms the thyroid gland. The second pair of pouches (in mammals) form the tonsils, the third pair the thymus, and both the third and fourth pairs the parathyroids. In the ventral midline opposite the fourth pouches form the laryngotracheal groove which becomes the trachea, and produces paired buds which generate the bronchi and tissue of the lungs. Much of the floor of the pharynx becomes the tongue, which is mostly composed of muscle from the head mesenchyme.

Regional differentiation of the gut tube is indicated in Fig. 16.5. After the pharyngeal region comes the esophagus, which is initially lined with columnar and later with stratified squamous epithelium. Then comes the stomach, where the epithelium becomes glandular and specialized to secrete hydrochloric acid and pepsin. At the exit of the stomach is the pyloric sphincter leading into the small intestine where the embryonic epithelium is columnar but on maturation differentiates into the crypt and villus arrangement described in Chapter 13. In the mature animal the junctions between the esophagus and stomach and the stomach and intestine are two of a very small number of places where there is a sharp discontinuity of epithelial type which has arisen *in situ* from a common cell sheet. The first section of small intestine is called the duodenum and this forms a loop to which are attached the liver and the pancreas. The liver is a ventral outgrowth of the endodermal epithelium that expands into the adjacent ventral mesentery shortly after formation of this part of the foregut. This region of the mesoderm is called the **septum transversum** in mammals. The pancreas arises from a large dorsal bud and a small ventral one (paired ventral buds in the chick). These move together and fuse in later development to form a single organ. The ventral pancreas is closely associated with the liver and the ventral pancreatic duct fuses with that of the liver to form the common bile duct. The subsequent parts of the small intestine are called the jejunum and the ileum. The junction between small and

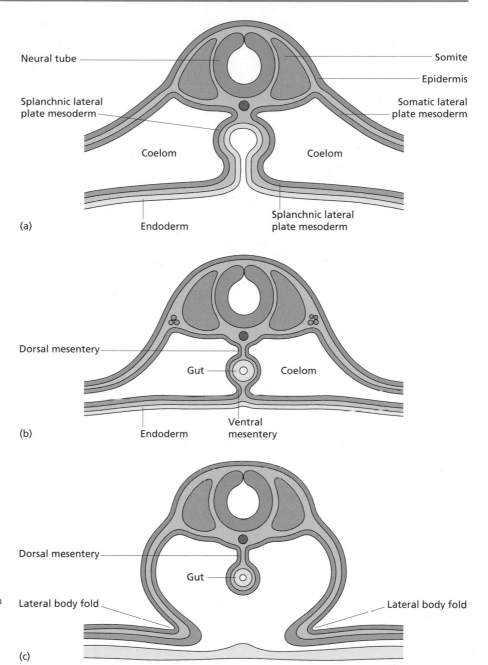

Neural tube

Somite

Epidermis

Splanchnic lateral plate mesoderm

Somatic lateral plate mesoderm

Coelom

Coelom

(a)

Endoderm

Splanchnic lateral plate mesoderm

Dorsal mesentery

Gut

Coelom

(b)

Endoderm

Ventral mesentery

Dorsal mesentery

Gut

Lateral body fold

Lateral body fold

(c)

Fig. 16.2 Enclosure of the gut and ventral closure of the body in the chick embryo. (a)–(c) represent transverse sections between midgut and hindgut levels at three days of incubation. (After Bellairs and Osmond 1998. *The Atlas of Chick Development*. Academic Press, figures 31, 32 and 33.)

large intestine (or colon) is often marked by the presence of outgrowths called caeca, which are blind-ended sacs lined with intestinal epithelium. The mature large intestinal epithelium resembles that of the small intestine but there are no villi and no Paneth cells. At the posterior end the large intestine becomes a rectum and in mammals joins to the exterior through the anal sphincter.

In the mature gut the terms fore-, mid-, and hindgut are still used, but they no longer relate to the gut segments enclosed by the anterior and posterior intestinal portals which are moving boundaries during the formation of the gut tube. Convention-

ally the **foregut** at late stages is taken to extend as far as the pancreas and liver, and the **midgut** as far as the junction between small and large intestine.

This account is generic for most amniote vertebrates, but experimental work is largely performed on the chick or the mouse, and they are not quite the same. The following checklist is provided to keep track of the differences:

1 The mouse has tonsils in the second pharyngeal pouch, the chick does not.

2 The chick forms parathyroids from the third and fourth pharyngeal pouch, the mouse only from the third.

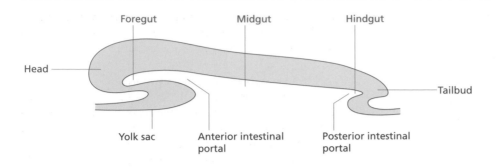

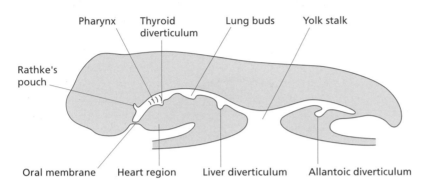

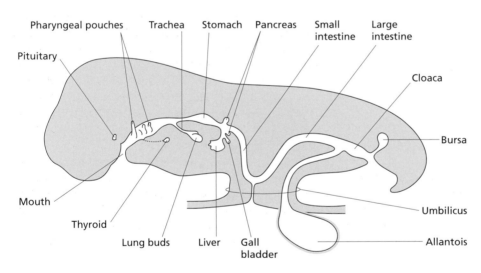

Fig. 16.3 Formation of the regions of the gut in an amniote embryo. (After Hildebrand 1995. *Analysis of Vertebrate Structure.* 4th edn. Wiley, figure 12.1.)

3 The mouse incorporates the ultimobranchial body into the thyroid as C (calcitonin) cells, the chick keeps them separate.

4 The mouse lungs have alveoli, the chick lungs have anastomosing channels running into large distal air sacs.

5 The chick has a specialization of the esophagus to form the crop, the mouse does not.

6 The chick has a division of the stomach into a glandular **proventriculus** and a muscular **gizzard**, the mouse has no gizzard.

7 The mouse embryo liver is a major hematopoietic organ while the chick liver is not.

8 The chick has paired ventral pancreatic buds, the mouse has one.

9 The mouse has one large cecum between small and large intestines, the chick has two small ones.

10 The chick has a bursa of Fabricius, which produces B lymphocytes, near the cloaca, and the mouse does not.

11 The chick has a **cloaca** while the mouse has separate urinary/reproductive and alimentary orifices.

Fate map of the endoderm

In the chick the endoderm has been fate mapped by several methods, including **orthopic grafting** of quail tissue and small extracellular injections of **DiI**, followed by *in vitro* culture of the

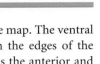

Arch 1
Pouch 1
Arch 2
Pouch 2
Arch 3
Pouch 3
Arch 4
Pouch 4
Arch 5
Pouch 5
Arch 6
Rudimentary pouch 6

Stomatodeum
Thyroid
Slit 2
Lumen of pharynx
Parathyroids
Thymus
Ultimobranchial body
Lungs
Esophagus

Fig. 16.4 Location of outgrowths of the endoderm in the pharyngeal region of the chick embryo.

whole embryo until the gut tube has closed. Comparable studies have also been made in mouse and *Xenopus*. As might be expected there is a broad conservation of anterior–posterior relationships, so the anterior endoderm becomes the foregut, the middle section becomes the midgut, and the posterior endoderm becomes the hindgut (Fig. 16.6a). Because the early endoderm is an open sheet, its midline becomes the dorsal midline of the gut tube and therefore structures like the dorsal pancreatic

bud are represented on the midline of the fate map. The ventral midline structures, like the liver, arise from the edges of the definitive endoderm, which come together as the anterior and posterior portals move towards one another – this is why there are often paired buds for ventral structures.

One very important aspect of the fate map is the relationship between prospective regions in the endoderm and those in the **splanchnic mesoderm** (Fig. 16.6b). The prospective regions in the mesoderm are quite different from those in the endoderm, tending to be arranged longitudinally. This shows that there is a considerable relative movement of the two germ layers during the process of closure of the gut and closure of the ventral body wall. An important theme in gut development is the inductive interactions between mesoderm and endoderm and the fate map shows clearly that a particular region of endoderm experiences contact with different regions of the mesoderm as the process of gut tube formation takes place. This means that the state of specification of the endoderm may change as it contacts different mesenchymes.

Determination of the endoderm

The initial formation of the endoderm has been investigated mainly in *Xenopus* and zebrafish. In *Xenopus* the endoderm forms from the tissue roughly corresponding to the eight most vegetal blastomeres at the 32-cell stage (see Fig. 7.9). The endoderm is formed because of the presence of mRNA for the transcription factor VegT. This activates the expression of various further transcription factors including Sox17α and β, Mix1, Mixer, and GATA1-4. Some of these are activated directly but others require intercellular signaling. In particular the expression of *mixer* and *GATA4* require nodal signaling. In zebrafish there is more dependence on nodal signaling as the endoderm is not formed at all in the *cyclops/squint* double mutant that lacks both of the *nodal* genes. A sox family gene called *casanova* and a POU domain homolog of *oct4* called *spiel-ohne-grenzen* are controlled by nodal and loss of function causes loss of the whole

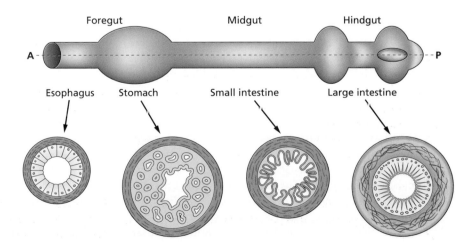

Foregut
Midgut
Hindgut
A
P
Esophagus
Stomach
Small intestine
Large intestine

Fig. 16.5 Cytodifferentiation of regions of the chick gut tube.

Classic Experiments

MASTER GENES OF THE
PANCREAS

Pdx1 (then called IPF1) was discovered as a transcription factor controlling insulin expression. But when its gene was knocked out it was found to be necessary for development of the whole pancreas (first paper). How a fraction of the endodermal cells was diverted to become endocrine cells was mysterious until the discovery of a lateral inhibition system similar to that controlling neurogenesis. The second paper

shows how neurogenin 3 is essential for endocrine cell formation and how Notch signaling represses endocrine development among the neighbors of neurogenin 3-expressing cells.

Jonsson, J., Carlsson, L., Edlund, T. & Edlund, H. (1994) Insulin-promoter-factor-1 is required for pancreas development in mice. *Nature* **371**, 606–609.

Apelqvist, A., Li, H., Sommer, L., Beatus, P., Anderson, D.J., Honjo, T., de Angelis, M.H., Lendahl, U. & Edlund, H. (1999) Notch signaling controls pancreatic cell differentiation. *Nature* **400**, 877–881.

endoderm. Likewise in mice, mutants that are defective in nodal signaling do not form the definitive endoderm.

The various endodermal transcription factors show a considerable mutual interdependence. They will mostly activate the other factors if overexpressed in *Xenopus* animal caps, and the introduction of **dominant negative** constructs made with the engrailed repression domain will generally suppress endoderm development. A similar set of transcription factors are expressed in the early endoderm of zebrafish and mouse, and mutants show endodermal defects of various degrees of severity, for example the mouse knockouts of the *sox17* genes lack the entire mid- and hindgut.

Regional specification

The transcription factors mentioned above seem to specify endoderm as a germ layer rather than any specific type of epithelium or organ. There are many other transcription factors that are expressed at specific levels within the endoderm (Fig. 16.7). Some of these are known to be important for regional specification because of their mutant phenotype. *pax9* (paired domain) is expressed in all the pharyngeal pouches and the knockout mouse has no thymus, parathyroid or ultimobranchial body. *Nkx2.1* (homeodomain, =*TTF1*) is expressed in the thyroid bud and in the tracheo-esophageal septum. The knockout mouse has no thyroid or lungs. *pdx1* (homeodomain, =*XlHbox8* in *Xenopus*) is expressed in a region of the proximal intestine including the future pancreatic buds and the knockout mouse has no pancreas (see below). *cdx2* (homeodomain, =*cdxC* in chick) is expressed in the posterior part of the gut corresponding to most of the intestine. The knockout of *cdx2* is an early lethal because it is also needed in the trophectoderm, but complete loss of function in clones arising within the intestine of heterozygous animals can give rise to transformation to an

esophagus-like epithelium, suggesting that *cdx2* distinguishes anterior and posterior endodermal types of differentiation. The three factors of the FoxA (forkhead domain) class, FoxA1–3, are expressed in a nested arrangement with staggered anterior boundaries. The knockout of *foxA2* shows a loss of fore- and midgut, although it is the one expressed in the whole endoderm. Interestingly, several of these transcription factors are expressed again at a later stage and have functions during the terminal differentiation of the organs concerned. For example, Pdx1 not only controls formation of the whole pancreas, but later on it is specifically expressed in β cells and helps to control *insulin* expression. Nkx2.1 not only controls formation of the whole thyroid, but later controls expression of the thyroid-specific genes *thyroglobulin* and *thyroperoxidase*.

Some transcription factors are regionally expressed in the splanchnic mesoderm rather than the endodermal epithelium and may affect differentiation of the mesoderm directly. For example, loss of the whole HoxD cluster (except d1 and d3) will provoke loss of the ileocecal sphincter, a muscular structure formed form the mesoderm. Others are probably important in controlling regional inductive signals from the mesoderm that control regional pattern in the endoderm (see below). For example viral misexpression of Hoxd13 in the midgut mesenchyme causes the midgut epithelium to develop as large intestine rather than small intestine. The original cause of the patterns shown in Fig. 16.7 must lie in the anteroposterior patterning system for the whole body in early development. As we have seen in Section 2 this is dependent on a posteriorizing signal during gastrulation which involves FGFs, Wnts, and retinoic acid. One of the important target gene groups is the *cdx* family and these in turn activate the more posterior Hox genes. *cdxA* is expressed in the posterior primitive streak of the chick from about stage 3. The early expression domain, which controls overall body pattern, fades by stage 10, after which *cdxA* is expressed again in the posterior endoderm in the region corresponding to most of the future intestine.

Epithelial–mesenchymal interactions

It is clear that there is a complex and continuous interchange of inductive signals between the endoderm and mesoderm. By this stage the splanchnic mesoderm is referred to as mesenchyme, since it has become a layer of undifferentiated mesenchyme surrounding the epithelial tube. The whole endodermal tube expresses Sonic hedgehog (Shh). This forms a radial gradient

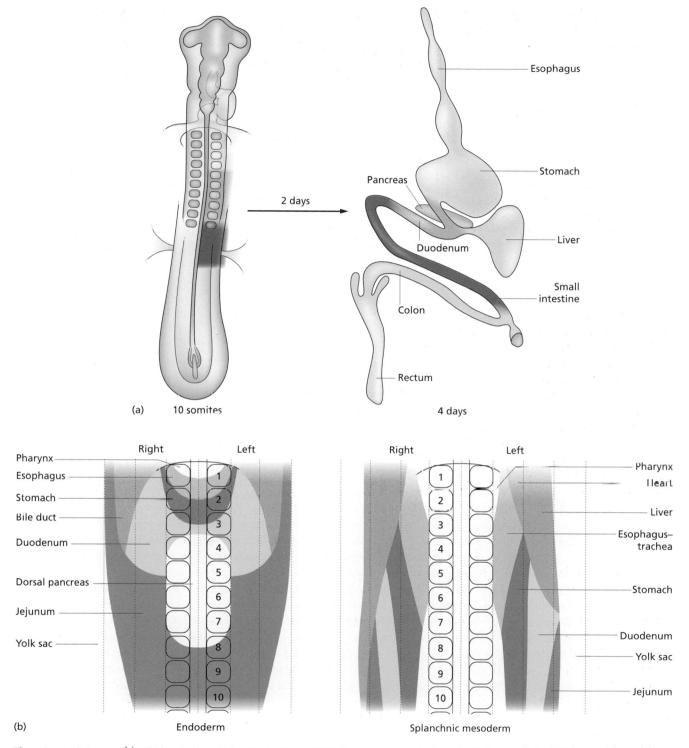

Fig. 16.6 (a) Fate map of the chick endoderm. (After Grapin-Botton.) (b) Mismatch between the fates of endoderm and splanchnic mesoderm at 1.5 days. (After Matsushita.)

across the mesenchyme and controls its differentiation into the various layers of the mature gut: lamina propria, muscularis mucosa, submucosa, and smooth muscle. Application of Shh from secreting tissue culture cells or via viral infection will induces this pattern locally, while the inhibitor of Shh, cyclopamine, will block its development.

Apart from this Shh signal from the endoderm attention has mostly been focused on signals going in the opposite direction,

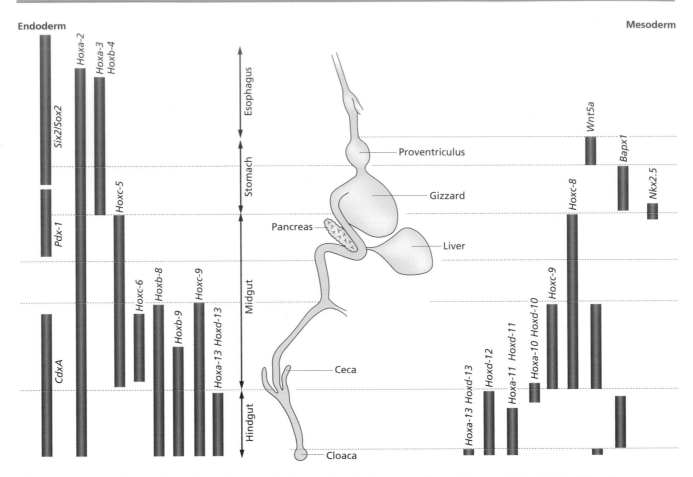

Fig. 16.7 Expression domains of transcription factor genes in the endodermal and mesodermal layers of the early chick gut. (After Roberts.)

from the mesoderm to the endoderm. It is generally found that mesenchyme provides a **permissive** signal that is required for maturation and differentiation of endodermal explants. In some circumstances it is found that when mesenchyme and epithelium from different regions of the chick embryo gut are recombined, the differentiation of the endoderm can be reprogrammed to that corresponding to the position of origin of the mesoderm. These experiments can be done by grafting the recombinants to a neutral site *in ovo* or by allowing them to differentiate in organ culture (Fig. 16.8). The degree of reprogramming can appear exaggerated if just the overall tissue architecture is scored, as this may be affected in a temporary and nonspecific way by the extracellular matrix associated with the mesenchyme, but if the cell differentiation and tissue architecture are considered together then the following results are found. From an early stage the chick embryo gut exhibits a "posterior dominance," meaning that anterior endoderm can be posteriorized by contact with posterior mesoderm, but posterior endoderm cannot be reprogrammed by anterior mesoderm (Fig. 16.9). In particular the future intestinal region, expressing CdxA, appears fixed in its commitment from before 1.5 days of incubation, while the prospective esophagus and stomach regions remain competent to form intestinal epithelium up to

about 4.5 days of incubation. Expression of CdxA can be provoked by treatment with several growth factors including activin A, FGF4, BMP7, and retinoic acid, but mesenchyme is also required in these experiments, so the effects are not direct but operate through some as yet unknown mesenchymal signal.

The chick stomach, unlike that of the mammal, is divided into two quite different parts. The anterior **proventriculus** has a glandular epithelial lining similar to the mammalian stomach, while the posterior **gizzard** has a **squamous** epithelial lining and a thick muscular wall, designed for churning and macerating the food. These two areas of epithelium remain interconvertable until about 9 days of incubation if they are cultured with mesenchyme from the other. The mesenchyme of the proventriculus expresses several BMPs while that of the esophagus and gizzard does not, and there is good evidence of the importance of BMP as the inductive signal. Treatment of either esophagus or gizzard epithelium with BMP will cause them to differentiate proventriculus-like glands, and this effect is inhibited by noggin. The transcription factor Nkx2.3 (homeodomain = Bapx1) is normally expressed in the gizzard mesenchyme. It prevents Shh from the endoderm from activating BMP synthesis in the mesenchyme. Introduction of this factor into the proventriculus mesenchyme will cause the adjacent epithelium to develop as

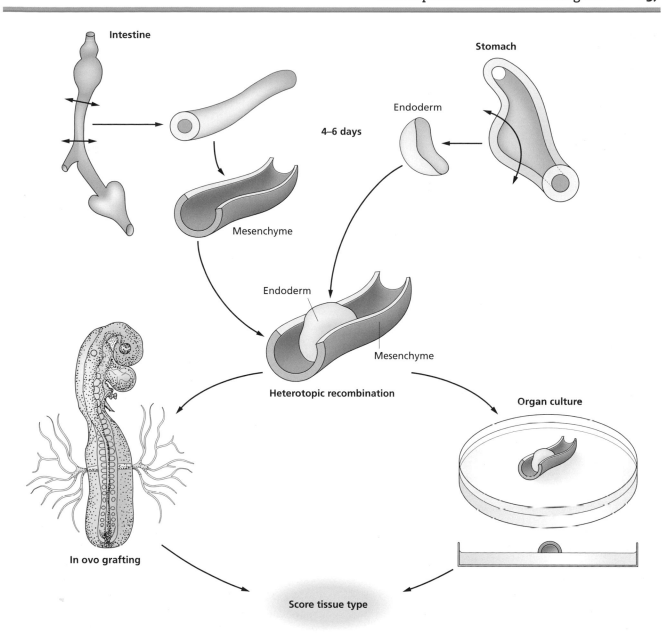

Fig. 16.8 Protocols for recombination experiments between endoderm and mesenchyme of the chick embryo gut.

gizzard rather than proventriculus. At the distal end of the gizzard the BMP signaling causes a ring of mesenchyme to express Nkx2.5 (the same transcription factor that controls heart development), and this controls formation of the pyloric sphincter.

Another example that is becoming understood is the formation of the liver and its distinction from the ventral pancreas. The liver arises from a ventral diverticulum of the foregut that proliferates into the ventral mesenchyme, this mesenchyme being called the **septum transversum** in mammalian embryos. But the liver diverticulum will not form unless it receives a signal from the adjacent cardiogenic mesoderm. This signal is FGF (Fig. 16.10). The cardiogenic mesoderm makes FGFs, FGF

inhibitors will block the signal, and FGFs will mimic the effect on isolated hepatic endoderm. Although FGFs are probably also involved in the early regionalization of the endoderm as posteriorizing factors, this interaction occurs at the primitive streak stage which is very much earlier than the local interaction involved in the formation of the liver, so it is possible to use the same signal again for a different purpose. In addition, the liver-inducing signal probably includes BMP and also something from the blood vessels. The evidence for this is that in recombination experiments normal hepatic endoderm will not develop well if the mesenchyme comes from an embryo lacking BMP4 or blood vessels (*flk1* knockout embryo). Once the liver bud is formed its later growth requires Wnt activity, as shown by

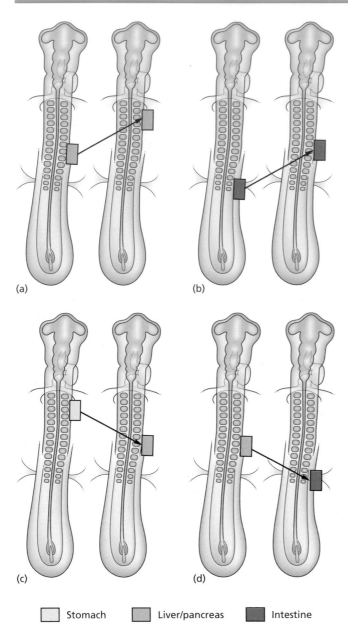

Stomach ▢ Liver/pancreas ▨ Intestine ▤

Fig. 16.9 *In ovo* grafting experiments showing posterior dominance. The colors indicate the structures as shown in Fig. 16.6(a).

the fact that viral overexpression of β-catenin in the chick liver will promote growth, and overexpression of Wnt inhibitors will reduce growth. Doubtless many other factors besides the Wnts are also involved in liver outgrowth.

FGF10 is known to be involved in the outgrowth of the lung buds and the cecum of the intestine. The *fgf10* and *fgfr2b* knock-outs both lack lungs and cecum as well as limb buds (see Chapter 15). The **branching morphogenesis** of the developing lungs seems to involve a lateral inhibition-type system whereby new tips produce FGF10 and suppress the formation of other tips in their immediate neighborhood.

The pancreas

The development of the pancreas has attracted substantial research interest in recent years because of the considerable importance of diabetes and of pancreatic cancer for human health. The mature pancreas (Fig. 16.11) is an organ with two quite distinct physiological functions. One is the synthesis of digestive enzymes. These are made in the **exocrine** cells which make up most of the mass of the pancreas. They are bunched into **acini** (singular: **acinus**) and secrete their products into a duct system that runs into the duodenum. The second is the synthesis of hormones that are released into the blood stream. These are made by **endocrine** cells which are mostly grouped into the islets of Langerhans, although some may be found isolated or in small clusters. The majority of the islet cells are β-**cells** which secrete insulin. In addition are smaller numbers of α-cells that secrete glucagon, δ-cells that secrete somatostatin, PP cells that secrete pancreatic polypeptide, and ε-cells that secrete ghrelin. In the embryo the proportion of endocrine cells is about 10% but it falls to 1–2% in the adult pancreas.

All the epithelial cell types of the pancreas arise from the endoderm (see below), while the blood vessels and the rather small amount of pancreatic connective tissue come from the mesenchyme. The pancreatic mesenchyme also forms the spleen. Development of the pancreas is fairly similar in all vertebrates (Fig. 16.12). It originates from the duodenum as a large dorsal bud and a smaller ventral bud (paired ventral buds in the chick). These buds expand and move towards each other until they fuse together. Each of the original buds has a duct and in humans the duct of the ventral bud becomes the principal duct and that of the dorsal bud an accessory duct. In the zebrafish the dorsal bud forms mostly endocrine cells and the ventral bud mostly exocrine cells.

Most experimental work has been done in the mouse, with some in the chick and a little in *Xenopus* and zebrafish. In the mouse the dorsal bud appears at 9.5 days and the ventral bud about 2 days later. Cells containing endocrine hormones are visible in the early buds and often make more than one hormone, unlike the mature cells. But it is now known that they are not precursors of the mature endocrine cells (see below). Growth of the buds is rapid in the period 14.5–18.5 days, during which there is a rapid increase of endocrine cell numbers and differentiation of much of the rest of the pancreas into exocrine cells.

Induction of pancreatic buds

Although both the dorsal and the ventral buds give rise to the same range of pancreatic cell types, they seem to arise in different ways. The dorsal bud appears in a region where the notochord contacts the gut roof. In this region the expression of Shh and Ihh, which is all over the rest of the endoderm, becomes suppressed. The effect of the notochord can be mimicked by administration of activin or FGF so these may be the inducing

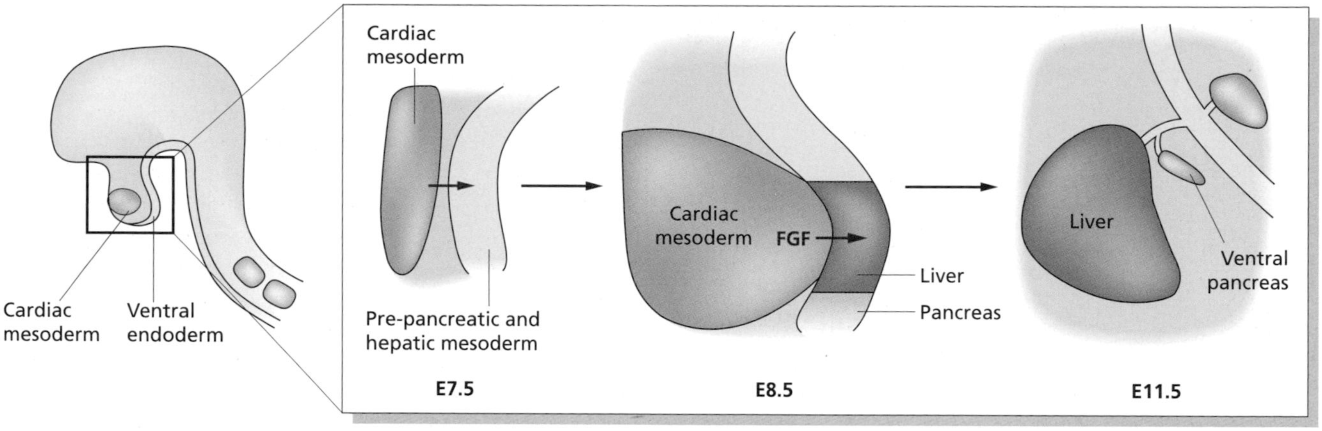

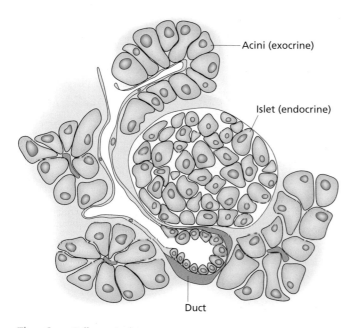

Fig. 16.10 Induction of the liver in the mouse embryo by FGF from the cardiac mesoderm.

Fig. 16.11 Cell types in the mature pancreas.

factors responsible. *Activin receptor* knockout mice have small pancreases and fewer endocrine cells, although have many other defects to the gut as well. Application of the Shh inhibitor cyclopamine to the endoderm will provoke the formation of ectopic pancreatic buds from the zone expressing Pdx1. However, neither contact with the notochord nor application of cyclopamine can induce ectopic pancreas from more posterior endoderm, and so the role of the notochord is considered **permissive** rather than **instructive**.

The ventral pancreas is formed from the adjacent region of the foregut floor to the liver and in the absence of the FGF signal the liver prospective region will also become ventral pancreas. The FGF seems to have the function of maintaining Shh expression in the endoderm that would otherwise be turned off. Normally the region that forms the ventral pancreas is too far

from the cardiac mesoderm to receive the FGF signal. However in mouse knockouts of *hex* (encoding a homeodomain transcription factor) the ventral pancreas is missing. This is not because the hex factor codes for pancreas; it is because hex is needed for the cell growth and movement that normally extends the ventral foregut, such that the region forming the ventral pancreas is not exposed to FGF. So it appears that the dorsal bud may be induced by FGF wheras the ventral bud develops because of an absence of FGF, although the common factor for the formation of both buds is suppression of Shh expression in the endoderm.

Experiments in both zebrafish and *Xenopus* indicate that retinoic acid is also needed at an early stage for formation of the pancreatic buds. The inhibitor BMS453 will prevent the buds forming and exogenous retinoic acid will promote endocrine rather than exocrine development.

Once the buds are formed, their continued outgrowth and exocrine differentiation depend on close proximity of the pancreatic mesenchyme. Again this is a permissive effect and mesenchyme from other parts of the gut can be successfully substituted. The lim domain transcription factor Islet-1 is expressed in the pancreatic mesenchyme and in the knockout mouse for *islet-1* the dorsal pancreas does not grow. But if the endoderm from the knockout is cultured with mesenchyme from a normal mouse then the bud will grow normally. The factors from the mesenchyme probably include FGFs. FGF10 is expressed in the mesenchyme; FGFs will stimulate proliferation of isolated pancreatic endoderm; and the knockout of *fgf10*, or the gene for its receptor *fgfr2b*, show reduced pancreatic growth. Although the mesenchymal factors are permissive they can also affect the proportions of the cell types that differentiate from the endoderm. In organ culture experiments treatment with TGFβ tends to produce more endocrine cells, and treatment with follistatin, an inhibitor of activin, tends to suppress endocrine differentiation.

Endocrine development is also favored by some signal from blood vessels. One environmental feature that the pancreatic

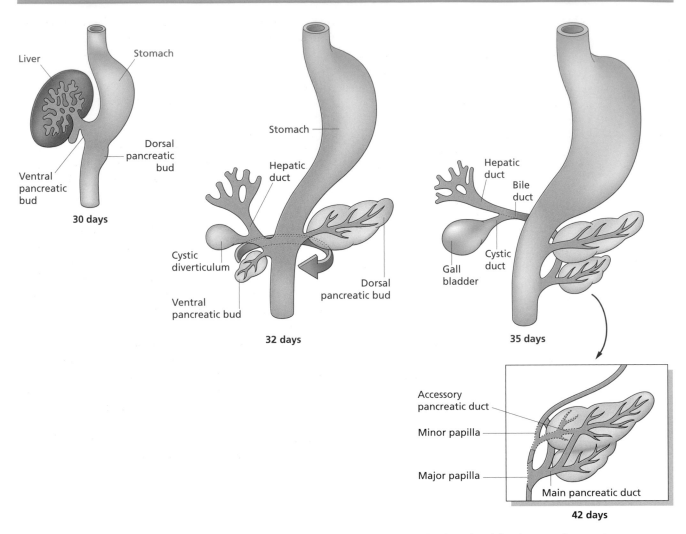

Fig. 16.12 Pancreatic bud development in the human embryo. (After Larsen 1993. *Human Embryology*. Churchill Livingstone, figure 9.6.)

New Directions in Research

Research on pancreatic development is dominated by the search for a method of creating and growing human β-cells. This is because diabetes, in which β-cells are either destroyed or insufficiently active, is such a massive international health problem. Insulin therapy is very effective but many diabetics still suffer serious and life-threatening complications from their disease. Islet cell transplantation has become a successful method for treating some types of diabetes but at present the only source of human islets is cadaveric organ donors, and the supply of these will always be hopelessly inadequate. So it is hoped to use an understanding of β-cell development to design new methods to make β-cells for transplantation.

Somewhat similar arguments apply to the liver, as the supply of human livers for transplantation is way short of demand. Although the liver will regenerate following damage, it is not possible to cultivate human hepatocytes *in vitro* as they rapidly lose their differentiated properties. Perhaps if an endodermal stem cell could be cultivated *in vitro* then large quantities of hepatocytes for transplantation could be produced.

buds have in common is the proximity of a major blood vessel: the dorsal aorta near the dorsal bud and vitelline veins near the ventral bud. Combinations of pancreatic buds with blood vessels can increase the proportion of endocrine cells and removal of the vessels reduces the proportion of endocrine cells. However, in the zebrafish, removal of all blood vessels in the *cloche* (*vegfr2*) mutant does not seem to affect pancreatic endocrine cell development.

Pancreatic transcription factors

Pdx1 (lim-homeodomain) is expressed in a region of the duodenum including the prospective territories of the pancreatic buds and it is essential for formation of the pancreas. The knockout mouse forms just rudimentary buds which do not grow or differentiate. Pdx1 is expressed in the whole of the early buds and is then downregulated to a low level with persistent high expression in β-cells where it serves as a transcription factor for *insulin* itself. A postnatal knockout of Pdx1, using the a tetracycline-regulated gene in place of the normal one, shows loss of insulin from the β-cells. Despite its apparent status as "master control gene" for the pancreas, Pdx1 does not cause the formation of pancreas from other parts of the gut when misexpressed, although an activated **domain swap** version, Pdx1-VP16, is capable of reprogramming the developing liver to pancreas in *Xenopus*.

Hb9 (encoded by *Hlxb9*, homeodomain) is expressed in a wider region than Pdx1 and at a slightly earlier stage. The knockout mouse has no dorsal pancreas and a ventral bud that is normal, except that β-cells do not mature properly, since Hb9 is also needed in differentiated β-cells.

PTF1 is a three subunit transcription factor required to activate various exocrine genes. It is expressed in the pancreatic buds themselves. When the p48 subunit is knocked out this prevents formation of the ventral bud. The dorsal bud forms but develops poorly, lacking exocrine tissue.

Pax4 and Pax6 (paired box) are both needed for endocrine cell development. The *pax4* knockout mouse lacks β- and δ-cells, while the *pax6* knockout lacks α-cells. The homeodomain factors Nkx6.1 and Nkx2.2 are also required for β-cell development but at later stages.

Hox11 (homeodomain, but not part of the Hox clusters despite its name) is expressed in pancreatic mesenchyme and the knockout mouse has a normal pancreas but completely lacks the spleen.

Just as lateral inhibition mechanisms control cell differentiation in neurogenesis and in the differentiation of the intestinal epithelium, so they do also in the pancreas. Early-stage endocrine cells make Delta, and repress endocrine differentiation by their neighbors by stimulating the Notch pathway. The key transcription factor promoting endocrine development is neurogenin 3 (bHLH transcription factor). This is expressed in scattered cells of the pancreatic buds that are not currently making the hormones, with maximal level at 13.5–15.5 days. The mouse knockout of *ngn3* loses all endocrine cells. Overexpression of *ngn3* in transgenic mice using the *Pdx1* promoter will drive a high proportion of pancreatic cells to become endocrine, mostly α-cells. Knockouts of components of the Notch pathway, including *delta like 1, RBPJκ*, and *hes1*, all have a similar effect to the overexpression of *ngn3* and increase the proportion of endocrine cells. Neurogenin 3 seems to drive mostly the formation of α-cells and it is not known what normally controls the relative proportions of the various different endocrine cell types.

The course of pancreatic development deduced from both the embryological experiments and the mouse knockouts is shown in Fig. 16.13.

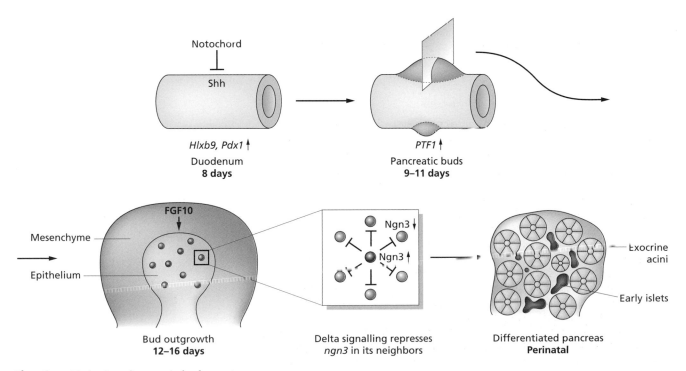

Fig. 16.13 Mechanism of pancreatic development.

Pancreatic cell lineage

The pancreatic endocrine cells are very neuron-like in morphology, behavior, and gene expression patterns. At one time they were thought to be derived from the **neural crest**, but quail to chick neural crest grafts showed that this was not the case. That they are indeed formed from the endoderm may be shown by culturing explants of isolated endoderm *in vitro*, which results in the formation of a few endocrine cells, or by recombining labeled endoderm with mesenchyme in order to get good development in organ culture, and then showing that all the resulting cell types are endoderm-derived. It has also been shown by retroviral labeling that a single endoderm cell can produce both exocrine and endocrine progeny.

To understand the lineage at higher resolution several **Cre-lox** lineage experiments have been carried out on the pancreas (see Chapter 10 for explanation of the Cre-lox system). In this type of application a transgenic line is made in which a promoter of interest is used to drive the *cre*. This is mated to a reporter line, such as R26R (see Chapter 10), and in the offspring this should produce permanent labeling of the cells that have activated the *cre* promoter at any stage in development. When the *pdx1* promoter is used to drive the *cre*, all pancreatic cell types become labeled. If the Cre is fused to the estrogen receptor hormone binding domain (CreER), then its activity can be induced by addition of a suitable **agonist** (usually the drug tamoxifen). This enables particular time windows to be examined as cells will only be labeled if they are both making the CreER and if tamoxifen is added. Exocrine and endocrine cells are labeled if the CreER is induced from 8.5 days of development onwards. But ducts are only labeled in the interval 9.5–11.5 days suggesting some

special mode of formation of this lineage at this stage. When the *neurogenin 3* promoter is used to drive the *cre*, then all endocrine cells are labeled although the expression of the *ngn3* itself is only transient. So these results confirm that Pdx1 is needed for pancreas formation and Ngn3 is needed for endocrine cell formation. When the *p48* promoter is used to drive the *cre*, it is found that all pancreatic cells are labeled, not just the exocrine cells. So this factor may have an early function in pancreas specification as well as the function controlling exocrine differentiation.

Because of the existence in the early buds of multihormone cells expressing two or three hormones simultaneously, it was considered that these might be precursors for all the later endocrine cells. But when the *insulin* promoter is used to drive *cre* the later formed α-cells are not labeled, showing that they had not expressed *insulin* in the past. Likewise when the *glucagon* promoter was used, the β-cells were not labeled, showing that they had not expressed glucagon in the past. However β-cells are labeled if the *pancreatic polypeptide* promoter is used, so cells expressing this hormone may indeed be precursors.

Another issue that has been addressed by Cre labeling is whether β-cells are formed from undifferentiated stem cells during postnatal life or only from pre-existing β-cells. When the *insulin* promoter is used to drive CreER, and the activity of the Cre is induced after birth, then it is found that it is possible to label virtually all the β-cells. This label is then retained indefinitely, showing that there is no recruitment of new β-cells from ducts or from other cells in the islets that were not themselves making insulin at the time of labeling.

A consensus pancreatic cell lineage based on these experiments is shown in Fig. 16.14.

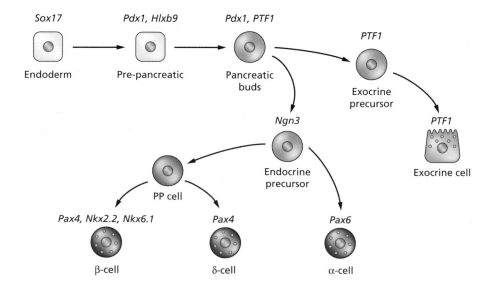

Fig. 16.14 Pancreatic cell lineage.

Key Points to Remember

• The endoderm of amniotes starts as a flat sheet of cells below the ectoderm and mesodermal layers. The gut tube is formed through the folding of the blastoderm to create a headfold and tailfold, which respectively contain pockets of fore- and hindgut. Subsequently the anterior and posterior intestinal portals move towards each other until they join. The gut tube and ventral body wall are closed by lateral body folds which meet each other on the ventral side of the embryo.

• The basic pattern of the gut is the same in all vertebrates with a pharynx, esophagus, stomach, liver, pancreas, small and large intestine. Each is lined with a characteristic epithelium derived from the endoderm.

• In the fate map the endoderm and mesenchyme of each part of the gut come from different positions in the embryo. They move together in the course of gut closure.

• Nodal signaling is necessary to activate expression of the early endodermal transcription factors which include Sox17 and mixer.

• Regional specification depends on inductive signals from the mesoderm. The posterior is determined early and the anterior later in development. Formation of the liver requires an FGF signal from the cardiac mesoderm.

• The pancreas consists of exocrine cells that secrete digestive enzymes into the intestine, and endocrine cells that secrete hormone, including insulin, into the bloodstream. Both cell types arise from the endoderm.

• The pancreas originates as dorsal and ventral buds from the duodenum. Suppression of Shh expression in the endoderm is necessary for bud formation and FGF is necessary for continued growth.

• Pdx1 is essential for the development of all pancreatic cell types. Neurogenin 3 is essential for the development of endocrine cells. Endocrine precursors expressing neurogenin 3 suppress endocrine development of their neighbors by Delta-Notch signaling.

Further reading

General endoderm

Yasugi, S. (1993) Role of epithelial–mesenchymal interactions in differentiation of epithelium of vertebrate digestive organs. *Development Growth and Differentiation* **35**, 1–9.

Wells, J.M. & Melton, D.A. (1999) Vertebrate endoderm development. *Annual Reviews of Cell and Developmental Biology* **15**, 393–410.

Grapin-Botton, A. & Melton, D.A. (2000) Endoderm development from patterning to organogenesis. *Trends in Genetics* **16**, 124–130.

Roberts, D.J. (2002) Molecular mechanisms of development of the gastrointestinal tract. *Developmental Dynamics* **219**, 109–120.

Shivasani, R.A. (2002) Molecular regulation of vertebrate early endoderm development. *Developmental Biology* **249**, 191–203.

Ober, E.A., Field, H.A. & Stanier, D.Y.R. (2003) From endoderm formation to liver and pancreas development in the zebrafish. *Mechanisms of Development* **120**, 5–18.

Liver and lung

Hogan, B.L.M. (1999) Morphogenesis. *Cell* **96**, 225–233.

Warburton, D., Schwartz, M., Tefft, D., et al. (2000) The molecular basis of lung morphogenesis. *Mechanisms of Development* **92**, 55–81.

Zaret, K.S. (2000) Liver specification and early morphogenesis. *Mechanisms of Development* **92**, 83–88.

Cardoso, W.V. (2001) Molecular regulation of lung development. *Annual Reviews of Physiology* **63**, 471–494.

Duncan, S.A. (2003) Mechanisms controlling early development of the liver. *Mechanisms of Development* **120**, 19–33.

Pancreas

Herrera, P.L. (2002) Defining the cell lineages of the islets of Langerhans using transgenic mice. *International Journal of Developmental Biology* **46**, 97–103.

Gu, G., Brown, J.R. & Melton, D.A. (2003) Direct lineage tracing reveals the ontogeny of pancreatic cell fates during mouse embryogenesis. *Mechanisms of Development* **120**, 35–43.

Murtaugh, L.C. & Melton, D.A. (2003) Genes, signals and lineages in pancreas development. *Annual Reviews of Cell and Developmental Biology* **19**, 71–89.

Hebrok, M. (2003) Hedgehog signaling in pancreas development. *Mechanisms of Development* **120**, 45–57.

Lammert, E., Cleaver, O. & Melton D.A. (2003) Role of endothelial cells in early pancreas and liver development. *Mechanisms of Development* **120**, 59–64.

Chapter 17

Drosophila *imaginal discs*

Metamorphosis

Drosophila belongs to one of the families of insects (Diptera) that undergo a complete **metamorphosis** (Fig. 17.1). After hatching, the larva passes through three **instars**, separated by molts. During this period it feeds voraciously and grows substantially in size, but without much change in body form. The third-instar larva becomes a **pupa** and during pupation most of the larval body is resorbed and replaced by adult structures derived from the **imaginal discs**. The contributions are shown in Table 17.1.

Imaginal discs are infoldings of the epidermis (Fig. 17.2). They are one cell thick, and are covered with a peripodial membrane. Associated with them are mesoderm-derived adepithelial cells which later differentiate into muscle cells. During larval life the discs grow, and each has a distinctive shape, but they do not become visibly differentiated. During pupation they evaginate to produce the appropriate appendage, and differentiate to give

Table 17.1 Origin of adult body parts from imaginal discs and histoblasts.

Disc/histoblast	Body part
Clypeolabral disc	Some mouthparts
Eye–antenna disc	Eyes, antennae, rest of head
Labial disc	Proboscis
Dorsal prothoracic disc	Dorsal prothorax
Prothoracic leg disc	First leg and ventral prothorax
Mesothoracic leg disc	Second leg and ventral mesothorax
Metathoracic leg disc	Third leg and ventral metathorax
Wing disc	Wing and dorsal mesothorax
Haltere disc	Haltere and dorsal metathorax
Dorsal abdominal histoblasts	Tergites (dorsal abdominal cuticle)
Ventral abdominal histoblasts	Sternites (ventral abdominal cuticle)
Genital disc*	Genital apparatus

*There is a single genital disc, but the others are paired on right and left sides.

the cuticle with its characteristic appendage-specific pattern of hairs and bristles. Unlike the situation in vertebrates, imaginal discs produce their own sensory neurons whose axons grow inwards to connect with the central nervous system.

Molting in insects is controlled by a surge of the steroid hormone ecdysone (of which the active form is 20-hydroxyecdysone). Neurosecretory cells in the brain release a peptide hormone called prothoracicotropic hormone (PTTH). This stimulates the prothoracic gland to secrete ecdysone. A pair of glands called the corpora allata, attached to the brain, secrete the terpenoid juvenile hormone (JH). When JH is high, a rise in ecdysone causes a larval molt; when JH is low, ecdysone initiates metamorphosis. The ecdysone receptor is a typical nuclear hormone receptor (see Appendix) and among its targets is its own gene, so following stimulation the responding tissues become still more responsive to the hormone.

The growth of discs can be prolonged beyond the normal duration of larval life by implanting them into the abdomen of an adult fly. If they are transplanted repeatedly they will grow indefinitely without differentiating. On the other hand, discs or disc fragments can be caused to differentiate by implantation into a larva about to metamorphose (Fig. 17.3), or by culture *in vitro* in the presence of ecdysone.

Genetic study of larval development

Many of the genes affecting the development of the imaginal discs are also important in embryonic development. Therefore null mutants often die as embryos and never display a disc-specific phenotype. Mutations allowing survival to adulthood are usually **hypomorphic** rather than **null**. They are often **temperature sensitive**, meaning that the protein product becomes inactivated when the organism is moved to the nonpermissive temperature. This makes it possible to find the time of action of the gene during larval life or metamorphosis. Groups of larvae are subjected to pulses of the nonpermissive temperature at different times. The temperature pulse will inactivate the gene

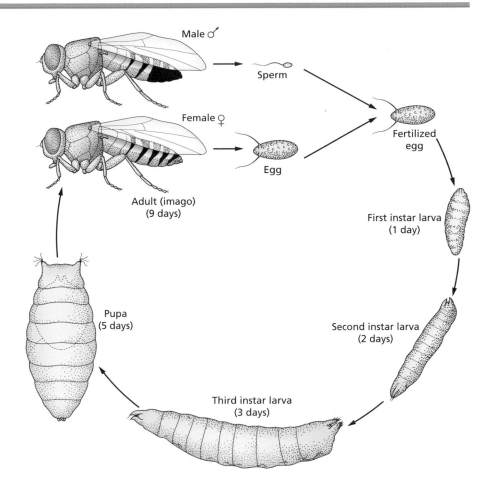

Fig. 17.1 Life cycle of *Drosophila*.

product and if the gene is required at that time then the mutant phenotype will develop.

Overexpression experiments in *Drosophila* are often done by introducing a transgene driven by a heat-shock promoter. Heat-shock proteins exist in all cells and function to assist in correct folding of other proteins. Their expression is greatly increased at elevated temperature, and this is a function of the promoter. When a heat-shock promoter is joined to another gene then this gene can be activated by raising the temperature. This technique should not, of course, be confused with the use of temperature-sensitive mutants.

Developmental genes often have several **enhancers**, each activated in a different expression domain. It is possible to take just one enhancer, link it to a **reporter gene**, such as *lacZ*, and make a transgenic fly line with this construct. X-Gal staining (see Chapter 5) of embryos, larvae, or discs will then reveal the activation domain of this particular enhancer, usually representing one of several expression domains of the normal gene. The use of such reporters is both more specific and more sensitive than looking at the messenger RNA of the endogenous gene by *in situ* hybridization.

Enhancer trap lines (see also Chapter 10) are transgenic flies in which a reporter gene is expressed with a particular spatial pattern that reflects the activity of an endogenous enhancer, but in this case one identified by random insertion. They are made by injecting a construct in which the reporter is under the control of a weak promoter. The promoter is too weak to drive expression unless the construct happens to integrate in the genome near to an enhancer sequence, in which case it is expressed with a pattern characteristic of the enhancer. Using this method, numerous lines have been created having particular patterns of *lacZ* expression, many resembling expression patterns of known genes but also many with novel patterns. They can be useful experimentally because particular cell populations may be highlighted and easy to observe simply using X-Gal staining.

A variation of the enhancer trap method is used to drive **ectopic** expression of any gene of interest. This is based on the GAL4 regulatory system from yeast. GAL4 is a transcriptional activator of the zinc-finger class and can drive expression of any gene sequence cloned downstream of the "upstream activating sequence" (*UAS*) to which it binds. Enhancer trap lines are made in which the gene introduced is not *lacZ*, but *GAL4*. Another line of transgenic flies is made in which the gene of interest is linked to *UAS*. When individuals of the two lines are crossed together, the GAL4 is expressed under the control of its enhancer and it will activate the target gene via the *UAS*. The spatial pattern of expression of the target gene will depend on the enhancer used

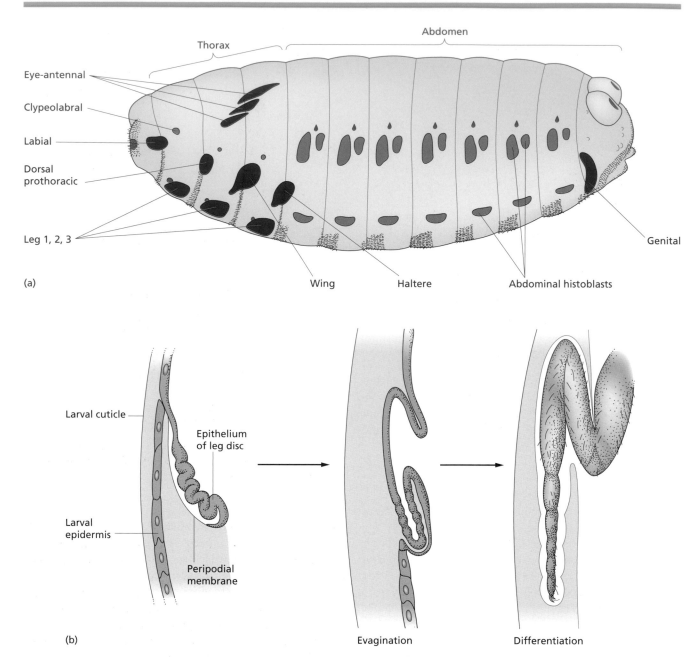

Fig. 17.2 Imaginal discs: (a) position in the late embryo; (b) eversion of a leg disc during pupation. ((a) After Hartenstein 1993. *Atlas of Drosophila Development*. Cold Spring Harbor Laboratory Press, p. 25)

to drive the *GAL4* (Fig. 17.4). Along with "FLP-out" (see below), this is now the most common way of bringing about regionally specific ectopic expression.

Mitotic recombination

Although recombination normally occurs at meiosis, it also possible to force recombination between homologous chromosomes at mitosis. If recombination is induced in a cell heterozygous for a mutation (+/−) then it can generate a pair of cells of which one is +/+ and the other is −/− (Fig. 17.5). These grow to form two adjacent **clones**, called a twin spot. The +/+ clone is not usually visible but the ability to generate the mutant −/− clones is important for at least four purposes, all of which will be discussed below:

1 The **clonal analysis** of determination in imaginal discs. See Chapter 4 for principle of clonal analysis.

2 Examination of the **autonomy** of mutations. If the mutation affects the surroundings as well as the clone itself then it is said to be nonautonomous. This implies that it affects some intercellular signaling mechanism, either directly or indirectly.

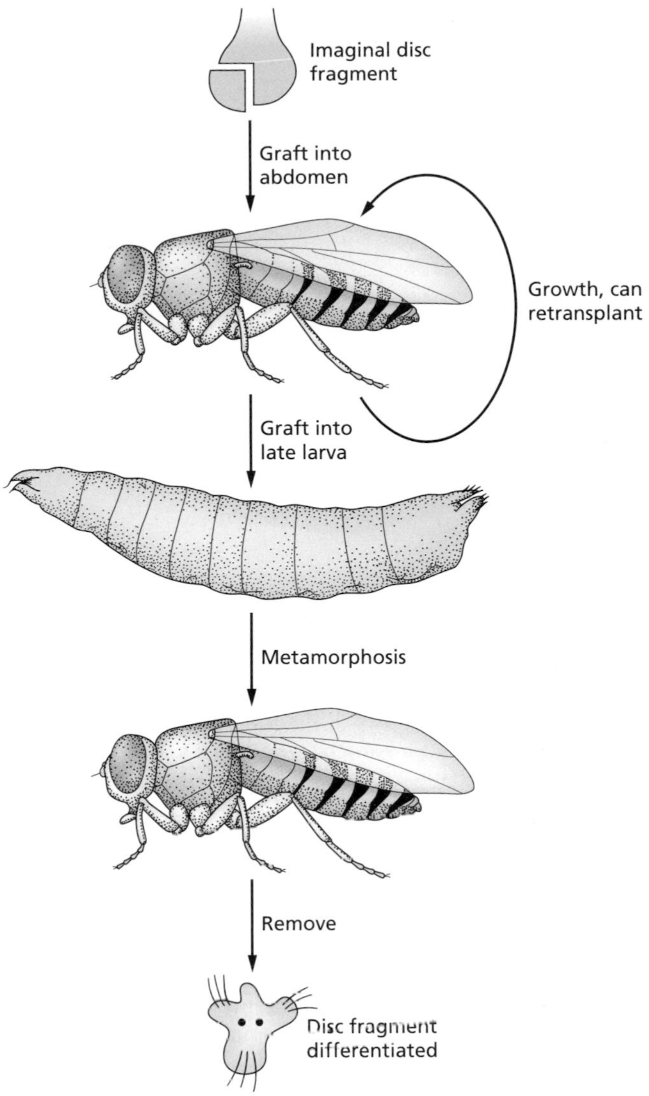

Imaginal disc fragment

↓ Graft into abdomen

↺ Growth, can retransplant

↓ Graft into late larva

↓ Metamorphosis

↓ Remove

Disc fragment differentiated

Fig. 17.3 Experimental methods for propagation of discs in an adult host and differentiation in late larva.

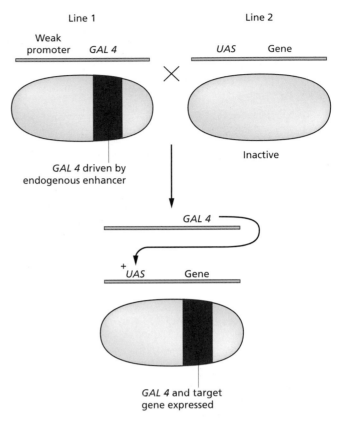

Fig. 17.4 GAL4 method for ectopic overexpression of a transgene. When the two transgenic lines are mated, the offspring will show tissue-specific expression of the target gene.

3 The study of late phenotypes of mutations that are lethal during embryonic development. It is often the case that a mutation that arrests development of the whole embryo is not lethal at the cell level. Therefore a mutant clone in an otherwise normal larva or adult will probably be viable. Such a clone may display morphological defects indicating a late function which could not have been observed otherwise.

4 Localized overexpression of individual genes. This is complementary to the GAL4 method described above. GAL4 generates one specific pattern of overexpression depending on the promoter that is used to drive it, while mitotic recombination can generate a random set of overexpressing clones. The two methods are both heavily used in research on disc development.

If the object of induced mitotic recombination is just to visualize clones in normal development, then the mutation should not perturb development but simply change the appearance of the cells. Mutations originally used for this purpose were cuticular markers such as *yellow* and *multiple wing hairs*, although these have the limitation of only being visible in the adult and not during embryonic or larval stages. More recently a number of transgenic labels, such as green fluorescent protein (*GFP*), have been introduced and are of general utility.

In early experiments mitotic recombination was induced using X-irradiation, which causes chromosome breaks. But because of its low efficiency this has now been superseded by the **FLP** (pronounced "Flip") system. The basic elements are the FLP recombinase, which is a yeast enzyme, and the FRT sites in the target DNA. The FLP recombinase recognizes two FRT sites and catalyzes a recombination event between them. A line of flies carrying the FRT sites is crossed to one carrying the FLP recombinase, driven by a heat-shock promoter. Recombination is initiated by a brief temperature shock, which causes transient transcription of the recombinase, and hence brings about recombination in just a few cells. If there is a marker transgene such as *GFP* or *lacZ* further from the **centromere** than the FRT site, then this will be lost in the recombinants, generating clones that are identifiable by the absence of the marker but are otherwise developmentally normal.

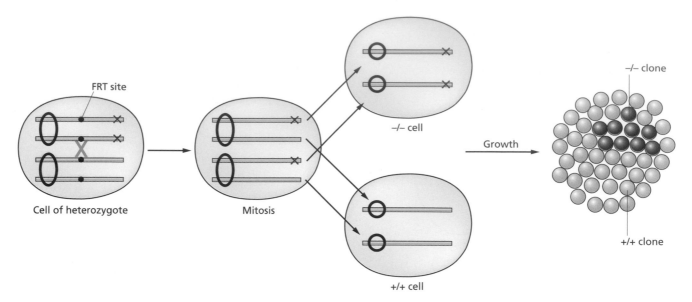

Fig. 17.5 Mitotic recombination generates a twin spot consisting of +/+ and a –/– clone. Position of mutation is indicated by the red cross, and of centromere by the black circle. Recombination takes place at the FRT site, catalyzed by FLP recombinase.

In this way labeled clones can be produced, but the more usual application is to make clones that are not just labeled but in which a particular functional gene is ablated or ectopically expressed. If the intention is to ablate the gene in the clones, then the FRT sites need to lie on the centromeric side of the functional mutation so that recombination generates a –/– clone as in Fig. 17.5. A transgene intended for ectopic expression would contain a constitutive promoter to drive the gene in question and a stop codon early in the coding region causing the product to be truncated and functionless. This stop codon is flanked by two FRT sites. When induced, the recombinase can excise the stop codon by recombination between these adjacent FRT sites, leading to the production of clones in which the full-length gene is expressed (Fig. 17.6). This variant of the method is known as "FLP-out." It is not always necessary to have a separate marker gene for identification of the clone because, if the gene under study is a transgene, an identification sequence can be incorporated into its coding region. For example the "myc tag" is a short peptide that can be recognized by immunostaining with a specific antibody and enables the mutant clone to be visualized.

Classic Experiments

DISCOVERY OF DEVELOPMENTAL COMPARTMENTS

The idea of the developmental compartment has been a basic concept in developmental biology that is by no means confined to *Drosophila*. Compartments were discovered by clonal analysis of imaginal discs and were defined as units of body structure that were produced by the progeny of a small number of cells. Through work on the properties of *engrailed* mutants they also became identified as domains of activity of homeotic genes.

The first paper describes the behavior of labeled clones and the second introduces the *Minute* technique allowing production of

larger and more informative clones. The third paper is actually a review but is unusual because of the messianic certainty of the authors that the work of the Spanish school was so important that it just had to receive wider publicity.

Garcia-Bellido, A., Ripoll, P. & Morata, G. (1973) Developmental compartmentalization of the wing disk of *Drosophila*. *Nature New Biology* **245**, 251–253.

Morata, G. & Ripoll, P. (1975) *Minutes*: mutants of *Drosophila* autonomously affecting cell division rate. *Developmental Biology* **42**, 211–221.

Crick, F.H.C & Lawrence P.A. (1975) Compartments and polyclones in insect development. *Science* **189**, 340–347.

Disc development

Origin of discs

The first sign of the thoracic imaginal discs is at about 5 hours of embryonic development when the germ band is maximally extended. They are not visible by ordinary microscopy but can be visualized in embryos transgenic for *lacZ* driven by an early enhancer of the gene *distalless*. *distalless* codes for a homeodomain transcription factor and is often expressed in appendages. Early disc development, and activation of the *distalless* enhancer, is dependent on the action of the *wingless* gene. As discussed in Chapter 11, *wingless* is required for embryonic segmentation,

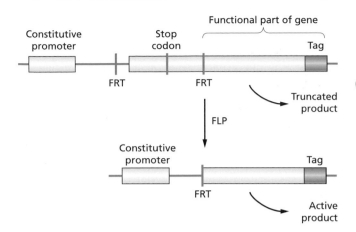

Fig. 17.6 Activation of a transgene in a clone produced by the FLP-out method.

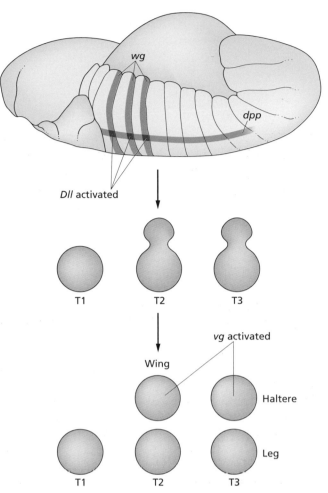

Fig. 17.7 Formation of the thoracic imaginal discs in the embryo.

and null mutants are lethal. However, there are also viable temperature-sensitive mutations of *wingless*, and use of these shows several sensitive periods for wing development. The first of these is just before 5 hours of development, when *wingless* is required to turn on the *distalless* enhancer.

In the embryo, *wingless* is initially expressed in circumferential stripes at the posterior margin of each parasegment. Another gene required for disc formation is *decapentaplegic (dpp)*. Early on, *dpp* is expressed in the dorsal 40% of the embryo, but later it resolves into several longitudinal bands. The three thoracic discs arise at the intersections of the circumferential bands of *wingless* with a longitudinal band of *dpp* expression (Fig. 17.7). During germ-band retraction a group of cells separates from two of the disc rudiments and moves dorsally. By 10 hours these dorsal clumps express *vestigial* which codes for a transcription factor of a novel class. These dorsal cell clusters are the wing and haltere discs while the three ventral clusters are the leg discs. At about this stage, all the imaginal discs and abdominal histoblasts begin to express the *escargot* gene, coding for a zinc-finger transcription factor. At least in the histoblasts, its function is to prevent **polyteny** of the DNA, and hence maintain the cells in a state capable of division.

Compartments and selector genes

The progressive formation and subdivision of the discs can rather crudely be followed by **clonal analysis**. If mitotic recombination is induced at the blastoderm stage, individual clones do not cross segment boundaries nor an invisible anteroposterior boundary running through each of the thoracic imaginal discs. This line of clonal restriction is called the anteroposterior **compartment** boundary and is now known to be the **parasegment** boundary of the early embryo, defined by the *cubitus* and *engrailed* systems (see Chapter 11). This boundary is inherited through cell divisions and maintained in each of the thoracic imaginal discs.

Although confined to a compartment, blastoderm-stage clones in the thorax can cross between wing and leg disc, whereas clones induced in larval life are confined to either wing or leg. This is now known to be because each thoracic disc rudiment originates as a single cell clump that divides into two (Fig. 17.7).

The anteroposterior compartment boundary can be visualized more easily if the labeled clone is very big so that it abuts the boundary over much of its length. This can be achieved by the use of *Minute* mutants which are a class of dominant mutants of ribosomal proteins that grow slowly during larval life because of a defect in protein synthesis. By inducing +/+ clones on a *Minute*/+ background, it is possible to give the clones a growth advantage over the surrounding tissue, with the result that a single clone can fill a large fraction of the final organ (Fig. 17.8). Remarkably this differential growth of clones does not disturb the final size or proportions of the organ. In the case of the wing disc (but not the leg disc), there is also a dorsal–ventral compartment boundary, which becomes established in the second larval instar. In other words, clones induced in the first instar will cross between dorsal and ventral, whereas those induced in late second instar are confined to either dorsal or ventral.

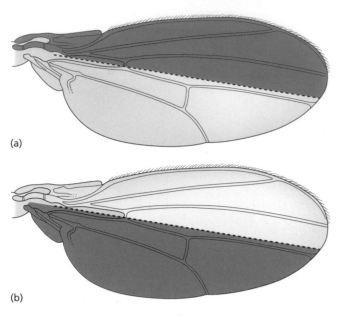

(a)

(b)

Fig. 17.8 Visualization of the anterior and posterior compartments using the *Minute* technique. A normal clone is induced on a *Minute* background, and fills up the whole compartment: (a) anterior compartment; (b) posterior compartment. (After Lawrence 1992. *The Making of a Fly*. Blackwell, figure 4.3.)

Each of the compartmental subdivisions depends on the expression of a particular homeotic gene, often called a **selector gene** in this context. The posterior compartment is defined by expression of the *engrailed* gene. *engrailed* activity defines the anterior stripe of cells in each embryonic parasegment, but since discs arise at the parasegment boundary, this becomes the posterior part of the disc. Null mutations of *engrailed* are embryo lethal, but there are also hypomorphic mutations that are viable and give a partial transformation of posterior to anterior wing. Clones experimentally induced by mitotic recombination may fall, at random, either in the anterior or the posterior compartment. If clones lacking *engrailed* function are induced in the anterior compartment, they remain confined to the anterior and produce a normal anatomy. If such clones are produced in the posterior, they produce anatomical defects, and the affected cells may cross over into the anterior compartment (Fig. 17.9). These experiments show that *engrailed* function defines the character of the posterior compartment, and lack of *engrailed*, by default, defines the anterior. The actual barrier to cell movement is based on a difference of cell adhesion between the compartments due to integins and other components whose expression depends on *engrailed*.

The dorsal compartment of the wing disc is defined by expression of *apterous*, a gene coding for a transcription factor of the Lim class. *apterous* expression starts in the second instar,

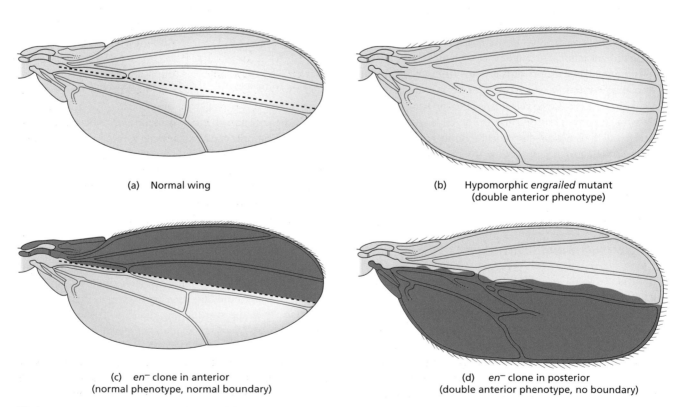

(a)   Normal wing

(b)   Hypomorphic *engrailed* mutant (double anterior phenotype)

(c)   *en⁻* clone in anterior (normal phenotype, normal boundary)

(d)   *en⁻* clone in posterior (double anterior phenotype, no boundary)

Fig. 17.9 Function of the *engrailed* gene. (a,b) A viable but hypomorphic mutant of *engrailed* gives a wing with structure tending toward double anterior. (c,d) Clones lacking *engrailed* respect the compartment boundary on the anterior but not on the posterior side. (After Lawrence 1992. *The Making of a Fly*. Blackwell, figure 4.3.)

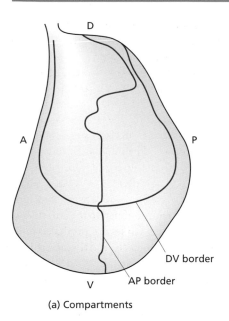
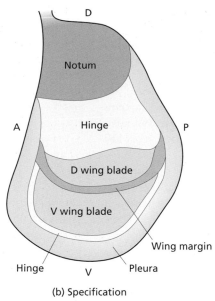

Fig. 17.10 The wing disc. (a) Compartment boundaries. (b) Specification map showing what develops from each part. (c) Regeneration behavior. So long as the "field center" is included, then missing parts are regenerated. Otherwise, existing parts are duplicated.

coincident with the time of dorsoventral clonal restriction, and is found in the dorsal but not the ventral compartment. It is activated by a neuregulin-like signal encoded by the *vein* gene. Neuregulins act through the EGF receptor and loss of function of the *EGFR* gene will prevent the activation of *apterous*. If clones are induced lacking *apterous* function, they develop normally on the ventral side, but on the dorsal side an ectopic wing margin is formed around the boundary of the clone. The cells lacking *apterous* lie within the clone and produce the ventral bristles and the cells around the clone which express *apterous* form the dorsal bristles. Thus, the ectopic dorsal–ventral boundary, like the normal one, arises at the boundary of *apterous* activity. Figure 17.10a shows the anterior–posterior and dorsal–ventral compartment boundaries drawn on the third-instar wing disc.

Regeneration

When fragments of third-instar discs are implanted into metamorphosing larvae, they will differentiate and form all of the structures formed in normal development. Different parts of the disc will produce different structures and this fact has been used to construct "**fate maps**" of the discs, although strictly these should be called **specification** maps (see Chapter 4). The epidermis of the adult *Drosophila* is very rich in various types of specialization such as hairs and bristles that can be identified by the skilled observer, and therefore the maps can have quite high resolution. In the wing disc the dorsal part forms the notum (dorsal cuticle), the ventral part the pleura (cuticle between wing and leg), and the central part the wing hinge and wing blade (Fig. 17.10b).

If a disc fragment is cultured in the abdomen of an adult female fly it will grow and regenerate. If it is then transferred to a metamorphosing larva, the structures that it produces on differentiation will show how the specification map has changed during the growth period. Over about a week in an adult host, a disc fragment will either regenerate the missing part or it will regenerate a copy of what is already present, a type of regeneration known in this context as "duplication."

In the wing disc there is a "field center" in the prospective wing blade region, which represents the zone with ability to regenerate all other parts (Fig. 17.10c). If fragments possess more than half the circumference of the disc they will reform a field center, if less they will not. It is thought that the cut edges of the fragments heal shortly after cutting and that the interactions that occur at the apposed edges determine whether or not a field center will be reconstituted.

The regenerative behavior of discs is still not entirely understood, although the ability to regenerate structures probably requires the inclusion or creation of compartment boundary regions as these have a special significance in determining the pattern of the remainder of the disc (see below).

Transdetermination

Although the disc type (wing, leg, genital) is usually conserved on short-term culture, it can switch on longer-term culture. This is called **transdetermination**, and is an all-or-none conversion of part of the regenerating structure into another disc type. Genetic labeling of the implant shows that the transdetermined structures really do arise from the implants and are not due to contamination by cells from other discs of the larval host. It is thought that transdetermination, unlike homeotic mutation, can occur simultaneously in a group of cells. The evidence for this is a classic set of experiments on the eye–antennal disc in which the discs were transplanted from larvae in which labeled clones had just been initiated. When the regenerated discs were recovered, some had regions of wing tissue formed by trans-determination. In some of these the wing tissue was of mixed (labeled + unlabeled) composition, showing that it must have originated from more than one cell.

Like disc regeneration, the molecular basis of transdeter-mination is still not fully understood, although it is presumed that the underlying cause is a change in expression of a homeotic gene. One candidate is the *vestigial* gene mentioned above, which codes for a transcription factor and is normally required for wing development. Ectopic expression of *vestigial* has been shown to cause wing structures to develop from the leg or haltere discs. This fits in with the homeotic effects of *Ultrabithorax (Ubx)* mutations. Ubx is a member of the Hox cluster and loss-of-function mutations transform haltere to wing, while ectopic expression transforms wing to haltere. The reason is that Ubx suppresses the activation of *vestigial* expression by both the wingless and the Notch pathways (see below), hence the absence of Ubx allows expression of *vestigial* and development of a wing.

Regional patterning of the wing disc

Anteroposterior pattern

The patterning mechanism for several of the imaginal discs in normal development is now reasonably well understood. In the wing disc, the basic model can be briefly summarized as fol-lows: the transcription factor engrailed, expressed in the poster-ior compartment, activates the *hedgehog* gene. The extracellular hedgehog protein stimulates the anterior cells along the antero-posterior (AP) compartment boundary and causes them to activate the *decapentaplegic (dpp)* gene. The dpp protein forms an extracellular gradient with the high concentration along the anterior–posterior boundary and the low concentration at the disc margins. This gradient brings about patterning of both compartments by activation of transcription factors at different concentration thresholds.

The evidence for this model comes from a variety of experi-ments mainly involving the ablation of gene activity in clones, or the ectopic expression of the gene, either in FLP-out clones or by the GAL4 method. The anterior–posterior compartment bound-ary arises in early embryogenesis because *engrailed* is on in the anterior of each parasegment and off in the remainder. The discs form around the parasegment boundaries such that *engrailed* activity defines the posterior compartment. Ectopic expression of *engrailed* in the posterior gives normal clones, but ectopic expression in the anterior gives abnormal clones. These produce

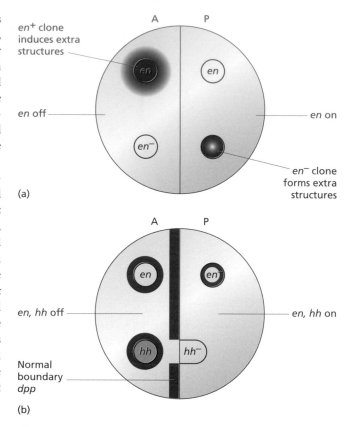

Fig. 17.11 (a) Behavior of clones with altered *engrailed* expression. Clones overexpressing *engrailed* in the anterior compartment induce pattern duplications in their surroundings; clones lacking *engrailed* activity in the posterior compartment themselves form ectopic structures. (b) Effects of clones with altered *engrailed* or *hedgehog* expression on activity of the *dpp* enhancer. Posterior clones lacking *engrailed* activate the *dpp* enhancer but only where they abut surrounding posterior tissue. Anterior clones expressing *engrailed* activate the *dpp* enhancer in the surroundings. Posterior clones lacking *hedgehog* prevent *dpp* activation across the normal boundary. Anterior clones expressing *hedgehog* activate *dpp* both internally and in the surroundings.

pattern duplications in the surrounding tissue (Fig. 17.11a). Since engrailed is a transcription factor, the nonautonomy of these clones shows that a signal must be emitted from *engrailed*-overexpressing cells. If *engrailed* is removed by mitotic recom-bination, the anterior clones are normal but posterior clones show excess proliferation and pattern duplication (Fig. 17.11a).

As in embryogenesis, the signal emitted by the *engrailed*-expressing cells is hedgehog, a signaling molecule with a short diffusion range. Cells of the posterior compartment are not competent to respond to hedgehog but those of the anterior com-partment are. Therefore the effect of the hedgehog signal is to stimulate a band of cells running along the anterior–posterior compartment boundary. As described in Chapter 11, the effect is to activate a Gli-type transcription factor, Cubitus interruptus (Ci). The target genes of Ci include *dpp* and *patched*, which encode the hedgehog receptor. Thus, the response to the hedge-hog signal is for the responding cells to secrete dpp and also to

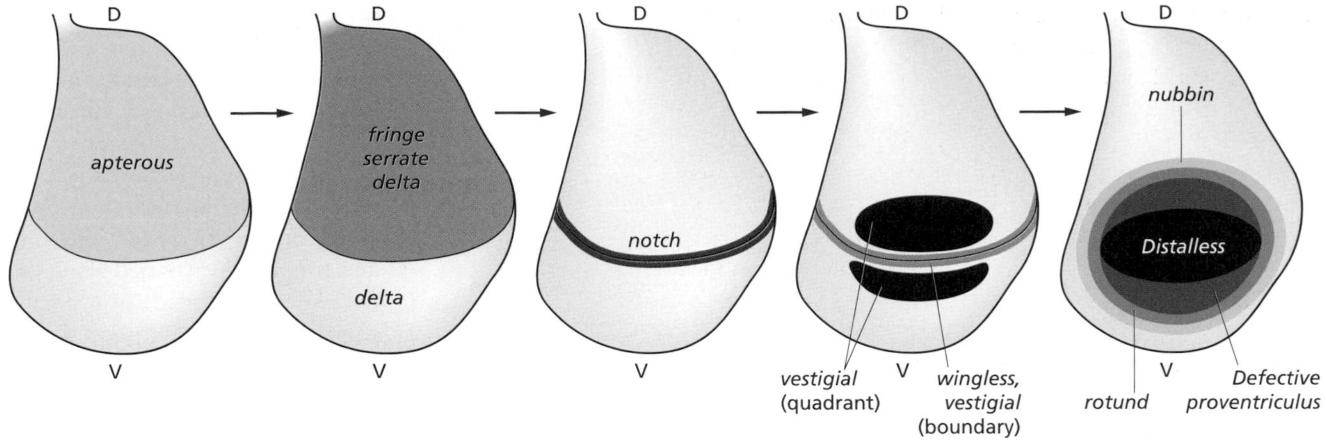

Fig. 17.12 Dorsoventral patterning in the wing disc. *apterous* in the dorsal compartment activates *fringe*. Notch becomes activated at the *fringe* boundary activates its own expression, and turns on *wingless* and *vestigial*. *vestigial* is also turned on by Dpp through the quadrant enhancer. Finally expression of a concentric set of transcription factors become activated that control the differentiation of the proximodistal pattern of the wing.

increase their own sensitivity to hedgehog. Ci activity is antagonized by protein kinase A (PKA), the cAMP-dependent protein kinase (see Appendix). Therefore ablating PKA function has a similar effect to activation of hedgehog signaling, and a number of experiments in this area are carried out using clones in which PKA activity has been removed.

Evidence for the process as described comes from the study of ectopic *hedgehog*-expressing clones and of clones lacking *hedgehog* activity. Clones producing hedgehog are normal if they lie in the posterior. If they lie in the anterior they cause pattern duplications in tissues surrounding the clone, just like *engrailed*-expressing clones. If *hedgehog*-expressing clones are induced in hosts carrying a *lacZ* reporter coupled to the *dpp* promoter, it can be seen that clones in the anterior activate the *dpp* promoter, while those in the posterior do not (Fig. 17.11b). By comparison, *engrailed*-expressing clones in the anterior compartment do not themselves activate *dpp*, but do cause an activation of *dpp* in the surrounding cells (Fig. 17.11b). Clones lacking *hedgehog* function only have a phenotype if they abut the anterior–posterior border. In this case there is no activation of *dpp* in neighboring anterior cells and there are local defects in the resulting pattern (Fig. 17.11b).

The dpp protein forms a gradient from the anterior–posterior border which patterns both anterior and posterior compartments. Evidence that dpp is the principal effector of patterning comes from the study of ectopic expression of *dpp*. In the anterior compartment, this has a similar effect to *hedgehog*, while in the posterior compartment it also induces some pattern duplication and overgrowth.

Dorsoventral pattern

In the wing disc there is an independent signaling system responsible for establishing the regional pattern in the dorsoventral axis. It depends on the stimulation of Notch along the dorsal–ventral compartment border, which leads to activation of a transcription factor gene, *vestigial*, and a signaling factor gene, the familiar *wingless* (Fig. 17.12).

The genes responsible for this process, *wingless*, *notch*, *delta*, *serrate*, and *fringe*, all have broadly similar effects to each other when ectopically expressed, or when ablated from clones. Ectopic expression tends to give ectopic patches of wing margin, while ablation tends to lead to defective wing margin or even complete absence of the wing blade. The wing margin of the adult is of developmental significance because it derives from the junction between the dorsal and ventral compartments in the larval wing disc.

Notch has two ligands, Serrate, which is expressed just in the dorsal compartment, and Delta, which is present all over the disc. However, Notch is mostly stimulated along the dorsal–ventral border, rather than in the remainder of the disc. This is achieved by two properties of the system. Firstly there is a positive feedback built in since Notch signaling activates the expression of genes encoding components of the Notch pathway, in particular *delta*, *serrate*, and *notch* itself. Secondly there is the action of Fringe which is expressed only in the dorsal compartment. Fringe encodes a glycosyl transferase which modifies Notch by addition of sugar residues and makes it more sensitive to Delta and less sensitive to Serrate. Because of the action of Fringe, Notch on the dorsal side is more sensitive to Delta so there is modest activation of Notch all over the dorsal compartment, resulting in production of slightly more Delta and Serrate. But the Serrate has no effect on the dorsal compartment because of the presence of Fringe. It has most effect on the first file of cells in the ventral compartment, which lack Fringe. So this file of cells produces more Serrate and Delta. This then becomes autocatalytic because the localized increase of Serrate and Delta cause increased expression of the genes encoding components of the Notch pathway. The effect is that both gene

New Directions in Research

Like early *Drosophila*, the imaginal discs are potentially useful for research into a variety of cell biology topics because of the powerful genetic methods that are available. Areas of current interest include the study of intracellular trafficking, of planar polarity, and of growth control (see also Chapter 18).

expression and signaling activity of Notch pathway components are upregulated just along the dorsoventral boundary. The evidence for this mechanism was based on the use of the GAL4 system to misexpress Delta and Serrate. It was shown that Delta could drive an increase in *serrate* transcription in the dorsal compartment while Serrate could drive an increase in *delta* transcription in the ventral compartment. Furthermore the key function of the dorsal compartment selector gene *apterous* is shown to be to drive *fringe*. If a transgene which drives *fringe* off the *apterous* promoter is introduced into *apterous* mutants then this will rescue wing formation.

vestigial codes for a transcription factor required for proliferation of wing cells and is normally on in the whole prospective wing blade. *vestigial* has the status of a "master control gene"

because if it is activated in other imaginal discs then it can force the formation of wing blade from these discs. *vestigial* has two enhancers each of which is responsive to one of the signaling systems active in the wing disc. The boundary enhancer is turned on by Notch signaling and is normally activated along the dorsal–ventral border. The quadrant enhancer is activated by dpp signaling and this activates *vestigial* in the remainder of the prospective wing blade.

The end result of the intercompartmental interactions is the activation of *wingless* along the prospective wing margin and of *vestigial* over the whole prospective wing blade. The proximodistal pattern of the wing is represented by a nested set of expression domains of transcription factors arranged concentrically from distal (central) to proximal (circumferential): distalless (homeodomain), Defective proventriculus (homeodomain), rotund (POU domain), nubbin (Zn finger). These are activated partly by wingless and Notch and partly by an additional unidentified signal from the *vestigial* domain. They lead to the final proximodistal pattern of visible structures consisting of hinge, proximal wing, and wing blade.

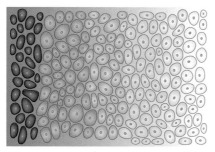

(a) Normal gradient

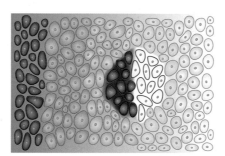

(b) Dpp receptor lof clone

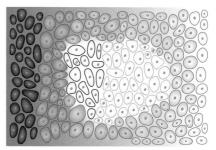

(c) Dynamin lof clone

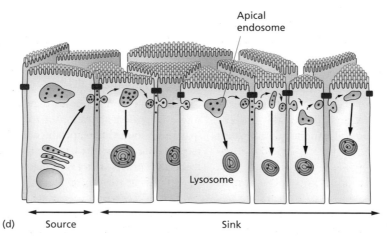

(d)

Apical endosome

Lysosome

Source   Sink

Fig. 17.13 Dpp gradient in the wing disc: (a) normal; (b) effect of a clone lacking the Dpp receptor; (c) effect of a clone lacking dynamin; (d) model of planar transcytosis. (After González-Gaitán 2003. *Mechanisms of Development* 120, 1265–82.)

The experiments on disc regeneration can now be explained to some extent since the production of wingless and dpp gradients will require the presence of the compartment boundaries. However, the "field center" is not, as might be expected, exactly at the point of intersection of the two compartment boundaries, and it is still not understood exactly how the disc reacts to injury in terms of resetting the gradients.

Morphogen gradients and polarity

The *Drosophila* imaginal discs provide the best evidence for the existence of extracellular **gradients** of **morphogens**. In particular the gradients of dpp and wingless have been observed directly by antibody staining or by the visualization of GFP fusion proteins in transgenic flies. But studies on the dpp gradient in the wing disc have shown that transport is not by simple diffusion of protein in the extracellular space, as formerly supposed, but actually involves receptors and the intracellular trafficking systems (Fig. 17.13). A gradient of dpp–GFP can be initiated by the use of a heat shock-regulated transgene. The gradient builds up over about 7 hours and has a roughly exponential form across about 25 cell diameters, as would be predicted by the local source-dispersed sink model (see Chapter 4). The phosphorylation of MAD (the *Drosophila* smad) occurs in concert, indicating biological activity. However, it can be seen that much of the protein constituting the gradient is intracellular rather than extracellular. Also, no gradient is formed if the transgene-encoded dpp is a construct unable to bind to its receptor. If a clone lacking the dpp receptor is induced then the dpp–GFP does not diffuse across it but accumulates on the source side of the clone (Fig. 17.13b). Internalization of ligand–receptor complexes requires the activity of the motor protein dynamin and the small GTPase Rab5 (see Appendix). The gradient is also unable to cross clones lacking dynamin (Fig. 17.13c), and mutants of *rab5* will alter the range of the gradient: loss of function reducing the range, and gain of function increasing it. So it seems that the gradient is set up and maintained by an active process of receptor binding, internalization, and re-release. It is presumed that vesicles carry the receptor–dpp complexes across the cell and then release the dpp on the other side in a process called transcytosis (Fig. 17.13d).

There is another aspect to the pattern of any body part, which is well displayed in *Drosophila* appendages such as the wing: the **polarity** of individual cells (Fig. 17.14). Wing cells each have one actin prehair which is oriented such that it points distally. This cell polarity uses a number of components of the Wingless signaling pathway which behave in a symmetry-breaking manner. Frizzled, the wingless receptor, recruits dishevelled. This both stabilizes the accumulation of Frizzled and also attracts Prickle-Spiney leg (a LIM domain protein) on the adjacent cell, which repels dishevelled to the other side of its cell. The accumulation of dishevelled on the other side of the second cell encourages the formation of another complex with Frizzled at this location and the attraction of Prickle on the same side of the next

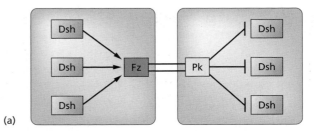

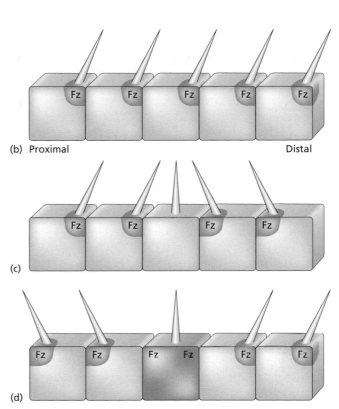

Fig. 17.14 Planar cell polarity system. (a) The formation of complexes involving Frizzled and Prickle-Spiney Leg leads to a propagating polarization of cells. (b) Normal orientation of wing hairs. (c) Orientation of hairs around a cell lacking Frizzled. (d) Orientation of hairs around a cell overproducing Frizzled.

neighbor. The net result is that, starting from a homogeneous situation, the polarity can spontaneously develop and propagate from cell to cell. The main evidence for this mechanism comes from studying the effects of mutants of *frizzled*. If clones lacking *frizzled* are induced then the hairs on the surrounding normal cells tend to point towards the clone. On the other hand if clones overexpressing *frizzled* are produced the hairs on the surrounding cells tend to point away. The accumulations of the various proteins can also be seen within the cells by immunostaining using confocal microscopy. The linkage to the global wingless gradient is thought to occur via fat and dachsous, which are types of cadherin that bias the system slightly to give it an overall polarity. Similar **planar cell polarity** systems also operate in other animals, for example in the control of the orientation of cells during *Xenopus* gastrulation.

Key Points to Remember

- The *Drosophila* larva undergoes two moults and then forms a pupa. In the pupa the majority of the larval body becomes replaced by growth and differentiation of a number of imaginal rudiments. The cuticle of the head, thorax, and genitalia is formed from imaginal discs that arise in the embryo and grow in size during larval life, but remain visibly undifferentiated until metamorphosis.
- Mitotic recombination has been extensively used to analyze disc development. It can be used for clonal analysis or to ablate or overexpress particular genes of interest in random clones.
- The Gal4 method is used to overexpress genes of interest in particular ectopic domains.
- The thoracic imaginal discs become subdivided by an invisible anteroposterior border, which can be detected by clonal analysis, and corresponds to the parasegment border of the embryo. The posterior compartment in each disc is defined by activity of the *engrailed* gene.

- The wing disc also becomes subdivided into dorsal and ventral compartments in the second instar larva. The dorsal compartment is defined by activity of the *apterous* gene.
- The anteroposterior pattern of the wing disc arises by engrailed activating *hedgehog* along the anteroposterior boundary, and hedgehog activating *decapentaplegic (dpp)*. This generates a gradient of dpp protein from the boundary towards both margins.
- The dorsoventral pattern of the wing disc is controlled by activation of Notch expression and signaling along the dorsoventral border. This activates expression of *wingless*. The combination of dpp, Notch, and wingless signals generates the pattern of cell differentiation over the whole disc.
- Study of the dpp gradient in imaginal discs have shown that it depends on receptor-mediated endocytosis for its establishment and maintenance.

Further reading

Website for data on *Drosophila* genes: http://flybase.bio.indiana.edu/allied-data/lk/interactive-fly/aimain/1aahome.htm

General

Cohen, S.M. (1993) Imaginal disc development. In: Bate, M. & Martinez-Arias, A., eds. *The Development of* Drosophila melanogaster, pp. 747–841. Cold Spring Harbor, NY: Cold Spring Harbor Laboratory Press.

Brook, W.J., Diaz-Benjumea, F.J. & Cohen, S.M. (1996) Organizing spatial pattern in limb development. *Annual Reviews of Cell and Developmental Biology* 12, 161–180.

Methods

Brand, A.H. & Perrimon, N. (1993) Targeted gene expression as a means of altering cell fates and generating dominant phenotypes. *Development* 118, 401–415.

Harrison, D.A. & Perrimon, N. (1993) Simple and efficient generation of marked clones in *Drosophila*. *Current Biology* 3, 424–433.

Blair, S.S. (2003) Genetic mosaic techniques for studying *Drosophila* development. *Development* 130, 5065–5072.

Disc regeneration and transdetermination

Bryant, P.J. (1978) Pattern formation in imaginal discs. In: Ashburner, M. & Wright, T.R.F., eds. *Biology of Drosophila*, vol. 2c, pp. 229–335. New York: Academic Press.

Maves, L. & Schubiger, G. (1999) Cell determination and transdetermination in *Drosophila* imaginal discs. *Current Topics in Developmental Biology* 43, 115–151.

Wei, G., Schubiger, G., Harder, F. & Muller, A.M. (2000) Stem cell plasticity in mammals and transdetermination in *Drosophila*: Common themes? *Stem Cells* 18, 409–414.

Gibson, M.C. & Schubiger, G. (2001) *Drosophila* peripodial cells, more than meets the eye? *Bioessays* 23, 691–697.

Maves, L. & Schubiger, G. (2003) Transdetermination in *Drosophila* imaginal discs: a model for understanding pluripotency and selector gene maintenance. *Current Opinion in Genetics and Development* 13, 472–479.

Compartments

Lawrence, P.A. & Struhl, G. (1996) Morphogens, compartments and pattern: lessons from *Drosophila*? *Cell* 85, 951–961.

Serrano, N. & O'Farrell, P.H. (1997) Limb morphogenesis: connection between patterning and growth. *Current Biology* 7, R186–R195.

Blair, S.S. (2003) Lineage compartments in *Drosophila*. *Current Biology* 13, R548–R551.

Wing

Zecca, M., Basler, K. & Struhl, G. (1995) Sequential organizing activities of engrailed, hedgehog and decapentaplegic in the *Drosophila* wing. *Development* 121, 2265–2278.

Cohen, S.M. (1996) Controlling growth of the wing: vestigial integrates signals from the compartment boundaries. *Bioessays* 18, 855–858.

Strigini, M. & Cohen, S.M. (1999) Formation of morphogen gradients in the *Drosophila* wing. *Seminars in Cell and Developmental Biology* **10**, 335–344.

Vervoort, M. (2000) Hedgehog and wing development in *Drosophila*: a morphogen at work? *Bioessays* **22**, 460–468.

Eaton, S. (2003) Cell biology of planar polarity in the *Drosophila* wing. *Mechanisms of Development* **120**, 1257–1264.

González-Gaitán, M. (2003) Endocytic trafficking during *Drosophila* development. *Mechanisms of Development* **120**, 1265–1282.

Growth, regeneration, and evolution

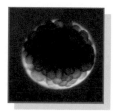

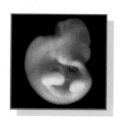

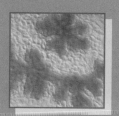

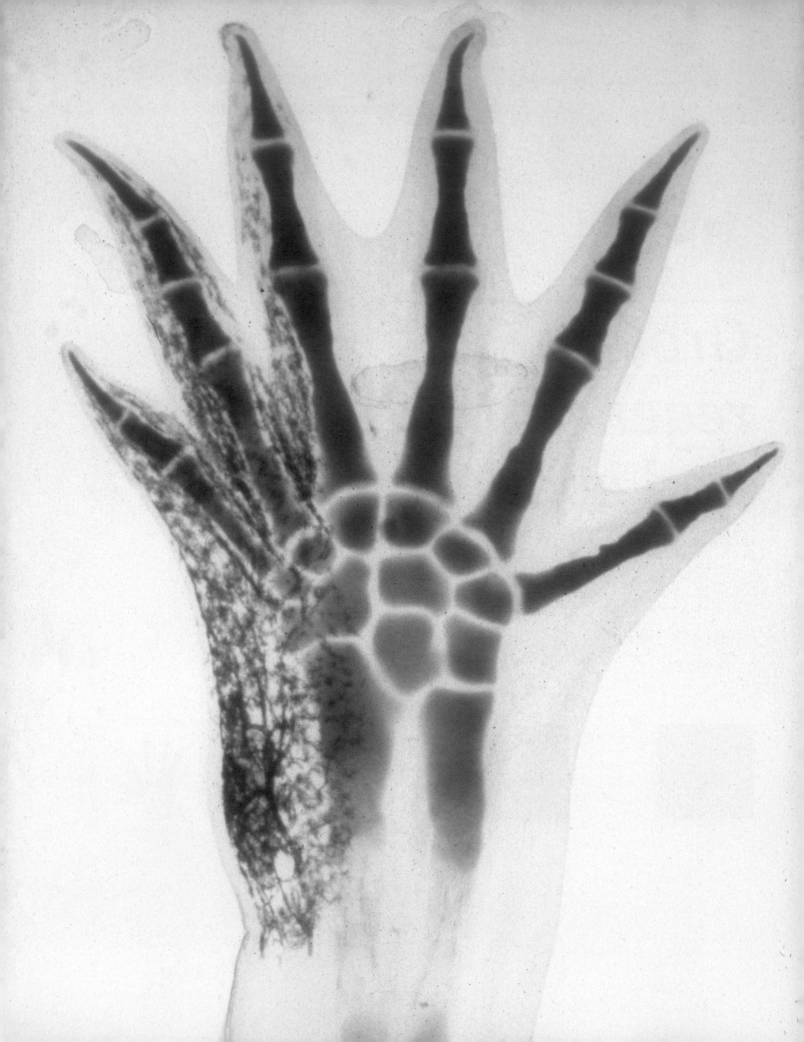

Chapter 18

Growth, aging, and cancer

Size and proportion

Although superficially familiar to all, growth is one of the least well understood aspects of development. By the time the general body plan is formed the whole embryo will be about 1 mm long and each organ is present as a small rudiment. In a large animal, such as the human, there will be an approximately 10^9-fold increase in volume from this stage to that of the full grown adult. This enormous change is driven largely by cell division, but also by secretion of extracellular matrix and by some increase in cell size. The intracellular events controlling cell division are now fairly well understood (see Chapter 2), as is the biochemical mode of action of the growth factors that act on cells from their environment (see Appendix), but the overall control of growth is still largely mysterious.

One problem relates to final size, or why the animal stops growing at a particular point. All animals do grow to reach a finite size, although many fish take such a long time to reach it that they may not do so in a normal lifetime and so their growth appears to be indeterminate. The growth of cells in tissue culture is typically exponential in the presence of excess nutrients. However exponential growth is rarely found in animals. The growth curve for an animal, or a part of an animal, will normally resemble Fig. 18.1, with a rate that is rapid at the beginning and tails off to zero as the final size is reached. Various equations have been proposed that give a mathematical description of overall growth but none is entirely satisfactory because of the large number of individual processes that all contribute in some way to increase of size. One general explanation for the existence of the final size is that the need for nutrients will be proportional to volume whereas the nutrient supply will always be limited by the number of terminal capillaries, and the number of termini of a branched distribution system cannot increase as fast as the volume of a three-dimensional object. But this insight does not currently explain, for example, the 80-fold range of size of different breeds of domestic dog.

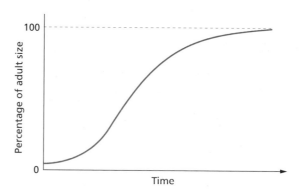

Fig. 18.1 Typical growth curve of an animal.

The other main issue is the coordination of growth between parts of the body. Small differences in growth rate between body parts over a 10^9-fold expansion would lead to very large changes in relative proportions. This suggests that there must exist mechanisms that can sense the overall size of the organism and regulate the growth of individual parts accordingly. In fact relative proportions do usually change systematically during growth, for example the human head becomes smaller relative to the body. The relationship between the growth of two parts of an organism, or the relationship of one part to the whole, can often be represented by the equation $y = bx^\alpha$ where y is the size of one part, x the size of the other, and b and α are constants. This is called "**allometry.**" However there is no real theoretical basis to the equation and it is often not followed, so should be regarded as a convenience for representing data rather than an explanation of the growth process.

Biochemistry of growth

Although cell division is fundamental to growth it is not the only factor driving increase in size. *Drosophila* wing discs have been much used for the investigation of growth control because of the ease of overexpressing genes in clones by the FLP-out method, or removing them from clones by mitotic recombination (see Chapter 17). It is possible to drive a local increase in cell division rate by overexpressing the transcription factor E2F, which activates expression of a number of genes required for the S phase of the cell cycle. Conversely the cell divison rate can be slowed by overexpressing the Retinoblastoma (RB) protein, which inhibits the endogenous E2F. But neither of these manipulations have much effect on overall size and proportion of the wing disc because the rate of cell size increase does not change. In other words the faster cell cycle generates a clone containing more, smaller, cells and the slower cell cycle generates a clone containing fewer, larger, cells. In order to change the rate of growth it is necessary also to change the rate of cell size increase. In normal circumstances, but not necessarily in experimental situations, the rate of cell division tends to be coupled to cell size such that a change in growth rate leads to altered volume of tissue composed of normal sized cells (Fig. 18.2).

The key mediator of cell size is the insulin/IGF signaling system, and the associated intracellular TOR (target of rapamycin) system (Fig. 18.3). These are essentially similar in all animals although the number of parallel components can vary, for example C. elegans has no fewer than 37 insulin-like genes compared with three or four in mammals. TOR gets its name because it is the site of action of the drug rapamycin, used as an immunosuppressant to support human patients who have received organ grafts. The operation of the system is best understood by considering the sequence of events following binding of insulin or IGF to its receptor. The receptors are heterotetramers of two

α and two β subunits. The β subunits are transmembrane proteins with internal tyrosine kinase domains. When activated by ligand binding they transphosphorylate and activate their downstream targets via adapter proteins. There are two key signal transduction pathways activated. The familiar ERK pathway is coupled to the receptor via the adapter proteins Shc, Grb2, and SOS. These lead to activation of the GTP exchange protein Ras, this activates the kinase Raf by recruitment to the cell membrane, and this phosphorylates MEK which phosphorylates ERK. Active ERK enters the nucleus and turns on target genes in cooperation with other transcription factors. The ERK pathway is activated by many other growth factors besides the insulin group and has effects on a large number of processes, often including stimulation of cell division.

The other main pathway activated by the insulin-like receptors is the PI3 kinase (PI3K) pathway. This is more concerned with metabolism, and particularly control of the rate of protein synthesis, which is the main determinant of cell size. PI3K is activated via an adapter protein called the insulin receptor substrate (IRS). When activated, it phosphorylates the membrane lipid phosphatidyl inositol-4,5-diphosphate to phosphatidyl inositol-3,4,5-trisphosphate (PIP_3). This then activates a serine-threonine kinase called protein kinase B (PKB or Akt). PIP_3 does not accumulate indefinitely because it becomes dephosphorylated by a phosphatase called PTEN. The PI3K pathway unleashes a number of well-known metabolic consequences, in particular the increased uptake of glucose, of fatty acids, and increased synthesis of glycogen. In terms of growth regulation the active PKB phosphorylates, and thereby inactivates, a complex of two proteins encoded by the *tuberous sclerosis* genes 1 and 2, which in turn normally inactivate TOR. Because two inhibitions equals one activation this means that insulin signaling activates TOR. TOR is an intracellular serine-threonine kinase which phosphorylates another kinase, S6 kinase, which in turn phosphorylates and activates the ribosomal protein S6. This enables ribosomes to translate a class of mRNAs encoding ribosomal proteins, thus increasing the overall protein synthesis capacity of the cell. It also stimulates protein synthesis by another route: inhibiting the protein 4EBP which is an inhibitor of the protein synthesis initiation factor eF1a. The pathway also has an effect on gene expression via the transcription factor FoxO (forkhead class) which is inactivated by PKB. In the absence of insulin signaling FoxO becomes dephosphorylated, goes to the nucleus, and activates transcription of certain components (including 4EBP) that reduce protein synthesis and others (including the Cdk inhibitor p27[kip1]) that reduce cell division.

The operation of all these components have been analyzed using the methods of *Drosophila* genetics, particularly using the wing disc. In essence loss or reduction of function of any of the positive components on the insulin pathway (the insulin receptor, PI3K, PKB, TOR) will reduce both the size and number of cells, hence reducing the size of the structures they make up. Conversely loss or reduction of function of the inhibitory components (PTEN, TSC1 and 2, FoxO) will increase the size and

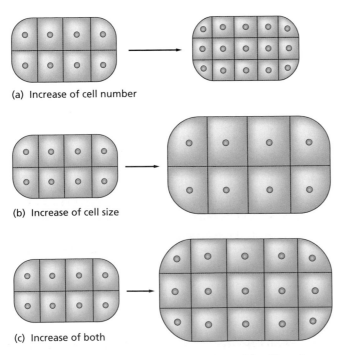

(a) Increase of cell number

(b) Increase of cell size

(c) Increase of both

Fig. 18.2 Effect of increasing the cell division rate and the cell number.

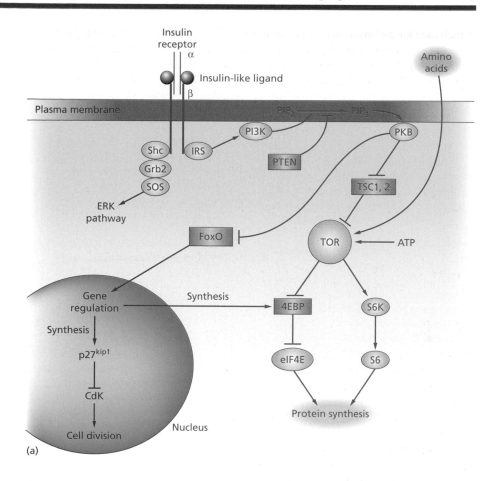

Fig. 18.3 The insulin/TOR pathway.
(a) The components promoting growth are shown as ovals and those inhibiting growth as rectangles. (b) State of each component in the presence or absence of insulin-like ligand.

(b)

	Receptor	PI3K	PKB	TSC	TOR	FoxO	Protein synthesis
+ Ligand	Active	Active	Active	Inactive	Active	Inactive	Up
– Ligand	Inactive	Inactive	Inactive	Active	Inactive	Active	Down

number of cells and increase the size of resulting structures. This biochemical system ensures that the amount of growth will be determined by the appropriate growth promoting signals, the available nutrients, and the energy supply. The point at which all these factors are brought together is TOR, because it is activated by ATP and by amino acids as well as by insulin signaling. It thus integrates the sensing of growth factors, energy supply, and nutrient availability.

The same pathway is important for growth control in mammals. The two insulin-like growth factors, IGF1 and IGF2, are of central importance in growth control. They both operate through a receptor called the "IGF1 receptor" which is similar to the generic insulin receptor described above. IGF2 is more important for supporting growth during the fetal period, largely through its effects on the placenta. IGF1 is more important in supporting postnatal growth. It is made in a wide variety of

tissues and is under the control of the growth hormone (GH or somatotrophin) secreted by the somatotrophs of the anterior pituitary gland. Animals or humans lacking sufficient growth hormone grow and mature slowly and their final size is reduced. Hence deficiencies of growth hormone or GH receptor lead to a type of dwarfism. Conversely individuals suffering a pituitary tumor that secretes excessive growth hormone will grow faster during childhood and adolescence and reach a larger than normal final size, or gigantism. Pituitary-deficient animals can be restored to near-normal growth rates by administration of either growth hormone or IGF1. There are normal hereditary differences in these components, for example there is a considerable difference of IGF1 levels in the serum of large and small breeds of dog.

Mouse knockouts of *igf1* or *igf2* are about 60% normal size at birth, and the double knockout is only 30% normal size and dies

Classic Experiments

MOLECULAR BASIS OF CANCER

There have been many discoveries building our picture of the molecular biology of cancer but these papers focus on two key genes with developmental relevance. In the early 1980s it was found that DNA from human tumors could cause oncogenic transformation of tissue culture cells. When isolated the genes responsible were found to be similar to the mysterious "oncogenes" already discovered in oncogenic retroviruses. The first three papers show that a single base change was all that distinguished an oncogene from the normal homolog. This oncogene was later found to encode the Ras protein in the key ERK signaling pathway.

Reddy, E.P., Reynolds, R.K., Santos, E. & Barbacid, M. (1982) A point mutation is responsible for the acquisition of transforming properties by the T24 human bladder-carcinoma oncogene. *Nature* **300**, 149–152.

Tabin, C.J., Bradley, S.M., Bargmann, C.I., Weinberg, R.A., Papageorge, A.G., Scolnick, E.M., Dhar, R., Lowy, D.R. & Chang, E.H. (1982) Mechanism of activation of a human oncogene. *Nature* **300**, 143–149.

Taparowsky, E., Suard, Y., Fasano, O., Shimizu, K., Goldfarb, M. & Wigler, M. (1982) Activation of the T24 bladder-carcinoma transforming gene is linked to a single amino-acid change. *Nature* **300**, 762–765.

The other two papers below deal with a tumor suppressor gene, one that is normally required to suppress growth and so the loss-of-function mutant is oncogenic. The gene, *APC*, is defective in the human hereditary disease of adenosis polyposis coli. Individuals with this condition have one good copy of the gene, and whenever this is lost in one cell by somatic mutation an adenoma is initiated. These papers identified the gene in mice and showed that loss of the good copy was an early step in development of cancer. APC is now known to be a component of the Wnt pathway, needed for targeting β-catenin for degradation.

Powell, S.M., Zilz, N., Beazerbarclay, Y., Bryan, T.M., Hamilton, S.R., Thibodeau, S.N., Vogelstein, B. & Kinzler, K.W. (1992) *APC* mutations occur early during colorectal tumorigenesis. *Nature* **359**, 235–237.

Su, L.K., Kinzler, K.W., Vogelstein, B., Preisinger, A.C., Moser, A.R., Luongo, C., Gould, K.A. & Dove, W.F. (1992) Multiple intestinal neoplasia caused by a mutation in the murine homolog of the *APC* gene. *Science* **256**, 668–670.

the gene entirely. The result is numerous chaotic outgrowths of bone and connective tissue, and this condition may have been experienced by the famous "Elephant Man," Joseph Merrick. The *Tuberous sclerosis* genes are also tumor suppressor genes and get their name from a dominant hereditary disease in which loss of one copy of the gene predisposes to the formation of numerous benign tumors, again because of occasional somatic mutations in the good copy.

Control of relative proportion

This is a somewhat neglected but very important problem in growth control. Even if the overall size of the organism is correct, how can it be guaranteed that different body parts expand in proportion to one another? Or, in cases where there are allometric deviations from proportional growth, how is this controlled?

The data relating to these questions are limited, although in general there is a tendency for grafts to behave in an autonomous way, not adapting to the size of the host, nor affecting the relative growth of corresponding parts within the host. This is actually very surprising because it is hard to think of a mechanism to control relative proportions that does not involve a feedback inhibition of each body part on itself. For example, if each body part produces an inhibitor that is diluted into the total blood volume of the organism then its concentration will measure the size of the part relative to the size of the whole. If the part gets a bit too big, the inhibitor will slow it down, and if the part is too small the inhibitor level will be low, so its growth will accelerate (Fig. 18.4). Such a hypothetical negative feedback signal is called a **chalone** (pronounced "kaylone").

Although this model is simple and hypothetical, there are some real examples of chalone-like systems. For example, it is known that removal of thoracic imaginal discs from metamorphosing butterflies will increase the size of the structures formed from the other thoracic discs, but not affect the size of structures in the head or the abdomen. The best characterized example of chalone action in vertebrates is the control of muscle growth exerted by the TGFβ superfamily member myostatin (= growth and differentiation factor 8; GDF8). This is secreted by developing

at birth. Likewise, the knockout of the *IGF1 receptor*, or of two of the three *PKB* genes, are about 50% normal size and die at birth. The knockout of *igf2* alone grows normally after birth, which is why it is thought that its functions are mostly exerted in fetal life. Conversely transgenic overexpression of either IGF1 or of growth hormone, which stimulates the synthesis of IGF1, can increase size up to about twofold.

In addition to pituitary dwarfism and gigantism, components of the same pathway appear in some types of disproportionate growth found in humans. Pituitary tumors in adult life do not lead to gigantism because the growth zones of the long bones are already shut down (see below). But they do lead to a disproportionate enlargement of the jaws, hands, and feet, known as acromegaly. *PTEN* is often mutated to inactivity in a variety of tumor types. It is a **tumor suppressor gene** (see below), so the absence of one copy of the gene in the germline leads to a high risk of cancer because cells may lose the other copy if it acquires a somatic mutation during postnatal life. Complete loss of function of *PTEN* leads to "Proteus syndrome." This is exceedingly rare but can occur as a very early somatic mutation in embryonic development, creating a mosaic in which part of the body lacks

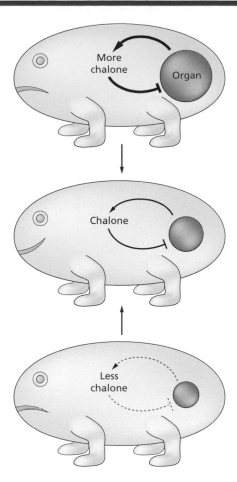

Fig. 18.4 Operation of a hypothetical chalone system that regulates the size of an organ by negative feedback control of growth.

muscle cells and it reduces both the cell division of myoblasts and the enlargement of myofibers. The *myostatin* knockout mouse has a two- to three-fold increase in muscle mass, arising from the presence of both more and larger myofibers. Some breeds of domestic cattle, such as the Belgian Blue, have been shown to carry hypomorphic alleles for *myostatin*, indicating that traditional animal breeding has already anticipated rational genetic modification in this area. In the embryo, *myostatin* is expressed in the central part of the dermamyotome, in between the myoblast-generating areas, and later in the limb buds both by the myoblasts and some other mesenchymal tissue. The early expression in tissues neighboring the developing muscle indicates that it may have a role in defining the boundaries of the myogenic areas as well as the later negative feedback regulation.

Another well documented chalone is GDF11 (growth and differentiation factor 11) in the olfactory system. This is expressed in olfactory neurons and inhibits further neurogenesis by inducing the Cdk inhibitor $p27^{kip1}$. The knockout mouse for *gdf11* has an increased number of progenitor cells and neurons in the olfactory epithelium. Interestingly, both myostatin and GDF11 belong to the same subgroup of TGFβ-like factors, and both are inhibited by follistatin. This suggests that there could be global controls that modulate the behavior of a number of tissue-specific negative feedback mechanisms.

Growth in stature

The height of a human being, or overall stature of a vertebrate animal, depends to a large extent on the length of the long bones in the limbs. These long bones, like most of the embryonic skeleton, are initially formed as a cartilage **model** which become replaced by bone in later development (Fig. 18.5). Cartilage consists of **chondrocytes** embedded in holes (lacunae) in a clear extracellular matrix without blood vessels, nerves, or lymphatics. The matrix is characterized by the presence of type II collagen and aggrecan, a specific cartilage-type proteoglycan. Sox9 serves as a key transcription factor in cartilage development, whether from mesoderm or neural crest, and activates transcription of the specific cartilage-type collagens. The human genetic disease campomelic dysplasia is a dominant haploinsufficient condition of cartilage defects arising from the loss of one copy of *sox9*. The periphery of the cartilage, or **perichondrium**, is composed of fibroblast-like cells that divide to generate **chondroblasts**. These divide a few more times and then differentiate to become chondrocytes embedded in the matrix.

The bones of the face and cranium are known as **membrane bones** as they are not formed from a cartilage model but differentiate directly from the dermis, which in the head is of neural crest origin. Bone formation by either method requires the action of the bone morphogenetic proteins (BMPs) which are normally expressed in differentiating bone and in healing fractures. Because the BMPs are also required for various earlier developmental events, BMP knockout mice may not develop far enough to form a skeleton. However the knockout of *bmp7* is viable and does have skeletal defects, as do naturally occurring null mutants of *bmp5* and *growth and differentiation factor 5* (*gdf5*), also a member of the BMP family.

The shaft of a long bone is called the **diaphysis** and the knobs at either end are called the **epiphyses**. Primary ossification starts in a collar around the center of the diaphysis where the perichondrium becomes converted to a **periosteum** containing **osteoblasts**. Combinations of perichondrium from **Rosa26** mice with unlabeled cartilage, grown under the kidney capsule of **nude mice**, shows that the osteblasts are derived from the perichondrium. The key transcription factor driving osteogenesis is called Cbfa1 (homeodomain, = Runx2). The mouse knockout of *cbfa1* has a fully cartilaginous skeleton with a complete absence of bone. The heterozygote is known in the human syndrome of cleidocranial dysplasia, and has defects of membrane bone ossification. During ossification, the osteoblasts invade the cartilage in association with blood vessels and secrete a matrix (osteoid) mainly composed of type I collagen. As this thickens, some cells become trapped in it and they, now called **osteocytes**, secrete calcium phosphate which precipitates in the matrix as hydroxyapatite. Unlike chondrocytes, the osteocytes remain in

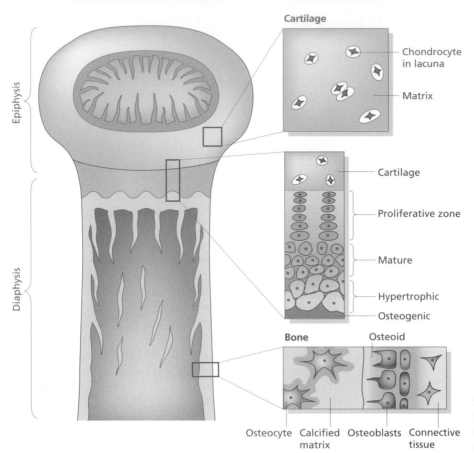

Fig. 18.5 A developing long bone, showing the cartilaginous epiphysis, the epiphyseal growth plate, and the ossification of the diaphysis.

contact with one another by means of thin cellular processes running through channels called canaliculi. The bone continues to grow at the surface and becomes eroded from the inside by **osteoclasts**, which are giant multinucleate cells formed in the bone marrow from the hematopoietic lineage (see Chapter 13). Osteoclasts also make channels in the bone matrix for blood vessels, nerves, and lymphatics. Additional ossification centers appear in the epiphyses, but growth continues for some time at the ends of long bones in a zone between the remaining cartilage of each epiphysis and the bone of the diaphysis. This region is called the epiphyseal **growth plate**. Within the growth plate the most distal part consists of actively dividing chondrocytes, and more proximally the division ceases and the cartilage cells become **hypertrophic** (enlarged) and the matrix becomes calcified. The hypertrophic chondrocytes eventually become replaced by osteocytes and the matrix becomes converted to bone. When full adult stature has been attained, the ossified epiphysis fuses with the diaphysis and the growth plate disappears. The ultimate size of an individual depends both on the activity of the growth plates and on the length of time that they remain active. Because of the timing of the shutdown of the growth plates, excess growth hormone given during childhood will increase stature, but after epiphyseal fusion it leads to acromegaly.

The most common form of human dwarfism, achondroplasia, is rather different from pituitary dwarfism as it leads to a disproportion in which the long bones are abnormally short. It is caused by a dominant gain-of-function mutation of *fgf receptor 3* (*fgfr3*) that reduces proliferation of the growth plate. Interestingly the mouse knockout of *fgfr3* has the opposite phenotype, namely abnormally long bones caused by a prolongation of activity of the growth plate. These results are explained because in this context the FGF system acts as a growth inhibitor rather than a growth promoter. This inhibitory activity is exerted on the chondrocytes of the growth plate via the STAT signal transduction pathway, rather than the ERK pathway normally activated by the FGFs. Stronger gain-of-function mutants of FGFR3 lead to the more severe and fatal thanatophoric dysplasia, in which there is a truncation of growth of much of the skeleton other than the skull. Gain-of-function mutations of other FGF receptor genes cause defects in membrane bone formation, particularly resulting in craniosynostosis, in which there is premature fusion of the bones of the skull.

Within the growth plate, proliferation of chondrocytes is maintained by a feedback loop involving a member of the hedgehog family of signaling molecules, Indian hedgehog (Ihh), and parathyroid hormone-related protein (PTHrP), a protein similar to parathyroid hormone (Fig. 18.6). Ihh is produced by the cartilage cells that have finished dividing and become committed to hypertrophic differentiation. It acts on the perichondrium to stimulate expression of *pthrp*. PTHrP

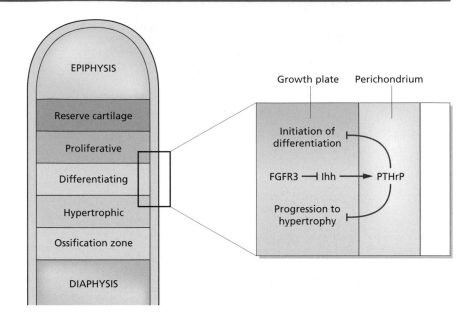

Fig. 18.6 Feedback loop for control of chondrocyte hypertrophic differentiation. Ihh from the committed, but nonhypertrophic, cartilage represses differentiation via PTHrP.

then acts on the dividing chondrocytes to maintain division and hence inhibit their progression to the hypertrophic compartment. Knockouts of genes for either factor, or of the *pthrp receptor*, lead to reduced proliferation of the growth plates, premature ossification and hence another type of short-limbed dwarfism. Overexpression of Ihh prolongs the phase of chondrocyte division and hence suppresses the formation of hypertrophic cartilage. This also leads to a type of dwarfism because the hypertrophic differentiation itself normally drives elongation of the limbs to some extent. However, the phenotype is actually the opposite of that found in the *pthrp* knockout, because the formation of hypertrophic cartilage is delayed instead of advanced. The overexpression of Ihh does not rescue the phenotype of the *pthrp* knockouts, but a constitutive mutation of *pthrp receptor* will do so, confirming that the PTHrP acts downstream of the Ihh. Therefore in essence this is another chalone-like system in which the rate of production of a cell type (the hypertrophic cartilage) is regulated by the size of the cell population committed to this pathway.

Aging

There has long been debate about whether aging is really a developmental process or not. The argument in favor is that senescence and death arise from pre-programmed events which are under developmental control. The argument against is that such events are simply due to progressive mutational and other damage, which have an inherently random character. In fact both positions have proved correct to some extent. Aging is developmental insofar as it responds to genes that can be manipulated, and remarkably it turns out that the system controlling it is the same as the system regulating growth, namely the insulin

signaling pathway. On the other hand the genes regulated by this pathway mostly encode components that minimize the consequences of oxidative stress, indicating that the actual destruction of the organism does depend on random and nondevelopmental damage.

The role of the insulin pathway in aging was discovered in *C. elegans*. If conditions are bad in terms of limited nutrients or overcrowding during the first larval stage then it can form a **dauer larva** instead of the third larval stage. This is a nonfeeding, impermeable, dispersal phase. The dauer larva can survive for months but can still develop to an adult if conditions improve. When mutants were screened to find what caused this developmental switch, it was found that loss-of-function mutants of components of the insulin pathway will cause dauer formation. Furthermore, hypomorphic alleles in the same genes do not spontaneously form dauers but do live longer than the normal 2–3 weeks. The *C. elegans* insulin receptor gene is called *daf2*, the *PI3kinase* gene is *age1/daf23*, and the PKB gene is *Akt*. The homolog of FoxO is called *daf16*. It was found that the insulin pathway would not exert its effect unless FoxO was present suggesting that, unlike growth control, the aging pathway mostly operates through transcriptional regulation. FoxO promotes transcription of genes for a variety of products that reduce oxidative stress, such as the enzymes catalase and superoxide dismutase. Many other hypomorphic mutants of *C. elegans* that extend lifespan are in components of the mitochondria, suggesting that aerobic metabolism, with its concomitant production of free radicals, causes the damage that leads eventually to death. So FoxO prolongs life by causing production of products that reduce oxidative stress. Because the insulin pathway represses FoxO it follows that conditions which normally stimulate insulin signaling, namely an abundance of food, will tend to reduce lifespan. Treatment of *C. elegans* with RNAi against the

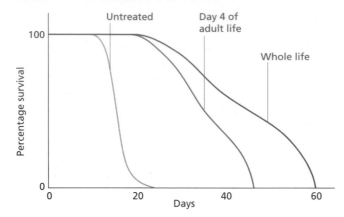

Fig. 18.7 Effects of downregulating the insulin pathway on the lifespan of *C. elegans*. Feeding with RNAi to insulin receptor will extend life substantially if started before the last molt, but the effect falls off the later in adult life the treatment is started.

insulin receptor makes it possible to determine when the protective effect is exerted, and it is found that the receptor needs to be downregulated soon after the final molt in order to get the maximum degree of prolongation of life (Fig. 18.7).

The general significance of the results with *C. elegans* is suggested by comparisons with other animals. *Drosophila* adults have a life span of about 3–4 months and this can be extended in hypomorphic mutants of the insulin receptor. However these are not normal but tend to be dwarfed and subfertile. Mice live 2–3 years but the Snell and the Ames dwarf mice can live 40% longer. The Snell dwarf mouse lacks the transcription factor Pit1 (POU domain) and the Ames dwarf mouse the transcription factor Prop1 (paired homeodomain). Both of these strains fail to form three cell types in the anterior pituitary: the somatotrophs, lactotrophs, and thyrotrophs. As described above, the reduction of growth hormone arising from the absence of somatotrophs leads to reduction of IGF1 production and so reduction of insulin pathway signaling.

It has also been known for many years that the lifespan of mammals can be extended significantly by "caloric restriction," in other words supplying just enough food to avoid symptoms of starvation. Such individuals mature slowly and live a long time. This may be another way of reducing insulin signaling although there may also be some parallel or synergistic effects as the pituitary dwarf mice can have their lives extended even longer by imposing caloric restriction.

Another type of long-lived mouse is the knockout of the gene for the Shc adapter protein. Although shown coupling Ras to the insulin receptor in Fig. 18.3, it also couples Ras to many many other receptors and is generally responsible for intracellular signal transduction in response to "stress" signals, such as treatment with hydrogen peroxide or ultraviolet light.

Studies on mammalian cells grown in culture show that they tend to slow down and eventually stop growing after a certain number of passages. This is known as the **Hayflick limit**. Permanent cell lines that grow indefinitely have undergone mutational changes similar to those undergone by cancers which enable them to evade the limit (see below). Although various explanations have been proposed for the Hayflick limit, it is now thought that the main reason is the erosion of chromosome ends. Usually at each DNA replication cycle the chromosomes ends, or **telomeres**, become slightly shorter and eventually the genes near the telomere become damaged. The telomeres consist of thousands of repeats of a six-base-pair sequence, of which 50–100 base pairs are lost in each cycle. However, the ends can be repaired by an enzyme–RNA complex called **telomerase**, normally present in germ cells and, perhaps, in stem cells. This includes a piece of RNA encoding the telomere sequence and a reverse transcriptase to make the DNA. It has been debated for many years whether the Hayflick limit has anything to do with the aging of whole animals. Recently the advent of whole-animal cloning has raised the interesting question of whether the telomeres might be eroded more rapidly in clones because the cells have already been through a lifetime of divisions. In fact the recent studies on both cloned mice and cattle have shown that the telomeres are normal and there is no reason to expect short life span from this cause.

Postnatal disorders of growth and differentiation

Cancer (see below) is well known and feared. It is fundamentally a growth of the body's own tissue to an inappropriate extent or in an inappropriate place. But there are also many other disorders of growth and differentiation that may be less life threatening but still have biological interest.

Hyperplasias represent an excessive cell production but with fairly normal cell differentiation and turnover. Some hyperplasias represent normal physiological events, for example the growth of the mammary epithelium at puberty or in pregnancy. Others are pathological, for example a goiter is a hyperplasia of the thyroid due to a lack of iodine, or other thyroid hormone insufficiency. An inadequate level of thyroid hormones causes production of more thyroid-stimulating hormone (TSH) by the pituitary and this causes increased growth of the thyroid. Although not usually described as such this is another **chalone**-like system. Where hyperplasia affects one of the "expanding" tissues, which are the type that grows while the organism is growing (see Chapter 13), it can lead to an increase in final size. In a renewal tissue, hyperplasia may not increase size, but instead causes a faster flux of cells through the system, as for example in psoriasis, a skin disease in which there is a massive continuous overproduction of keratinocytes associated with an elevated level of TGFβ.

Metaplasia means the conversion of one tissue type into another. It occurs quite commonly, usually in association with tissue regeneration provoked by trauma or infection. It represents a switch in developmental commitment of a stem cell from one tissue type to another. Many glandular epithelia can develop

squamous metaplasia, which means that they acquire patches of tissue organized as a stratified squamous epithelium (see Chapter 13). For example squamous metaplasia of the bronchus is common in smokers. Patches of intestinal metaplasia can arise in the stomach, often in association with ulcers. These may be "complete metaplasia," in which the ectopic patches are exactly similar to normal small intestine, or "incomplete," in which the ectopic tissue resembles the intestine but is not normal. Metaplasias can also arise in connective tissues. Patches of cartilage or bone are often formed in ectopic locations, for example in surgical scars or in traumatized muscle. They are thought to derive from fibroblasts that have been caused to enter the skeletal development pathway. **Transdifferentiation** is a term originally coined to describe direct conversions of one type of differentiated cell into another, thus being a subdivision of metaplasias. It has more recently been applied to transformations from stem cells to differentiated cells of a different type, for example the rare occurrence of donor-derived myofibers following graft of hematopoietic stem cells to an irradiated host. Because of the controversy about whether such transformations really happen at all (see Chapter 13), the well-established examples of transdifferentiation in other areas of biology have tended to be unnecessarily discredited. For example, Wolffian regeneration of the lens in some amphibians is a well-documented phenomenon that involves transformation of the pigmented cells of the iris to replace a lens which has been removed.

Cancer

The terms **neoplasm**, meaning "new growth," and tumor, originally meaning "swelling," are more or less synonymous, as a tumor is any type of neoplastic growth. Tumors may be benign or malignant (Fig. 18.8). A benign tumor will be well differentiated, localized to the site of origin, and often surrounded by a fibrous capsule. A malignant tumor sends cells invading into the locality and, via the bloodstream or lymphatic system, to establish secondary tumors, or **metastases**, at distant sites in the body. It is also likely to be less well differentiated and faster growing. The spread to other parts of the body makes malignant tumors particularly difficult to treat by surgery. The term **cancer** tends to be reserved for malignant tumors, although benign tumors can also, on occasion, cause death.

Most cancers in the human body arise ultimately from a single cell that has acquired several somatic mutations in its growth and differentiation machinery. As it takes a long time to collect several relevant mutations in one cell, most cancers increase in frequency with age and are commonest in older people. In many cases there exist visible precursor lesions that have some altered growth control but are not in themselves dangerous. These include some of the metaplasias, for example most lung cancer arises within areas of squamous metaplasia, and many gastric cancers arise within areas of incomplete intestinal metaplasia. A general term for regions of altered growth and differentiation is

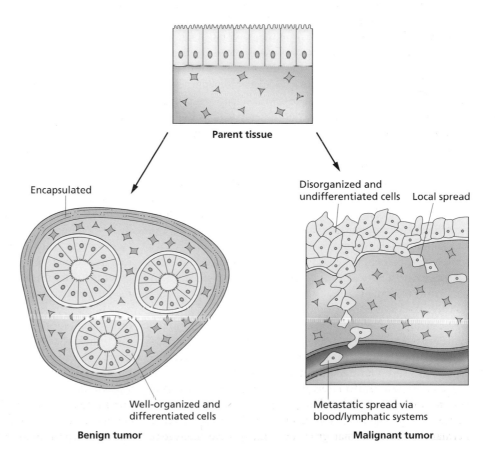

Parent tissue

Encapsulated

Disorganized and undifferentiated cells   Local spread

Well-organized and differentiated cells

Metastatic spread via blood/lymphatic systems

Fig. 18.8 Tumors: benign and malignant.   **Benign tumor**   **Malignant tumor**

New Directions in Research

Growth control is a very rich area for future research because some of the basic questions remain unanswered. In particular:
1 What determines the final size of an organism?
2 What regulates the proportions between its parts?
A variety of biological systems or methods of approach might prove fruitful in this area.

dysplasia. Some dysplasias are precursor lesions to cancer but others may regress spontaneously. An example is the dysplasia of the uterine cervix detected by cervical smear cytology tests. This is now usually called cervical intraepithelial neoplasia (CIN) and may, but does not necessarily, progress to cervical cancer.

Even malignant tumors frequently retain a good deal of the differentiated histological characteristics of the parent cell, and the histological classification of tumors is based on this fact. This means that there should be at least as many kinds of cancer as there are cell types in the body. In fact at a molecular level there are many more, because there are several ways in which each cell type can lose its normal growth control capability.

Cancers of epithelial tissues are known as **carcinomas**. If they are glandular in histology they are called adenocarcinomas, and if they are squamous in histology they are called squamous cell carcinomas. Tumors of the connective tissues (in the broad sense) are called **sarcomas**. Thus, an osteosarcoma is a malignant tumor of bone, and a liposarcoma of adipose tissue. The corresponding benign tumors carry the suffix -oma, for instance an adenoma is a benign neoplasm of glandular epithelium, a papilloma of squamous epithelium, an osteoma of bone, or a lipoma of adipose tissue. If a tumor is so undifferentiated that its parent tissue cannot be recognized, it is described as anaplastic.

Leukemias are cancers of the hematopoietic tissue which result in an overproduction of one or more types of blood cell. In general one of the **transit amplifying** cell types will have acquired indefinite multiplication ability. The type of leukemia depends on where in the pathway of hematopoietic differentiation the defect is located, so granulocytic leukemias are diseases of a **granulocyte** progenitor cell and lymphocytic leukemias of a **lymphocyte** progenitor cell. Lymphomas are solid overgrowths of lymphocytes, although this may also be accompanied by circulating malignant cells.

There are, in addition, various miscellaneous types of cancer. These include some that probably only need one or two mutations for their causation as they are most commonly found in children, for example neuroblastoma is a tumor of immature neuroblasts and Wilm's tumor an overgrowth of stem cells in the kidney. A tumor of particular interest to developmental biologists is the **teratocarcinoma**, which arises from germ cells and closely resembles the embryonic stem cells of mice. Like ES cells, teratocarcinomas can form a wide variety of differentiated tissues and may contain a jumbled mixture of structures in addition to the stem cells themselves (see also Chapter 10).

Molecular biology of cancer

The molecular biology of cancer has become moderately well understood in recent years. It is known that it is usually necessary for a cell to acquire defects in six different systems to become a full-blown cancer, although precursor dysplasias will require fewer changes. The defects may individually arise through somatic mutation, through inherited germ-line mutation, or through changes in gene expression. Somatic mutation is probably the most important, and may accelerate as the cancer develops because the cells become more liable to chromosomal losses and rearrangements. Although most human cancer is due to random somatic mutation acquired throughout life, predispositions to various types of cancer will exist where there are already **germ-line mutations** of the same target genes. For example, if six mutations are normally needed for the formation of a tumor, then an individual with one pre-existing germ-line mutation will only need to collect five more and this will represent an hereditary predisposition, increasing the lifetime risk of cancer.

The first type of defect is independence from growth factors for continued division. The mutations bringing this about are those that generate **oncogenes**, which are genes that can cause changes in tissue-culture cells that resemble cancerous behavior, such as loss of contact inhibition. Oncogenes are often elements of the signaling systems that normally control growth, either the growth factors themselves, or growth factor receptors, or components of the signal transduction pathways, or transcription factors activated by these pathways. Gain-of-function mutations may create constitutively active forms of the proteins, for example human tumors often contain constitutive mutations of the GTP exchange protein Ras, a component of the ERK pathway activated by many receptor tyrosine kinases (see Appendix). Oncogenic mutations may also act by causing an inappropriate expression of otherwise normal gene products. For example many tumors secrete growth factors, and these may be factors normally supplied by neighboring tissues in order to create a stem-cell niche. When the tumor itself makes the factors, or activates the pathway independent of the presence of the factors, then its growth will no longer be controlled by the neighboring tissue.

The next change is the insensitivity to growth inhibitors. The genes involved here are known as **tumor-suppressor genes**, and normally code for inhibitors of growth. When they are inactivated by mutation the normal inhibition is lifted and this can contribute to the development of a cancer. An example is the product of the *retinoblastoma (Rb)* gene. This is an inhibitor of the cell cycle (see Chapter 2), normally activated by factors of the TGFβ family. If both copies of the *Rb* gene are lost from a retinal cell then it will develop into a retinoblastoma. Under normal circumstances this is an exceedingly rare event since it requires an independent loss-of-function mutation of both copies of the gene in the same cell. However there is a hereditary form of retinoblastoma in which affected individuals are highly susceptible to development of the disease and often suffer from multiple foci forming in both eyes. This is because they are heterozygotes who have already inherited one defective copy of the gene through the germ line. In such individuals the disease will be initiated in any retinal cell that suffers a single somatic mutation inactivating the remaining good copy, and since there are tens of thousands of retinal cells it is highly probable that a few cells will experience this mutation. Retinoblastoma is unusual because the loss of function of a single gene is enough to initiate formation of the tumor, but it exemplifies very clearly the nature of tumor suppressor genes.

The third change is the ability to avoid apoptosis. This often depends on a loss of function of the gene coding for p53, a protein that detects various conditions normally leading to cell death. If this control is removed then many cells that have acquired defects of various kinds will fail to die and will instead survive and proliferate. The mutations in *p53* are often **dominant negative**, so just a single dominant mutation is enough to reduce cell death substantially.

The fourth change is the ability to proliferate without limit. This is due to an upregulation of **telomerase** to evade the consequences of chromosome shortening (see above).

The next element that is essential for the growth of a tumor is the ability to promote **angiogenesis**. Unless the tumor can attract the inward growth of capillaries from the surroundings it cannot obtain nutrients and cannot grow. Many tumors secrete angiogenesis factors such as vascular endothelial growth factors (VEGFs) or fibroblast growth factors (FGFs) and are thereby able to attract their own abundant blood supply.

Finally there are the changes required for invasion and metastasis. These include the downregulation of various cell adhesion molecules and the secretion of proteases that degrade the surrounding extracellular matrix and enable cells to move more freely.

Colon cancer

An example where the mutational sequence is reasonably well known is the development of carcinoma of the colon (Fig. 18.9). The histological structure of the colon is similar to the small intestine described in Chapter 13, but without the villi or the Paneth cells. The sequence described is a common pathway, but is not the only possible one.

The first step in the development of cancer is the loss of the tumor-suppressor gene *APC (Adenomatous Polyposis Coli)*, which leads to the formation of a small adenoma. *APC* encodes a cytoplasmic protein that forms the complex responsible for the phosphorylation of β-catenin by GSK3, which marks β-catenin for degradation. APC is therefore an inhibitor of Wnt signaling. As discussed in Chapter 13, the Wnt pathway is normally

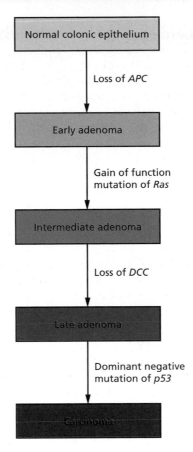

Fig. 18.9 Sequence of mutational changes involved in the formation of a carcinoma of the colon.

involved in controlling the size of the proliferative zone in the intestinal crypts. Individuals suffering from the hereditary disease familial adenomatous polyposis coli are heterozygous for loss-of-function mutations of *APC*. They suffer multiple colonic adenomas due to repeated somatic loss of the good allele, and have a very high lifetime risk of cancer. The next step is the acquisition of a constitutive mutation of *ras*, involved in the ERK signal transduction pathway. This increases the size of the adenoma. Then there is the loss of the netrin receptor called *DCC (Deleted in Colon Carcinoma)*, which leads to a large adenoma. Finally there is a dominant negative mutation of *p53*, which increases net growth by the suppression of apoptosis in the tumor. Having collected all these changes, the tumor finally becomes invasive and hence malignant.

This sequence of events as currently understood is not complete, and not all colon carcinomas show all four mutations, or all in the same order. However, it does represent the most common mutational pathway for generating this type of cancer, and this example illustrates that multiple developmental events involving growth, differentiation, and cell adhesion need to be perturbed in order for a cancer to develop. It also points toward a variety of potential targets for new therapies, which could only have been identified by knowing the molecular and developmental processes involved.

Key Points to Remember

- Growth normally requires increase of cell number, often accompanied by increase of cell size and the amount of extracellular matrix.
- Growth is controlled to a large extent by the insulin/IGF signaling pathway. Loss-of-function mutations will reduce growth and gain of function procedures will increase growth.
- TOR (target of rapamycin) integrates insulin signaling with the availability of nutrients and energy.
- In the postnatal life of mammals growth hormone from the pituitary controls synthesis of IGF1, and loss of any component of this system can lead to dwarfism.
- There is some evidence for chalone-like negative feedback systems controlling the growth of individual tissues.
- The growth of stature in vertebrates depends mainly on the elongation of long bones. This depends on the activity of the terminal cartilage growth plates. Their growth is under negative control from FGFs and their differentiation from Indian Hedgehog.
- The insulin pathway also affects lifespan, by negative regulation of genes for proteins that reduce oxidative stress.
- There is a variety of pathological abnormalities of normal growth and differentiation, classified as hyperplasias, dysplasias, neoplasias, and metaplasias.
- The distinction between benign and malignant tumors mostly depends on their propensity to invade surrounding tissues, and establish metastases elsewhere in the body.
- Development of a cancer normally requires several mutational changes to occur in a single cell.

Further reading

Growth

Robertson, E.J. (1995) Insulin-like growth factors, imprinting and epigenetic growth control. *Seminars in Developmental Biology* **6**, 293–299.

Lee, S.J. & McPherron, A.C. (1999) Myostatin and the control of skeletal muscle mass. *Current Opinion in Genetics and Development* **9**, 604–607.

Day, S.J. & Lawrence, P.A. (2000) Measuring dimensions: the regulation of size and shape. *Development* **127**, 2977–2987.

Lupu, F., Terwilliger, J.D., Lee,K. et al. (2001) Roles of growth hormone and insulin like growth factor 1 in mouse postnatal growth. *Developmental Biology* **229**, 141–162.

Saucedo, L.J. & Edgar, B.A. (2002) Why size matters: altering cell size. *Current Opinion in Genetics and Development* **12**, 565–571.

Kozma, S.C. & Thomas, G. (2003) Regulation of cell size in growth, development and human disease: PI3K, PKB and S6K. *Bioessays* **24**, 65–71.

Nijhout, H.F. (2003) The control of body size in insects. *Developmental Biology* **261**, 1–9.

Skeletal development

Erlebacher, A., Filvaroff, E.H., Gitelman, S.E. & Derynck, R. (1995) Toward a molecular understanding of skeletal development. *Cell* **80**, 371–378.

Wallis, G.A. (1996) Bone growth: coordinating chondrocyte differentiation. *Current Biology* **6**, 1577–1580.

Vortkamp, A. (1997) Skeleton morphogenesis: defining the skeletal elements. *Current Biology* **7**, R104–R107.

Karsenty, G. & Wagner, E.F. (2002) Reaching a genetic and molecular understanding of skeletal development. *Developmental Cell* **2**, 389–406.

Aging

Kenyon, C. (2001) A conserved regulatory system for aging. *Cell* **105**, 165–168.

Carter, C.S., Ramsey, M.M. & Sonntag, W.E. (2002) A critical analysis of the role of growth hormone and IGF-1 in aging and lifespan. *Trends in Genetics* **18**, 295–301.

Helfand, S.L. & Rogina, B. (2003) Genetics of aging in the fruit fly *Drosophila melanogaster*. *Annual Reviews of Genetics* **37**, 329–348.

Cancer

Bishop, J.M. (1991) Molecular themes in oncogenesis. *Cell* **64**, 235–248.

Vogelstein, B. & Kinzler, K.W. (1993) The multistep nature of cancer. *Trends in Genetics* **9**, 138–142.

Kinzler, K.W. & Vogelstein, B. (1996) Lessons from hereditary colorectal cancer. *Cell* **87**, 159–170.

Garcia, S.B., Novelli, M. & Wright, N.A. (2000) The clonal origin and clonal evolution of epithelial tumors. *International Journal of Experimental Pathology* **81**, 89–116.

Hanahan, D. & Weinberg, R.A. (2000) The hallmarks of cancer. *Cell* **100**, 57–70.

Frank, S.A. & Nowack, M.A. (2004) Problems of somatic mutation and cancer. *Bioessays* **26**, 291–299.

Chapter 19

Regeneration of missing parts

Distribution of regenerative capacity

The ability to regrow missing parts is something that has fascinated humanity for millennia. But only when the molecular basis is understood will it be possible to know whether we humans could ever regenerate missing limbs, or defective kidneys or heart muscle. Like the other topics in this section, regeneration is at the frontier of developmental biology and involves many unsolved problems.

The most dramatic type of regeneration is the ability to regrow the main body **axis** following a severing of the main body. This is shown most profoundly by nemertean worms, which can grow a whole new body from a small fragment. Better studied are the **planarian** worms which can often regrow a new head from an anterior-facing cut surface and a new tail from a posterior-facing cut surface. Similarly, **hydroids**, including the familiar freshwater *Hydra*, can often regenerate new hydranths from a distal-facing surface and new basal parts from a proximal-facing surface. Although often supposed to have this ability, annelid worms show it to a much lesser degree. Tail regeneration is quite common, but head regeneration is less common, and a high level of regeneration of both heads and tails is confined to a few polychaete species

Whole-body regeneration is usually associated with the type of asexual reproduction in which the body can spontaneously fragment into two parts which each develop to reconstitute two new individuals. The requirements of asexual reproduction by fission and of whole-body regeneration are obviously very similar. Whole-body regeneration is also associated with a continuous turnover of cells in which there is a flux from undifferentiated stem cells to the principal body parts. This is found in both planarians and hydroids. An essential feature of whole-body regeneration is that it is bidirectional. The same cut surface can regenerate either a head or a tail, and normally what it forms depends on its relationship to the underlying tissues. This means that there is an essential polarity control that makes a decision about whether the regenerate is to be a head or a tail and does so

before the process of new part formation itself begins. Neither vertebrates nor insects, which are the animals best understood by developmental biologists, show any bidirectional regeneration at all, and regeneration of the whole body axis is confined to the tail of some vertebrates. They do, however, show some monodirectional (i.e. distal) regeneration of appendages, an ability that is not linked to asexual reproduction in the same way as whole-body regeneration. By contrast, the nematodes, well known to developmental biologists because of *Caenorhabditis elegans*, cannot regenerate at all.

Regeneration of external appendages in insects is confined to the **Hemimetabola**, those families which show a gradual progression to adulthood through a series of larval forms. If a leg or antenna is removed from a locust or cockroach, it will grow again and will become visible at the next molt when the old cuticle is shed. It is not found in the **Holometabola** which are those, like *Drosophila*, showing an abrupt metamorphosis in a pupal phase. Accordingly, *Drosophila* does not show any regeneration of adult structures, although the imaginal discs can regenerate if they are damaged prior to metamorphosis (see Chapter 17). Regeneration in vertebrates is displayed mainly by the amphibians. Urodeles (newts and salamanders) are usually able to regenerate limbs, tails, and jaws, both before and after metamorphosis. Anurans (frogs and toads) can often do so in the tadpole stage but lose the ability at metamorphosis. Lizards are well known for their ability to regenerate tails but the new tail does not include all of the tissues and structures of the original one.

Among mammals, the only regeneration of significant external appendages is the regrowth of antlers in deer. However, there are some interesting phenomena which are usually called regeneration but more properly referred to as hyperplastic growth. If a portion of the liver is removed, the remaining part rapidly grows back to the original volume. Liver regeneration is a hyperplasia rather than a true regeneration phenomenon because, although the volume of the organ is restored, its shape is not. A somewhat similar hyperplasia is shown by the single remaining

Classic Experiments

THEORY OF REGENERATION OF PATTERN

These classic papers in regeneration research are most unusual as they do not describe the result of a particular experiment, and were not even entirely correct in their conclusions. But they were very influential, particularly the first, which attempts to derive some general phenomenological rules for the regeneration of imaginal discs, limbs of hemimetabolous insects, and amphibian limbs. The second article is corrected in relation to some experimental results and focuses on the amphibian limb.

French, V., Bryant, P.J. & Bryant, S.V. (1976) Pattern regulation in epimorphic fields. *Science* **193**, 969–981.

Bryant, S.V., French, V. & Bryant, P.J. (1981) Distal regeneration and symmetry. *Science* **212**, 993–1002.

Although not "classic papers," the following provide a modern molecular account of the phenomena in insect legs that French, Bryant, and Bryant were trying to explain.

Campbell, G. & Tomlinson, A. (1995) Initiation of the proximodistal axis in insect legs. *Development* **121**, 619–628.

Mito, T., Inoue, Y., Kimura, S., Miyawaki, K., Niwa, N., Shinmyo, Y., Ohuchi, H. & Noji, S. (2002) Involvement of hedgehog, wingless, and dpp in the initiation of proximodistal axis formation during the regeneration of insect legs, a verification of the modified boundary model. *Mechanisms of Development* **114**, 27–35.

kidney following removal of the other kidney. This involves a rapid increase in the cell number but, except in very young animals, no increase in the number of nephrons.

This quick survey of regeneration shows that one of the problems for research is that extensive regenerative abilities are found mainly in animals that are not the familiar laboratory model species for developmental biology. This means that there are not any good pre-existing collections of cloned genes, or methods for introducing or removing genes, or methods for cell labeling or transplantation. New experimental approaches have needed to be introduced for individual experiments and this has been a factor reducing the rate of progress. However, important results have been obtained in regeneration research in recent years and the intrinsic interest of the phenomena ensures that they will stay high on the agenda.

Another important aspect of the distribution of regenerative capacity is that it often varies considerably between otherwise quite similar species. Such variations have led regeneration specialists to conclude that the ability to regenerate is probably an intrinsic biological property that is easily lost in evolution, presumably because it is associated with some significant cost. The alternative view, that each example of regenerative behavior is a specific adaptation to a specific life history, is not favored, although is hard to disprove altogether.

Planarian regeneration

Planarians are free-living freshwater worms belonging to the phylum Platyhelminthes. They are the simplest animals showing bilateral symmetry. They have three germ layers, ectoderm, mesoderm, and endoderm, but no coelom and no segmentation.

They have a head with eyes, but they do not have any circulatory or respiratory system. The gut opens at a muscular pharynx in the central part of the body and is a bilobed blind-ended sac, with no anus. Although simple, planarians display the basic features of animals, with Hox genes expressed in a nested way from posterior to anterior during regeneration, an *otx* homolog expressed in the regenerating head, and a *pax6* homolog expressed in the eyes (Fig. 19.1). A number of different species of planarian have been used for regeneration work. It is possible to graft cells and tissue between similar strains differing in **ploidy** and this enables some cell-lineage experiments to be undertaken, as the DNA content of donor and host tissues can be distinguished by staining. Planaria are not suitable for genetic experiments but the **RNAi** (**RNA interference**) method for ablation of specific messenger RNA (see Chapter 3) works well. The dsRNA is encoded in a plasmid with two T7 promoters facing each other. Bacteria containing this plasmid, and also the phage T7 polymerase, will transcribe the sequence in both directions and the two complementary RNA strands will hybridize to form double-stranded RNA. The bacteria can be mixed with the food fed to the planarians and when the worms eat it some of the dsRNA will be absorbed into their cells.

Planarians are in a state of continuous cell turnover and can grow or shrink on a daily basis depending on the food supply. There is a population of cells called **neoblasts** that are small with large nuclei and characterized by expression of an ATP-dependent RNA helicase similar to the *Drosophila* vasa protein. Vasa homologs are generally associated with germ-cell development and therefore with developmental **totipotency**. Neoblasts can be separated from the other cell types by **FACS**. They are the only cells that normally divide, and they are considered to comprise, or include, the **stem cells** that produce the 12–15 histologically distinguishable cell types making up the tissues of the worm. In the steady state, a worm contains about 20% neoblasts and 80% differentiated cells. It is not known whether the neoblast population contains subpopulations committed to particular fates, or whether there is a small stem-cell pool feeding a large transit amplifying pool, as is generally the case in higher animals (see Chapter 13).

Planarians have long been known for their high regenerative capacity. If cut in half, the anterior half will regenerate a new tail region from its posterior cut surface, while the posterior half will regenerate a new head region from its anterior cut surface (Fig. 19.2). Following transection, there is a muscular contraction

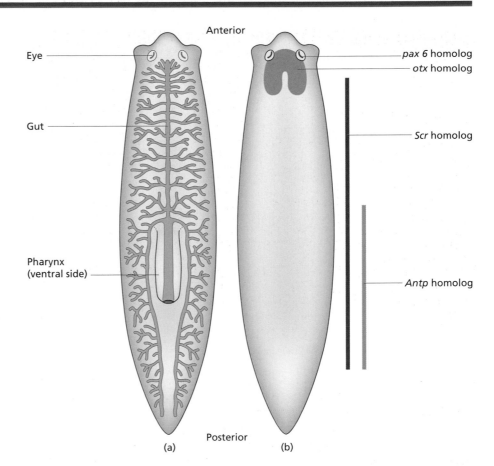

Fig. 19.1 The planarian worm.
(a) Anatomical features. (b) Aspects of the
zootype displayed during regeneration.

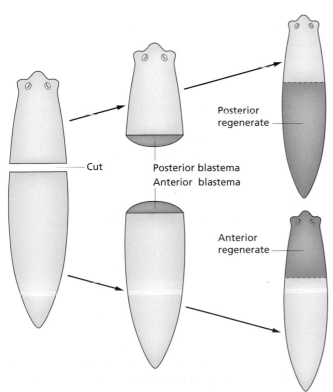

Fig. 19.2 Planarian anterior and posterior regeneration.

limiting the area of the cut, and a rapid formation of a thin wound epithelium across the cut surface. There is an accumulation of undifferentiated cells under the wound epithelium which makes up a regeneration blastema. **Blastema** is a general term for a regeneration bud containing undifferentiated proliferating cells, although in planarians the term confusingly refers to an unpigmented distal region in which there is no cell division. The cells populating the planarian blastema come from the dividing cells in the proximal region. The blastema enlarges by cell recruitment and redifferentiates over a few days to form the missing structures. The evidence that neoblasts give rise to the regenerate comes from two sources. Firstly, BrdU labeling of neoblasts before amputation leads to a blastema containing many BrdU-labeled cells. Secondly, the ability of worms to regenerate is destroyed by X-irradiation, which is followed by a rapid disappearance of neoblasts. There is no further production of new cells and the worms will die a few weeks later. It is possible to some extent to restore the ability to regenerate to X-irradiated worms by injection of purified neoblasts from an unirradiated donor. In unirradiated worms, although the bulk of the regenerate is formed from neoblasts, there is also considerable observational evidence for some dedifferentiation of cells near the cut, which enter the blastema.

The first decision that the cells of the blastema have to make is that of **polarity**: whether to be a head or a tail blastema. The

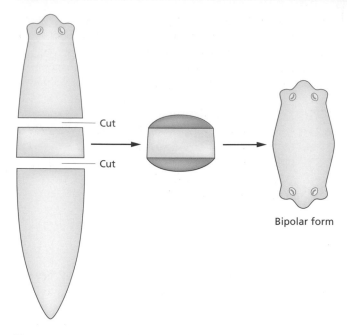

Fig. 19.3 Regeneration of a double anterior bipolar form from a short body section.

causes of this are not known, although it has been suggested that the origin of the wound epithelium is important, with an origin from dorsal epidermis in an anterior blastema and an origin from ventral epidermis in a posterior blastema. Certainly a surgically created dorsal–ventral confrontation is sufficient to initiate the formation of a new blastema. The polarity is unlikely to be encoded by Hox gene expression, as all of the Hox genes become activated in the first hour of regeneration in blastemas of both types. The existence of some sort of mechanism controlling polarity is demonstrated by the fact that errors can occur. In several species of planarian it is quite common for short segments from the central part of the worm to form anterior blastemas at both ends and regenerate into a Janus-headed **bipolar form** (Fig. 19.3). In such cases the posterior blastema must be lacking something normally derived from the anterior region that causes it to become a tail blastema.

Although events in different regions of a worm can occur with some autonomy, there is also evidence for some signaling processes associated with regeneration (Fig. 19.4). If a second head is grafted close to the first head, and then the first head is removed, its regeneration is now suppressed because of the presence of the second. If a head is grafted into the posterior,

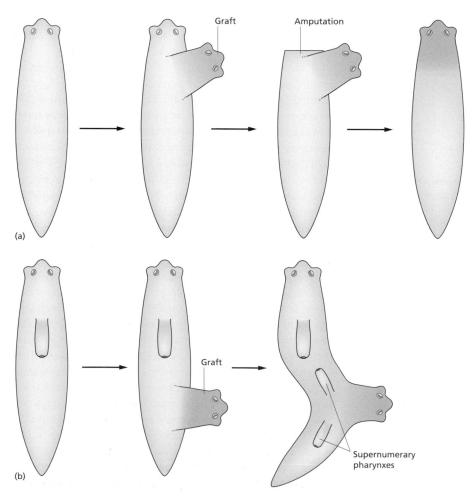

Fig. 19.4 Evidence for long-range interactions in planarian regeneration. (a) Suppression of head regeneration by grafted head. (b) Induction of extra pharynxes by a grafted head.

postpharyngeal region, then a pair of new pharynxes can regenerate from the junction. The nature of the signals involved in these interactions is presently unknown.

Vertebrate limb regeneration

Among the vertebrates only certain amphibian species can regenerate limbs after surgical removal. **Anuran** tadpoles can often regenerate limbs during the period of outgrowth and differentiation but the ability is lost during **metamorphosis**. By contrast, many **urodele** species can regenerate limbs throughout larval and adult life, and newts or axolotls (a type of salamander that never metamorphoses and remains in the larval condition for life) are normally used as experimental models for regeneration.

The process of regeneration

After amputation of a limb, the stump rapidly forms a wound epithelium by migration of epidermal cells over the cut surface. Then the internal tissues dedifferentiate to a depth of about 1 mm from the cut. The cells from this region form a **blastema** consisting of loose-packed mesenchymal type cells surrounded by a thick epidermal jacket. The blastema proliferates for a while and then the structures of the limb redifferentiate in a proximal–distal sequence. The regenerate is completed in miniature and then undergoes a long period of growth to get back to its original size. The stages of regeneration are referred to as dedifferentiation, cone, palette, early digit, and late digit (Fig. 19.5). The time required to regenerate the pattern in miniature varies with the age and size of the animal but is typically 2–3 weeks for small larvae and a few months for large adults.

Although apparently homogeneous, the blastema cells are of at least two types, distinguishable in newts by the use of a monoclonal antibody called 22/18. 22/18-positive cells are not found in the normal larval limb bud, but are found in the blastema following amputation of a mature limb. If larval buds are amputated at successively later stages, the 22/18 cells only appear in the blastema if the amputation takes place after the early digit stage. This is the time at which nerve fibers normally enter the limb bud, and, in fact, the 22/18 cells in the blastema are usually found in association with the nerves.

Because of their importance in limb development (see Chapter 15) the expression of Hox genes of the *a*- and *d*-paralog groups have both been studied in amphibian limb development and regeneration. During larval limb-bud development the expression of the genes is similar to higher vertebrates. Expression of the *a*-genes is nested from distal to proximal, and of the *d*-genes is nested from posterior to anterior. In regeneration they are expressed very early following amputation. Initially genes *a9* and *a13* come on all over the blastema, then as the blastema grows their patterns resolve into a nested arrangement

with *a13* distal. Some of the *d*-genes are activated as part of a wound response. This can occur in any part of the body and is not specific to regeneration. The Hox genes may be responsible for the **positional values** of blastemal cells which have been much investigated by grafting experiments (see below) although there is no direct evidence for this at present. Fibroblast growth factors (FGFs) are found to be expressed in the epidermal cap of the regeneration blastema. They are also mitogenic for isolated blastemas, although whether this reflects an apical ridge-like function or a neurotrophic function (see below) is unclear. Sonic hedgehog is expressed at the posterior margin of the blastema, as in the larval limb bud.

The source of cells for regeneration

When considering the source of cells for regeneration, there are three distinct issues that are often confused with one another:
1 Are progenitor cells local or distant in origin?
2 Are they formed by dedifferentiation of functional cells or from undifferentiated "reserve cells"?
3 Are they uni-, pluri-, or totipotent?

In relation to the first question it is known that the origin of cells is local to the amputation surface. This may be shown by X-irradiation which can completely suppress regeneration following a dose that inhibits most, but not all, cell division. The radiation beam can be shielded so that only a small part of a limb is irradiated. If the level of the future amputation site is irradiated, there is no regeneration, while if the region of the future amputation is protected, but neighboring tissues are irradiated, regeneration proceeds normally. This indicates that the blastema must arise from cells located no further than a few millimeters from the amputation surface, rather than from stem cells in the bone marrow or other distant sites.

The second question relates to the possible existence of "reserve cells" which are supposed to be pluripotent cells left over from early embryonic development and scattered throughout the tissues. Such hypothetical cells are often called **neoblasts**, following the planarian usage. However, neoblasts are clearly visible in planarians but are not visible at all in the amphibian limb. While it is never easy to prove the nonexistence of an entity like the reserve cell, there is no positive evidence that they do exist. By contrast, there is no doubt that dedifferentiation of functional cells can occur in regeneration. If newt **myofibers** are labeled by introduction of a marker gene and grafted into a limb, which is then amputated, they will break up into mononucleate cells. These can still be identified by the expression of the marker gene and they can be shown by BrdU labeling to undergo cell division. If the myofibers are kept in culture for some days they will spontaneously start to break up into mononuclear cells. This is accompanied by an increase in expression of the homeodomain transcription factor Msx1, and the change can be antagonized by treatment with a **morpholino** against Msx1. Similar changes can also be induced in mammalian **myotubes**

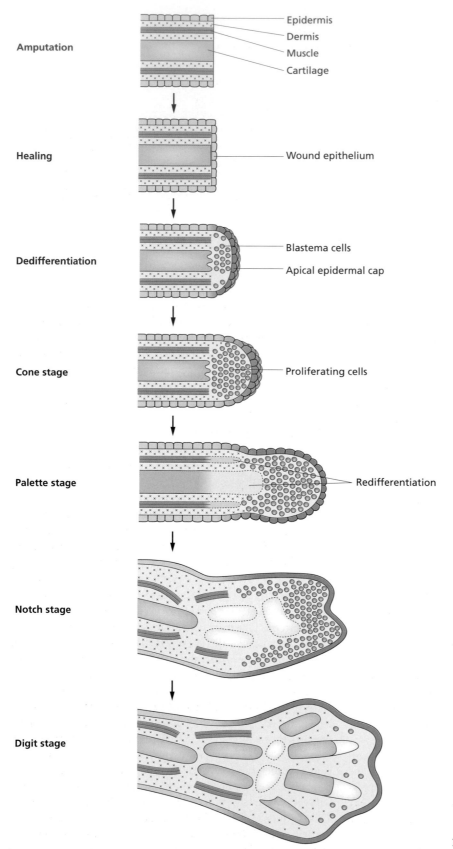

Fig. 19.5 Stages of amphibian limb regeneration.

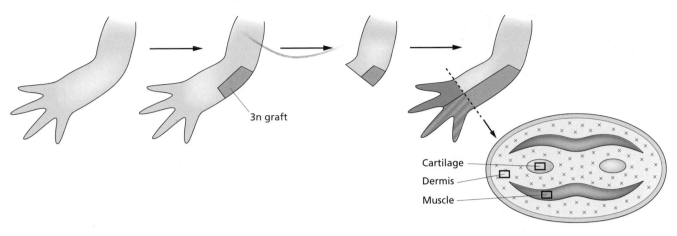

Fig. 19.6 Metaplasia during regeneration. A skin graft is made from a triploid donor, then the limb is amputated through the graft. The regenerate is analyzed for the presence of triploid cells in different tissues. Typically the dermis and the cartilage would contain 3n cells.

by overexpression of Msx1. In addition, newt myotubes in tissue culture will re-enter **S-phase** on addition of serum. In this regard they differ from the myotubes of mammals which are entirely postmitotic. The S-phase re-initiation is associated with inactivation by phosphorylation of the cell-cycle inhibitor Rb (retinoblastoma protein; see Chapters 3 & 18), and it depends on the activity of thrombin, the protease that unleashes the blood clotting cascade. It is thought that a product of thrombin proteolysis is responsible, and this offers a linkage between the events of wounding and those of regeneration.

The final question relates to the occurrence or otherwise of **metaplasia** during regeneration. Metaplasia was mentioned in Chapter 18 and means a conversion of one tissue type into another. It has been studied in regeneration using triploidy as a cell marker (Fig. 19.6). In most amphibian eggs it is possible to induce triploidy by administering a pressure shock soon after fertilization. The effect of this is to drive the second polar body back into the egg. It fuses with the male and the female pronuclei to yield a 3n, or triploid, zygote. Such embryos develop normally to adulthood except that they are sterile and their gonads are vestigial. The cell volume is greater in 3 : 2 proportion to diploid cells, and in axolotls (but not in most other species) the triploid cells are readily identifiable because they have three nucleoli while the diploid ones have only two. If a graft is made of a pure tissue from a triploid to a diploid, allowed to heal, and the limb is amputated through the graft, then it can be asked which cell types in the regenerate carry the triploid marker. Such experiments show that a graft of labeled epidermis can only generate more epidermis and not internal tissues. By contrast, a graft of labeled cartilage can populate not only cartilage but also all the connective tissues: dermis, tendons, ligaments, and fibrous capsules. Furthermore, dermis can readily convert to cartilage. A graft of labeled myofibers will label predominantly muscle in the regenerate, with a few percent of other cell types including cartilage. Thus, these experiments suggest that there are probably three tissue lineages, epidermal, muscle, and connective

tissue, although extensive metaplasia can occur between the different types of connective tissue. This is similar to the situation in the developing limb (see Chapter 15) where there are also three lineages: the connective tissue types derived from the limb-bud mesenchyme, the muscle cells from the somites, and the epidermal cell types from the limb-bud epidermis.

The "neurotrophic" factor

Limb regeneration is absolutely dependent on a nerve supply to the limb. In animals in which the nerve supply has been transected, dedifferentiation occurs as normal resulting in the formation of a blastema, but this blastema fails to grow and regeneration is aborted. The function of the nerves is to release mitogenic factors, often referred to as the **neurotrophic factor(s)**. Note that the term "neurotrophic factor" normally means a growth factor required for the growth or survival of neurons, whereas in the regeneration context it means a mitogenic activity derived from neurons, required for the growth of the blastema. The neurotrophic factors probably include FGFs, which are abundant in nerve axons and which are mitogenic for isolated blastemas. They probably also include a neuregulin formerly called glial growth factor. Neuregulins are similar to epidermal growth factor and are ligands for the receptor tyrosine kinase ErbB2. Only the 22/18 subpopulation of blastemal cells respond to neuregulin. Although these factors satisfy the expression and activity criteria of the three criteria for proof (see Chapter 4), it has not so far been possible to ablate their activity in the sort of animal capable of limb regeneration, so the case that they really are the neurotrophic factors remains unproved.

A remarkable fact about the neurotrophic effect is that it is possible to make limbs that lack almost all innervation and these "aneurogenic" limbs do not require a neurotrophic factor for regeneration. One way of making them is to remove the neural tube from the limb region and to join the operated embryo to

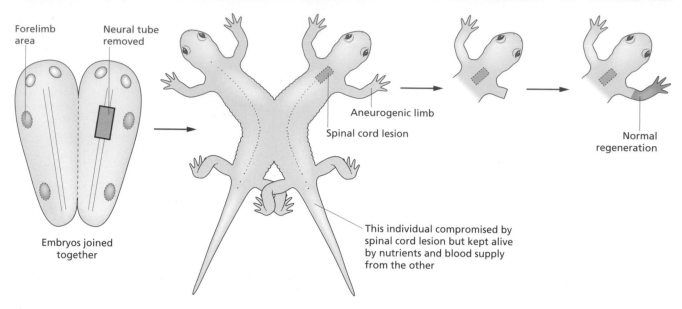

Fig. 19.7 Aneurogenic limb regeneration. The limb lacks innervation because of the removal of part of the neural tube from the embryo. Despite this, it regenerates normally.

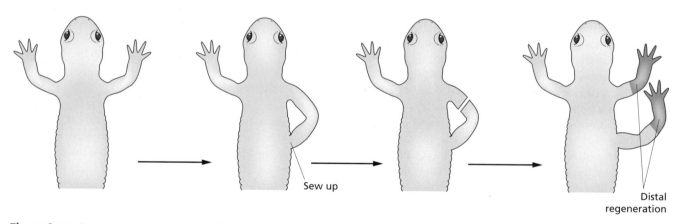

Fig. 19.8 Distal regeneration occurs even when the cut surface is proximal facing, as shown in this classic experiment.

another in **parabiosis** (joint blood circulation) to ensure survival (Fig. 19.7). A limb will then grow out in the operated region in which the innervation is limited or nonexistent. These limbs are otherwise normal although there is a degeneration of muscle at a later stage. Despite the lack of nerves, they can regenerate normally. The blastemas of aneurogenic limbs do not include the 22/18 cell population that responds to neuregulin. So it seems that in the absence of these cells, the other cells can form a complete functional blastema, but if 22/18 cells are present then no blastema can be formed without the neurotrophic factor.

Regeneration of regional pattern

The pattern of the regenerate conforms with the stump such that all parts **distal** to the cut are replaced. Proximal regeneration does not occur from an amputation surface even when the cut surface is proximal facing, as achieved by the procedure shown in Fig. 19.8. This fact is generalized as the "law of distal transformation," which holds for all vertebrate and arthropod appendages capable of regeneration. In the amphibian limb it also holds for **intercalary regeneration**, which means regeneration occurring at a tissue junction between parts that are not normally neighbors. When blastemas are transplanted distal to proximal, the potential gap in the pattern of the regenerate is filled in by intercalary regeneration from the stump side. However, if a proximal blastema is grafted to a distal stump, there is no reverse intercalation to fill in the gap, instead each component regenerates what it normally would, leaving a discontinuity in the pattern (Fig. 19.9). In such experiments the tissue contributions can be determined by using grafts between triploid and diploid animals.

The results of many experiments on regeneration suggest that the tissues of the limb carry cryptic codes for regional identity, often called **positional values**, which specify what structures shall be formed on differentiation. This may be illustrated by the

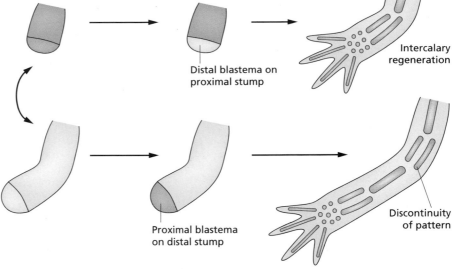

Fig. 19.9 Intercalary regeneration in the proximodistal axis. The stump can generate more distal tissue, but the blastema cannot generate more proximal tissue.

Distal blastema on proximal stump

Intercalary regeneration

Proximal blastema on distal stump

Discontinuity of pattern

following type of experiment: a limb is irradiated to suppress regeneration; a cuff of skin is grafted from the upper arm of an unirradiated animal to the lower arm of the irradiated one; then the composite limb is amputated through the graft. A regenerate will form, derived from the graft, and the regenerated pattern will start at the upper arm rather than at the lower arm, since this represents the proximodistal code of the graft. The molecular basis of the positional value remains unclear, but it is associated with differences in cell adhesion. This may be shown by grafting a blastema from one limb to the junction of the blastema and stump on another. Following the outgrowth and differentiation of this graft–host combination, the donor blastema will form an ectopic limb branching off the host limb. Remarkably, the proximodistal level at which the ectopic limb branches off corresponds to its own proximodistal character. In other words if it is a proximal blastema then it will end up proximally, while if it is a distal blastema it will end up distally (Fig. 19.10). One possible molecular component of positional value is the newt homolog of CD59, an inhibitor of complement activation. This is more abundant in the proximal than the distal blastema and is induced by retinoic acid (see below). It is a cell surface protein with a GPI anchor. Normally a proximal blastema will engulf a distal one if they are cultured in contact, indicating that distal cells adhere stronger to each other than do proximal cells (see Chapter 3). But a neutralizing antibody to CD59 will prevent this, suggesting that the reduced adhesivity of the proximal cells is due to the higher level of CD59. Transcription factors associated with positional value are the Meis proteins, which are expressed in the proximal segment of the vertebrate limb during development, and also in the proximal part of the *Drosophila* leg disc. If *Meis* genes are introduced by electroporation into cells of an axolotl blastema, the cells containing them will move to more proximal positions than control cells.

However, the limb is asymmetrical in three dimensions and

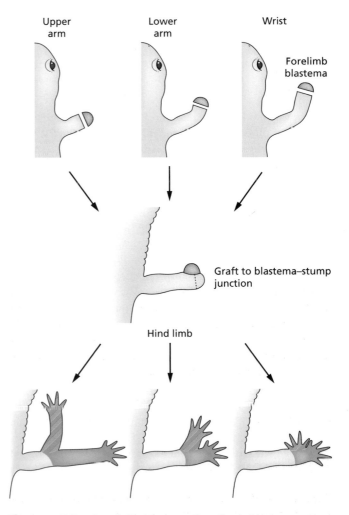
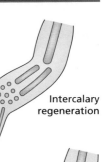

Upper arm

Lower arm

Wrist

Forelimb blastema

Graft to blastema–stump junction

Hind limb

Fig. 19.10 Migration of a blastema, transplanted to the blastema–stump junction of the hindlimb. Proximal blastemas end up in proximal positions, distal blastemas end up in distal positions.

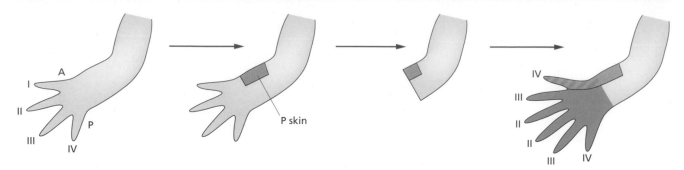

Fig. 19.11 Intercalary regeneration in the anteroposterior axis. A posterior skin graft provokes the formation of the intervening digits from host tissue.

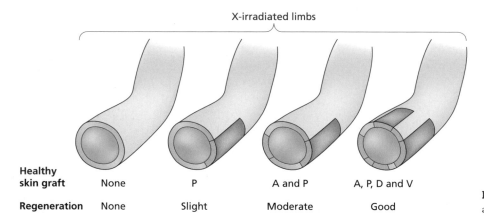

Fig. 19.12 Need for pattern discontinuity at the cut surface for distal regeneration.

so in principle should require three sets of codes to specify the pattern of differentiation in the three anatomical axes: proximodistal, anteroposterior, and dorsoventral. In some respects it does behave accordingly. The cryptic codes can only be detected when dedifferentiation and blastema formation are provoked by amputation. A common experimental protocol involves rearranging a limb surgically and sewing it together so that it heals up. This in itself does not provoke regeneration. But if the modified limb is then amputated, a blastema will form at the cut surface and the pattern of structures regenerated depends both on the representation of codes at the cut surface and on the nature of interactions between them.

There are two reasonably well-established principles that summarize the operation of these interactions in the transverse axes (anteroposterior and dorsoventral). First is the principle of **intercalation**. This states that where a blastema is generated from two tissue regions with a discontinuity between their codes, the regenerate will fill in the gap with structures that would normally form in between the two regions. This may be shown by grafting skin from one part of the limb to another and then amputating through the graft. If neither component was irradiated, the intervening structures are derived from both graft and host (Fig. 19.11). Such experiments can also reveal which tissues in particular carry the regional codes. The answer is that they are carried by all the connective tissues, but not by the muscle or by the epidermis. Hence, although skin grafts are often used in such experiments, it is only the **dermis** that is

active. This is shown by the fact that grafts of pure dermis have a similar effect whereas grafts of pure epidermis are without effect.

The other principle is that some degree of discontinuity of codes at the cut surface is necessary to initiate distal regeneration. For example, if a small piece of skin is grafted to an irradiated limb followed by amputation through the graft, then only a very limited regenerate will be formed. But if two similar-sized pieces from opposite sides of the limb are grafted, then the regenerate will be much more substantial and contain many pattern elements (Fig. 19.12). The same applies to the formation of extra regenerates following nerve deviation. If the brachial nerve is surgically deviated through the skin of the upper arm then this will provoke the formation of a blastema, but this then regresses. If the nerve deviation is accompanied by a posterior to anterior skin graft then a complete limb will form at the junction. It is not, however, necessary that *all* codes need be present at the cut surface, as partial limbs, or abnormal limbs lacking some structures, can both regenerate well. For example, double posterior limbs made by ZPA (**zone of polarizing activity**) grafting in the embryo (see Chapter 15) will regenerate double posterior regenerates.

Both principles are seen at work in a spectacular type of experiment that has gripped the imagination of all students of regeneration. This is the graft of a blastema to a stump on the contralateral side of the animal. Because limbs are asymmetrical in three dimensions, it is not possible to superimpose a right

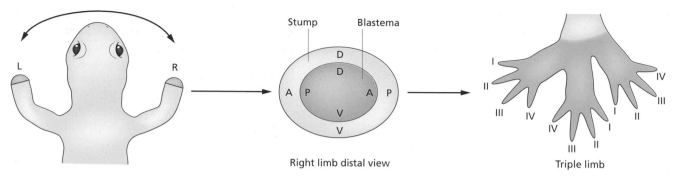

Fig. 19.13 Axial inversion of a blastema onto a stump produces a triple limb with one central member and two supernumeraries.

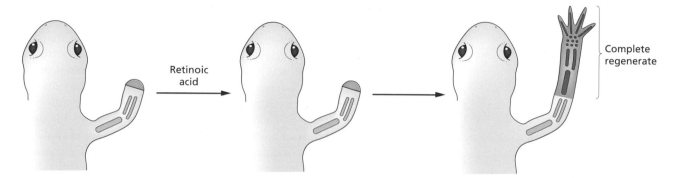

Fig. 19.14 Proximalization of blastema by retinoic acid treatment.

limb on a left limb. If a blastema is grafted from right to left, or left to right, side, then it will inevitably have one axis inverted in relation to the host stump (Fig. 19.13). This graft reliably gives a pair of supernumerary regenerates at the points of maximum disparity. Therefore in total there are three limbs regenerated. One arises from the blastema, with its original polarity, and two arise from the discordant blastema–stump junctions, with polarity inverted in relation to the central limb (Fig. 19.13). A simple rotation of the blastema on its own stump can also give rise to supernumeraries, but the number, position, and anatomical completeness is much more variable because the disparity of codes is similar all round the circumference.

Again, there is almost no information about the molecular identity of the positional values around the circumference of the limb. Shh is expressed in the posterior part of the blastema and introduction of *Shh* into the anterior side of a blastema using vaccinia virus will produce a double posterior duplication. This shows at least some conservation of mechanism from limb development, but the positional values are presumably among the unknown targets of Shh signaling.

Retinoic acid effects

Several different types of respecification of code can be achieved by treating cone-stage blastemas with retinoic acid. These include effects on each of the three pattern axes, leading to proximaliza-tion, posteriorization, and ventralization. Since in these experiments the whole animal is immersed in a solution of retinoic acid the blastema is probably exposed to a uniform dose of the reagent during treatment.

When an amputated but otherwise anatomically normal limb is treated, the result is proximalization, such that the further development of the blastema leads to serial duplication of structures. So for example a wrist-level blastema treated with retinoic acid can regenerate a complete arm (Fig. 19.14). This of course violates the normal law of distal transformation. There is some dose–response effect, with higher doses giving more proximal-ization, but above a certain dose a growth inhibition is found, causing a total suppression of regeneration.

The posteriorization effect is shown when blastemas from half limbs are treated. Normally a posterior half limb will show more regenerative capacity than an anterior half limb, the former often regenerating a complete distal limb, and the latter little or nothing. But this is altered by retinoic acid, which causes normal limb regeneration from anterior halves and suppresses regeneration from posterior halves. If limbs are surgically created with double-anterior or double-posterior morphology, then normally the double posteriors produce some distal regen-erate while the double anteriors produce nothing. Again, in retinoic acid this is reversed and the double-anterior limbs pro-duce two complete regenerates while the double-posterior limbs produce nothing (Fig. 19.15). These results can be explained on the assumption that retinoic acid posteriorizes the blastema.

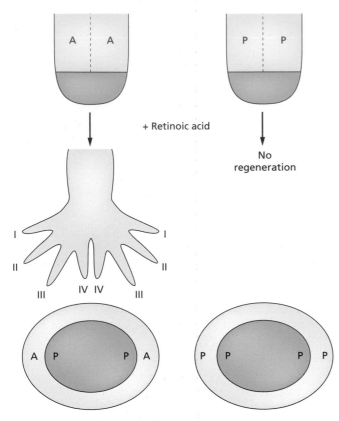

Fig. 19.15 Effect of retinoic acid on surgically produced double-anterior and double-posterior limbs. The double-anterior limb is caused to form a double regenerate because of interaction between the posteriorized blastema and the unmodified stump.

The uniform treatment with retinoic acid makes the blastema posterior in character all over. Interactions between the blastema and newly dedifferentiated tissue from the stump then give rise to a limb wherever there is a discontinuity between anterior and posterior codes.

Comparable results are obtained when retinoic acid is used to treat dorsal and ventral half limbs. After retinoic acid treatment the dorsal halves regenerate and double-dorsal halves form

double limbs. This suggests that the retinoic acid acts to ventralize the blastema, and the regenerates arise after interaction between blastema and newly dedifferentiated stump tissue. Evidence that this type of interpretation is correct comes from the fact that late treatment with retinoic acid, after dedifferentiation has finished, leads to pattern truncation. This is because by this stage the stump can no longer provide further dedifferentiated tissue to enable the interactions to occur that are needed to initiate distal regeneration.

Thus, the overall effects of retinoic acid are rather complex, involving a simultaneous proximalization, posteriorization, and ventralization of the blastema. The molecular basis of this remains unclear except for the proximalization process, where it is known that both CD59 and *Meis* genes are upregulated by retinoic acid. The initial step in its action requires binding of the retinoic acid to its receptors, which belong to the nuclear receptor class. The urodele limb contains several types of retinoic acid receptor, including some forms specific to urodeles called δ-receptors. To identify which receptor was responsible for the recoding effects, **domain swaps** were made with thyroid receptor, such that the chimeric molecules had the thyroid hormone-binding domain but the retinoic acid receptor domains for DNA binding and gene activation (Fig. 19.16). This means that the genes normally regulated by retinoic acid can now be regulated by thyroid hormone, but only if the particular receptor introduced into the limb is the one that can transduce the signal. Genes for the receptors, together with a gene for cell identification, *alkaline phosphatase*, were introduced into retinoic acid-treated distal blastemas by biolistic particle bombardment. The blastemas were then grafted to proximal stumps. Proximalization is revealed by the presence of alkaline phosphatase-positive cells in the intercalary regenerate, which is normally formed from the stump tissue alone. These experiments showed that proximalization was only mediated by one out of the five possible receptors, one of the urodele-specific types called retinoic acid receptor δ2.

Although the spectacular effects of experimentally applied retinoic acid are well described, this does not necessarily mean that retinoic acid also has an endogenous role in the control of regeneration. Levels of endogenous retinoic acid have been measured in one of two ways, either by chemical extraction and separation, or by observing the response to the amphibian tissues of mammalian reporter cells containing a retinoic acid response element (RARE) coupled to a reporter gene. These methods have shown that retinoic acid is present in the wound epidermis of regeneration blastemas. It is actually the 9-*cis* isomer of retinoic acid, which is more potent than the all-*trans* form. It is found to be enriched

New Directions in Research

The sky is the limit in regeneration research. Some key questions are:
1 Why can some structures regenerate and other apparently similar ones not do so (e.g. urodele and anuran limbs)?
2 What is the molecular basis of dedifferentiation?
3 What are the rules about the degree of metaplasia that can occur in regeneration?
4 What is "positional value"?
5 How can cells regenerate a particular pattern when this may be quite different from what is present at the cut surface (e.g. the pattern of a hand versus the upper forelimb)?

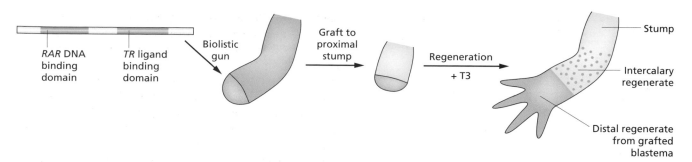

Fig. 19.16 Experiment to find which retinoic acid receptor causes proximalization. A domain-swapped receptor is transfected into a distal blastema. This is grafted to a proximal stump and the animal treated with thyroid hormone. The presence of donor cells in the intercalary regenerate is an indication of proximalization induced by the thyroid hormone.

in proximal over distal blastemas, and in the posterior over anterior region within a blastema. So retinoic acid is present in the wound epidermis and it has biological activity, but proof that it has a role *in vivo* must await an ablation experiment. This is one of many experiments on regeneration that has yet to be carried out.

Key Points to Remember

- Some invertebrates can regenerate whole bodies from small tissue fragments. In such cases a transected body can form a new head from the anterior-facing cut surface and a new tail from the posterior-facing cut surface. This is known as bidirectional regeneration. Insects and vertebrates can only regenerate appendages, and this is monodirectional, always proceeding distally from the cut surface.
- Planarians are animals capable of bidirectional regeneration. Regenerative capacity is associated with the existence of neoblasts, which are small undifferentiated mitotic cells making up about 20% of the worm's body. Neoblasts act as stem cells for normal growth and for regeneration.
- Most urodele amphibians can regenerate limbs throughout life. The regenerate is formed from the cells near the cut surface. There is some dedifferentiation, in particular

of multinucleate myofibers which re-enter the cell cycle and break up into mononucleate cells.
- A regeneration blastema of undifferentiated cells is formed. It grows and differentiates in proximal to distal sequence to reform the missing parts. The miniature organ then grows to the normal size.
- There is some metaplasia on regeneration, in particular the dermis and cartilage interconvert extensively and there is limited conversion of myofibers to other cell types.
- Growth factors secreted by nerves are needed for the growth of the blastema.
- The limb behaves in response to grafting experiments as though it contains a system of positional values, labeling each part, which control the pattern of structures formed during regeneration. Some discontinuity of positional value is necessary for distal regeneration.
- Retinoic acid can proximalize, posteriorize, and ventralize the blastema.

Further reading

General

Goss, R.J. (1969) *Principles of Regeneration*. New York: Academic Press.

Scadding, S.R. (1977) Phylogenetic distribution of limb regeneration capacity in adult amphibia. *Journal of Experimental Zoology* **202**, 57–68.

Brockes, J.P., Kumar, A. & Velloso, C.P. (2001) Regeneration as an evolutionary variable. *Journal of Anatomy* **199**, 3–11.

Planarian regeneration

Baguña, J., Saló, E., et al. (1994) Regeneration and pattern formation in planarians: cells, molecules and genes. *Zooloogical Science* **11**, 781–795.

Agata, K. (2003) Regeneration and gene regulation in planarians. *Current Opinion in Genetics and Development* **13**, 492–496.

Agata, K., Tanaka, T., Kobayashi, C., et al. (2003) Intercalary regneration in planarians. *Developmental Dynamics* **226**, 308–316.

Sanchez-Alvarado, A. (2003) The freshwater planarian *Schmidtea mediterranea*: embryogenesis, stem cells and regeneration. *Current Opinion in Genetics and Development* **13**, 438–444.

Amphibian limb regeneration

General
Iten, L.E. & Bryant, S.V. (1973) Forelimb regeneration from different levels of amputation in the newt *Notophthalamus viridescens*: length, rate and stages. *Wilhelm Roux's Archives of Developmental Biology* **173**, 263–282.

Mescher, A.L. (1996) The cellular basis for limb regeneration in urodeles. *International Journal of Developmental Biology* **40**, 785–795.

Brockes, J.P. (1997) Amphibian limb regeneration: rebuilding a complex structure. *Science* **276**, 81–87.

Nye, H.L.D., Cameron, J.A., Chernoff, E.A.G. & Stocum, D.L. (2003) Regeneration of the urodele limb: a review. *Developmental Dynamics* **226**, 280–294.

Dedifferentiation and metaplasia
Namenwirth, M. (1974) The inheritance of cell differentiation during limb regeneration in the axolotl. *Developmental Biology* **41**, 42–56.

Brockes, J.P. & Kumar, A. (2002) Plasticity and reprogramming of differentiated cells in amphibian regeneration. *Nature Reviews in Molecular and Cellular Biology* **3**, 566–574.

Tanaka, E.M. (2003) Cell differentiation and cell fate during urodele tail and limb regeneration. *Current Opinion in Genetics and Development* **13**, 497–501.

Positional information and retinoic acid
Tank, P.W. & Holder, N. (1981) Pattern regulation in the regenerating limbs of urodele amphibians. *Quarterly Reviews of Biology* **56**, 113–142.

Bryant, S.V. & Gardiner, D.M. (1992) Retinoic acid, local cell–cell interactions, and pattern formation in vertebrate limbs. *Developmental Biology* **152**, 1–25.

Stocum, D.L. (1996) A conceptual framework for analysing axial patterning in regenerating urodele limbs. *International Journal of Developmental Biology* **40**, 773–783.

Maden, M. (1997) Retinoic acid and its receptors in limb regeneration. *Seminars in Cell and Developmental Biology* **8**, 445–453.

Chapter 20

Evolution and development

The interface between developmental biology and evolutionary biology has a long history. At the beginning of the nineteenth century the German embryologist Von Baer noticed that the early stages of different types of vertebrate embryo were very similar and proposed that the more general features of animals develop before the more specific. In the mid-nineteenth century Ernst Haeckel formulated the theory that "Ontogeny recapitulates Phylogeny," in other words the developmental stages of an individual organism resemble the sequence of its evolutionary ancestors. This view made sense in the context of the then current Lamarkian theory of evolution in which heritable changes could arise from life experience and would therefore be "added on" to the end of a developmental sequence. In the early twentieth century Neodarwinism became generally accepted. This is the synthesis of Darwin's theory of natural selection, Mendel's genetics, and the quantitative mathematical theory of genetics produced by Fisher, Wright, and Haldane. The mechanism of evolutionary change is considered to arise from mutations each of which confers a reproductive advantage on the individuals carrying them such that the mutation spreads through the population to become the wild-type allele. According to Neodarwinism, morphological change will occur gradually and result from the action of a number of mutations each having a small effect. Change arising from natural selection is called **adaptative evolution** or adaptation. In the second half of the twentieth century studies in molecular biology made it clear that a great deal of change in the primary sequence of DNA was not adaptive but **neutral evolution**, consisting of an accumulation of mutations of no selective consequence which spread through the population by the effects of random sampling of alleles from one generation to the next (genetic drift).

The molecular understanding of development has had an important impact on evolutionary biology because it has answered certain questions that could not be answered before. The most impressive result is the ability to reconstruct a credible model of remote animal ancestors way beyond the reach of comparative morphology. Developmental biology has also contributed to attempts to understand the true phylogeny of animals. In addition it raises certain questions for Neodarwinism, concerning the possible role of developmental constraints and of mutations of large effect.

Macroevolution

Organisms are classified in a hierarchical way, and the subdivisions of life are called **taxa** (singular **taxon**). Species are grouped into genera, families, orders, classes, and the highest-grade animal taxa are the **phyla** (singular **phylum**) of which there are about 35, the exact number depending on the author. They vary in size from large phyla such as the Mollusca containing tens of thousands of species, to tiny phyla such as the Loricifera, with only a handful of species. The vertebrates and insects, which are the most important groups from the point of view of developmental biology, are not actually phyla but classes. The vertebrates make up by far the largest class of the phylum Chordata, and the insects are the largest class of the phylum Arthropoda.

A classification system (= **taxonomy**) can be completely arbitrary and still be useful in the sense that it allows for the unambiguous identification of specimens. However there has long been an attempt in biology to make the taxonomy congruent with the actual evolutionary history of the organisms concerned. According to this tradition each taxon should be a **clade** consisting of the complete set of descendants from a common ancestor (Fig. 20.1). Certain well-known groups are not clades but are tolerated for reasons of familiarity. For example the reptiles are not a clade because they do not contain the birds, which are descendants from within the reptiles and not a separate group derived form the vertebrate stem line. Although any number of arbitrary classification schemes are possible, there is only one "true tree" which follows exactly the bifurcating branches that actually occurred in evolution. This is called a **phylogenetic tree** whether it deals with the phyla themselves or with lower-level taxa. It follows that each node of a phylogenetic tree should

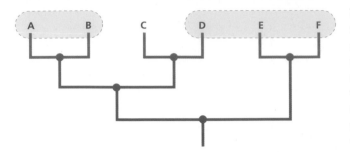

Fig. 20.1 Phylogenetic (= cladistic) taxonomy. A–F represent taxa arising from the phylogeny shown. A+B represents an acceptable higher taxon because it is a complete set of organisms derived from a common ancestor. D+E+F is not an acceptable higher taxon because some descendants of the common ancestor are omitted.

correspond to a population of real ancestral organisms that split to give rise to the two branches. Of course this will only be literally true at the species level of detail, because what actually evolves are species in the sense of interbreeding populations of organisms. Evolution above the species level is called **macroevolution**. But it must be remembered that there is no motor of evolution of higher taxa because this would involve parallel changes being selected simultaneously over all the constituent species, which is exceedingly unlikely. Macroevolution is simply the consequence of a lot of species-level evolution, including extinction events, which over long time periods shifts the overall composition of life on Earth. So a node on a phylogenetic tree representing higher taxa still corresponds to one specific ancestor, namely the individual species that split to give the two lines of descent, but there will be many other organisms of related taxa around at the same time and represented on the same tree.

The construction of phylogenetic trees is a complex business outside the scope of this book. But there are two approaches in principle. One involves scoring large numbers of characters, including continuous quantitative characters, and constructing the tree on the basis that more similar specimens should be closer together. This is called numerical, or phenetic, taxonomy. The other involves applying certain assumptions about the likelihood of different types of change to try to deduce an actual phylogenetic tree. This is called cladistic, or phylogenetic, taxonomy. Zoologists have long drawn a distinction between two types of similarity between different organisms. They are called **homology** and **analogy**. If two structures are homologous, it means not only that they look similar, but that this similarity is due to descent from a common ancestor possessing the ancestral version of the part in question. An example is the **tetrapod** limb. Many types of vertebrate, from humans to crocodiles, have limbs that are obviously similar in the number and arrangements of the bones and muscles, and in their general position on the body. The reason they are similar is because there was once an ancestral tetrapod, from which all existing tetrapods are descended, and its limbs looked like this. Because they share common descent, homologous structures are normally formed by similar developmental mechanisms. By contrast, if two structures are analogous it means that there is no common ancestry and the parts look similar because the pressure of natural selection has forced a convergence of structure to meet the need for a similar function. An example is the insect wing versus the wing of a bird. Obviously these cannot have a winged common ancestor, because we know from the fossil record that there was a long history both of vertebrates and of insects before any winged members appear in either group. Furthermore, the common ancestor of insects and vertebrates must have been a marine organism as it would have existed before colonization of the land by any animal group. Note that although the wings themselves cannot be homologous, various individual genes or genetic pathways involved in making the wings may still be homologous (see below). Characters that are thought to resemble those of ancestors are called primitive or **basal**, while those thought to result from adaptive evolution are called **derived**.

Molecular taxonomy

A great deal of modern taxonomy uses the primary sequences of genes as the raw data rather than morphological characters. Because only a fraction of the amino acids in a protein are necessary for its biological activity, the sequences of genes and proteins change gradually over evolutionary time. Most of the changes are neutral, that is they do not affect the survival or reproduction of the individual organism and they are neither favored nor discriminated against by natural selection. For any individual mutation there is a very small chance that it will eventually spread through the whole species and become the normal version of the gene or protein. Because evolutionary time is very long, even a very small chance eventually becomes a certainty and there are many substitutions of amino acids in those positions in the protein where they do not compromise the biochemical function. Population genetic theory predicts the existence of a **molecular clock**, indicating an approximately linear relationship between the time since the divergence of two lineages and the number of differences between the sequences. Phylogenetic trees can be constructed from primary sequence data using various different types of algorithm including ones that simply compare the number of differences and those that try to minimize the number of substitution events. The ideal genes to use for these studies are those that can clearly be identified as homologous and that show some variation over the range of organisms being studied, but not so much that substitutions are piled on top of each other. For long-range phylogeny much use has been made of the large and small ribosomal RNA genes as these have retained the same function since at least the divergence of prokaryotes and eukaryotes. Among protein-coding genes some favorites have been RNA polymerases, glycolytic enzymes, and cytochromes. Molecular phylogenetic methods have the attraction that they are completely independent of morphology and can in principle be applied to a set of sequences from present-day organisms without knowing anything about the organisms from which the sequences came. However, they do not necessarily predict a unique phylogenetic tree and the closest approach to the true tree is likely to be obtained by combining information from comparative morphology, molecular sequences, and fossil evidence.

The rise of molecular taxonomy has extended and refined our concepts of homology and analogy. Although many amino acids at inessential positions in a protein sequence will change because of neutral mutation, the amino acids at many inessential positions also remain the same. It is the presence of these inessential identities that shows that two genes from different organisms are really homologous. But it is not entirely straightforward identifying homologous genes in different organisms because of the widespread occurrence of **gene duplication** in evolutionary time. If a gene becomes duplicated then only one copy is required to perform the original function and the other copy is free to assume a different, although often related, function. In particular, among the vertebrates a high proportion of genes belong to multigene families. Two genes in the same organism which are the result of a gene duplication are called **paralogs**. Two genes in different organisms are called **orthologs** if they are directly related by lineage and preserve a common function. To make proper comparisons between species it is essential to identify the true ortholog, otherwise the comparison that is made will relate to the time of the gene duplication event and not the time of the splitting of the ancestral species.

If two genes converge to the same function starting from totally different sequences they may have no identities at all, or any they do have are confined to a small region of the molecule actually responsible for its catalytic or other biochemical activity. A molecular example of analogy is provided by the crystallins, which are the proteins making up the lens of the eye. In different types of vertebrate quite different sorts of protein have been brought into service to make a transparent lens, and although they serve the same physiological function they have no primary sequence in common.

Phylogeny of animals

Figure 20.2 shows a modern consensus phylogenetic tree for a few of the most important phyla. Because of the lack of

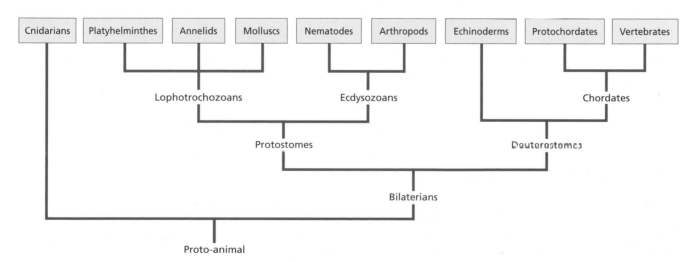

Fig. 20.2 Phylogenetic trees showing the principal animal taxa.

morphological characters in common between the adult animals of different phyla, information from embryos and larvae has traditionally been used to attempt to draw up the phylogenetic tree. The first of these characters is the number of germ layers. As we have seen, most animals have formed three tissue layers by the end of gastrulation, the ectoderm, mesoderm, and endoderm. They are called **triploblasts**. Only three phyla are not triploblastic, the Porifera (sponges), which do not have well-defined tissue layers, the Cnidaria (*Hydra*, jellyfish, sea anemones), and the Ctenophora (comb jellies), which have two layers. The Cnidaria and Ctenophora are called **diploblasts**. Animals of these phyla also, in the main, show radial symmetry whereas most of the triploblastic animals show bilateral symmetry and are therefore also known as the Bilateria.

The next important feature is the presence or otherwise of a **coelom**. This is the cavity formed within the mesoderm, lined with a mesoderm-derived epithelium, the peritoneum, and often the principal body cavity. Platyhelminthes (flatworms) and Nemertea (ribbon worms) do not have a coelom and are described as acoelomate. Many other invertebrate phyla, including the Nematoda, have a body cavity that is only partly surrounded by mesoderm, and are described as pseudocoelomate. Annelida (segmented worms), Mollusca, Arthropoda, Echinodermata (sea urchins, starfish, etc.), and Chordata all have a coelom.

The coelomate phyla were traditionally divided into two "super phyla" called the **Protostomia** and the **Deuterostomia**. The latter group contains the Echinodermata and Chordata. It was defined by **radial cleavage**, formation of the anus from the blastopore**,** and **enterocoely**, which is formation of the coelom by budding of the mesodermal rudiment from the gut. In fact, true enterocoely is not found in vertebrates but it is found in some protochordates and echinoderms and thus serves to link these two phyla together. The defining features of protostomes were **spiral cleavage**, early-acting **cytoplasmic determinants**, and **schizocoely** (formation of the coelom by splitting of the mesodermal layer). Older textbooks may also include formation of the mouth from the blastopore, but this is incorrect. In protostomes the blastopore typically narrows to a ventral slit from which both mouth and anus derive.

Recently the use of molecular taxonomy has changed this picture. The deuterostomes remain as a taxon, but the significance of the coelom has disappeared and there are now two new superphyla making up the protostomes. The annelids and molluscs, which share a type of larva called the **trochophore**, are grouped with some other phyla in the **Lophotrochozoa**, while the arthropods and nematodes, together with all other molting animals, comprise the **Ecdysozoa**.

The advantage of molecular taxonomy is that it can look back further in time than morphology and can in principle tell us when the last common ancestor of all animals was flourishing. In practice this has proved rather difficult as the estimates depend on the calibration dates derived from the fossil record. Although the fossil dates themselves are reasonably reliable, when applied to molecular trees it turns out that the vertebrates have been accumulating neutral mutations more slowly than invertebrates. The reason for this is not known but may be because of the larger overall gene number in vertebrates which may increase the number of protein–protein interactions and thereby reduce the proportion of neutral as opposed to slightly deleterious mutations. Because of the calibration uncertainty, estimates of the last common ancestor for all animals span quite a long period from about 1000 Myr ago to about 600 Myr ago. The latter date is obtained using invertebrate divergence times and does correspond quite well to the earliest probably metazoan fossils dating from the Vendian period.

The fossil record

The fossil record provides a real view of the past and is the only source of time calibration for phylogenetic trees. A fossil is any residue of an organism preserved in the rocks. Virtually all fossils are found in sedimentary rocks: those formed by accretion of sediments in the sea or fresh water. Most fossils consist of only the hard parts of organisms, for example the shells of marine invertebrates or the bones of vertebrates, and these are often mineralized such that all of the original organic material has been replaced by rock. Fossils may also be "trace fossils" which show evidence of activity but without the actual remains, for example burrows or tracks made by organisms. A very small number of geological sites contain organisms in an exceptional state of preservation in which soft parts are also visible, for example the well-known Burgess shale and Chengjiang deposits from the Cambrian period.

Figure 20.3 shows the geological periods and the first occurrence of fossils of various taxonomic groups. It may be noticed that most of these first occurrences are of vertebrates. Most, and possibly all, of the invertebrate phyla were already in existence during the Cambrian period which is the earliest geological period for which the rocks contain significant numbers of fossils. The "Cambrian explosion" during which the invertebrate phyla appeared occupies a geologically rather short period of 5–20 million years. This rapid diversification poses problems for attempts to find the true phylogenetic tree of the animal kingdom. Trees are constructed from characters of modern organisms, whether morphological or gene sequences, and almost all the changes that have happened were in the lines of descent following the Cambrian explosion. For neutral characters that should evolve in a clock-like manner only about 2% are likely to relate to the time of the explosion and the remaining 98% have accumulated since. This means that it is hard to find an informative signal among all the noise of later evolution.

The fossil record cannot prove a phylogeny because it is impossible to know what is an ancestor to what else. However some possibilities can be excluded because a late-arising group obviously cannot be an ancestor of an early-arising group. The fossil record also provides valuable time calibration for

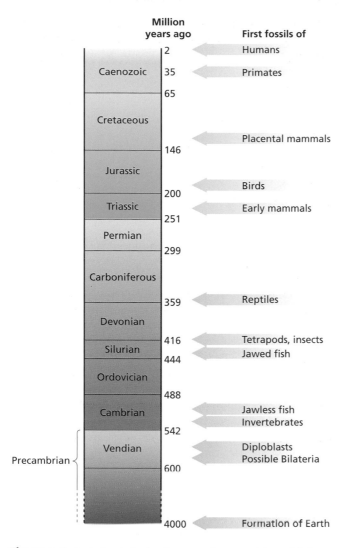

Fig. 20.3 Geological periods, showing currently accepted dates of their commencement and the first occurrence of fossils for major animal groups.

represent some other type of radially symmetrical life which had hard parts and which died out without leaving descendants. Evidence of recognizable metazoan animals has been hard to find in the Precambrian rocks although tiny 100–200 μm long worm-like creatures have recently been found in the Doushantuo formation of SW China which is of Vendian age. Sections of these indicate that they may be coelomate animals with a through gut and some anterior sense organ (see below and Fig. 20.4).

The primordial animal

The most important result of evolutionary developmental biology is the ability to uncover previously invisible homologies through the study of expression and function of the genes that actually make the body plan. In order to explain this it is necessary to consider first the concepts of the **body plan** itself and of the **phylotypic stage**. The body plan (or **Bauplan** in the original German) refers to the idea that it is possible to abstract the essential features of anatomical organization from a wide range of organisms. This is done in zoology textbooks whenever they show a diagram of a generalized vertebrate or mollusc. Evidence that the body plan is real rather than an arbitrary abstraction has come from developmental biology because of the fact that the key transcription factors and signaling molecules are found to have very similar expression domains within a taxon. For example the expression of the *brachyury* gene in vertebrates occurs in rather differently shaped regions in *Xenopus*, chick, and mouse, but they all correspond to what was previously thought to be the mesoderm, and indeed reinforce the concept of mesoderm as a real cell state definable by the activity of a combination of transcription factors.

The phylotypic stage

The **phylotypic stage** is the stage of development at which all members of a taxon show the maximum morphological similarity. This concept has been most widely applied to the insects and vertebrates although phylotypic stages can also be defined for other groups. Insects all look rather similar at the **extended germ band** stage. At this stage the segmental arrangement of the body is very obvious with three gnathal, three thoracic, and a variable number of abdominal segments (Fig. 20.5a). The extended germ band is reached before dorsal closure of the embryo and consists of a plate composed of ectodermal and mesodermal components. There is a ventral nerve cord but most details of the future epidermal structures are yet to appear, and the endoderm, in terms of the anterior and posterior midgut invaginations, has yet to form. Among vertebrates the phylotypic stage is the **tailbud** stage (Fig. 20.5b), when all vertebrates have a dorsal nerve cord with anterior specialization, segmented somites, a ventral heart, and a set of pharyngeal arches. Phylotypic stages for groups other than the insects and vertebrates

phylogenetic trees in the form of divergence dates, which are the points at which a new group appears for the first time. For example the divergence of birds and mammals is thought to be 300 Myr ago and of *Drosophila* from mosquitos 235 Myr ago. Of course these times cannot be deduced from the fossils themselves, as the fossils just provide a stratigraphy enabling the correlation of different geological sites into a coherent sequence. Absolute times can only be obtained by radioactive dating of the rocks, and this itself is a complex business since sedimentary rocks rarely contain radioactive elements with suitable half-lives for this sort of measurement. However, correlation of results from many sites around the world has led to the time scale shown on Fig. 20.3.

Preceding the Cambrian is a period called the Vendian, or Ediacaran, in which are found fossils that resemble cnidarians. There is some uncertainly about this since jellyfish are notoriously difficult to preserve as fossils and the Vendian fauna may

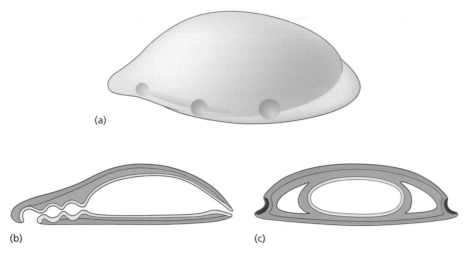

Fig. 20.4 Possible bilaterian fossil from Doushantuo formation of SW China. (a) Reconstruction. (b) Sagittal section. (c) Transverse section. The creatures appear to have three germ layers, an anteroposterior pattern, and a through gut.

might include the **veliger** larva of molluscs. This has a head, a posterior, a through gut, a dorsal shell gland, and a ventral muscular foot. A phylotypic annelid might be considered to be a stage at which a certain number of segments have arisen and the organism has a head, a ventral nerve cord, a through gut, and segments involving all the germ layers. The echinoderms are a remarkable phylum in which metamorphosis from the larva to the adult is particularly dramatic. Here it would seem that the **echinus rudiment** of sea urchins, or its equivalent in the other classes, should be considered phylotypic, since the earlier stage larvae are typically bilaterally rather than radially symmetrical and do not have echinoderm features such as a water vascular system or tube feet. Finally the **nauplius** larva which has three pairs of head appendages has been considered a phylotypic stage for crustaceans.

It is helpful to have some idea of why conserved body plan features should be displayed at one stage. Although it is often stated that the earliest stages of development cannot change in evolution because this would compromise later events, it is nonetheless a familiar fact that early embryonic morphology is very diverse. The main reason for this is the diversity of reproductive lifestyle and strategy. The eggs of even quite closely related animals can differ markedly in yolk content. This is the result of evolutionary trade-offs such that organisms come to occupy different niches in which they may produce a lot of small eggs with a poor chance of survival or a few large ones with a good chance of survival. The presence of more yolk drives various changes in early development including the disposition of cleavages (**meroblastic** rather than **holoblastic**) and the nature of gastrulation movements (**epiboly** rather than **invagination**). **Viviparity** imposes even more drastic changes on early embryonic life and is necessarily accompanied by the early formation of a variety of extraembryonic membranes and supporting structures arising from the zygote as well as the mother. So early development is necessarily diverse because reproductive behavior is diverse. Late development is also diverse, but for quite different reasons. By late development the embryo will be

becoming quite similar to the postembryonic organism, whether larval or adult. Free-living organisms are subject to selection and must acquire distinct niches for their survival to the reproductive stage, so the late embryos have to be diverse because the organisms to which they give rise are diverse. It follows inexorably that the stage of maximum similarity within a large taxon will be the early–middle period of embryonic development, after the constraints imposed by reproductive strategy and before the constraints imposed by the adaptation of the free-living organism. The phylotypic stages are early–middle stages that exist within egg cases, or jelly layers, or a uterus, and so will not be interacting with the environment to more than a minimal degree. According to this way of thinking there is no specific cause for the phylotypic stage, it is just the stage in the middle at which the selective pressures for change are minimized, hence is most likely to retain the features of the common ancestor. This idea is encapsulated in the "phylotypic egg timer" diagram formulated by Duboule (Fig. 20.5c).

The zootype

Animal phyla have been defined in such a way that each of them corresponds to a different body plan which means that it is inevitably difficult to find any morphological homologies between them apart from the developmental features mentioned above. However the discoveries of developmental biology now make it possible to compare the expression patterns of key developmental genes between phyla. Some of these seem to be conserved across the whole animal kingdom. Figure 20.6 is a diagram of an animal embryo showing the expression domains of various key genes involved in regional specification. They are active around the phylotypic stage for all the main animal groups that have been examined. The totality of common expression domains is called the **zootype**, to signify that this cryptic anatomy of developmental gene expression patterns defines what an animal actually is. The vertebrate nomenclature

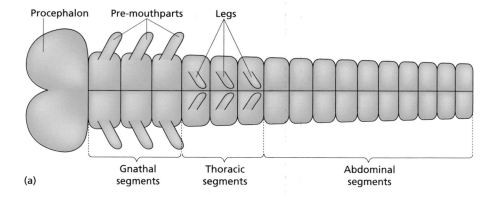

Procephalon Pre-mouthparts Legs

Gnathal segments Thoracic segments Abdominal segments

(a)

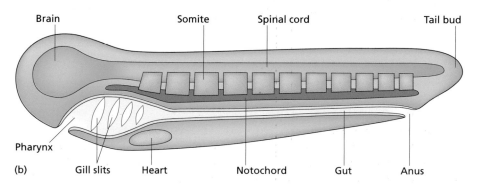

Brain Somite Spinal cord Tail bud

Pharynx Gill slits Heart Notochord Gut Anus

(b)

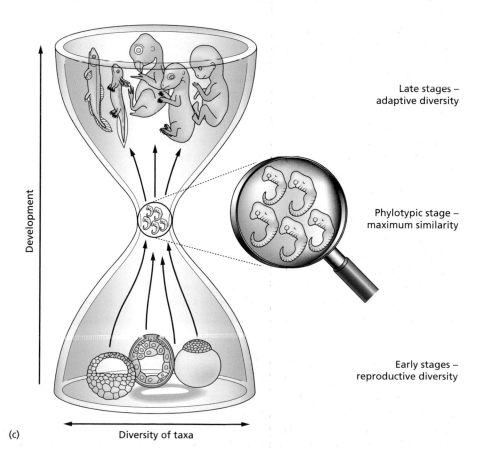

Late stages – adaptive diversity

Phylotypic stage – maximum similarity

Early stages – reproductive diversity

Development

Diversity of taxa

(c)

Fig. 20.5 The phylotypic stage.
(a) Phylotypic (extended germband) stage
of insects. (b) Phylotypic (tailbud) stage of
vertebrates. (c) "Phylotypic hourglass" for
vertebrates. (After Duboule 1994. In: The
evolution of developmental mechanisms.
Development Supplement 139.)

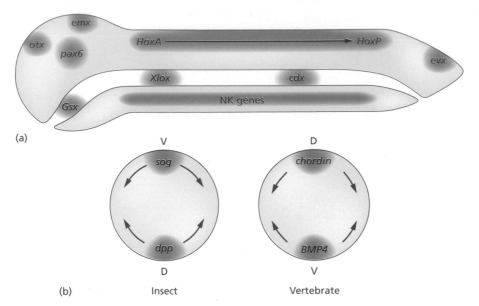

(a)

(b) Insect Vertebrate

Fig. 20.6 The "zootype," or common constellation of expression domains for developmental genes found in the animal body plan. (a) Transcription factors active at the phylotypic stage. (b) Homologous dorsoventral signaling systems in insects and vertebrates.

is used, but homologs of all these genes exist in *Drosophila* and several other invertebrates and perform comparable tasks. It is intriguing to speculate whether this constellation of gene expression domains, deduced from modern organisms, is actually what was active 600 Myr ago in the tiny creatures of the Doushantuo fomation (Fig. 20.4).

Hox genes

The Hox genes have already been described in Section 2 in the context of each of the model organisms. Current data suggest that all animals, except probably the sponges, use the same set of Hox genes to control anteroposterior pattern within the ectoderm-derived structures. From cnidarians to humans it has been possible to recover Hox genes that fall into homologous groups, and in those organisms where detailed genomic analysis has been possible they comprise a cluster on one chromosome, sometimes interrupted by other genes. Where expression patterns have been determined there is a general trend for each gene to be expressed in a different anteroposterior domain and for the sequence of anterior expression limits to be similar to the sequence of genes on the chromosome. In those systems where experimentation is possible it has been shown that loss of function of a Hox gene generally causes a homeotic transformation such that the affected body level is anteriorized, and gain of function generally leads to posteriorization.

From an evolutionary point of view the critical thing about Hox genes, and indeed almost all the components of the developmental toolbox, is that they function simply as switches, activating or repressing other genes. Any transcription factor could do this job given appropriate connections from the signaling pathways. If animals had evolved from a variety of different ancestral lower eukaryotes then it is virtually certain that different sets of transcription factors would have been

coopted to deal with anteroposterior pattern. The fact that the same set of genes does this job in all animals proves that they must have arisen from a single common ancestor.

Study of the vertebrate Hox genes has shown that a remarkable event must have occurred around the time of the vertebrate origin. The creature called Amphioxus is a **protochordate** and has long been regarded as the modern animal most similar to a putative vertebrate common ancestor. It has a dorsal nerve cord, notochord, somites, pharyngeal gill slits, and a through gut. Like other invertebrates it has a single Hox cluster, but unlike them it has increased the number of genes to 14 (Fig. 20.7). Vertebrates have four Hox clusters on different chromosomes (Fig. 20.7). Each cluster contains a subset of orthologs of the Amphioxus cluster, making it look as though the vertebrate common ancestor acquired four copies of the cluster by chromosome duplication, and then individual genes were lost from each cluster independently within the various vertebrate lineages. The vertebrate clusters are called a, b, c, d, and individual genes are numbered 1–13 starting from the 3′ end of the cluster. Because the expression domains in the embryo lie in the same order as the position on the chromosome, the numbers also reflect the sequence of anterior expression boundaries. The individual genes with the same number, for example *a4*, *b4*, *c4*, *d4*, are **paralogs** (as explained above), and the set of paralogs is called a **paralog group**. It has been suggested that the entire genome of the vertebrate ancestor may have undergone two duplications, increasing gene number from about 15,000 to about 60,000, followed by gene loss back to about 30,000. This is not proved and it may be that the increased gene number of vertebrates is due to a number of partial duplications. But either way the huge increase in useful genetic material probably helped provide the opportunity for the adaptive radiation of the vertebrates.

In vertebrates the Hox expression domains do not extend anterior to the hindbrain. The regional pattern of the fore- and

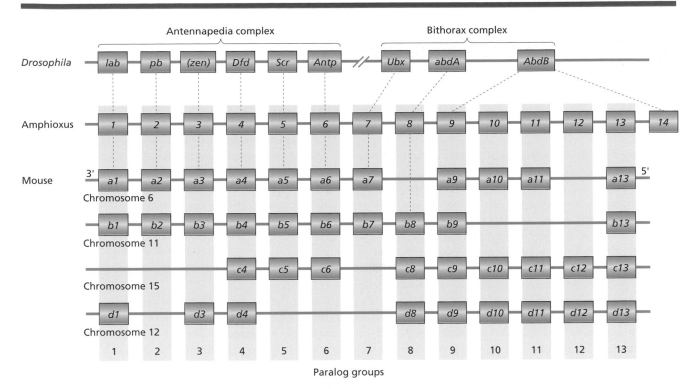

Fig. 20.7 Hox genes found in *Drosophila*, Amphioxus (a protochordate), and the mouse.

midbrain is controlled by other homoebox genes of the *otx* and *emx* groups. These also have orthologs in invertebrates that are expressed at the extreme anterior and so are considered to be part of the zootype.

Other clusters

Although Hox genes are expressed in all germ layers in vertebrates, they are confined to the ectoderm in most invertebrates and anteroposterior pattern in the other germ layers is controlled by other genes. The endoderm seems to be regionalized by a sister cluster of genes called the Parahox cluster. This may represent a very early duplication of the Hox cluster and contains three genes *gbx*, *xlox*, and *cdx*. The latter two are well known as being very important in vertebrate gut development (see Chapter 16). The *xlox* group includes *pdx1*, essential for pancreatic development, and the *cdx* group includes *cdx2*, essential for development of the intestine. The *gsx* group are expressed in the brain of vertebrates, but may be found in the anterior endoderm of invertebrates.

It is also possible that the regional

pattern of the mesoderm is controlled by the NK cluster. The group of homeobox genes known as Nkx in vertebrates form a single cluster in *Drosophila*. This cluster is broken into three pairs in Amphioxus and is even more dispersed in vertebrates. One piece of evidence for a role in regional patterning of the mesoderm is that one of these genes, called *Nkx2.5* in vertebrates and *tinman* in *Drosophila*, is especially associated with the development of the heart (see Chapter 15). The primordial bilaterian animal would probably not have had a heart because if it was as small as the organisms of the Doushantuo formation its tissues

New Directions in Research

EvoDevo is still a wide open field. We have very little idea what actually happened in evolution to create each of the key morphological novelties in animals. There are millions of animal species whose development has not been investigated with molecular tools and the methods are now available to make the necessary investigations on nonmodel organisms:
1 Homology cloning to isolate important genes.
2 In situ hybridization to analyze expression patterns.
3 RNAi to knock out gene activity.
4 Electroporation to introduce genes for overexpression.
Even better, the best science journals have an insatiable appetite for stories about evolution! Nevertheless, the funding bodies tend to be keener on medically oriented research than on pure biology, so the global budget available for EvoDevo is somewhat limited.

would have been oxygenated by diffusion. Only in the Cambrian, when animals became large, would the need for a circulatory system have been felt. So the primordial role of *Nkx2.5* was probably to specify an anteroventral mesodermal territory, rather than to specify a heart as such.

Sense organs and cell types

One of the defining characteristics of animals is that they have some sort of concentration of sense organs at one end, representing a "head," and referred to as the anterior end. The *pax6* gene is intimately associated with the formation of eyes. It is defective in the *small eye* mutant of the mouse; and in *Drosophila pax6* is called *eyeless* and its loss of function prevents eye development completely. *pax6* is also expressed in the eyes of cephalopod molluscs and planarians. Cephalopods are significant because they have image-forming eyes structurally quite similar to those of vertebrates except for the orientation of the retina. This had always been a textbook case of **analogy** because it was felt impossible for vertebrates and cephalopods to have a common ancestor with an image-forming eye. However the involvement of the same transcription factor in making the two eyes suggests that they really are homologous at least to the extent that the ancestor had some sort of photoreceptor made with the help of *pax6*. Planarians are significant because it is possible to do functional gene ablation experiments by treating regenerating planarians with RNAi. In fact it turns out that *pax6* itself is not required for eye regeneration but another homeobox gene, *sine oculis*, is required. *sine oculis* is immediately downstream of *pax6* in the *Drosophila* eye development pathway; and its vertebrate homolog *six3* is needed for eye development in mice. These results show a remarkable degree of conservation of function of a small group of transcription factors involved with the creation of light-sensitive organs across the whole animal kingdom.

There is also conservation at least between vertebrates and insects of many other developmental control genes required not for regional specification but for specific types of cell differentiation. For example the myogenic gene family, of which the prototype member is *MyoD*, exists in all animals and is needed for the differentiation of muscle cells. The Delta–Notch signaling system is responsible for controlling the differentiation of neurons from neurogenic epithelia, and, as we have seen in Chapter 16, also of specialized cell types within epithelia.

Dorsal–ventral pattern

Work on *Xenopus* showed that the main activity of the organizer was the secretion of bone morphogenetic protein (BMP) inhibitors such as chordin. This established a ventral to dorsal gradient of BMP signaling activity that controls the dorsoventral pattern in both the ectoderm and mesoderm. Work on *Drosophila* showed that there was a dorsal to ventral gradient of activity of the BMP homolog decapentaplegic (dpp). This was antagonized by Short-gastrulation which is a chordin homolog secreted from ventrolateral ectoderm. Expression data from regenerating planarians suggest that a *dpp* homolog is also expressed ventrally. This molecular information can be correlated with the long-known fact that vertebrates have their principal nerve cord on the dorsal side, while at least some invertebrates, such as arthropods and annelids, have it on the ventral side. In those invertebrate phyla that have hearts, this tends to be dorsal while it is ventral in vertebrates. This suggests that there is a primordial system of dorsoventral patterning that operates in all animals but it became inverted at some point such that the vertebrate polarity is now opposite to that of the invertebrate groups (Fig. 20.6b).

What is an animal ?

The constellation of developmental gene activities represented by the zootype is a very compelling set of characters which can be used to define what an animal actually is. However it is necessary to examine those organisms formerly considered to be the most basal animals in order to find whether this notion is congruent with traditional zoology. The phylum Cnidaria comprises sea anemones and jellyfish, and also the familiar freshwater polyp *Hydra*, used in regeneration research. Cnidarians have traditionally been considered as basal animals that differ from most others in that they possess only two germ layers (ectoderm and endoderm) instead of three; and that they are radially rather than bilaterally symmetrical. Cnidarians do have Hox genes and at least some of these form a linked cluster. Expression data show that there is some staggering of anterior expression limits just as in bilaterian animals (Fig. 20.8a). Moreover they do sometimes possess some elements of bilateral symmetry, and in at least one case that has been examined express a *dpp* homolog in a restricted manner suggesting that the BMP–chordin system may operate even here.

It is not even clear that the cnidarians are really diploblastic. The hydrozoan *Podocoryne carnea* shows a typical alternation of generation and will undergo the whole life cycle in laboratory conditions. This involves a **polyp** phase, the formation of asexual buds to form a **medusa**, sexual reproduction of the medusae, and embryonic development of a motile planula larva which settles down to become the new polyp. The medusae, like many jellyfish, possess striated muscle cells. These develop from a cell layer called the entocodon which arises in the asexual buds and expresses mesoderm-type genes including homologs of *twist*, *snail*, and *mef2* (Fig. 20.8b). In Cnidarians without a medusa phase the same genes may be expressed in regions of the endoderm (Fig. 20.8c), which has given rise to the suggestion that the mesoderm originated as a subdivision of the endoderm. In any case, the molecular analysis of cnidarians has advanced sufficiently to show that they correspond to the zootype and are definitely animals.

The most basal animals in modern phylogenies are the

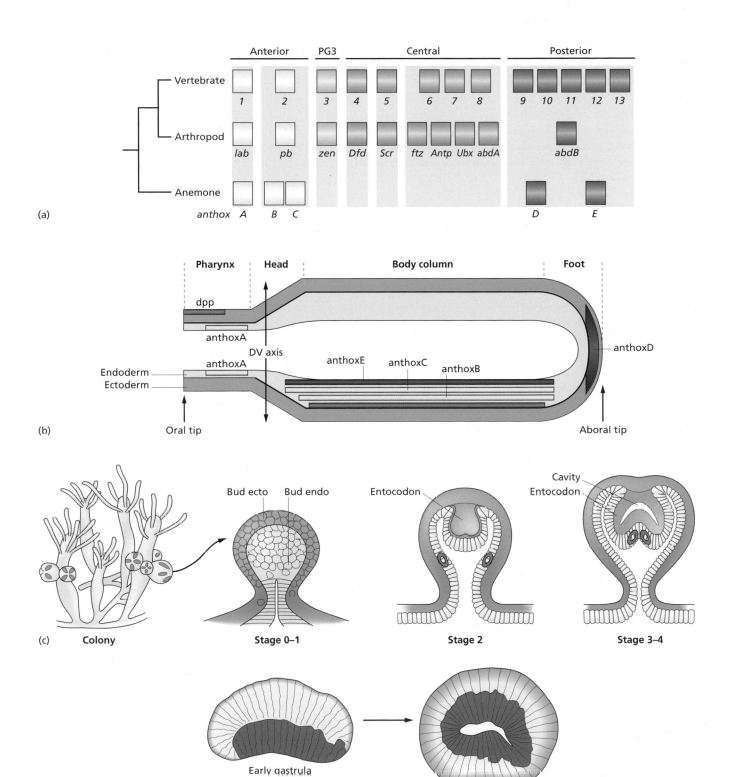

Fig. 20.8 Developmental gene expression in cnidarians. (a) Hox genes of *Nematostella* (anemone). (b) Expression domains in *Nematostella* of Hox genes and *dpp*. (c) Formation of the entocodon in *Podocoryne* (hydrozoan). (d) Expression of *snail* in the endoderm embryo of *Acropora* (coral). ((a,b) After Finnerty et al. 2004. *Science* **304**, 1335–7, figure 3. (c) After Ball et al. 2004. *Nature Reviews Genetics* **5**, 567–77, figure 1.)

Porifera or sponges. These are traditionally thought to have few cell types and no real tissue grade of organization. At present it is unclear whether they really contain orthologs of the Hox genes and there is no evidence for a Hox cluster or anteroposterior expression domains. It seems that if the zootype were officially adopted as the criterion for being an animal then the sponges would become an outgroup within the lower eukaryotes, rather as has already happened with unicellular creatures such as *Amoeba* or *Paramecium*.

What really happened in evolution?

The other main area where developmental biology can assist in the solution of evolutionary problems is to find out what actually happened in evolution. Some evidence about this can be obtained by comparing the expression patterns of key developmental genes that give rise to two different morphologies, one ancestral and one derived. However this is not sufficient. As when elucidating developmental mechanisms it is also necessary to establish the actual biological activity of the genes in question by performing overexpression and loss-of-function experiments. This may sometimes be difficult as the organisms under study will not be the standard laboratory models and so it may even be difficult to obtain embryos let alone perform successful experiments on them. It is also important to avoid the "fallacy of design," in other words once a developmental system is understood there is a tendency to say "extra legs can be produced very easily by allowing gene A to turn on gene B, so this must be what happened in evolution." What actually happened may have been a number of small changes, perhaps in different genes altogether, that had the effective consequence of correlating the states of genes A and B and the new morphology. In general developmental systems are much more complex than would seem necessary had they been designed by a conscious agent and they do not necessarily change from one behavior to another by a simple route.

This also relates to the issue of **developmental constraints**. Biologists not concerned with developmental mechanisms sometimes implicitly assume that there are no limits to possible mutational variation. This means that with enough mutants, enough time and enough selection it is possible to evolve anything from anything else. If you want to evolve a snail into an elephant: no problem. Developmental biologists often tend to fall into the other extreme of believing that the only mutational variation that is possible is that produced by gain or loss of function of each known component of a particular developmental system. This will confine the range of attainable morphologies to a relatively small set obtained by one-step changes away from the wild-type pattern. In one sense developmental constraints must exist because there are all sorts of mutations that are never observed even in mutagenesis screens, suggesting that they cannot arise at all, regardless of whether they would be advantageous or not. Also, it is possible to imagine all sorts of body plans

that would probably confer greater fitness but have not evolved, such as angels which have a useful pair of wings in addition to arms and legs. On the other hand, it would always be possible to evolve a snail into an elephant given sufficient time and resources. It could be done by selecting initially for the loss of all morphology to convert the organism to a primitive worm-like creature, and then applying a second selective regime designed to culminate in large size, four legs, tusks, and a trunk. The potential importance of developmental constraints to evolutionary theory is immense. If constraints are very limited then the course of evolution is determined almost entirely by adaptation to a changing environment. But if developmental constraints are substantial then the course of evolution is determined mostly by what is possible. At the present time the true extent and role of developmental constraints remains unknown.

In general the mutations of the known components of well-understood systems, such as those controlling segmental identity or limb patterning, will be **macromutations**: single mutations which bring about a significant morphological change in one step. Neodarwinists have always been uncomfortable with macromutations, believing that they will virtually always be deleterious and selected against. Developmental biologists are often more sympathetic to what has been called the "**hopeful monster**," the creature arising from a macromutation, because this type of mutational change is congruent with developmental explanations. Also some morphological characters seem intrinsically qualitative, and it is hard to imagine small imperceptible intermediate forms. For example if an organism adds a segment then the extra segment is either there or not there, and it seems more plausible that a mutant with an extra segment would arise in one step and then take its chances with natural selection rather than follow a trajectory of a new segment gradually increasing in size over many generations. But orthodox Neodarwinists prefer to believe that evolution occurs by small imperceptible steps and that many of the mutations that bring these about occur in "modifier genes," in other words genes outside the developmental machinery but which have quantitative effects on this machinery. Again, the truth is that we do not know the relative importance of macromutations versus those of small effect.

Segmented body plans

The arthropods are a very large phylum containing four classes: insects, crustaceans, myriapods (centipedes and millipedes), and chelicerates (spiders and scorpions). Each class is distinguished by a fundamentally different arrangement of segments in the body plan, but we know from both morphological and molecular phylogeny that the arthropods are a clade and evolved from a common ancestor. Since the discovery of the Hox cluster and its role in controlling segment identity it has been natural to think that the changes in body pattern arose from changes in Hox gene expression. In particular in *Drosophila* it is known that

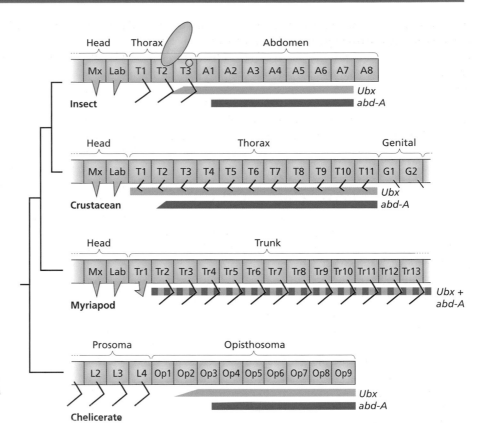

Fig. 20.9 Correlation of segmental pattern, appendage types, and Hox gene expression in arthropods. (After Carroll 2001. *From DNA to Diversity*. Blackwell, figure 5.2.)

the genes *Ubx* and *abdA* effectively determine the boundary between thorax and abdomen, as they repress formation of legs in the abdomen by repressing an enhancer of the *distalless* homeobox gene which is required for the outgrowth of all appendages. The expression of *Ubx* and *abdA* has been studied in a variety of arthropods (Fig. 20.9). This shows that there is some general association of Hox domains with body regions, but there is no specific association with the thorax or with the suppression of legs. In crustaceans these two Hox genes are expressed throughout the thoracic segments, all of which bear legs, but they are not expressed in head or genital regions. In some crustaceans that have maxillipedes (feeding appendage) instead of legs on the first thoracic segments, these segments do not express *Ubx*, suggesting that maybe *Ubx* represses maxillipedes but not legs. In myriapods *Ubx* and *abdA* are expressed in all but the first thoracic segment. In chelicerates the anterior boundary occurs within the ophisthosoma, the posterior body region which does not bear any appendages. The general conclusion from this is that there are some associations of Hox boundaries with boundaries of segment type, but that functions such as leg suppression are not conserved across large evolutionary distances. In crustaceans and myriapods, *distalless* is coexpressed in the early legs with *Ubx*, so cannot be repressed by *Ubx*. It follows that what the Hox genes do is to provide a universal coordinate scheme, or set of **positional values**, for the organism and that most of the evolutionary innovations occur in the regulatory connections downstream of them.

Further evidence for changes in downstream regulatory connections comes from an examination of wings. It is known that some early fossil insects had larvae (nymphs) with wings on all segments, and that the adults were wingless. In modern insects wings reach their full development in the adult. This type of shift of relative timing of events in the life history, with resulting change in adult morphology, is known as **heterochrony**, and is a common theme in evolution. In *Drosophila* the development of wings on all the non-wing-bearing segments is suppressed by Hox genes. In the first thoracic it is suppressed by *Sex combs reduced*, in the abdomen by *Ubx*, *abdA*, and *AbdB*. Since *Drosophila* is a dipteran, it also lacks wings on the third thoracic segment where the dorsal imaginal disc forms a haltere and not a wing. This is controlled by *Ubx* by preventing the activation of *vestigial* (see Chapter 17). Since the expression of Hox genes is fairly similar in all insects it seems that the wing-suppressing functions of modern insects must have arisen from the establishment of new regulatory connections and not through changes in Hox expression.

A mutation that yields a morphology characteristic of an ancestor is known as an **atavism**. Since much of the diversification of arthropods seems to have depended on suppression of appendages by Hox activity, it is not surprising that loss-of-function mutations of Hox genes will often have atavistic phenotypes. For example the four-winged *Drosophila* resulting from complete loss of *Ubx* from the haltere disc is an atavism. Atavisms may sometimes become established as the wild type

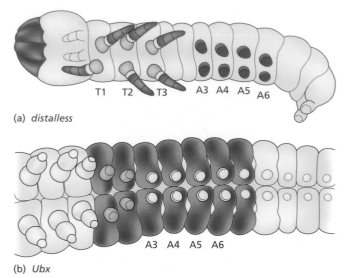

(a) *distalless*

(b) *Ubx*

Fig. 20.10 Abdominal legs of the butterfly caterpillar. (a) *Distalless* expression. (b) *Ubx* expression at the embryonic stage.

and therefore represent partial reversals of evolution. For example the caterpillars of butterflies have legs on four of their abdominal segments (A3–A6; Fig. 20.10). It is thought that the primordial arthropod probably had legs on most segments, like modern myriapods. When the expression of *Ubx*, *abdA*, and

distalless are examined in butterflies it can be seen that the Hox genes are actually turned off in the leg buds, and this presumably allows the expression of *distalless* and the formation of the atavistic legs.

Vertebrate limbs

The vertebrate limb has a distinguished history in evolutionary biology. It was early recognized as being a homologous structure across the tetrapods. Because of the presence of a complex skeleton it is often preserved in fossils. It provides some striking examples of **allometry**, or differential growth. For example the middle digit of the horse leg has elongated during evolution until the modern horse just has effectively one digit with the others reduced to vestiges. By contrast, all the digits of the bat forelimb have become greatly elongated to bear the skinflaps that constitute the wings.

It is believed from the fossil record that the first tetrapods arose in the Devonian period and that their ancestors resembled a creature called *Panderichthys* (Fig. 20.11a). These were lobe finned fishes belonging to the same order as the modern coelacanth. They are thought to have lived in shallow water and perhaps even been able to creep out of the water like some modern catfish. The paired fins possessed skeletal elements quite similar to the proximal parts of a modern limb, but with no

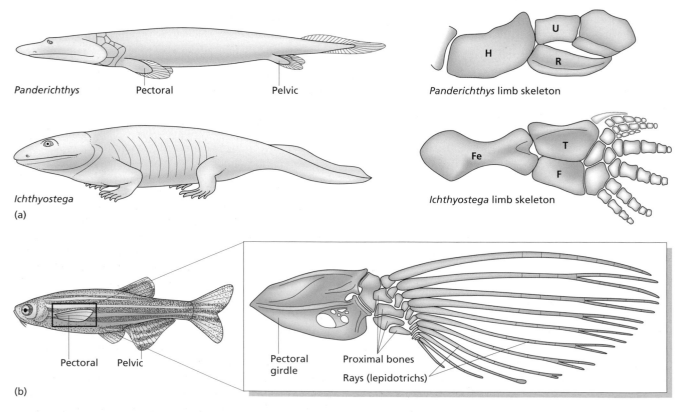

Fig. 20.11 Origin of legs. (a) *Panderichthys* and *Ichthyostega* with limb skeletons. (b) Zebrafish pectoral fin skeleton.

autopodium (hand or foot). In fact the paired fins of modern fish still possess the basic developmental machinery required to make a limb (see Chapter 15). Modern zebrafish fin buds have an early phase of outgrowth which is remarkably similar to tetrapod limb buds, with an apical ectodermal ridge expressing *fgfs* and a posterior region expressing *shh*. Furthermore the same genes seem to initiate limb development, since **morpholinos** against *wnt2b* or *tbx5* will both inhibit the pectoral fin bud formation. During this phase are laid down four parallel cartilage elements which later ossify and serve as attachments for the fin rays. These rays themselves are formed by dermal ossification and are quite unlike anything in the tetrapod limb (Fig. 20.11b). The four proximal bones are also a lot less like a modern limb than the bones of the *Panderichthys* fin, but it seems highly likely that the basic machinery for specifying outgrowth and three-dimensional pattern of the bud was present in primitive fish fins and predated the tetrapods by a considerable period.

The critical innovation was the **autopodium**. In this connection it is interesting that there exists in the mouse an enhancer region for the Hox d cluster that is required for the late phase of expression when all of these genes are activated in a distal region corresponding to the autopodium. This comes later than the nested posterior to anterior expression pattern shown in Fig. 15.20, and it could be the critical evolutionary innovation that brought about the formation of a limb that was capable of movement on land.

One of the most obvious features of modern tetrapod limbs is that the number of digits very rarely exceeds five. The very name **pentadactyl** limb indicates that the primitive limb has five digits and that all types with reduced digit numbers are derived. It has been speculated that there is a developmental constraint preventing more than five digits because this is the number of Hox d genes available to create distinctions between the digit territories. Unfortunately this theory does not stand up to the fossil evidence because there are some early Devonian tetrapods such as *Ichthyostega*, appearing soon after *Panderichthys*, having an autopodium which very obviously possesed more than five digits (Fig. 20.11a).

Limb presence and positioning

Some tetrapods have lost their limbs. These include snakes, whales, and the forelimbs of flightless birds. Often a rudimentary girdle or portion of the proximal limb skeleton survives as a vestige, indicating that the ancestors had more substantial limbs than their modern descendants. It is believed that the positions of the limbs on the lateral plate are specified by the antero-posterior patterning system for the whole body. This obviously includes the Hox genes although may include other genes as well, for example those of the ParaHox and NK clusters. Evidence of a role for Hox genes is provided by the effects of transgenic expression of *Hoxb8* at a more anterior level than usual in the mouse, which results in the formation of an extra ZPA and duplication of the forelimb. In the mouse and chick,

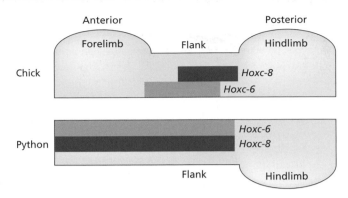

Fig. 20.12 Hox gene expression in the flank of chick and python embryos.

the Hox genes *c6* and *c8* are expressed in the lateral plate between the two limb buds and it is believed that they repress limb formation in this region. The python is a type of snake that has no trace remaining of the forelimbs but does have a rudimentary pelvic girdle and a proximal vestige of the hindlimbs. When its Hox expression domains are examined it is found that expression of *c6* and *c8* extend right up to the head, but posteriorly the expression border stops just short of the rudimentary hindlimbs (Fig. 20.12). In this case the nature of the critical mutations cannot be known but there is at least a correlation between the domain of the putative leg-suppressing genes and the absence of legs.

Vertebrate and insect wings

The vertebrate and the insect wing are, as mentioned above, obviously analagous and not homologous, since the common ancestor of the two groups must have lived in the sea and cannot have possessed wings. However there are a remarkable number of similarities in the expression patterns and the functions of transcription factors and signaling molecules in the two systems. Both have a dorsoventral pattern controlled by a homologous Lim transcription factor gene expressed on the dorsal side (*apterous* and *Lim1*). Both have an involvement of a *fringe* gene in setting the dorsoventral boundary. Both have hedgehog signaling controlling the anteroposterior pattern, and there is even an interaction of hedgehog with BMP/dpp in both cases. The distal outgrowth depends in both cases on the homeodomain factor distalless. With so much in common surely the insect and vertebrate wing have to be homologous! The solution to this mystery is not yet clear. But it is generally thought that the common features represent modules of developmental genetic circuitry. These may have evolved at an early stage, perhaps even before multicellularity, and could then be coopted for various purposes: in this case making appendages that needed to be asymmetrical in three dimensions. The lesson is that it is always essential to specify the level at which homology exists. The insect and vertebrate wings are not homologous as wings. It is possible that they are homologous as appendages. Some appendage made with these systems may have existed in the common ancestor but probably had very different functions. It is certain that the two types of wing contain active genes and genetic pathways that are homologous. This example shows the importance of maintaining clear thinking about difficult concepts. This is valuable throughout developmental biology, but nowhere more than in the complex borderland between development and evolution.

Key Points to Remember

- Evolution involves both adaptive changes brought about by natural selection and neutral changes established by genetic drift. Neutral changes are most apparent in DNA sequences.
- Classification may be arbitrary but usually attempts to represent the evolutionary history of the groups concerned. This is called a phylogenetic or cladistic taxonomy.
- Homologous features are similar because they derive from a common ancestor. Analogous features are similar because they are driven together by natural selection.
- A phylogenetic tree for the animal phyla themselves can be created using information from descriptive embryology, molecular taxonomy, and the fossil record. It is thought that the common ancestor of all animals lived between 1000 and 600 million years ago.
- The time calibration for phylogenetic trees comes ultimately from radioactive dating of rocks whose relative ages are deduced from the fossil record.
- Animal taxa often have a phylotypic stage at which all groups show maximum similarity.

- All animals share expression domains, and probably function, of a basic set of genes responsible for setting up the general body plan. This configuration of gene activity is called the zootype, comprising the Hox, ParaHox, and NK clusters, and certain other genes including *otx*, *emx*, and *pax6*.
- The importance in evolution of developmental constraints and of macro- as opposed to micromutations is still not known.
- Changes in the segmental organization of arthropods have mostly come about through changes in regulatory connections stemming from Hox genes rather than from changes in expression of Hox genes themselves.
- The key innovation in the evolution of the vertebrate limb seems to be the creation of an enhancer responsible for a late, autopodial, expression of Hox genes.
- Homology may exist at the level of genetic pathways and circuits even where there is no morphological homology. This indicates the extreme antiquity of the genetic pathways.

Further reading

General

Gould, S.J. (1977) *Ontogeny and Phylogeny*. Cambridge, MA: Harvard University Press.

Raff, R.A. (1996) *The Shape of Life: genes, development and the evolution of animal form*. Chicago: University of Chicago Press.

Gerhart, J. & Kirschner, M. (1997) *Cells, Embryos and Evolution*. Malden, MA: Blackwell Science.

Carroll, S.B., Grenier, J.K. & Weatherbee, S.D. (2001) *From DNA to Diversity*. Malden, MA: Blackwell Science.

Davidson, E.H. (2001) *Genomic Regulatory Systems in Development and Evolution*. New York: Academic Press.

Wilkins, A.S. (2001) *The Evolution of Developmental Pathways*. Sunderland, MA: Sinauer.

Hall, B.K. & Olson W.M., eds (2003) *Keywords and Concepts in Evolutionary Developmental Biology*. Cambridge, MA: Harvard University Press.

Taxonomy

Holland, P.W.H., Garcia Fernández, J., Williams, N.A. & Sidow, A. (1994) Gene duplications and the origins of vertebrate development. *Development* (suppl) 125–133.

Bolker, J.A. & Raff, R.A. (1996) Developmental genetics and traditional homology. *Bioessays* **18**, 489–494.

Aguinaldo, A.M.A. & Lake, J.A. (1998) Evolution of the multicellular animals. *American Zoologist* **38**, 878–887.

Hedges, S.B. (2002) The origin and evolution of model organisms. *Nature Reviews Genetics* **3**, 838–849.

Baldauf, S.L. (2003) Phylogeny for the faint of heart: a tutorial. *Trends in Genetics* **19**, 345–351.

Paleontology

Conway Morris, S. (1993) The fossil record and the early evolution of the metazoa. *Nature* **361**, 219–225.

Valentine, J.W., Jablonski, D. & Erwin, D.H. (1999) Fossils, molecules and embryos: new perspectives on the Cambrian explosion. *Development* **126**, 851–859.

Conway Morris, S. (2000) Evolution: bringing molecules into the fold. *Cell* **100**, 1–11.

Chen, J.Y., Olivieri, P., Davidson, E. & Bottjer, D.J. (2004) Small bilaterian fossils from 40 to 55 million years before the Cambrian. *Science* **305**, 218–222.

Body plans

Slack, J.M.W., Holland, P.W.H. & Graham, C.F. (1993) The zootype and the phylotypic stage. *Nature* **361**, 490–492.

Holland, P.W.H., Garcia Fernández, J. (1996) Hox genes and chordate evolution. *Developmental Biology* **173**, 382–395.

Arendt, D. & Nübler-Jung, K. (1997) Dorsal or ventral: similarities in fate maps and gastrulation patterns in annelids, arthropods and chordates. *Mechanisms of Development* **61**, 7–21.

Ferrier, D.E.K. & Holland P.W.H. (2001) Ancient origin of the Hox gene cluster. *Nature Reviews Genetics* **2**, 33–38.

Erwin, D.H. & Davidson, E.H. (2002) The last common bilaterian ancestor. *Development* **129**, 3021–3032.

Arthropods

Akam, M., Averof, M., Castelli-Gair, J., et al. (1994) The evolving role of Hox genes in arthropods. *Development* (Suppl.) 209–215.

Weatherbee, S.D. & Carroll, S.B. (1999) Selector genes and limb identity in arthropods and vertebrates. *Cell* **97**, 283–286.

Akam, M. (2000) Arthropods: developmental diversity within a superphylum. *Proceedings of the National Academy of Sciences USA* **97**, 4438–4441.

Limbs and eyes

Ahlberg, P.E. & Milner, A.R. (1994) The origin and early diversification of tetrapods. *Nature* **368**, 507–514.

Sordino, P. & Duboule, D. (1996) A molecular approach to the evolution of vertebrate paired appendages. *Trends in Ecology and Evolution* **11**, 114–119.

Shubin, N., Tabin, C. & Carroll (1997) Fossils, genes and the evolution of animal limbs. *Nature* **388**, 639–648.

Gehring, W.J. (2002) The genetic control of eye development and its implications for the evolution of the various eye types. *International Journal of Developmental Biology* **46**, 65–73.

Key molecular components

This appendix contains a brief summary of the principal types of macromolecule and cell component involved in development. It is not intended as a substitute for a grounding in cell biology or molecular biology or biochemistry, but can be referred to if molecules or processes mentioned in the main text are unfamiliar.

Genes

A gene is a sequence of DNA that codes for a protein, or for a nontranslated RNA, and it is usually considered also to include the associated regulatory sequences as well as the coding region itself. The vast majority of eukaryotic genes are located in the nuclear chromosomes, but there are also a few genes carried in the DNA of mitochondria and chloroplasts. The genes encoding RNA include those for ribosomal or transfer RNAs, and also an indeterminate number of microRNAs that are probably involved in translational control. It is normal practice to italicize in lower case names of genes or their primary transcripts and to write the name of the protein product in Roman type, e.g. the *wingless* gene codes for the *wingless* messenger RNA and the wingless protein.

DNA is normally complexed into **chromatin** by the binding of basic proteins called histones. Protein-coding genes are transcribed into messenger RNA (mRNA) by **RNA polymerase II**. Transcription commences at a transcription start sequence and finishes at a transcription termination sequence. Genes are usually divided into several exons, each of which codes for a part of the mature mRNA. The primary RNA transcript is extensively processed before it moves from the nucleus to the cytoplasm. It acquires a "cap" of methyl guanosine at the 5′ end, and a polyA tail at the 3′ end, both of which stabilize the message by protecting it from attack by exonucleases. The DNA sequences in between the exons are called introns and the portions of the initial transcript complementary to the introns are removed by splicing reactions catalyzed by "snRNPs" (small nuclear ribonucleoprotein particles). It is possible for the same gene to produce several different mRNAs as a result of alternative splicing, whereby different combinations of exons are spliced together from the primary transcript. In the cytoplasm the mature mRNA is translated into a polypeptide by the ribosomes. The mRNA still contains a 5′ leader sequence and a 3′ untranslated sequence flanking the protein-coding region, and these untranslated regions may contain specific sequences responsible for translational control or intracellular localization.

The total number of genes in vertebrate animals is about 30,000. In invertebrates it is rather fewer, about 19,000 in *Caenorhabditis elegans* and 13,000 in *Drosophila*. Of these, perhaps 1–2% are intimately involved in the processes of development, although many of the others are necessary for the formation of viable cells capable of developing.

Control of gene expression

There are many genes whose products are required in all tissues at all times, for example those concerned with basic cell structure, protein synthesis, or metabolism. These are referred to as "housekeeping" genes. But there are many others whose products are specific to particular cell types and indeed cell types differ from each other because they contain different repertoires of proteins. This means that the control of gene expression is central to developmental biology. Control may be exerted at several points. Most common is control of transcription, and we often speak of genes being "on" or "off" in particular situations, meaning that they are or are not being transcribed. There are also many examples of translational regulation, where the mRNA exists in the cytoplasm but is not translated until some condition is satisfied. This can be particularly important in early embryos where mRNA present in the egg may not be translated until after fertilization. Control may also be exerted at the stage of nuclear RNA processing, or indirectly via the stability of individual mRNAs or proteins.

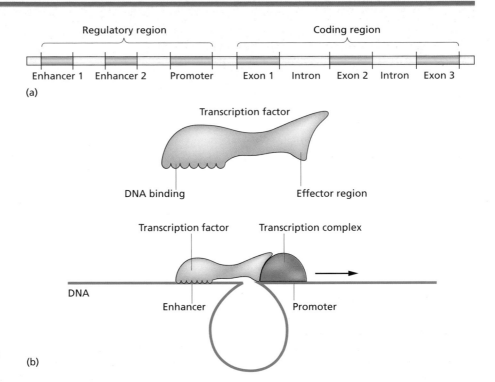

Fig. A.1 (a) Typical structure of a eukaryotic gene. (b) A transcription factor may bind to an enhancer and stimulate the transcription complex "at a distance."

Control of transcription depends on regulatory sequences in the DNA, and on proteins called **transcription factors** that interact with these sequences. The **promoter** region of a gene is the region just upstream from the transcription start site to which the RNA polymerase binds. It binds assisted by a set of general transcription factors, which together make up a transcription complex. In addition to the general factors required for the assembly of the transcription complex, there are numerous specific transcription factors that bind to specific regulatory sequences that may be either adjacent to or at some distance from the promoter (Fig. A.1). Transcription factors usually contain a DNA-binding domain and a regulatory domain, which will either activate or repress transcription. Looping of the DNA may bring these regulatory domains into contact with the transcription complex and either promotes or inhibits its activity.

Genes involved in development often have very complex regulatory regions consisting of several sites. Each site may be capable of binding both positive and negative transcription factors. Whether the gene is or is not transcribed will depend on the combination of factors that is present in the cell. The same transcription factors will bind to regulatory sites controlling many different genes, thus ensuring that target genes are controlled in a coordinated manner. Regulatory sites at a distance from the gene are often called **enhancers**, although they may sometimes repress as well as enhance transcription. The term **promoter** is frequently used not just for the region binding the transcription complex but for a larger region adjacent to the 5′ end of the gene containing several regulatory sites. A promoter in this broad sense may act independently of its surroundings.

If a specific promoter is joined to another gene and then introduced into the genome of an animal, it will often drive expression of the new gene with the appropriate tissue specificity, regardless of the position in the genome in which it integrates.

Transcription factors

Transcription factors are the proteins that regulate transcription. There are many families, classified by the type of DNA-binding domain they contain, such as the **homeodomain** or zinc-finger domain, and those most important in development are listed below. Most are nuclear proteins, although some exist in the cytoplasm until they are activated and then enter the nucleus. Activation often occurs in response to intercellular signaling (see below). One type of transcription factor, the nuclear receptor family, is directly activated by lipid-soluble signaling molecules such as retinoic acid or glucocorticoids.

Each type of DNA-binding domain in a protein has a corresponding type of target sequence in the DNA, usually 20 nucleotides or less. These can be identified by sequencing of the regulatory region of the target gene, although evidence that a particular transcription factor actually does regulate a given target gene *in vivo* needs to be established by experiment. Activation domains of transcription factors often contain many acidic amino acids forming an "acid blob," which accelerates the formation of the general transcription complex. Some transcription factors recruit histone acetylases which open up the chromatin by neutralizing amino groups on the histones, and allow access of other proteins to the DNA. In general, the molecular organization of transcription factors is modular, so new factors can be created in the laboratory by recombination of individual domains of existing factors. For example, a transcriptional activator may be converted to a repressor by swapping its acid blob for a repression domain. In the nuclear receptor family it is possible to combine the hormone-binding site of one factor with the DNA-binding and effector sites of another. For example if a retinoic acid-binding site in the receptor is replaced by a thyroid-binding site, then the genes normally responsive to retinoic acid would become responsive to thyroid hormone instead.

Although it is normal to classify transcription factors as activators or repressors of transcription, their action is also sensitive to context, and the presence of other factors may on occasion cause an activator to function as a repressor, or vice versa.

Transcription factor families

This lists some of the structural families of transcription factor most important in development. They are classified by DNA-binding domain.

Homeodomain factors

The **homeodomain** is an approximately 60 amino acid sequence containing many basic residues, and forms a helix-turn-helix structure that binds specific sites in DNA. The homeodomain sequence itself is coded by a corresponding **homeobox** in the gene. The homeobox was given its name because it was initially discovered in **homeotic genes** (see Chapter 4). However, there are many transcription factors that contain a homeodomain as their DNA-binding domain and although they are often involved in development, possession of a homeodomain does not guarantee a role in development, nor are mutants of homeobox genes necessarily homeotic. A very large number of homeodomain proteins have important developmental functions, e.g. Engrailed in *Drosophila* segmentation, Goosecoid in the vertebrate organizer, Cdx proteins in anteroposterior patterning. An important subset are the **HOX proteins** which have a special role in the control of anteroposterior pattern in animals. Homeobox genes are found in animals, plants, and fungi, but the **Hox** subset are found only in animals.

LIM-homeodomain proteins

The LIM domain is a cysteine-rich zinc-binding region responsible for protein–protein interactions, but is not itself a DNA-binding domain. LIM-homeoproteins possess two LIM domains together with the DNA-binding homeodomain. Examples are Lim-1 in the organizer, Islet-1 in motorneurons, Lhx factors in the limb bud, and Apterous in the *Drosophila* wing.

Pax proteins

These are characterized by a DNA-binding region called a paired domain with 6 α-helical segments. The name is derived from the paired protein in *Drosophila*. Many of the pax proteins also contain a homeodomain. Examples are Pax6 in the eye and Pax3 in the developing somite.

Zinc-finger proteins

This is a large and diverse group of proteins in which the DNA-binding region contains projections ("fingers") with Cys and/or His residues folding around a zinc atom. Some examples are the GATA factors important in the development of the blood and the gut, Krüppel in the early *Drosophila* embryo, Krox20 in the rhombomeres of the hindbrain, WT-1 in the kidney, and GAL4, the yeast transcription factor used for overexpression experiments.

Nuclear receptor superfamily

These are intracellular receptors that are also transcription factors. They are stimulated by lipophilic ligands including steroids, thyroid hormones, and retinoic acid. They have a hormone-binding domain, a DNA-binding domain, and a transcription-activation domain. The receptor is normally complexed with a heat shock protein called hsp90 and thereby retained in the cell cytoplasm. On binding of the hormone, the hsp90 is displaced and the factors can form active dimers which move to the nucleus and bind to the target genes. Examples include the retinoic acid receptors (RARs, also now called NR1Bs) and the ecdysone receptor.

Basic helix–loop–helix (bHLH) proteins

These transcription factors are active as heterodimers. They contain a basic DNA-binding region and a hydrophobic helix-loop-helix region responsible for protein dimerization. One member of the dimer is found in all tissues of the organism and the other member is tissue specific. There are also proteins containing the HLH but not the basic part of the sequence. These form inactive dimers with other bHLH proteins and so inhibit their activity. Examples of bHLH proteins include E12, E47 which are ubiquitous in vertebrates, the myogenic factor MyoD, and the *Drosophila* pair-rule protein hairy. An example of an inhibitor with no basic region is Id, which is an inhibitor of myogenesis.

Winged helix proteins

These have a 100 amino acid winged helix domain which forms another type of DNA-binding region and are known as "Fox" proteins. Examples are Forkhead in *Drosophila* embryonic termini and FoxA2 (formerly HNF3 β) in the vertebrate main axis and gut.

T-box factors

These have a DNA-binding domain similar to the prototype gene product known as "T" in the mouse and as brachyury in other animals. They include the endodermal determinant VegT and the limb identity factors Tbx4 and Tbx5.

High mobility group (HMG)-box factors

These factors differ from most others because they do not have a specific activation or repression domain. Instead they work by bending the DNA to bring other regulatory sites into contact with the transcription complex. Examples are SRY, the testis-determining factor, Sox9, a "master switch" for cartilage differentiation, and the TCF and LEF factors whose activity is regulated by the Wnt pathway.

Other controls of gene activity

Some aspects of gene control are of a more stable and longer-term character than that exerted by combinations of positive and negative transcription factors. To some extent this depends on the remodelling of the **chromatin** structure, which is still poorly understood. Much of the DNA is complexed with histones into nucleosomes, and is then arranged as a 30-nm filament and higher structures. In much of the genome the nucleosomes are to some extent mobile, allowing access of transcription factors to the DNA. This is called **euchromatin**. In other regions it is highly condensed and inactive. This is called **heterochromatin**. In the extreme case of the nucleated red blood cells of nonmammalian vertebrates the entire genome is heterochromatic and inactive. Chromatin structure is regulated to some degree by **chromobox** proteins, for example Polycomb in *Drosophila*, which affect the expression of many genes but are not themselves transcription factors.

An important element of the chromatin remodelling is the control through acetylation of lysines on the N-termini of histones. This opens up the chromatin structure enabling a transcription complex to assemble on the DNA. The degree of histone acetylation is controlled, at least partly, by DNA methylation, because histone deacetylases are recruited to methylated regions. Methylation occurs on cytosine residues in CG sequences of DNA. Because CG on one strand will pair with GC

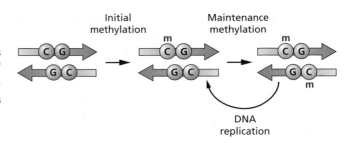

Fig. A.2 DNA methylation. Methylation is preserved through DNA replication by the action of the maintenance methylase.

on the other, antiparallel, strand, potential methylation sites always lie opposite one another on the two strands. There are several DNA methyl transferase enzymes, including *de novo* methylases that methylate previously unmethylated CGs, and maintenance methylases that methylate the other CG of sites bearing a methyl group on only one strand. Once a site is methylated, it will be preserved through subsequent rounds of DNA replication, because the hemimethylated site resulting from replication will be a substrate for the maintenance methylase (Fig. A.2). There is an overall deficit of the dinucleotide CG in the vertebrate genome compared to the other 15 possible dinucleotides. Many of the CGs are clustered into about 40,000 regions near the 5′ ends of genes. These so-called CpG islands are predominantly unmethylated and are thought to indicate the positions of the "housekeeping" genes.

The state of histone acetylation may also, independently of methylation, be transmitted through DNA replication, perhaps by permanent association of the transacetylases with acetylated regions of chromatin. So both DNA methylation and histone acetylation provide means for maintaining the state of activity of genes in development even after the original signals for activation or repression have disappeared.

Signaling systems

Development depends to a large extent on the effects of one group of cells on another. These effects are exerted by **embryonic induction**, which involves the release of an **inducing factor** from a signaling center and its action on a responding cell group. There are several important molecular families of inducing factors, and these molecules may also be known as growth factors, cytokines, or hormones in other contexts. Inducing factors are mainly proteins, although a few are small lipid soluble molecules such as retinoic acid. Most are secreted from the signaling cells, although some are integral membrane proteins that can only interact with immediately adjacent cells. They bind to specific **receptors** and activate a **signal transduction pathway** within the cell. This may lead to activation or repression of specific genes or to changes in cell behavior mediated through the cytoskeleton or through changes to cellular metabolism. The repertoire of responses that a cell can show depends on which receptors it possesses, how these are coupled to signal transduction pathways, and how these pathways are coupled to gene regulation. An inducing factor that can evoke more than one response at different concentrations is often called a **morphogen**. This is because a concentration gradient emitted from the signaling center will induce different responses at different distances from the source and thus create "morphology" where there was previously none (see Chapter 4).

Secretion of proteins

All secreted proteins, including the inducing factors, normally contain a signal peptide, which is a hydrophobic sequence of amino acids at the N-terminal end. During protein synthesis this is recognized by a signal recognition particle which binds the ribosome to a translocation channel in the endoplasmic reticulum. As synthesis proceeds the growing polypeptide is posted through this channel into the endoplasmic reticulum lumen. Finally the signal peptide is cleaved off and the protein released. Once in the endoplasmic reticulum, the protein becomes further processed. Any disulfide bonds that are required for the secondary structure are formed. These may be intramolecular or intermolecular and may lead to the formation of either homo- or heterodimers. Most cell-surface and extracellular proteins also become glycosylated by the addition of carbohydrate chains to serine, threonine, or asparagine residues. This usually does not affect the biological activity of the molecule but may affect its stability or interaction with other components. In some cases specificity is affected, for example in the case of Notch (see below). The proteins are carried in vesicles to the Golgi apparatus where further processing of the carbohydrates occurs, and then in exocytotic vesicles to the cell surface where they are released.

Signal transduction

The small molecule lipid soluble inducing factors, such as retinoic acid, can enter cells freely by diffusion. The retinoic acid receptor is itself a transcription factor, and ligand binding causes translocation to the nucleus where the receptor complex can activate its target genes (Fig. A.3a).

Protein-inducing factors cannot diffuse across the plasma membrane and so work by binding to specific cell-surface receptors. There are three main classes of these: enzyme-linked receptors, which are particularly important in development, G-protein-linked receptors, and ion channel receptors.

The main enzyme-linked receptors of importance in development are tyrosine kinases or Ser/Thr kinases (Fig. A.3b). All have a ligand-binding domain on the exterior of the cell, a single transmembrane domain, and the enzyme active site on the

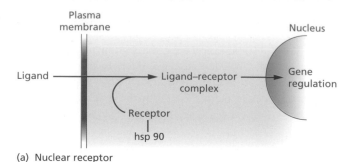

(a) Nuclear receptor

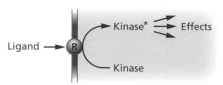

(b) Enzyme–linked receptor

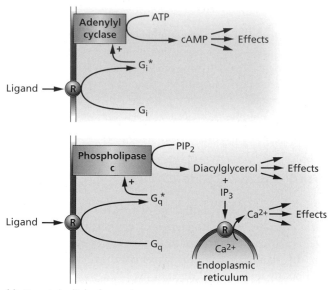

(c) G protein–linked receptors

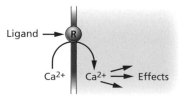

(d) Ion channel receptor

Fig. A.3 General modes of signal transduction. (a) Nuclear receptor mechanism used by steroids and retinoic acid. (b) Enzyme-linked receptor. (c) G-protein-linked receptor. (d) Ligand gated ion channel.

cytoplasmic domain. For receptor tyrosine kinases, the ligand-binding brings about dimerization of the receptor and this results in an autophosphorylation whereby each receptor molecule phosphorylates and activates the other. The phosphorylated receptors can then activate a variety of targets. Many of these are transcription factors that are activated by phosphorylation and move to the nucleus where they activate their target genes. In other cases there is a cascade of kinases that activate each other down the chain, culminating in the activation of a transcription factor. Roughly speaking, each class of factors has its own associated receptors and a specific signal transduction pathway; however, different receptors may be linked to the same signal transduction pathway, or one receptor may feed into more than one pathway. The effect of one pathway upon the others is often called "cross talk." For example, activated ERK (extracellular signal regulated kinase) can inhibit bone morphogenetic protein (BMP) signaling by phosphorylation of smad1 in its linker region. The significance of cross talk can be hard to assess from biochemical analysis alone, but is much easier using genetics. This is why the model organisms used for studying development are also useful for functional analysis of signal transduction and thereby useful for screening libraries of potential new drugs affecting signal transduction pathways. The inducing factor families and the signal transduction pathways most important in development are listed below.

There are several classes of G-protein-linked receptor (Fig. A.3c). The best known are seven-pass membrane proteins, meaning that they are composed of a single polypeptide chain crossing the membrane seven times. They are associated with trimeric G proteins composed of α-, β-, and γ-subunits. When the ligand binds, the activated receptor causes exchange of guanosine diphosphate (GDP) bound to the α-subunit for guanosine triphosphate (GTP), the activated α-subunit is released and can interact with other membrane components. The most common target is adenylyl cyclase, which converts adenosine triphosphate (ATP) to cyclic adenosine monophosphate (cAMP). Cyclic AMP activates protein kinase A (PKA) which phosphorylates various further target molecules affecting both intracellular metabolism and gene expression.

Another large group of G-protein-linked receptors use a different trimeric G protein to activate the inositol phospholipid pathway (Fig. A.3c). Here the G protein activates phospholipase C β which breaks down phosphatidylinositol bisphosphate (PIP_2) to diacylglycerol (DAG) and inositol trisphosphate (IP_3). The DAG activates an important membrane-bound kinase, protein kinase C. Like protein kinase A this has a large variety of possible targets in different contexts and can cause both metabolic responses and changes in gene expression. The IP_3 binds to an IP_3 receptor (IP_3R) in the endoplasmic reticulum and opens Ca channels which admit Ca ions into the cytoplasm. Normally cytoplasmic Ca is kept at a very low concentration of around 10^{-7} M. An increase caused either by opening of an ion channel in the plasma membrane, or as a result of IP_3 action, can again have a wide range of effects on diverse target molecules.

Ion channel receptors (Fig. A.3d) are not so prominent in development as the other types, but they play a key role in fertilization and their importance elsewhere may have been underestimated. They open on stimulation to allow passage of Na, K, Cl, or Ca.

Inducing factor families

Some important classes of protein-inducing factor are listed here. Most of these factors exist in both vertebrates and invertebrates but they often have different names if discovered by different routes. The normal vertebrate name is used here. Usually an invertebrate such as *C. elegans* or *Drosophila* will contain one gene for a particular factor, while the vertebrate species will contain a corresponding family of genes coding for proteins with similar, although not necessarily identical, biological activity. The usually accepted signal transduction mechanisms are shown in simplified form in Fig. A4, but the true picture is much more complex with many interactions between pathways.

Transforming growth factor β superfamily

Transforming growth factor (TGF) β was originally discovered as a mitogen (i.e. a factor promoting cell growth in tissue culture) secreted by "transformed" (cancer-like) cells. It has turned out to be the prototype for a large and diverse superfamily of signaling molecules, all of which share a number of basic structural characteristics. The mature factors are disulfide-bonded dimers of approximately 25 kDa. They are synthesized as longer pro-forms which need to be proteolytically cleaved to the mature form in order for biological activity to be shown.

The TGFβs themselves are in fact often inhibitory to cell division and promote the secretion of extracellular matrix materials. They are involved mainly in the organogenesis stages of development. The activin-like factors include the nodal-related family, which are all involved in induction and patterning of the mesoderm in vertebrate embryos. The bone morphogenetic proteins (BMPs) were discovered as factors promoting ectopic formation of cartilage and bone in rodents. They are involved in skeletal development, and also in the specification of the early body plan.

There are a number of receptors for the TGFβ superfamily. Their specificity for different factors is complex and overlapping, but in general different subsets of receptors bind to the TGFβs themselves, the activin-like factors, and the BMPs. In all cases the ligand binds first to a type II receptor and enables it to form a complex with a type I receptor. The type I receptor is a Ser Thr kinase and becomes activated in the ternary complex. Activation causes phosphorylation of smad proteins in the cytoplasm. Smads 1, 5, and 8 are targets for BMP receptors; smads 2 and 3 for activin receptors. Smad 4 is required by both pathways, and smad 6 is inhibitory to both by displacing the binding of smad 4. Phosphorylation causes the smads to migrate to the nucleus where they function as transcription factors, regulating target genes (Fig. A.4a).

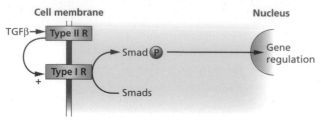

(a) TGFβ pathway

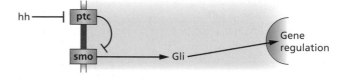

(b) FGF pathway

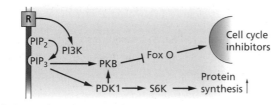

(c) Insulin/IGF pathway

(d) Hedgehog pathway

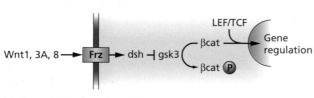

(e) Canonical Wnt pathway

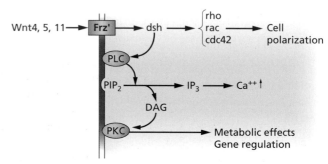

(f) Other Wnt pathways

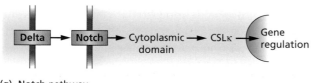

(g) Notch pathway

(h) Cytokine system

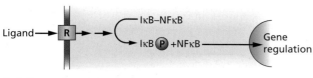

(i) Inflammatory system

Fig. A.4 Various specific pathways of signal transduction.

Fibroblast growth factor family

As their name suggests, fibroblast growth factors (FGFs) were first identified as mitogens for fibroblasts in tissue culture. They form a large family of factors with many important biological activities. They are monomeric polypeptides that bind tightly to heparan sulfate, a glycosaminoglycan found on all cell surfaces. FGFs have a very wide range of developmental functions including early anteroposterior patterning, regionalization of the brain, and promotion of limb outgrowth.

The FGF receptors are of the tyrosine kinase class. In vertebrates there are four receptors encoded by different genes and these each have splice forms that may show ligand specificity. In particular the R2b splice form is selective for FGF 7 and 10, while the R2c splice form is specific for FGF 4 and 8.

The complex of FGF and heparan sulfate binds to the receptors and brings about formation of receptor dimers. This leads to autophosphorylation and the activation of the extra-cellular signal regulated kinase (ERK) signal transduction pathway. This pathway involves a cascade of events. The first step is the activation of the GTP binding protein Ras, by exchange of a bound GTP for GDP. The activated Ras then activates the Raf protein by causing it to associate with the cell membrane. Raf is a kinase that phosphorylates another kinase called MEK (mitogen-activated, ERK-activating, kinase), which in turn phosphorylates ERK. Activated ERK enters the nucleus and activates various transcription factors by phosphorylation (Fig. A.4b). ERK used to be synonymous with MAP kinase, but this term now refers to a group of kinases which includes JNK and p38 as well as ERK itself.

Insulin family

Factors such as insulin and the insulin-like growth factors are important in growth control. They bind to tyrosine kinase-type receptors which are tetramers of two α and two β chains joined by disulfide bridges. These phosphorylate adapter proteins including IRS1 (insulin receptor substrate 1), which leads to activation of phosphatidylinositol 3-kinase (PI3 kinase). This acts in the membrane to phosphorylate various phosphatidyl-inositol substrates to generate a group of 3-phosphoinositides. These act as intact phospholipids and should not be confused with the free inositol 1,4,5-trisphosphate involved in the G-protein-linked pathway. The 3-phosphoinositides activate protein kinase B (also called Akt), and PDK1, which phosphorylate further substrates that have the effect of boosting protein synthesis and growth (Fig. A.4c).

The phosphatase PTEN dephosphorylates PKB and antagonizes insulin signaling.

Other tyrosine kinase-linked factors

Epidermal growth factor (EGF) was originally identified as a protein causing premature tooth eruption in newborn mice. It is structurally similar to TGFα, a factor isolated at the same time as TGFβ and hence with a similar name, although it is biochemic-ally quite distinct. EGF/TGFα factors are very important in the maternal stage patterning of Drosophila. Like FGF receptors, EGF receptors are Tyr kinases and they also activate the ERK pathway.

A similar group of factors are the neuregulins, all formed by differential splicing from a single gene. These bind a different group of receptor tyrosine kinases (e.g. ErbB2) and are often secreted by neurons.

Platelet-derived growth factor (PDGF) is the major growth factor in fetal serum. There are A and B forms of the factor and α and β forms of the receptor. Knockout or inhibition of this system causes a number of developmental process to fail.

Hepatocyte growth factor (HGF) is a ligand distinct from the FGF and EGF classes. It binds to another tyrosine kinase type receptor called c-met. It is important in liver regeneration and for migration of myoblasts into the limb buds. It is also called scatter factor because its first discovered activity was to cause epithelial to mesenchymal transition of epithelial cells.

Neurotrophins are principally, although not exclusively, of importance in the nervous system. They are essential for survival of neurons and are often secreted by the target cells to which the axons project. The neurotrophins comprise nerve growth factor (NGF), brain-derived neurotrophic factor (BDNF), and neurotrophins 3 and 4 (NT3 and 4). The receptors are called Trk (pronounced "track") proteins, which are tyrosine kinase type receptors. Glial-derived neurotrophic factor (GDNF) has a different receptor, ret, also a tyrosine kinase.

Ephrins are ligands for the Eph class of tyrosine kinase receptors. The ephrin A subgroup remains attached to the producing cell by a glycerophosphoinositol (GPI) anchor and binds predominantly to the Eph A receptors. Members of the ephrin B subgroup are themselves transmembrane proteins and bind mainly to the Eph B receptors. So the ephrin system must operate between cells making contact with each other and signaling may occur in both directions where B-type ephrins are involved. The ephrins are known to be involved in gastrulation, the establishment of segmentation in the hindbrain, and establishment of topographic maps in the nervous system.

Hedgehog family

The hedgehogs were first identified because mutations of the gene in Drosophila disrupted the segmentation pattern and made the larvae look like hedgehogs. Sonic hedgehog is very important for the dorsoventral patterning of the neural tube and for anteroposterior patterning of the limbs. Indian hedgehog is important in skeletal development. The full-length hedgehog polypeptide is an autoprotease, cleaving itself into an active N-terminal and an inactive C-terminal part. The N-terminal fragment is normally modified by covalent addition of a fatty acyl chain and of cholesterol, which are needed for full activity.

The hedgehog receptor is called patched, again named after the phenotype of the gene mutation in Drosophila. This is of the G-protein-linked class. It is constitutively active and is repressed by ligand binding. When active it represses the activity of another cell membrane protein, smoothened, which in turn represses the proteolytic cleavage of Gli-type transcription factors. Full-length Gli factors are transcriptional activators that can move to the nucleus and turn on target genes, but the constitutive removal of the C-terminal region makes them into repressors. In the absence of hedgehog, patched is active, smoothened inactive, and Gli inactive. In the presence of hedgehog, patched is inhibited, smoothened is active, and Gli is active (Fig. A.4d). Activation of protein kinase A also represses Gli and hence antagonizes hedgehog signaling.

Wnt family

The founder member of the Wnt family was discovered through two routes, as an oncogene in mice and as the *wingless* mutation in *Drosophila*. Wnt factors are single-chain polypeptides containing a covalently linked fatty acyl group which is essential for activity and renders them insoluble in water.

The Wnt receptors are called frizzleds after another *Drosophila* mutation. There are several classes of receptor for different ligand types and they do not necessarily cross-react. Wnt 1, 3A, or 8 will activate frizzleds that cause the repression of a kinase, glycogen synthase kinase 3 (gsk3) via a multifunctional protein called dishevelled. When active, gsk3 phosphorylates β-catenin, an important molecule involved both in cell adhesion and in gene regulation. When gsk3 is repressed, β-catenin remains unphosphorylated and in this state can combine with a transcription factor, Tcf-1, and convey it into the nucleus (Fig. A.4e). This pathway is important in numerous developmental contexts, including early dorsoventral patterning in *Xenopus*, segmentation in *Drosophila*, and kidney development.

Other Wnts, including Wnts 4, 5, and 11, bind to a different subset of frizzled that activate two other signal transduction pathways. In the **planar cell polarity** pathway a domain of the dishevelled protein interacts with small GTPases and the cytoskeleton (see below) to bring about a polarization of the cell. In the Wnt-Ca pathway phospholipase C becomes activated by a frizzled. This then acts to generate diacylglycerol and inositol 1,4,5 trisphosphate, with consequent elevation of cytoplasmic calcium, as described above under G-protein-coupled receptors (Fig. A.4f).

Notch system

For the Delta–Notch system both the ligands (Delta, Jagged) and the receptor (Notch) are integral membrane proteins. Their interaction can therefore only take place if the cells making them are in contact, as for the ephrin-Eph system above. Binding of ligand to Notch causes cleavage of the cytoplasmic portion of Notch by an intramembranous protease, γ-secretase, and this causes release into the cytoplasm of a transcription factor, CSL-κ (**CBF1**, **S**uCH, and **L**ag1 group, = suppressor of hairless in *Drosophila*). This migrates to the nucleus and activates target genes (Fig. A.4g). The γ-secretase is the same protease that generates the peptide whose accumulation in the brain leads to Alzheimer's disease.

Notch can carry O-linked tetrasaccharides and the presence of this carbohydrate chain can affect its specificity, increasing sensitivity to Delta and reducing sensitivity to Jagged (= Serrate in *Drosophila*). Control is often exercised through the activity of the glycosyl transferase Fringe, which adds a GlcNAc to the O-linked fucose.

The Delta–Notch system is important in numerous developmental situations, including neurogenesis, somitogenesis, and imaginal disc development.

Cytokine system

A somewhat heterogeneous group of signaling molecules including some classical cytokines (interferons), the hematopoietic growth factors, growth hormone, and leukemia inhibitory factor (LIF), work through this type of system. The receptors, which may be homodimeric or heterodimeric, bind intracellular tyrosine kinases called JAKs (Janus kinases). When the ligand brings receptor molecules together the JAKs phosphorylate and activate each other, and the receptor, and the active complex can then phosphorylate STAT-type transcription factors. Phosphorylation causes them to dissociate from the receptor and dimerize and they can then enter the nucleus and activate target genes (Fig. A.4h).

Inflammatory system

The proinflammatory cytokines comprising the interleukin 1 (IL1) and tumor necrosis factor families have receptors that are coupled to the activation of a transcription factor NFκB, which is normally complexed in an inactive form in the cytoplasm by another protein called IκB. When ligand binds receptor it activates a cytoplasmic kinase that phosphorylates IκB and causes it to dissociate from the complex, allowing NFκB to move to the nucleus and turn on target genes (Fig. A.4i). IκB can also be phosphorylated by protein kinase C.

IL1 and TNF are not prominent in developmental biology, but the same system acts in the dorsoventral patterning of *Drosophila* where the receptor called Toll is the homolog of the vertebrate IL1 receptor.

Axonal guidance systems

Diffusible factors primarily responsible for axonal guidance include the netrins and the semaphorins. The receptors for semaphorins are called neuropilins, and the receptors for netrins called DCC ("deleted in colon carcinoma").

Slit proteins are ligands for roundabout (robo)-like receptors. These mostly display axonal repulsion activity. They need heparan sulfate for activity and the intracellular effects work through the small GTPases (see below).

Cytoskeleton

The cytoskeleton is important in development for four distinct reasons. Firstly, eggs often have a significant internal regionalization, for example in the localization of particular mRNAs. Secondly, the orientation of cell division may be important. Thirdly, animal cells move around a lot, either as individuals or as part of moving cell sheets. Finally, the shape of cells is an essential part of their ability to carry out their functions. All of these activities are functions of the cytoskeleton.

The three main components of the cytoskeleton are:

1 **microfilaments**, made of actin;
2 **microtubules**, made of tubulin;
3 **intermediate filaments**, made of:
   (a) cytokeratins in epithelial cells
   (b) vimentin in mesenchymal cells
   (c) neurofilament proteins in neurons
   (d) glial fibrilliary acidic protein (GFAP) in glial cells.

Microtubules and microfilaments are universal constituents of eukaryotic cells, while intermediate filaments are found only in animals.

Microtubules

Microtubules (Fig. A.5) are hollow tubes of 25 nm diameter composed of tubulin. Tubulin is a generic name for a family of globular proteins which exist in solution as heterodimers of α- and β-type subunits and is one of the more abundant cytoplasmic proteins. The microtubules are polarized structures with a minus end anchored to the centrosome and a free plus end at which tubulin monomers are added or removed. Microtubules are not contractile but exert their effects through length changes based on polymerization and depolymerization. They are very dynamic, either growing by addition of tubulin monomers, or retracting by loss of monomers, and individual tubules can grow and shrink over a few minutes. The monomers contain GTP bound to the β-subunit and in a growing plus end this stabilizes the tubule. But if the rate of growth slows down, hydrolysis of GTP to GDP will catch up with the addition of monomers. The conversion of bound GTP to GDP renders the plus end of the tubule unstable and it will then start to depolymerize. The drugs colchicine and colcemid bind to monomeric tubulin and prevent polymerization. Among other effects this causes the disassembly of the mitotic spindle. These drugs cause cells to become arrested in mitosis and are often used in studies of cell kinetics.

The shape and polarity of cells can be controlled by locating capping proteins in particular parts of the cell cortex which bind the free plus ends of the microtubules and stabilize them. The positioning of structures within the cell also depends largely on microtubules. There exist special **motor proteins** that can move along the tubules, powered by hydrolysis of ATP, and thereby transport other molecules to particular locations within the cell.

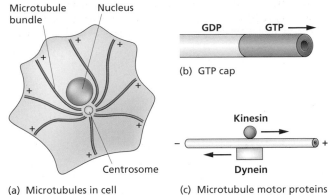

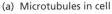

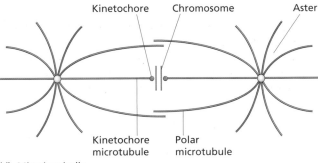

(a) Microtubules in cell

(b) GTP cap

(c) Microtubule motor proteins

(d) Mitotic spindle

Fig. A.5 Microtubules. (a) Arrangement of microtubules in cell. (b) GTP cap at the plus end of microtubule. (c) Motor proteins associated with microtubules. (d) Arrangement of tubules in a mitotic spindle.

The kinesins move towards the plus ends of the tubules while the dyneins move towards the minus ends. In a developmental context they are important for the localization of substances in the egg, for example the dishevelled protein in *Xenopus*, or the bicoid mRNA in *Drosophila*.

Microtubules are prominent during cell division. The minus ends of the tubules originate in the centrosome, which is a microtubule-organizing center able to initiate the assembly of new tubules. In mitotic prophase the centrosome divides and each of the radiating sets of microtubules becomes known as an aster. The two asters move to the opposite sides of the nucleus to become the two poles of the mitotic spindle. The spindle contains two types of microtubules. The polar microtubules meet each other near the center and become linked by plus directed motor proteins. These tend to drive the poles apart. Each chromosome has a special site called a kinetochore which binds another group of microtubules called kinetochore microtubules. At anaphase the kinetochores of homologous chromosomes separate. The polar microtubules continue to elongate while the kinetochore microtubules shorten by loss of tubulin from both ends and draw the chromosome sets into the opposite poles of the spindle.

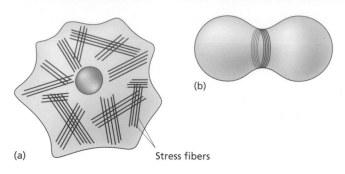

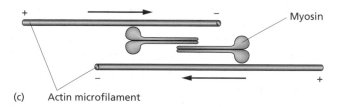

Fig. A.6 Microfilaments. (a) Arrangement of microfilaments in a fibroblast in tissue culture. (b) Contractile ring in a dividing cell. (c) Contraction of filament bundle driven by myosin.

Microfilaments

Microfilaments (Fig. A.6) are polymers of actin, which is the most abundant protein in most animal cells. In vertebrates there are several different gene products of which α actin is found in muscle and β/γ actins in the cytoskeleton of nonmuscle cells. For all actin types the monomeric soluble form is called G-actin. Actin filaments have an inert minus end, and a growing plus end to which new monomers are added. G-actin contains ATP and this becomes hydrolysed to ADP shortly after addition to the filament. As with tubules, a rapidly growing filament will bear an ATP cap which stabilizes the plus end. Microfilaments are often found to undergo "treadmilling" such that monomers are continuously added to the plus end and removed from the minus end while leaving the filament at the same overall length. Microfilament polymerization is prevented by a group of drugs called cytochalasins, and existing filaments are stabilized by another group called phalloidins. Like microtubules, microfilaments have associated motor proteins that will actively migrate along the fiber. The most abundant of these is myosin II, which moves toward the plus end of microfilaments, the process being driven by the hydrolysis of ATP. To bring about contraction of a filament bundle, the myosin is assembled as short bipolar filaments with motile centers at both ends. If neighboring actin filaments are arranged with opposite orientation then the motor activity of the myosin will draw the filaments past each other leading to a contraction of the filament bundle.

Microfilaments can be arranged in various different ways depending on the nature of the accessory proteins with which they are associated. Contractile assemblies contain microfilaments in antiparallel orientation associated with myosin. These

are found in the contractile ring which is responsible for cell division, and in the stress fibers by which fibroblasts exert traction on their substratum. Parallel bundles are found in **filopodia** and other projections from the cell. Gels composed of short randomly orientated filaments are found in the cortical region of the cell.

Small GTPases

There are three well-known GTPases that activate cell movement in response to extracellular signals: Rho, Rac, and cdc42. They are activated by numerous tyrosine kinase-, G-coupled-, and cytokine-type receptors. Activation involves exchange of GDP for GTP and many downstream proteins can interact with the activated forms. Rho normally activates the assembly of stress fibers. Rac activates the formation of **lamellipodia** and ruffles. Cdc42 activates formation of filopodia. In addition all three promote the formation of focal adhesions, which are integrin-containing junctions to the extracellular matrix. These proteins can also affect gene activity through the JNK and p38 MAP kinase pathways.

Cell adhesion molecules

Organisms are not just bags of cells, rather each tissue has a definite cellular composition and microarchitecture. This is determined partly by the cell-surface molecules by which cells interact with each other, and partly by the components of the **extracellular matrix (ECM)**. Virtually all proteins on the cell surface or in the ECM are glycoproteins, containing oligosaccharide groups added in the endoplasmic reticulum or Golgi apparatus after translation and before secretion from the cell. These carbohydrate groups often have rather little effect on the biological activity of the protein but they may affect its physical properties and stability.

Cells are attached to each other by adhesion molecules (Fig. A.7). Among these are the cadherins, which stick cells together in the presence of Ca, the cell adhesion molecules (CAMs), that do not require Ca, and the integrins that attach cells to the extracellular matrix. When cells come together they often form **gap junctions** at the region of contact. These consist of small pores joining the cytosol of the two cells. The pores, or connexons, are assembled from proteins called connexins. They can pass molecules up to about 1000 molecular weight by passive diffusion.

Cadherins

This is a family of single-pass transmembrane glycoproteins which can adhere tightly to similar molecules on other cells in the presence of calcium. Cadherins are the main factors

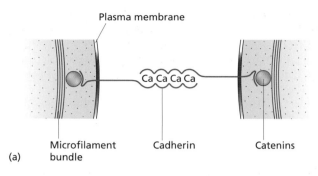

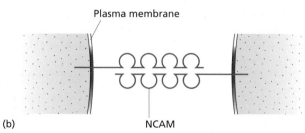

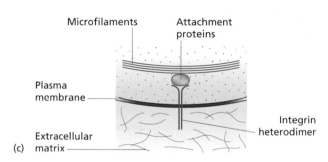

Fig. A.7 Cell adhesion molecules (CAMs). (a) Calcium-dependent adhesion via cadherins. (b) Calcium-independent adhesion via CAMs. (c) Adhesion to extracellular matrix via integrins.

attaching embryonic cells together, which is why embryonic tissues can often be caused to disaggregate simply by removal of calcium. The cytoplasmic tail of cadherins is anchored to actin bundles in the cytoskeleton by a complex including proteins called catenins. One of these, β-catenin, is also a component of the Wnt signaling pathway (see above), providing a potential link between cell signaling and cell association. Cadherins were first named for the tissues in which they were originally found, so E-cadherin occurs mainly in epithelia and N-cadherin occurs mainly in neural tissue.

Immunoglobulin superfamily

These are single-pass transmembrane glycoproteins with a number of disulfide-bonded loops on the extracellular region, similar to the loops found in antibody molecules. They also bind to similar molecules on other cells, but unlike the cadherins they do not need calcium to do so. The neural cell adhesion molecule (NCAM) is composed of a large family of different proteins formed by alternative splicing. It is most prevalent in the nervous system but also occurs elsewhere. It may carry a large amount of polysialic acid on the extracellular domain, and this can inhibit cell attachment because of the repulsion between the concentrations of negative charge on the two cells. Related molecules include L1 and ICAM (intercellular cell adhesion molecule).

Integrins

The integrins are cell-surface glycoproteins that interact mainly with components of the extracellular matrix. They are heterodimers of α- and β-subunits, and require either magnesium or calcium for binding. There are numerous different α and β chain types and so there is a very large number of potential heterodimers. Integrins are attached by their cytoplasmic domains to microfilament bundles, so, like cadherins, they provide a link between the outside world and the cytoskeleton. They are also thought on occasion to be responsible for the activation of signal transduction pathways and new gene transcription following exposure to particular extracellular matrix components.

Extracellular matrix components

Glycosaminoglycans

Glycosaminoglycans (GAGs) are unbranched polysaccharides composed of repeating disaccharides of an amino sugar and a uronic acid, usually substituted with some sulfate groups. GAGs are constituents of proteoglycans, which have a protein core to which the GAG chains are added in the Golgi apparatus before secretion. One molecule of a proteoglycan may carry more than one type of GAG chain. GAGs have a high negative charge and a small amount can immobilize a large amount of water into a gel. Important GAGs, each of which have different component disaccharides, are heparan sulfate, chondroitin sulfate, and keratan sulfate. Heparan sulfate, closely related to the anticoagulant heparin, is particularly important for cell signaling as it is required to present various growth factors, such as the FGFs, to their receptors. Hyaluronic acid differs from other GAGs because it occurs free, and not as a constituent of a proteoglycan. It consists of repeating disaccharides of glucuronic acid and N-acetyl glucosamine, and is not sulfated. It is synthesized by enzymes at the cell surface and is abundant in early embryos.

Collagens

Collagens are the most abundant proteins by weight in most animals. The polypeptides, called α chains, are rich in proline and glycine. Before secretion, three α chains become twisted around each other to form a stiff triple helical structure. In the extracellular matrix, the triple helices become aggregated together to form the collagen fibrils visible in the electron microscope. There are many types of collagen, which may be composed of similar or of different α chains in the triple helix. Type I collagen is the most abundant and is a major constituent of most extracellular material. Type II collagen is found in cartilage and in the notochord of vertebrate embryos. Type IV collagen is a major constituent of the basal lamina underlying epithelial tissues. Collagen helices may become covalently cross-linked through their lysine residues, and this contributes to the changing mechanical properties of tissues with age.

Elastin

This is another extracellular protein with extensive intermolecular cross-linking. It confers the elasticity on tissues in which it is abundant, and also has some cell signaling functions.

Fibronectin

Fibronectin is composed of a large disulfide-bonded dimer. The polypeptides contain regions responsible for binding to collagen, to heparan sulfate, and to integrins on the cell surface. These latter, cell-binding, domains are characterized by the presence of the amino acid sequence Arg-Gly-Asp (= RGD). There are many different forms of fibronectin produced by alternative splicing.

Laminin

Laminin is a large extracellular glycoprotein, found particularly in basal laminae. It is composed of three disulfide-bonded polypeptides joined in a cross shape. It carries domains for binding to type IV collagen, heparan sulfate, and another matrix glycoprotein, entactin.

Further reading

General
Brown, T.A. (2001) *Gene Cloning and DNA Analysis: an introduction*, 4th edn. Oxford: Blackwell Science.
Alberts, B., Johnson, A., Lewis, J., Raff, M., Roberts, K. & Walter, P. (2002) *Molecular Biology of the Cell*, 4th edn. New York: Garland Publishing.
Primrose, S.B., Twyman, R.M. & Old, R.W. (2002) *Principles of Gene Manipulation*, 6th edn. Oxford: Blackwell Science.
Darnell, J.E. (2003) *Molecular Cell Biology*, 5th edn. New York: W.H. Freeman.

Gene regulation
Cross, S.H. & Bird, A.P. (1995) CpG islands and genes. *Current Opinion in Genetics and Development* **5**, 309–314.
Gould, A (1997) Functions of mammalian Polycomb group and trithorax group related genes. *Current Opinion in Genetics and Development* **7**, 488–494.
Wolffe, A. (1998) *Chromatin: structure and function*, 3rd edn. San Diego: Academic Press.
Latchman, D.S. (2003) *Eukaryotic Transcription Factors*. New York: Academic Press.

Signaling
Hancock, J.T. (1997) *Cell Signaling*. Harrow: Longman.
Hunter, T. (2000) Signaling – 2000 and beyond. *Cell* **100**, 113–127.
Downward, J. (2001) The ins and outs of signaling. *Nature* **411**, 759–762.
Heath, J.K. (2001) *Principles of Cell Proliferation*. Oxford: Blackwell Science.

Specific signaling pathways
Cadigan, K.M. & Nusse, R. (1997) Wnt signaling: a common theme in animal development. *Genes and Development* **11**, 3286–3305.
Holder, N. & Klein, R. (1999) Eph receptors and ephrins: effectors of morphogenesis. *Development* **126**, 2033–2044.
Szebenyi, G. & Fallon, J.F. (1999) Fibroblast growth factors as multifunctional signaling factors. *International Review of Cytology* **185**, 45–106.
Massagué, J. & Chen, Y.G. (2000) Controlling TGFβ signaling. *Genes and Development* **14**, 627–644.
Munn, J.S. & Kopan, R. (2000) Notch signaling: from the outside in. *Developmental Biology* **228**, 151–165.
Ingham, P.W. & McMahon, A.P. (2001) Hedgehog signaling in animal development: paradigms and principles. *Genes and Development* **15**, 3059–3087.
Von Bubnoff, A. & Cho, K.W.Y. (2001) Intracellular BMP signaling regulation in vertebrates: pathway or network? *Developmental Biology* **239**, 1–14.
Tata, J.R. (2002) Signaling through nuclear receptors. *Nature Reviews Molecular Cell Biology* **3**, 702–710.

Cytoskeleton, adhesion molecules, and extracellular matrix
Adams, J.C. & Watt, F.M. (1993) Regulation of development and differentiation by the extracellular-matrix. *Development* **117**, 1183–1198.
Kreis, T. & Vale, R. (1999) *Guidebook to the Cytoskeletal and Motor Proteins*, 2nd edn. Oxford: Oxford University Press.
Kreis, T. & Vale, R. (1999) *Guidebook to the Extracellular Matrix and Adhesion Proteins*, 2nd edn. Oxford: Oxford University Press.
Beckerle, M.C. (2002) *Cell Adhesion*. Oxford: Oxford University Press.
Thiery, J.P. (2003) Cell adhesion in development: a complex signaling network. *Current Opinion in Genetics and Development* **13**, 365–371.
Wheelock, M.J. & Johnson, K.R. (2003) Cadherins as modulators of cellular phenotype. *Annual Reviews of Cell and Developmental Biology* **19**, 207–235.

Glossary

To keep the glossary to a reasonable length it does not include the names of most animal taxa or individual genes or gene products. However the accounts of these in the text can easily be located from the index

acini (singular **acinus**): clusters of cells in a glandular **epithelium**

acrosome: large secretory vesicle found in sperm

adaptive evolution: evolutionary changes produced by natural selection

adipose tissue: tissue storing and synthesizing fats

adult stem cell: = **tissue stem cell**

aggregation chimera: mouse made by aggregating two embryos of different **genotypes**

AGM region: aorta-gonad-mesonephros region of vertebrate embryo mesoderm

agonist: a drug that will stimulate a particular cellular receptor or response, often mimicking the action of an endogenous substance

allantois: extraembryonic structure of **amniotes** arising from **posterior mesoderm**

allele: genes with alternative DNA sequences that occupy the same **locus** in the **genome**

allelic series: a set of **alleles** showing a progressively stronger **phenotype**

allograft: graft from one mature individual to another: likely to provoke an immune reaction unless the individuals are of the same inbred strain

allometry: differential growth between two parts of an organism, or between homologous parts of different organisms

amniocentesis: sampling of cells from within the amniotic cavity of a human embryo. This enables the early detection of chromosome abnormalities

amnion: **extraembryonic membrane** characteristic of **amniotes**, consisting of **ectoderm** on the side facing the embryo and **mesoderm** on the side facing the chorion

amniote: animal whose embryo has an **amnion**, i.e. a mammal, bird, or reptile

analogy: structures that resemble each other for reasons other than **homology**

anchor cell: cell acting as a **signaling center** for the *C. elegans* vulva

androgenetic: an embryo whose genetic material derives only from the sperm

angiogenesis: formation of new blood vessels by growth from pre-existing vessels

animal cap: explanted region from the animal pole of an amphibian embryo, used to assay **inducing factors**

animal hemisphere: the upper hemisphere of an **egg** or **oocyte**; after fertilization usually bears the **polar bodies**

animal model: normally a transgenic or **knockout** mouse that is designed to have a defect similar to a human disease

animal pole: the pole of an egg that normally lies uppermost, contains relatively yolk-free cytoplasm, and usually bears the **polar bodies**

animal–vegetal axis: line running from animal to vegetal pole of an **egg** or **oocyte**

antagonist: a substance that inhibits or blocks the action of another substance

anterior intestinal portal: opening of the **foregut** into the subgerminal (**midgut**) cavity of an **amniote** embryo

anterior midgut: the part of an insect embryo gut invaginationg from the **anterior** end

anterior: the head end of an animal

anteroposterior: the direction towards, or line joining, the **anterior** and **posterior** extremities of an animal

antibody: immunoglobulin protein produced by B lymphocytes of the immune system, which can bind specifically and with high affinity to a specific substance of interest. Antibodies are made by immunizing an animal with the substance of interest, which is called the **antigen**

antigen: any substance used to raise an **antibody** by immunization of an animal

antimorphic: = **dominant negative**

antisense: sequence of DNA or RNA complementary to a target sequence and thus capable of hybridizing with it

anurans: tailless amphibians: the frogs and toads

apical ectodermal ridge (AER): epidermal ridge on limb bud, needed for distal outgrowth

apical: of **epithelia**, the side away from the **basement membrane**, usually facing a **lumen**

apoptosis: = **programmed cell death**

appositional induction: induction between two cell sheets in contact

archenteron: cavity formed as a result of **gastrulation** movements; becomes or contributes to the later gut **lumen**

area opaca: outer region of the avian **blastoderm** which expands to form the **yolk sac**

area pellucida: central region of an avian **blastoderm** from which the embryo itself develops

asexual reproduction: reproduction without the fusion of two **gametes** to form a **zygote**. This can takes several forms including budding, fragmentation, or **parthenogenesis** from eggs

aster: the microtubule array which serves to separate the chromosomes during cell division

astrocyte: a type of **glial cell** in the central nervous system

asymmetrical division: a situation where the two daughter cells resulting from a cell division are different from each other

atavism: a morphological character, usually caused by a **mutation**, that resembles an ancestral organism

autologous graft: graft from one part of the mature individual to another; avoids any immune rejection

autonomous: of a process not requiring any input from surrounding cells

autopod (= autopodium): the hand or foot of a vertebrate limb

autoradiography: method for detecting radioactivity on a gel or a histological section by exposure to a photographic emulsion

autosomes: chromosomes that are not sex chromosomes

AVE region (= anterior visceral endoderm): a **signaling center** in the mouse **conceptus**

axis: (1) a line or direction, as in **anteroposterior** axis; (2) the midline structures of a vertebrate embryo: **notochord**, **somites**, and **neural tube**

B

β cells: insulin-secreting **endocrine** cells of the islets of Langerhans, in the pancreas

β-galactosidase: product of the *lacZ* gene. β-galactosidase enzymatic activity is easily detected histochemically and therefore the *lacZ* gene is often used as a reporter of gene expression

β-geo: a fusion protein of **β-galactosidase** and **neomycin** resistance protein

balancer chromosome: a chromosome that will not recombine with its **homolog**, and usually carries a recessive lethal mutation and a dominant marker. Especially used for simplifying mutagenesis screens and for maintaining stocks of recessive lethal mutants in heterozygous form

band shift assay: gel electrophoretic technique for demonstrating the binding of a protein to a nucleic acid

basal: (1) of **epithelia**, the side next to the **basement membrane**; (2) of **taxa**, those which are most similar to the common ancestor

basement membrane: extracellular matrix layer underlying an epithelium

bipolar form: a **mirror symmetrical** duplication of the whole body, especially encountered in the regeneration of **planaria** or **hydroids**

birth defects (= congential defects): defects of a mammalian embryo apparent at the time of birth. These may arise from genetic causes, or chromosome abnormalities in the embryo, or **teratological** stimuli

bistable switch: molecular mechanism which has two stable steady states that can be interconverted by some external signal

bivalent: (1) the meiotic chromosomes containing four chromatids, formed by pairing of paternal and maternal chromosomes; (2) of a molecule, having two binding sites

blastema: an undifferentiated bud of cells, dividing except in **planaria**, that gives rise to a structure

blastocoel: fluid-filled cavity in the center of a **blastula**

blastocyst: mammalian embryo after cavitation; differs from **blastula** because it contains differentiating extraembryonic structures

blastoderm: similar to a **blastula** but arranged as a sheet rather than a ball of cells

blastomeres: the large cells arising from the fertilized egg by **cleavage** divisions

blastopore: depression or slit through which cells move to the interior during **gastrulation**

blastula: early developmental stage; a ball of similar cells arising from repeated **cleavage** of a fertilized egg

blot: a blot is made by transferring the products of a gel electrophoretic separation onto a membrane that is suitable for biochemical analysis by nucleic acid hybridization or antibody binding (see **Northern, Southern, Western blot**)

body plan (= Bauplan): the essential features of the whole body of a species or higher taxonomic group

border cells: specialized ovarian follicle cells that control the **polarity** of the *Drosophila* **oocyte**

brachial plexus: arrangement of nerves formed by the fusion and rebranching of the **spinal nerves** supplying the vertebrate forelimb

branchial arches: (= pharyngeal arches) segmental structures between the pharyngeal pouches each comprising a cartilaginous arch, blood vessel, and cranial nerve

branching morphogenesis: formation of a branched structure by cell movement and/or growth of an **epithelium**

bromodeoxyuridine (BrdU): DNA nucleotide analog used for labeling cells in **S phase**

C

cancer: uncontrolled growth of the body's own cells; often displaying local invasive behavior and **metastasis**

capacitation: of mammalian sperm, a process of conditioning in the female reproductive tract which makes them capable of fertilization

carcinoma: a **cancer** derived from an **epithelium**

cardia bifida: congenital abnormality: having two hearts

carpals: bones of the wrist or ankle

cartilage model: see **model**

caudal: relating to the tail, often eqivalent to **posterior**

cavitation: formation of an internal cavity in a cell mass

CCD (= charge coupled device): the optical sensor in a digital camera

cell biology: approach to biology that focuses on the structure, function, and behavior of cells

cell lineage: the "family tree" of a group of cells, showing which cell is the descendant of which other cell

cellular blastoderm: stage of insect development at which the **syncytial blastoderm** becomes divided into cells

centromere: the region of a chromosome that attaches to the spindle at cell division, attaches sister chromatids following DNA replication, and attaches homologous chromosome during meiosis

cerebellum: part of the vertebrate brain controlling movement, formed from **rhombomere 1**

chalone: a circulating, tissue-specific, inhibitor of growth

checkpoint: time during the cell division cycle at which control is exerted on whether the cell shall continue in the cycle or not

chemokine: class of extracellular signaling molecule particularly important for activation of white blood cells

chim(a)era: (1) = **genetic mosaic**, especially a mammalian embryo in which cells of one **genotype** have been injected into an embryo of a different genotype; (2) also used of molecules assembled from more than one source by **molecular cloning:** see **fusion protein** and **domain swap**

chondroblast: mitotically active cartilage cell, not yet a mature **chondrocyte**

chondrocyte: mature cartilage cell

chordates (= Chordata): phylum comprising the vertebrates together with the **protochordates**

chorioallantoic membrane (CAM): extraembryonic membrane of avian embryos formed by fusion of the **allantois** with the **chorion**

chorion: (1) **extraembryonic membrane** of **amniotes**, similar to the **amnion** but forming an outer layer; (2) an extracellular layer surrounding an insect or fish egg

chromatin: DNA combined with chromosomal proteins in a maner which helps regulate its expression and behavior

chromobox genes: genes encoding homologs of the *Polycomb* gene that regulate **Hox** gene expression via chromatin structure

clade: a **taxon** which consists of all the descendants of a common ancestor

cleavage: a type of cell division that occurs without growth, such that daughters are smaller than the mother; typically found in the early stages of development

cloaca: common urogenital and anal aperture

clonal analysis: information obtained about developmental mechanisms by studying the position and differentiation of the progeny of a single labeled cell

clone: (1) population of cells derived from a single progenitor cell, either *in vivo* or *in vitro*; (2) DNA molecule prepared by **molecular cloning**

cloning: (1) assembling a DNA sequence with the use of restriction enzymes and ligases followed by insertion into a cloning vector and growth to a useful quantity (= **molecular cloning**); (2) growth of a colony of cells from a single cell (= cell cloning); (3) formation of a whole organism from one cell

(= whole animal cloning, or **reproductive cloning** of humans)

clonogenic: of cells, the ability to divide and form a **clone** in an appropriate assay system

cnidarians (= Cnidaria): animal phylum including jellyfish and sea anemones

coelom: body cavity lined with mesoderm

combinatorial chemistry: methods for making large libraries of related compounds simultaneously instead of one at a time

combinatorial: situation where different states are defined not by single elements but by a combination of elements. E.g. different states of cell **commitment** generally depend on the combination of **transcription factors** that is present, rather than a single factor

commisural neurons: neurons which connect one side of the spinal cord to the other

commitment: of a cell or tissue region indicates that it is programmed to follow a particular developmental pathway or **fate**

compaction: process whereby cells of a mammalian **morula** become more adhesive and pack together more tightly

compartment: a region of an embryo within which a **clone** of cells may move around but whose boundaries it does not cross

competence: ability to respond to an **inducing factor**

complementation test: genetic test to find if two recessive mutants lie in the same gene by introducing one mutation on the maternal and the other on the paternal chromosome. If they are in different genes then the wild-type alleles should complement and show no phenotype. But if they are in the same gene they will fail to complement and a phenotype will result

compound microscope: type of visible light microscope used for looking at sections or other very thin objects

conceptus: a mammalian embryo together with all the **extraembryonic membranes** formed from the **zygote**

condensation: (1) a dense patch of cells within a **mesenchyme**; (2) formation of such a patch

conditional knockout: a **knockout** mouse in which the ablation of the gene occurs under a particular set of circumstances controlled by the experimenter

confocal (scanning) microscope: a microscope that scans the specimen with a laser to build up a digital image of a particular optical section

connective tissue: strictly refers to tissues containing a substantial collagen-rich **extracellular matrix** secreted by **fibroblasts.** Sometimes used more widely to refer to all tissues that are not **epithelia**

constitutive: describes a gene or gene product that is active continuously

convergent extension: morphogenetic movement in which a cell sheet elongates and narrows because of active movements of the constituent cells to alter the overall packing arrangement

coronal section: = frontal section

corpus luteum: structure formed in the ovary after **ovulation** of mammalian oocyte

cortex: outer region of a cell, especially an **egg** or **oocyte**, comprising a few microns thickness beneath the plasma membrane

cortical granules: secretory vesicles containing **vitelline membrane** components, released from the **egg** at fertilization

cortical rotation: rotation of the **zygote** cortex relative to the internal cytoplasm

cranial nerves: nerves coming from the brain

cranial: relating to the head, often equivalent to **anterior**

Cre-lox: system used to bring about DNA recombination or excision events at particular sites under controlled circumstances

crossing over: recombination of paternal and maternal chromatids at meiosis; can also be artificially induced at four-strand stage of normal mitotic cycle

cryostat (= cryotome): microtome operated below freezing point to cut frozen sections of a specimen

cumulus: ovarian **follicle cells** released with a mammalian **oocyte** on maturation

cystoblast: precursor cell to the **oocyte+15 nurse cells** in *Drosophila*

cytoplasmic determinant: see **determinant**

cytoskeleton: system of filaments and tubules that makes up the structural support for the cytoplasm of a cell

D **dauer larva:** resistant dormant state of *C. elegans*

deciduum: the swelling of a uterine crypt produced by its reaction to an implanted embryo

deep cells: cells in the interior of an early embryo, especially in fish embryos

definitive endoderm: the embryonic **endoderm** of an **amniote** embryo, as opposed to **extraembryonic** endoderm

dehydration: removal of water

denticles: small cellular appendages on the ventral side of a *Drosophila* larva

derived: of taxa, those which are most different from the common ancester

dermal papilla: mound of dermal cells, especially in hair follicle

dermamyotome: part of the **somite** forming the **dermatome** and **myotome**

dermatome: part of the **somite** forming the **dorsal dermis**

dermis: the **connective tissue** layer of the skin

determinant (= cytoplasmic determinant): substance localized to part of an **egg** or **blastomere** that causes the cells that inherit it to acquire a particular developmental **commitment**

determination: type of developmental **commitment** of cells or tissue explant that is irreversible following grafting to any different location in the embryo

Deuterostomia: group of animal **phyla** characterized by **radial cleavage, regulative** development, and formation of the anus from the **blastopore**

developmental constraint: aspect of the developmental mechanism that prevents the viability of some types of mutant

diaphysis: the shaft of a long bone

diencephalon: the **posterior** part of the vertebrate **forebrain**

differentiation: the acquisition during development of the properties of mature functional cells. The term usually refers to the last step of development rather than to earlier events of **commitment**

diI, diO: lipid-soluble **fluorescent vital dyes** used for **fate** mapping

diploblasts: animal **taxa** with only two **germ layers:** the cnidarians and ctenophores

diploid: having two sets of chromosomes, one from the father and one from the mother

dissecting microscope: type of microscope for looking at, and manipulating, relatively large solid objects

distal tip cell: cell acting as a **signaling center** for the *C. elegans* **germ line**

distal: further from the body

DNA sequencing: determination of the sequence of bases (A,T,C,G) in DNA

domain swap: a gene encoding a protein in which two functional domains of other proteins are combined, for example a DNA binding and hormone binding region from different proteins

dominant negative: type of **mutation** that antagonizes the effect of the wild-type **allele**

dominant: type of **mutation** that produces a phenotype in the presence of the wild-type **allele**

dorsal closure: morphogenetic movement of insect embryos leading to enclosure of the yolk mass by the ventrally located germ band

dorsal lip: dorsal part of the ring-shaped **blastopore** of an amphibian embryo. The tissue just above the dorsal lip is **Spemann's organizer**

dorsal root ganglia (DRG): the sensory ganglia on the **dorsal** branch of a **spinal nerve**

dorsal: the upper surface (or back) of an animal

dorsoventral: the direction towards, or line joining, the **dorsal** and **ventral** extremities of an animal

dsRNA: double-stranded RNA, used in **RNA interference**

ductus arteriosus: junction of aorta and pulmonary artery of mammalian fetus

dysplasia: altered pattern of growth and differentiation

E **Ecdysozoa:** a superphylum defined on the basis of primary sequence data; includes the arthropods and nematodes among other groups

echinus rudiment: the bud in the sea urchin larva that gives rise to the adult during **metamorphosis**

ectoderm: the outer of the three embryonic **germ layers**

ectopic: not in its usual position; as in ectopic structures or ectopic gene expression

egg chamber: structure in the female *Drosophila* consisting of the **oocyte** and its **nurse cells** surrounded by ovarian **follicle cells**

egg cylinder: stage of rodent embryo in which the embryo consists of two layers arranged in a cup shape

egg: strictly the female gamete after completion of the second meiotic division (= **ovum**). Also a fertilized egg (= **zygote**). Also often loosely used for secondary **oocytes** and early embryos

electron microscope: microscope that forms an image using a beam of electrons and therefore has much higher resolving power, and possible magnification, than a light microscope

electroporation: introduction of substances, usually DNA, into cells by means of electric field pulses

embryoid bodies: structures resembling embryos which are formed by **ES cells** or **teratocarcinoma** cells when they are removed from a growth-promoting medium

embryonic induction: the process whereby the development of one group of cells, called the **competent** region, is altered by an **inducing factor** from another group, called a **signaling center** or **organizer**

embryonic shield: thickened dorsal part of the **germ ring** of a fish embryo

embryonic stem cells (= ES cells): cells grown from early mammalian embryos that are developmentally totipotent

emission: the light of a particular characteristic wavelength that is emitted from a **fluorochrome**

endocardial cushions: structures contributing to the **ventricular septum** and the atrioventricular valves of the heart

endocardium: the inner, **endothelial**, layer of the heart

endocrine: of a ductless gland, secreting substances into the blood

endoderm: the inner of the three germ layers

endothelial cells: the cells lining blood vessels and forming capillaries

enhancer trap: a transgenic line containing a **reporter gene** whose activity is regulated by an endogenous **enhancer**

enhancer: regulatory region of DNA that controls the expression of a gene; often independently of its precise position in relation to the gene

enterocoely: formation of the coelom by budding of the mesodermal rudiment from the gut

enterocytes: absorptive cells of small intestinal epithelium

enteroendocrine cells: endocrine cells of the gut epithelium

enveloping layer (EVL): outer layer of an early fish embryo, composed of flattened cells

ependyma: the **ventricular** lining cells of the vertebrate central nervous system

epiblast: upper layer of a mammalian or avian **blastoderm**, or fish **germ ring**

epiboly: active spreading and increase in area of a cell sheet

epidermis: the epithelial part of the skin

epigenetic: may be used for anything to do with development, but nowadays more usually refers to mechanisms of gene control based on DNA methylation or **chromatin** structure

epiphysis (plural **epiphyses**): (1) the expanded terminal regions of long bones; (2) = pineal gland

epistasis: in general an effect exerted on the phenotype of one gene by the activity of another. In particular used to describe the suppression of one **mutant** phenotype by a **mutation** in a different gene

epithelium (plural **epithelia**): a tissue type in which cells are arranged as a single or multilayered sheet lying on a **basement membrane**

epitope: part of a molecule recognized by an antibody

equivalence group: set of cells with the same **competence**

erythrocyte: red blood cell

ES cells: see **embryonic stem cells**

euchromatin: decondensed, active **chromatin**

excitation: the input of energy required to evoke **fluorescence** from a **fluorochrome**

exocrine: of a gland secreting substances into a duct

exogastrula: abnormal **gastrula** in which the **endoderm** + **mesoderm** separates from the **ectoderm** instead of **invaginating** within it

experimental embryology: study of development by microsurgical methods

extended germ band: the **phylotypic stage** of insect development

extracellular matrix (ECM): material filling the space between cells, and also such structures as **basement membranes**, **vitelline membranes**, etc.

extraembryonic: relating to structures derived from the **zygote** which are not part of the embryo itself but serve to support or nourish it

F **FACS:** fluorescence-activated cell sorting. Enables separation of cell populations based on the attachment of different fluorescent labels

fasicle: bundle of nerve axons, often surrounded by **connective tissue**

fate map: diagram of an embryo or organ rudiment showing where each part will move and what it will become in the course of normal development

fate: indicates what will happen to a region of the embryo. The fate can often be altered by experimental manipulation

feeder layer: **tissue culture** cells treated to prevent division, which provide an environment enabling the growth of other cells that cannot be grown in standard media

femur: bone of the vertebrate upper hindlimb (also a segment of insect leg)

fertilized egg: the egg from the time of sperm entry to first **cleavage**

fibroblast: stellate cell secreting collagen, the main cell type of **connective tissue**

fibula: posterior bone of vertebrate lower hind limb

filopodium (plural **filopodia**): long thin finger-like extensions from cells

fixation: treatment of a specimen to preserve it and enable it to withstand further treatments such as staining or sectioning

fixative: chemical substance used for **fixation**, e.g. formaldehyde or glutaraldehyde

floor plate: midventral part of the vertebrate embryo spinal cord

floxed: a DNA sequence including loxP sites, which is a substrate for Cre-mediated recombination

FLP system: method for inducing recombination at desired genomic sites, particularly used in *Drosophila*

fluorescence: the emission of light of one wavelength following absorption of light of another, shorter, wavelength

fluorochrome: fluorescent substance that can be attached to proteins, antibodies, etc.

follicle cells: somatic epithelial cells of the ovary

foramen ovale: gap in the interatrial septum of the fetal mammalian heart

forebrain: region of the vertebrate embryonic brain that will become the **telencephalon** and **diencephalon**

foregut: region of the gut of a vertebrate embryo that lies anterior to the **anterior intestinal portal**. Later the region running from the pharynx to the common bile duct

forward genetics: analysis of a biological phenomenon starting from a mutant phenotype

foster mother: in mice, a female recipient to which a preimplantation embryo is transferred, to enable its further development

frontal: relates to plane or **section** separating the **dorsal** and **ventral** parts of the body

functional genomics: understanding gene function, especially in the context of total genome sequences where the function of most of the genes is not known

functional proteomics: understanding protein function, especially in the context of comparison of complex populations of proteins

fusion protein: protein composed of two or more functional domains, which is produced by molecular **cloning** and expression

G **gain of function:** of a **mutation**, confers additional or altered activity on the gene product

gametes: haploid reproductive cells; sperm or **eggs**

gametogenesis: the development of **gametes** from **germ cells**

ganciclovir: drug used to select for desirable recombination events in **tissue culture**

ganglion (plural **ganglia**): a structure in the peripheral nervous system containing neurons

gap genes: genes whose expression defines a body region in the *Drosophila* embryo

gap junctions: cell contacts that allow the passage of low molecular weight substances between the cells

gastrula: stage of development in which **gastrulation** takes place

gastrulation: phase of morphogenetic movements in early development that brings about the formation of three **germ layers**

gene duplication: the appearance of a copy of an existing gene in the **genome**

gene therapy: therapy involving the introduction of a gene into a patient

gene trap: type of insertional mutation in which a **reporter gene** is introduced into the **locus** of an endogenous gene. The function of the endogenous gene is usually destroyed, but its regulatory elements drive the reporter in the normal expression pattern

genetic marker: see **marker gene**

genetic mosaic: an organism composed of two or more types of cell that are genetically different

genital ridge: region of **mesoderm** of a vertebrate embryo that forms the **gonads**

genome: the whole of the nuclear DNA of an organism

genotype: relating to a single organism or to a genetic line, the genotype means the particular **alleles** present at specific genetic loci

germ cells: cells belonging to the **germ line**

germ layers: the **ectoderm, mesoderm,** and **endoderm**

germ line: the cell lineage that will form the **gametes**, comprising the **primordial germ cells, spermatogonia,** and **oogonia,** together with other clonal progeny such as **nurse cells**. The germ line is often formed by the action of **cytoplasmic determinants**

germ plasm: visible cytoplasmic determinant causing formation of the **primordial germ cells**

germ ring: the thickened margin of the **blastoderm** of a fish embryo

germinal vesicle: the nucleus of a primary **oocyte**

GFP: see **green fluorescent protein**

gizzard: muscular region of the avian stomach

glial cells: the non-neuronal cells of the central nervous system and peripheral **ganglia**

gnathal segments: the segments of an insect embryo lying in the **posterior** part of the head and bearing the mouthparts

goblet cells: secretory cells containing a large mucinous vesicle: found in intestinal **epithelium**

gonad: the somatic structure containing the **germ cells**

gradient: a continuous change in some property with position. Often used to refer to concentration gradient of a **morphogen**

graft: a piece of tissue transplanted from one place to another, or the act of carrying out a graft

granulocytes: cells of the blood with characteristic granules, comprising eosinophils, neutrophils, and basophils

gray crescent: region of intermediate pigmentation on the dorsal side of an amphibian egg following the **cortical rotation**

green fluorescent protein (GFP): a **fluorescent** protein often used as a **reporter**

growth cone: structure at the leading tip of a growing nerve axon

growth factors: generic name for biologically active extracellular proteins active at short range, includes many **embryonic inducing factors**

growth plate: the zone of mitotic cartilage between the **epiphysis** and **diaphysis** of a developing vertebrate long bone

gynogenetic diploid: an organism that is diploid by virtue of a duplication of the **genome** of the **egg,** and contains no paternal **genome**

H ^{3}**H-thymidine (tritiated thymidine):** radioactively labeled thymidine, used for labeling cells in **S phase**

halteres: small balancing organs found on the **metathorax** of flies in place of the second pair of wings found in other insects

haploid: having a single set of chromosomes, as after meiosis

haploinsufficient: type of genetic **dominance** arising because the loss of one gene copy reduces the level of product enough to produce a **phenotype**

Hayflick limit: the number of passages in **tissue culture** that a primary cell line can successfully be grown

hemangioblast: pluripotent cell that forms both **endothelial** cells and **hematopoietic stem cells**

hematopoietic (=hemopoietic): blood-forming

Hemimetabola: the group of insects that undergo a gradual **metamorphosis** with nymphal stages separated by molts

hemolymph: the "blood" of an arthropod

Hensen's node: a condensation of cells at the anterior end of the **primitive streak** of an avian embryo

hermaphrodite: individual organism producing both male and female **gametes**

heterochromatic: condensed, inactive **chromatin**

heterochrony: an evolutionary shift of the relative timing of events in development

heterotopic: type of graft that is performed to a different position in the host to that from which it came in the donor

hindbrain: region of the brain of a vertebrate embryo that will become the cerebellum and medulla oblongata

hindgut: region of the gut of a vertebrate embryo that lies posterior to the **posterior intestinal portal**

histological sections: thin slices of biological specimens that can be stained with a variety of techniques and enable visualization of cell arrangement and structure

holoblastic: type of **cleavage** where the whole **zygote** becomes subdivided into **blastomeres**

Holometabola: group of insects undergoing a complete **metamorphosis**

homeobox: DNA sequence encoding a **homeodomain**, which is a 60-amino-acid DNA-binding domain defining an important class of **transcription factors**

homeodomain: protein sequence encoded by a **homeobox**

homeotic gene (= selector gene): a gene whose expression distinguishes two body parts. If mutated then one body part will be converted into the other

homolog: (1) structures or genes which resemble each other because they are **homologous,** i.e. descended from a common ancestor; (2) the maternally and paternally derived chromosomes that pair with each other at meiosis

homologous recombination: recombination of a **transgene** into a **locus** in the **genome** which has exactly the same sequence

homology: related by descent from a common ancestor

homophilic: binding of a molecule to others of the same type

hopeful monster: individual organism arising as the result of a **macromutation**

horseradish peroxidase (HRP): an enzyme used for cell labeling and neuronal tracing

Hox genes: subset of **homeobox** genes that controls anteroposterior specification in animals

humerus: bone of the upper vertebrate forelimb

hybridization: (1) of nucleic acids, forming a double helix from two complementary sequences; (2) of animals, crossing one species with another to produce hybrid offspring, which are normally inviable or sterile

hybridoma: cell line producing a particular **monoclonal antibody**

hydroids: class of animal belonging to the phylum **Cnidaria,** usually sessile and colonial with a number of **polyps** connected by a stolon

hyperdorsalized: having an excessive representation of dorsal-type structures

hypermorph: mutant **allele** showing **gain of function**

hyperplasia: excessive growth

hypertrophy: excessive growth

hypoblast: the lower layer of an avian or other amniote **blastoderm,** or of the **germ ring** of a fish embryo

hypodermis: syncytial outer layer of *C. elegans*

hypomorph: mutant **allele** showing partial **loss of function**

imaginal discs: structures in the larva of *Drosophila* and other **holometabolous** insects that form the adult cuticle at the time of **metamorphosis**

immunoprecipitation: isolation of a particular substance by incubating the sample with a specific **antibody** and then recovering the antibody–substance complex as a solid phase

immunostaining: procedure to make visible the location of an **antigen** in a **section** or **wholemount** by using an **antibody** and a detection system

imprinting: the expression of a gene from only one of the parental genomes

in situ **hybridization:** detection of a specific RNA, or occasionally DNA, in a biological specimen by hybridizing to a specific **probe** and then visualizing the **probe** with a suitable detection method

in vitro **fertilization (IVF):** creation of a fertilized egg outside the mother, typically by mixing eggs and sperm in a dish

inducible: relating to a gene or gene product whose activity can be controlled by some experimental method

inducing factor: a signal substance responsible for an embryonic **induction**

induction: (1) = **embryonic induction;** (2) can refer to the regulated activation of transcription of a gene

inner cell mass: the cells within a mammalian **blastocyst** that form the entire embryo and most of the **extraembryonic membranes**

insertional mutagenesis: creation of mutations by means of a DNA element, such as a **transposon** or **retrovirus,** that will integrate at random into the genome and disrupt endogenous genes in the process

instar: period in the life cycle of an organism in between two molts

instructive induction: a process of induction that increases the complexity of the embryo because the responding cells can develop along two or more possible pathways depending on the concentration of the signal

intercalary regeneration (= intercalation): regeneration at a junction between two body parts of the structures that would normally lie in between them

intermediate filaments: components of the **cytoskeleton;** intracellular filaments of a size between **microfilaments** and **microtubules** and composed of various types of protein

intermediate mesoderm: region of **mesoderm** in a vertebrate embryo lying between the **somites** and the **somatopleure**

invagination: infolding of a cell sheet to form an internal protrusion or pocket

involution: internalization of a cell sheet by movement led by a free edge

isomerism: (1) loss of the normal left–right asymmetry of an animal; (2) in chemistry isomerism refers to two substances having the same molecular formula but different structures

isthmus: constriction between vertebrate embryonic **midbrain** and **hindbrain**

Keller explant: dorsal explants from the *Xenopus* **gastrula** used to study morphogenetic movements

keratinocyte: principal cell type of the **epidermis**

knock-in: introduction of a transgene to a specific **locus** in the **genome** by **homologous recombination**

knockout: null mutation made by targeted **mutagenesis**

Kollar's sickle: thickened posterior region of **area pellucida** of early avian **blastoderm**

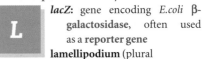

lacZ: gene encoding *E.coli* β-galactosidase, often used as a **reporter gene**

lamellipodium (plural **lamellipodia**): large flat extensions at the leading edge of a motile cell

lampbrush chromosomes: large **bivalent** chromosomes in the **germinal vesicle** of an amphibian **oocyte**

lateral inhibition: regional specification mechanism that involves isolated cells or cell clusters differentiating in a particular direction, and emitting an **inducing factor** that inhibits the surrounding cells from doing the same

lateral plate: region of **mesoderm** of a vertebrate embryo lying lateral to the **somites**

left–right asymmetry: differences between left and right sides of the body in terms of gene expression, morphogenetic movements, or differentiation

Leydig cells: endocrine cells of the testis

lineage label: method of labeling a cell or group of cells that will enable identification of all their descendants

lineage: see **cell lineage**

locus (plural **loci**): position in the **genome** occupied by a particular gene

long terminal repeat (LTR): strong promoter of **retroviruses** often used to drive **transgenes**

longitudinal section: a section made parallel to the long axis of the organism, often but not necessarily in the **sagittal** plane

Lophotrochozoa: a superphylum defined on the basis of primary sequence data; includes the annelids, molluscs, and platyhelminths

loss of function: usually describes a **mutation** which causes the protein product of the **gene** to be inactive or of lower activity than the **wild type**

luciferase: enzyme catalyzing a phosphorescent reaction and often used as a **reporter**

lumbosacral plexus: arrangement of nerves formed by the fusion and rebranching of the **spinal nerves** supplying the vertebrate hindlimb

lumen: the cavity within an organ, often lined with an **epithelium**

lymphocytes: cells responsible for specific immunity; T and B cells

M

macroevolution: evolution of **taxa** above the species level

macromeres: large **blastomeres**

macromutation: mutation that produces a significant morphological effect

macrophage: cell concerned with immunity and inflammation; originates from **hematopoietic stem cell** but found in tissues

mantle layer: cell-rich layer adjacent to the **ventricular zone** of the embryonic vertebrate central nervous system

marginal zone: (1) in an amphibian embryo the ring around the **blastula** that invaginates during **gastrulation**; (2) in a chick embryo the junction of the **area pellucida** and **area opaca**; (3) in the embryonic vertebrate central nervous system the cell-poor region near the **pial surface**

marker gene: a **transgene**, or **allele** of endogenous gene, allowing visual identification of the cells possessing it

mass spectrometry: technique for determining the molecular weight of a molecule, or of decomposition fragments of the molecule, to very high precision

maternal effect: situation where the phenotype of the embryo corresponds to the **genotype** of the mother rather than its own **genotype**. This arises from defects in the assembly of the **oocyte**

matrix: (1) the region of proliferating epidermal cells in the hair bulb; (2) = **extracellular matrix**

medial: relating to the midline of the organism

mediolateral: the direction from, or line joining, the midline to the lateral extremities of an animal

medusa: the jellyfish stage of the life cycle of a **cnidarian**

megakaryocytes: bone marrow cells that form blood platelets

meiosis: the final division leading to the formation of **gametes**, in which the chromosome complement is halved

melanocytes: cells making the melanin pigments

membrane bone: bone that develops directly from the **dermis**, with no cartilage **model**

meroblastic: type of **cleavage** where only part of the **zygote** cleaves and the remainder, usually a yolk mass, does not

mesenchyme: tissue type in which cells lie scattered within an **extracellular matrix**

mesoderm induction: formation of the **mesoderm** in response to **inducing factors**

mesoderm: the middle of the three **germ layers**

mesometrium: membrane attaching the uterus of a mouse to the body wall

mesonephros: the middle segments of the developing kidney

mesothorax: the second thoracic segment of an insect

metacarpals: **proximal** bones of the hand

metamorphosis: major remodeling of the body at a late stage of development

metanephros: the posterior region of the developing kidney, becomes the definitive kidney in mammals

metaplasia: change of one tissue type into another

metastasis (plural **metastases**): implantation and growth of **cancer** cells at remote sites in the body

metatarsals: **proximal** bones of the vertebrate (hind)foot

metathorax: the most posterior of the three thoracic segments of an insect

microarray: a small slide containing a grid of dots of specific cDNAs or oligonucleotides. Used for establishing gene expression profiles by hybridization with cDNA from the cell types of interest

microfilaments: components of the **cytoskeleton**; thin intracellular filaments composed of actin

micromeres: small **blastomeres**

microtome: machine for making thin **sections** of a biological specimen

microtubule organizing center: a place in the cell, such as a centrosome or basal body, from which microtubules grow

microtubules: components of the **cytoskeleton**; intracellular rod-like structures composed of tubulin, with a larger diameter than **microfilaments** or **intermediate filaments**

microvilli: small projections from a cell, usually found on the apical surface of absorptive epithelia

midblastula transition (MBT): set of coordinated changes in the late **blastula** including the onset of zygotic genome transcription

midbrain: region of the embryonic vertebrate brain that will become the **optic tecta**, or equivalent, and some other structures

midgut: region of an **amniote** embryo **endoderm** that lies between the **anterior** and **posterior intestinal portals**; eventually closes to the umbilicus

mirror-symmetry: two sets of parts related such that one could be the reflection of the other in a mirror

mitotic index: the proportion of cells in a specimen undergoing mitosis

model: (1) **model organism:** species used for experimental work because of favorable technical considerations. In developmental biology the results obtained on a model species are considered to have a significance well beyond the species itself; (2) an **animal model** of a human disease; (3) a **cartilage model** is a skeletal element formed in cartilage that later becomes replaced by bone

molecular clock: mechanism for the fixation in the population of **neutral mutations**: it leads to the prediction that the number of neutral mutations fixed in two lines of descent will be proportional to the time of divergence from a common ancestor

molecular cloning: insertion of a nucleic acid sequence into a cloning vector, usually a bacterial plasmid, such that it can be amplified to a useful quantity

monoclonal: consisting of or produced from a single clone of cells

monocytes: macrophage-like cells in the blood

morphogen: type of **inducing factor** to which competent cells can make at least two different responses at different **threshold** concentrations. This means that the responding cells will form a series of differently committed territories in response to a concentration **gradient** of the factor

morphogenesis: the aspects of development involving movement of cells or of cell sheets, with the associated formation of structures

morpholino: a chemical grouping that has given its name to a type of oligonucleotide analog which can **hybridize** with nucleic acids while being resistant to nuclease enzymes

morphology: the structure of the organism (= **anatomy**)

morula: multicellular pre-**blastula** embryonic stage, usually of a mammalian embryo

mosaic: (1) a type of embryo in which each part continues to develop in accordance with the **fate map** after separation from the rest of the embryo; (2) = **genetic mosaic**

motor proteins: proteins that translocate along **microtubules** or **microfilaments**

MSC: mesenchymal stem cell or marrow stromal cell

mucosa: a moist internal **epithelium** together with its immediately underlying **connective tissue**

Mullerian duct (= paramesonephric duct): duct that becomes the reproductive tract in females and regresses in males

mutagen: substance inducing genetic **mutations**

mutagenesis screens: the process for isolating a large number of **mutants** affecting some property of the organism. This may be the overall anatomy of the embryo, or some specific organ or cell type

mutant: organism carrying a **mutation**

mutation: change in the genomic DNA

myeloid cells: nonlymphocytic cells of the blood comprising **erythrocytes, monocytes, granulocytes,** and **megakaryocytes**

myoblast: mononucleate cell that is committed to differentiate into muscle

myocardium: the muscular wall of the heart

myoepithelial cell: epithelial cell capable of contraction

myofiber: the multinucleate fibers of skeletal muscle

myogenic: leading to formation of muscle

myotome: the part of the **somite** forming striated muscle

myotube: **myofiber**-like structures formed by **myoblasts** in **tissue culture**

nauplius: a type of larva characteristic of crustacea

neoblasts: the only dividing cells in **planarian** worms; required for tissue renewal and regeneration

neocortex: additional layers of neurons in the cerebral cortex of mammals, as compared to lower vertebrates

neomycin: drug used to select for desired recombination events in **tissue culture**

neoplasm: any "new growth"; need not necessarily be malignant

nephric duct (= Wolffian duct): duct forming the collecting system of the kidney, and, in males, the vas deferens

nephrogenic: leading to formation of the kidney

neural crest: migratory cells from the dorsal **neural tube** of a vertebrate embryo that form a variety of cell types

neural plate: flat sheet of **neuroepithelium** that will roll up to form the **neural tube** of a vertebrate embryo

neural stem cell: stem cell in the central nervous system that can produce neurons

neural tube: the primordium for the vertebrate central nervous system

neuroblast: a cell that divides to produce neurons

neuroenteric canal: transient connection between the **neural tube lumen** and the **hindgut** of a vertebrate embryo

neuroepithelium: the tissue comprising the **neural tube**

neurogenesis: formation of neurons

neurospheres: structures derived from the embryonic vertebrate central nervous system that contain neural **stem cells** and can be grown in suspension culture

neurotrophin or neurotrophic factor: (1) class of **growth factor** that is particularly relevant to neuronal growth and survival; (2) substances released from nerves that are required for regeneration of other structures

neurula: stage of vertebrate development during which the **neural tube** is formed

neurulation: the morphogenetic movements of vertebrate embryos that create the **neural plate** and **neural tube**

neutral evolution: evolution arising from the gradual accumulation of **mutations** which are neither beneficial nor deleterious

neutral mutations: mutations which are neither beneficial nor deleterious

neutralizing antibody: antibody that does not just bind to its target but also inhibits its biological activity

Nieuwkoop center: region of the early amphibian embryo that induces **Spemann's organizer**

node: condensation of cells at the anterior end of the **primitive streak** of a mouse embryo, similar to **Hensen's node** in the chick

nonautonomous: process that is affected by events in surrounding cells

nonpermissive: temperature at which a **ts mutation** causes a **mutant phenotype**

normal development: events that occur in the course of development of the embryo undisturbed by any experimental manipulation

Northern blot: method of detecting a specific mRNA by **hybridization** with a labeled **probe** following fractionation of the sample by gel electrophoresis, and transfer of the sample from the gel to a membrane

notochord: cartilage-like rod which comprises the dorsal-most part of the **mesoderm** of a vertebrate or invertebrate **chordate** embryo

nucleic acid hybridization: association of DNA or RNA strands through base complementarity (A binds to T/U and C binds to G). The specificity of **hybridization** enables the location or quantity of a particular sequence to be measured using a complementary labeled probe

nucleus (plural nuclei): (1) of a cell; (2) a condensation of neurons in the central nervous system

nude mouse: mouse with no thymus that does not reject tissue grafts

null: relating to a **mutation** that shows a complete **loss of function**

nurse cells: sister cells of the insect **oocyte**; 15 **nurse cells** and one **oocyte** are formed from female **oogonia** by four successive mitotic divisions

oligodendrocyte: a type of **glial cell** in the central nervous system that forms myelin sheaths around axons

oncogene: gene encoding a product that can confer cancer-like behavior if overexpressed in tissue culture cells

oocyte: the female **germ cell** after completion of its mitotic divisions is a **primary oocyte**, and after completion of the first meiotic division is a **secondary oocyte**

oogenesis: the development of **oocytes**

oogonia: the mitotic germ cells that become **oocytes**

optic cups = optic vesicles

optic tecta (singular tectum): regions of the **midbrain** where the axons from the optic nerve terminate

optic vesicles: outgrowths of the **forebrain** of a vertebrate embryo that form the eyes

organ culture: the growth of embryonic organ rudiments *in vitro*

organizer (= signaling center): group of cells emitting an **inducing factor**

organogenesis: the process, or stage, of formation of individual organs in vertebrate development

ortholog: a gene is an ortholog of a gene in another species if it is the direct **homolog**, as opposed to a **homolog** to a duplicated **locus** that arose before the divergence of the two species

orthotopic graft: type of graft that is performed to the same position in the host to that from which it came in the donor

osteoblast: cell of the bone, strictly an immature mitotic cell

osteoclast: multinucleate cell type of **hematopoietic** origin that resorbs bone and is essential for bone remodeling

osteocyte: mature cell of the bone

otic placodes: epidermal structures forming the ear vesicles

oviduct: part of female reproductive tract: a tube conveying mature **oocytes** from the ovary to the uterus

ovulation: the release of an **oocyte**, usually accompanied by the completion of first meiotic division

P-element: transposable element of *Drosophila*

P-granules: (= polar granules)

pair-rule genes: genes controlling the segmentation pattern of *Drosophila* that are expressed as one stripe for each prospective pair of segments

Paneth cells: neutrophil-like cells at the base of small intestinal crypts that secrete a variety of antimicrobial substances

parabiosis: two animals joined together in such a way that they share a common blood circulation

paralog group: set of similar genes in the same **genome**, arising from a **gene duplication**. Especially used for the homologous sets of **Hox genes** within the vertebrate Hox clusters

parasagittal: a plane parallel to the **sagittal** plane but displaced towards the right or left side

parasegments: the initially formed segmental repeating structures in insect embryos

paraxial mesoderm: plate of **mesoderm** that forms the **somites** of a vertebrate embryo

parietal endoderm: extraembryonic endoderm lining the **trophectoderm** of a mouse embryo

parthenogenesis: asexual development from **oocytes**, without a genetic contribution from the sperm

pentadactyl: having five digits

perdurance: persistance of a gene activity after the gene expression has stopped. It is due to continued presence of the protein product

perichondrium: connective tissue sheath around a cartilage structure

periosteum: connective tissue sheath around a bone

permissive: (1) of **inductions**, a signal required for the continuation of a particular developmental pathway, but which does not control alternative developmental fates; (2) of temperatures, the temperature at which a **ts mutant phenotype** is not displayed

phalanges: bones of the digits in vertebrate fore- or hindlimb

pharyngeal arches: see **branchial arches**

pharyngeal pouches: segmentally repeated pits or openings to the exterior in the pharyngeal region of a vertebrate embryo

phenocopy: the same **phenotype** as that produced by a **mutation**, but produced by some other means

phenotype: the collected characteristics of an organism, particularly in relation to other members of the same species. It is mostly determined by the individual's **genotype**

phosphorescence: emission of light in the course of a chemical reaction

phylogenetic tree: classification scheme which depicts the **taxa** in terms of their actual evolutionary relationships

phylotypic stage: developmental stage at which members of an animal group most closely resemble each other

phylum (plural **phyla**): the highest subdivision in animal **taxonomy**

pial surface: the outer surface of the vertebrate central nervous system

pigment epithelium: see **retinal pigment epithelium**

placode: a thickening of the **epidermis** which is the rudiment for a structure, usually a sense organ

planar cell polarity: polarity displayed by neighboring cells within a cell sheet

Planaria: group of free-living flatworms belonging to the phylum **Platyhelminthes**. They have a blind-ended gut and no coelom

Platyhelminthes: phylum including free-living and parasitic flatworms and tapeworms

ploidy: indicates the number of parental chromosome sets per nucleus. Animals are normally **diploid** (two sets, one from each parent), but under special circumstances could be **haploid** (one set), **triploid** (three sets), **tetraploid** (four sets), etc.

pluripotent: able to form more than one type of differentiated cell

polar body: chromosome set arising from a meiotic division of the oocyte and expelled from the **oocyte** as a small vesicle

polar granules (= **P-granules**): associated with **germ plasm** of *C. elegans*

polarity: (1) of cells, a difference between one end and the other, e.g. apico-basal polarity; (2) of an embryo or part of an embryo, a difference in **commitment** between the regions along a particular axis, e.g. **animal–vegetal** polarity

pole cells: germ line cells of *Drosophila*, formed before cellularization of the rest of the embryo

pole plasm: associated with **germ plasm** of *Drosophila*

polyclonal: consisting of or formed from more than one **clone** of cells

polydactyly: having too many digits

polymerase chain reaction (PCR): amplification of a DNA sample by repeated cycles of strand separation, primer hybridization, and replication

polyp: (1) the sessile phase of a **cnidarian** organism; (2) a projecting differentiated **neoplasm**, usually benign

polyploidy: duplication or higher multiplication of the entire chromosome set

polyspermy: fertilization of an **egg** by more than one sperm

polytene, polyteny: repeated replication of DNA without separation of chromosomes; leads to giant chromosomes

positional cloning: cloning of a **gene** starting from a **mutation**, which is mapped to very high resolution to find the gene within which it lies

positional value: a hypothetical property of animal tissues that bears a one-to-one relationship to normal position in the body, and is called into play during regeneration

posterior (= caudal) intestinal portal: the opening of the hindgut into the subgerminal cavity (**midgut**) of an **amniote** embryo

posterior midgut: the part of an insect embryo gut invaginating from the **posterior** end

posterior: the tail end of an animal

potency: of a cell or tissue region indicates the full range of structures or cell types that it is capable of becoming if exposed to a range of different environments

prenatal screening: a test to detect a developmental abnormality, which may be of genetic or spontaneous origin, before birth

primary oocyte: see **oocyte**

primitive ectoderm: = **epiblast** of mammalian embryo

primitive endoderm: first formed layer of **endoderm** in mammalian embryo, contributes to **extraembryonic membranes**

primitive streak: the region of convergence and invagination of cells in a gastrulating **amniote** embryo

primordial germ cells (PGCs): germ line cells at a stage before they become **oogonia** or **spermatogonia**

probe: a labeled and antisense nucleic acid that can be used to detect the complementary nucleic acid by **hybridization**, either in a biochemical method such as a **blot**, or *in situ* in a specimen

procephalon: the part of an insect embryo forming the anterior head

proctod(a)eum: most posterior part of the gut, nominally lined by **ectoderm**

programmed cell death (= apoptosis): death of cells by a process involving the activation of caspase enzymes, which avoids the release of toxic products into the surroundings. Often occurs in embryonic development

progress zone: the undetermined mesenchymal cells at the **distal** tip of a vertebrate limb bud

promoter: the region of a gene 5′ to the coding sequence where the **RNA polymerase** binds. It may also contain regulatory sequences binding **transcription factors** that control gene expression

pronephros: anterior segments of the developing kidney

proneural cluster: group of cells which are competent to form neurons but of which only one or a few cells do so because of a **lateral inhibition** process

pronucleus: transient nucleus-like structure of a **fertilized egg** containing the sperm or **egg** chromosomes

proteomics: analysis of proteins, especially on a large scale using the techniques of 2D electrophoresis to resolve a complex mixture, followed by **mass spectrometry** to identify individual components

prothorax: the most **anterior** of the three thoracic segments of an insect

protochordate: member of the classes of the phylum **Chordata** that are not vertebrates, for example amphioxus is a cephalochordate

Protostomia: group of animal **phyla** characterized by **spiral cleavage**, **mosaic** development, and **schizocoely**

proventriculus: glandular region of the avian stomach

proximal: nearer to the body

pseudoalleles: genes that are so similar in sequence that they appear to be **alleles**, but actually belong at different genetic **loci**. Normally the result of **gene duplication**

pseudocoelom: body cavity differing from a true **coelom** because it is not fully lined with **mesoderm**

pseudopregnant: in mice, the hormonal state resulting from mating with a sterile male

pseudotetraploid: a genome that has undergone complete duplication to the **tetraploid** state followed by some evolutionary divergence between the duplicated copies

pupa: stage of **holometabolous** insect development at which **metamorphosis** occurs

R

radial cleavage: symmetrical type of **cleavage**

radial glia: cells spanning the full width of the developing brain, from the ventricular to the **pial surface**, now known to be neural **stem cells**

radius: **anterior** bone of the vertebrate lower forelimb

RCAS virus: replication-competent, avianspecific **retrovirus**, used for overexpression experiments in chick embryos

real time PCR: a quantitative type of PCR in which the formation of the product is continuously monitored during the amplification reaction

receptor: type of molecule recognizing and activated by a specific ligand, often an **inducing factor**

recessive: a mutation that produces no **phenotype** if it is present alongside the **wild type**

redundancy: partial or complete overlap of function of two or more genes

regeneration: regrowth of a missing body part at an adult or late developmental stage

regional specification: starting from a population of similar cells, the formation of a set of territories of cells each committed to become a different structure or cell type

regulative: in addition to its general meaning of one thing controlling another, a **regulative embryo** is one where isolated parts do not necessarily develop in accordance with the **fate map**, usually forming more structures than predicted by the **fate map**

renewal tissue: tissue undergoing continuous cell turnover

reporter gene: usually a **transgene** that produces a product that is very easy to detect and hence "reports" on the nature and activity of its regulatory environment

reproductive cloning: probably possible, but currently illegal, formation of a complete human from a single cell

retinal pigment epithelium: pigmented layer of the eye derived from the optic cup

retrovirus: a type of virus with an RNA genome that is reverse transcribed to DNA in the course of infection

reverse genetics: investigations on a known gene, especially by creation of **loss-of-function** mutants

reverse transcription polymerase chain reaction (RT-PCR): detection of a specific mRNA in a sample by reverse transcribing to DNA then amplifying the specific sequence of interest by the **polymerase chain reaction (PCR)**

rhombomeres: segmental structures in the **hindbrain** of vertebrate embryos

RNA interference (= RNAi): method of inactivating a specific mRNA by introducing a short complementary double-stranded RNA sequence

RNA polymerase II: the RNA polymerase responsible for transcription of protein-coding genes

RNAse protection: method of detecting a specific mRNA by **hybridization** of the sample to a labeled **probe**, followed by RNAse digestion of unhybridized RNA, and separation of the protected product by electrophoresis

roof plate: mid-dorsal part of the vertebrate embryo spinal cord

Rosa26: **gene trap** mouse line that expresses β-**galactosidase** in all tissues

S

S phase: the part of the cell cycle during which DNA is replicated

sagittal: the **medial** plane of an organism, separating left and right halves of the body

sarcoma: **cancer** derived from **connective tissue** (in broad sense)

satellite cells: mononuclear cells associated with vertebrate muscle that can become **myoblasts** and form new **myofibers** on appropriate stimulation

scaffold: three-dimensional **extracellular matrix** used in **tissue engineering**; often made of synthetic polymers rather than natural extracellular components

schizocoely: formation of the **coelom** by cavitation of a pre-existing mass of **mesoderm**

Schwann cells: cells that form the myelin sheaths of peripheral nerve axons

sclerotome: the part of the **somite** forming the vertebrae

secondary oocyte: see oocyte

sections: see histological sections

segment polarity genes: genes in *Drosophila* that are expressed in one stripe per prospective segment

segmental plate: = **paraxial mesoderm**

segmentation: subdivision of the body, or of an appendage, into repeating structural units, usually involving all **germ layers**. In older literature can also be a synonym for **cleavage**

selector gene: = **homeotic gene**

septum intermedium: part of the interventricular septum of the heart

septum transversum: region of **splanchnic mesoderm** into which the liver bud grows

Sertoli cells: supporting cells of the testis

sex determination: mechanism for ensuring that some individuals become male and some female

sex linked: of genes or **mutations** that lie on a sex chromosome

signal transduction pathways: metabolic pathway connecting a cell surface **receptor** to its intracellular effector (gene, cytoskeleton, secretory system). It often involves a sequence of protein phosphorylation events and movement of activated **transcription factors** into the nucleus where they can regulate gene activity

signaling center (= organizer): group of cells emitting an **inducing factor**

sink: region of cells destroying or otherwise removing an **inducing factor**

situs inversus: inversion of the normal left–right asymmetry

situs solitus: the normal asymmetrical arrangement of some organs on the left and right side of the body

somatic cells: all cells of an organism that are not the **germ line**; may be collectively referred to as the **soma**

somatic mesoderm: the layer of lateral **mesoderm** lying exterior to the **coelom**

somatic: refers to all tissues that are not part of the **germ line**

somatopleure: the lateral part of an **amniote** embryo consisting of **somatic mesoderm** overlain by **epidermis**

somites: segmented **mesodermal** structures in vertebrate embryos that give rise to vertebrae and striated muscles

source: region of cells producing an **inducing factor**

Southern blot: method of detecting a specific DNA sequence by **hybridization** with a labeled probe following fractionation of the sample by gel electrophoresis, and transfer of the sample from the gel to a membrane (invented by Ed Southern)

specification: type of developmental **commitment** of cells or a tissue explant that is manifested on culture in isolation, but is not irreversible

Spemann's organizer: the dorsal **blastopore** lip region of an amphibian embryo that emits BMP inhibitors and thereby dorsalizes the surrounding tissue

spermatogonia: proliferating cells that give rise to sperm

spinal nerves: segmental nerves coming from the spinal cord of a vertebrate

spiral cleavage: asymmetrical type of **cleavage** in which quartets of **micromeres** are cut off in alternating directions

splanchnic mesoderm: layer of lateral mesoderm lying internal to the **coelom**

splanchnopleure: the **splanchnic mesoderm** together with the apposed **endoderm**

squamous: describes a thin flat type of cell, usually found in a **squamous epithelium**

stage series: description of normal development of a species divided into a number of standardized stages which can be identified by externally visible features

stem cell niche: microenvironment that supports the survival and function of **stem cells**

stem cell: a cell that is undifferentiated; survives a long time; divides to produce more copies of itself; and also divides to produce progeny destined to differentiate

superficial cleavage: cleavage of nuclei without formation of cells

symmetry breaking: process whereby two or more discrete cellular states arise starting from a uniform situation

syncytium: a mass of cytoplasm containing many nuclei

T

tailbud: (1) region at the **posterior** of a vertebrate embryo that generates part or all of the tail; (2) the **phylotypic stage** of vertebrate embryos

tarsals: bones of the ankle

taxon (plural **taxa**): a group of organisms

taxonomy: the science of classification of organisms

telencephalon: the **anterior** part of the vertebrate **forebrain**

telomerase: enzyme-RNA complex that restores the **telomeres** of chromosomes after DNA replication

telomeres: the ends of a chromosome, consisting of a short repeating sequence

telson: unsegmented **posterior** extreme of an insect embryo

temperature sensitive (ts): describes **mutations** that have an effect at one, usually high, temperature but not at another, usually lower, one

teratocarcinoma: **pluripotent** tumor derived from **germ cells**

teratogenic: leading to teratological effect

teratology: the study of developmental abnormalities arising from exposure of the embryo to noxious stimuli such as toxic substances, radiation, or infection

terminal system: system that activates certain genes at both termini of the *Drosophila* embryo

tet-off: a system of inducible gene expression which is activated by withdrawal of tetracycline

tetraploid: organism whose genome has become duplicated so that it contains two maternal copies and two paternal copies of every chromosome

tetrapods: four-legged vertebrates: amphibians, reptiles, birds, and mammals

thecal cells: endocrine cells of the ovary

therapeutic cloning: growth in **tissue culture** of differentiated cells for transplantation, starting from a cloned human **blastocyst**

therapeutic targets: gene products that may be suitable targets for the development of drugs to inhibit or enhance their activity

threshold response: sharp and discontinuous change in cell state occurring at a particular concentration of an **inducing factor**

tibia: anterior bone of the vertebrate lower hind limb (also a segment of insect leg)

tight junctions: band-like **apical** cell junctions within an **epithelium** that prevent substances crossing the **epithelium** through spaces between cells

tissue culture: growth of cells *in vitro*, usually as a cell monolayer

tissue engineering: the growth of tissues and organs *in vitro*, sometimes on a three-dimensional **scaffold**, either for transplantation or for temporary replacement of the function of a damaged organ

tissue stem cell (= **adult stem cell**): **stem cell** found in a **renewal tissue**; normally committed to forming just cell types found in that tissue

topographic: indicates a one-to-one mapping between parts of one structure and another, especially relevant to the formation of nerve connections

totipotent: indicates that a cell population is capable of differentiating into any cell type in the organism

tracheae: the respiratory system of insects

transcription factor: protein controlling the transcription of specific genes

transdetermination: change of the state of specification of a *Drosophila* **imaginal disc** to that of a different disc type

transdifferentiation: strictly the change of one **differentiated** cell type to another, without passing through undifferentiated intermediate states. But often used as synonym for **metaplasia**

transfilter: experimental setup whereby a signaling and responding tissue are separated by a membrane with defined permeability properties

transgene: cloned gene introduced into an organism

transgenesis: introduction of a new gene into an organism, normally into the **germ line**

transit amplifying cells: dividing cells in a **renewal tissue** that are not stem cells. Normally they can only divide a limited number of times

transplantation: the movement of tissue from one part of an organism to another, or from one individual to another. It can refer either to microscopic grafts carried out on embryos, or to organ transplants in adults

transposable element or **transposon:** a piece of DNA that may occasionally move from one part of the **genome** to another

transverse section: section made orthogonal to the **anteroposterior** axis of the organism

triploblasts: animal **taxa** with three **germ layers**, **ectoderm**, **mesoderm**, and **endoderm**; most animals are triploblasts

triploid: having three chromosome sets per nucleus

trochophore: type of larva found among both annelids and molluscs

trophectoderm: the outer layer of a mammalian **blastocyst**, forms part of the placenta

trophoblast: **extraembryonic** structure formed from the **trophectoderm**

truncoconal swellings: ingrowths of the outflow tract of the heart that separate it into pulmonary and systemic branches

ts: = **temperature sensitive**

tumor suppressor gene: a gene that encodes a growth inhibitor, such that its removal will provoke uncontrolled growth

turning: morphogenetic movement whereby a rodent head-fold stage embryo moves and/or twists such that the dorsal midline structures come to lie in a C-shaped curve along the outside

 ulna: posterior bone of the vertebrate lower forelimb

umbilical tube: the tube connecting an **amniote** embryo to the yolk mass or placenta

unfertilized: strictly an **egg** arising from the second meiotic division. In fact most vertebrates are fertilized at the **secondary** oocyte stage, so the term is also often used of secondary oocytes

ureteric bud: outgrowth of the nephric duct that forms the duct system of the **metanephros**

urodeles: the tailed amphibians: newts and salamanders

 vasculogenesis: formation of new blood vessels by *de novo* differentiation from **mesoderm**

vegetal hemisphere: the lower hemisphere of an **egg** or **oocyte**

veliger: a type of larva characteristic of molluscan classes except cephalopods

ventral furrow: invagination of prospective **mesoderm** into an insect embryo

ventral: the lower surface of an animal

ventricular septum: wall between the right and left ventricles of the heart

ventricular zone: the cells next to the ventricles of the vertebrate central nervous system

visceral endoderm: extraembryonic endoderm initially forming the lower layer of the **egg cylinder** stage embryo of rodents

vital dye (= vital stain): stain that can be applied to a living specimen

vitelline membrane: extracellular membrane around a **zygote**

vitellointestinal duct: persistent projection of the **midgut** into the **umbilical tube** of an **amniote** embryo

viviparous: where the embryo and fetus develops within the mother, nourished by a placenta

 Western blot: detection of a specific protein by use of an **antibody**, following fractionation of the sample by gel electrophoresis, and transfer of the sample from the gel to a membrane

wholemount: a specimen that is viewed as a three-dimensional object

wild type: the normally occurring **allele** at a genetic **locus**

Wolffian duct: see **nephric duct**

 X-chromosome: sex-determining chromosome, in mammals females are XX and males are XY

X-inactivation: shutting down of the activity of one of the two X chromosomes in female mammals

 yolk: granules of food reserve deposited in the **oocyte** and used for embryonic nutrition. Yolk granules contain a few principal proteins together with lipids

yolk sac: in the avian embryo the **extraembryonic membrane** that adheres to the yolk mass, consisting of extraembryonic **mesoderm** and **endoderm**; in the rodent embryo the corresponding structure is the middle of three membranes eventually surrounding the fetus, derived from the endoderm and mesoderm connecting the **egg cylinder** with the ectoplacental cone

yolk syncytial layer (YSL): syncytial region at the junction of the yolk cell and the cellular part of a fish embryo

zona pellucida: transparent extracellular layer surrounding a mammalian early embryo

zone of polarizing activity (ZPA): **signaling center** in the **posterior** region of the vertebrate limb bud that controls the **anteroposterior** pattern of **regional specification**

zootype: configuration of gene expression domains at the **phylotypic stage** that defines a common pattern for all animals

zygote: fertilized **egg**, strictly after the stage of fusion of male and female pronuclei

zygotic: relating to the **genome** of the embryo as opposed to that of the mother

Index